Sustainable 3D Printing for Innovative Biopolymer Production and Applications

Scrivener Publishing
100 Cummings Center, Suite 541J
Beverly, MA 01915-6106

Publishers at Scrivener
Martin Scrivener (martin@scrivenerpublishing.com)
Phillip Carmical (pcarmical@scrivenerpublishing.com)

Sustainable 3D Printing for Innovative Biopolymer Production and Applications

Edited by

Bhasha Sharma
Department of Chemistry, University of Delhi, India

Balaram Pani
Dean of Colleges at the University of Delhi, India

Shashank Shekhar
Department of Chemistry, Netaji Subhas University of Technology, New Delhi, India

and

Jude A. Okolie
St. Peter's College, University of Saskatchewan, Saskatoon, Canada

Scrivener Publishing

WILEY

This edition first published 2025 by John Wiley & Sons, Inc., 111 River Street, Hoboken, NJ 07030, USA and Scrivener Publishing LLC, 100 Cummings Center, Suite 541J, Beverly, MA 01915, USA

For more information about Scrivener publications please visit www.scrivenerpublishing.com.

Wiley Global Headquarters
111 River Street, Hoboken, NJ 07030, USA

For details of our global editorial offices, customer services, and more information about Wiley products visit us at www.wiley.com.

Library of Congress Cataloging-in-Publication Data

ISBN 978-1-119-79171-3

Front cover images courtesy of Adobe Firefly
Cover design by Russell Richardson

Set in size of 11pt and Minion Pro by Manila Typesetting Company, Makati, Philippines

Printed in the USA

10 9 8 7 6 5 4 3 2 1

Contents

Preface

Biopolymers are vastly available for several applications such as biomedicine, food, textile, and cosmetics due to their biodegradability, biocompatibility, and wide range of chemical composition. Recently, there has been an interest in the use of sustainable biopolymers in place of conventional ones due to environmental concerns. Biopolymers have attracted worldwide interest due to the obnoxious situation created by traditional plastic wastes and to protect the environment from further destruction. Techniques such as 3D printing are capable of producing materials with complex structures and varying functional groups. Therefore, 3D printing has been extended to the production and applications of biopolymers. Keeping this point into consideration, this book will cover a broad range of biopolymers and their applications in diverse fields by utilizing 3D printing. 3D printing technology has been extensively studied to fabricate reliable and efficient products for applications in tissue engineering applications. This book will provide a robust overview of inherent 3D printing and its properties in enhancing the fabrication of novel materials. This book will contain several chapters which will be contributed by eminent scientists who are effectively involved in technological research based on 3D printed materials and their applications by delineating their impact on the circular economy. This book will contain recent ideas, the latest research, and progress in biopolymer technology with challenges and future prospects. The main emphasis of this book is sustainable polymers fabricated by 3D printing. This handbook will consist of meticulous chapters contributed by the scientific community who are veterans in the field of science and technology and working on 3D printed biopolymers and their applications while addressing their limitations and fringe benefits. This book will combine several interdisciplinary areas including green biopolymers, involving nanotechnology and its functionalization, synthesis, and properties. Several processing techniques employed in 3D printing technology will also be discussed. Some major applications like water purification, energy storage, and biomedical implications will be discussed. This book is designed to meet the needs of several scholars including researchers from universities, early career researchers, and those working in the biopolymer industry.

The Editors
October 2024

1

The Framework of the Breakthroughs in the 3D Printing Technique

Bhasha Sharma* **and Shashank Shekhar**

Department of Chemistry, University of Delhi, New Delhi, India

Abstract

A rapidly emerging 3D printing technology was poised to revolutionize research and manufacturing and has become an imperative tool in numerous fields. There is plenty of contrivances in the 3D printing world to be exhilarated about- it's a veritable ferocious west of opportunities. The resolution, fabrication speed, and available matrix should be contemplated in the 3D printing process which depends on the end-use application. 3D printing has become unpretentiously dominant and is employed to refer to the combination of additive manufacturing techniques. This indagation will cast light on the evolution, development, and fundamentals of 3D printing technology. This exploration bestows elementary comprehension on novel 3D printing materials like ceramics, composites, biomaterials, polymers, and smart materials, etc. Nevertheless, this technology is an integral segment of the multi-process system to emulate the burgeoning of novel materials. 3D printing technology has consequently emanated in an enormous outburst of innovations in copious sectors. The world of tomorrow might be very feasible to be pioneered with 3D-printed, electronics, cars, drones, doorknobs, planes, and all categories of products we either never came across or were never affordable, available, and customizable.

Keywords: 3D printing, polymers, additive manufacturing, biopolymers

**Corresponding author*: sharmabhasha@gmail.com

Bhasha Sharma, Balaram Pani, Shashank Shekhar and Jude A. Okolie (eds.) Sustainable 3D Printing for Innovative Biopolymer Production and Applications, (1–24) © 2025 Scrivener Publishing LLC

Acronyms

ABS	Acrylonitrile Butadiene Styrene
AM	Additive Manufacturing
CAD	Computer Aided Design
CNT	Carbon Nanotubes
FDA	Food Development Authority
FDM	Fused Deposition Modelling
RP	Rapid Prototyping
PE	Polyethylene
PEEK	Polyetheretheketone
PC	Polycarbonate
PCL	Polycaprolactone
PGA	Polyglycolic Acid
PLA	Polylactic Acid
PMMA	Poly Methyl Methacrylate
PP	Polypropylene
PVA	Poly (vinyl alcohol)
SFF	Solid Free Form
SLS	Selective Laser Printing
STL	Stereolithography

1.1 Outlook: From Cradle to Grave

Despite the colossal progression that has been made, the diminution nature of conventional manufacturing processes has diverse limitations that have eventually navigated an avant-garde paradigm in manufacturing, i.e., additive manufacturing (AM), now generally referred as 3D printing technology. 3D printing has turned into a broad amalgamation of additive-based technologies that operate on the credo of fabricating objects from layer by layer, bit by bit, bottom-up in comparison to top-down by removing excess material to attain the final product. The maneuvering of 3D printing technology for manufacturing and rapid tooling has harbingered to generate complex-geometry components based on computer designs. The 3D printing basic functions were first described in 1964 by Arthur C. Clarke (sci-fi author). It is also referred to as rapid prototyping (RP), additive manufacturing, and solid free form (SFF), which is a process of amalgamation of materials to produce objects from 3D model data through layer-by-layer deposition [1], described by Charles Hull in 1986 [2] displayed in Figure 1.1. Hideo Kodama in 1981 at Nagoya Municipal Industrial Research Institute produced his functional rapid prototyping account employing photopolymers. The printed and solid material was fabricated in layers that correspond to the cross-sectional slice in a model. After 3 years, Charles Hull in 1984 invented stereolithography (history of 3D printing). Stereolithography generates 3D models by employing digital data which can be utilized to make tangible objects. In 1986, Hull established and developed the 3D systems. STL file format facilitates electronic handshake from computer-aided design (CAD) software by transferring files for printing of 3D components. Afterward, Hull and his team continued to innovate the first 3D printer system which was referred to as the "stereolithography apparatus" and the first 3D printer commercially available to the public which was SLA-250. In 1990, Scott crump at Stratasys developed and patented fused deposition modeling (FDM) [3]. The first apparatus in 1993 termed 3D printer patented by Michael Cima and Emanuel Sachs in MIT was able to print ceramics, metals, and plastics [4]. In 1999, the first 3D-printed organ in humans was implanted at Wake Forest Institute for Regenerative Medicine. Scientists printed a human bladder, a synthetic bladder which was coated with the patient's cells. This regenerative tissue was implanted into the body with extremely no chances of rejection into the immune system due to the attachment of the patient's body cells. Dr. Adrian Bowyer in 2005 launched an open-source project to produce a 3D printer which could fundamentally build itself or able to print most of its part. Presently, 3D printing is being employed in nearly every single industry, with innumerable applications in trade, food, toys art, implants, aircraft parts, fashion items, orbital transportation, etc. [5].

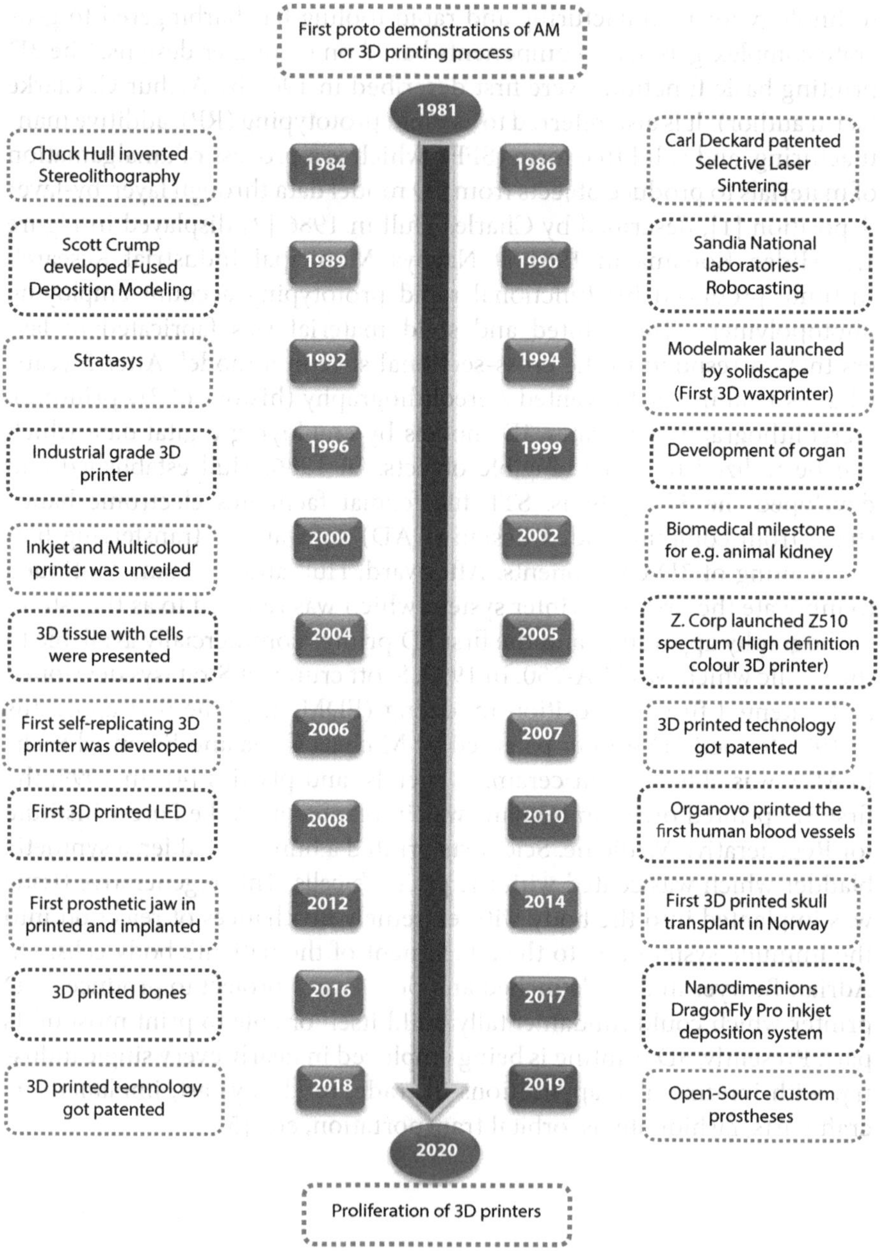

Figure 1.1 Evolution of 3D printing.

Box 1: History of additive manufacturing

The technique facilitates fabrication of structure layer by layer deposition process which is revolutionizing industry owing to its capability to acquire near net shape products with almost no material waste. A foremost key feature is based on solidification and melting is the elementary employment of elementary knowledge evolved by decades of research on welding technologies classified into laser and electron beam. The industrial applications utilize laser beam systems which produces parts with good dimensions, surface finish and precision under pertinent process control [6–8]. Nevertheless, electron beam systems tend to produce sound parts having remarkable mechanical properties as the process is evolved at a uniform high temperature under vacuum chamber which helps in decreasing the residual stresses formed during layer deposition process. Both laser and electron systems require high expensive costs. Therefore, additive manufacturing process with plasma and electric arc are allocated to produce manufacturing of large parts.

Additive manufacturing can be categorized into six diverse eras. The first era, i.e., late 1970s to early 1980s, is referred to frontrunner examples which could be defined as *Proto Additive Manufacturing*. The second era started in the middle of 1980s to 1990s which referred to the initial introduction and development of foundation 3D-printed technologies with the founding of the first 3D-printed technology which remains a major industry competitor until now. The third era, started from 1990 to 2005, designates the maturation of 3D printing within indigenous private firms, the conception of ancillary major 3D printing technologies together with tandem advancement and development of computational capabilities of 3D imaging. The fourth era (2005 to 2012) was not surprisingly aligned with the termination of original 3D printing patents, conventional acquisition of social media as well as initiation of maker movement which represents the beginning of 3D printing to be comprehensively introduced and adopted by spectators beyond the significant industrial players. The fifth era started from beginning of 2012 up to 2017 represents awakening of 3D printing. 3D-printable materials and novel additive technologies and even bioprinting begin to be acquired, being considerably funded, and prevailed, underpinned by prime federal government strive.

The ASTM Committee F42 in 2009 published a report containing standard terminologies of additive manufacturing. This acknowledged 3D printing as industrial manufacturing technology. The FDM patents expired in the same year with the production of the first low-cost 3D printer desktop by the RepRap project. Earlier, it costs $200,000 now all of a sudden and became available for less than $2,000. The adoption of 3D printing technology according to Wohlers kept proliferating. 3D desktop printers of more than 1 million have been accorded globally between 2015 to 2017, and it has been evinced that the industrial sale of metal printers get almost doubled in 2017 in contrary to the previous year. The locution "3D printing" is technically introduced as additive technology based on materials deposition. 3D printing of biomedical products has been turned omnipresent. The proliferation of advanced or novel biomaterials and biologics in addition to tissue and live cell bioprinting and associated fabrication technologies is a conception to be clinically broached. The prognostication of a future where 3D printing is not ordinarily employed to produce guides, models, or inert implants but also devilishly tissue regenerative and bioactive devices which assure the transformation of orthopedic medicine and healthcare immensely.

1.2 Understanding 3D Printing

3D printing is a technique of producing three-dimensional solid components from a digital file. The formation of a 3D-printed object is attained by employing additive processes. Here the object is formed through laying down successive layers of material. It is converse of subtracting manufacturing in which hollowing or cutting the piece of plastic or metal takes place through a milling machine. 3D printing circumscribes several materials and technologies as it is being employed in almost every industry one can contemplate. It is imperative to analyze it as a conglomerate of divergent industries with a myriad of fields—for instance, prosthetics, dental products, consumer products like furniture, eyewear, footwear, etc., replicating ancient artifacts, movie props, reconstructing fossils, industrial products (like prototypes, manufacturing tools, functional end-use parts), and in forensic pathology for evidence reconstruction. In this process, no specific tools are required like cutting tools with mold or geometries because the component is directly manufactured onto a layer-by-layer platform which facilitates a specific set of limitations and benefits. The procedure begins with the digital 3D model (blueprint of a physical object). The technique is sectioned by the printer software into two-dimensional or thin layers which turned into the set of instructions using machine language to execute for printing. Henceforth, the manner a 3D printer works differ by its procedures—for instance, desktop FDM printers which melt filaments of plastic and lay it down onto the printing platform *via* a nozzle (such as computer-controlled glue gum and high precision). Industrial SLS machines employ a laser to melt thin layers of plastic or metal. The materials available can be varied depending upon the process—for instance, plastics are the commonly employed materials, but metals are also considered for 3D printing. With regard to attributes to the size of the object and type of the printer, it has been analyzed that a print ordinarily takes 4 to 18 h for completion. Generally, parts of a 3D-printed item are hardly ready to use out of the machine though. They usually require post-processing methods to attain the desired level of the finished surface. The detailed section of different types of 3D printing technology will be discussed in Chapter 2. Currently, 3D printing has been employed for jewelry collections, artificial heart pumps, PGA rocket engines, steel bridges in Amsterdam, 3D-printed cornea, and other products in food as well as the aviation industry [9]. The acquisition of 3D printing technology will escalate the production speed while decreasing the cost. Simultaneously, consumers' behest will have more supremacy over production. Consumers have considerable reverberations in the resultant product and can solicit to tailor it according to the specifications. The amenities of 3D printing technology will be allocated closer to the end-user, according to a more responsive and flexible manufacturing process as well as substantial quality control. Additionally, when utilizing 3D printing technology, the exigency for global transportation is remarkably reduced. This is due to the nearer location of manufacturing sites to the end destination which results in saving time and energy due to the distribution through fleet tracking technology. Ultimately, the acquisition of 3D printing technology can alter the logistics of the firm which regulates the entire process from start to finish and more comprehensive services [10]. Currently, 3D printing is broadly employed globally. It has been increasingly utilized for production or mass production of different types of open-source designs in healthcare, agriculture, aerospace, and automotive industries for printing prototypes of airplane and car ancillaries [11]. The 3D printing applications in the government and private defense industry have been expeditiously perceived. Numerous applications in fashion and agriculture have been also transpired.

1.2.1 The Fundamental Process of 3D Printing Technology

There are several 3D printing technologies, but this segment will bestow emphasis on the elementary process from designing to the final part of the product. While every method of 3D printing fabricates components differently, there are five essential steps that are persistent throughout the process, namely:

(i) Creating a 3D file
The first crucial step in the 3D printing process is producing a digital model. The most commonly employed method for creating a digital model is CAD. Reverse engineering can also be utilized to create a digital model through 3D scanning. There are several considerations of design that depend on features such as materials, geometry limitations, and escape hole requirements.

(ii) STL creation and manipulation of file
A CAD model must be transformed into a format by which the 3D printer is capable to interpret the format to print a component in 3D printing. In the initial step is the conversion of the CAD model into the STL (stereolithography) file, also known as the Standard Triangle Language file. 3DP or OBJ are also allowable 3D printing files, but they are generally not used. STL employs polygons (triangles) to delineate the object surface, basically deciphering the often CAD complex model. CAD programs are competent for exporting a model as an STL file. After the generation of the STL file, the same has been imported into the slicer program which is followed by slicing of design into the several layers that will be employed to build up the component. Afterward, the slicer program transforms the STL file into G-Code. (G-Code is referred to as the numerically controlled programming language employed in CAM to control automated machines such as 3D printers and CNC machines.) The slicer program facilitates the 3D printer operator to define the build parameters of the 3D printer by specifying the support location, part orientation, and layer height. Ordinarily, slicer programs are proprietary to each type of 3D printer, even though there are various slicer programs such as simplify 3D, Netfabb, and Slic3r. Designers generally provide 3D printer operators with an STL file. The operator of the 3D

printer set the parameters to print and produce the G-Code file.

(iii) Printing
The printing process in 3D printing technologies depends on different additive manufacturing parts which will be discussed in detail in Chapter 2.

(iv) Print removal
Removing a print in some 3D printing technologies is as ordinary as splitting up the printed part from the build platform. In other 3D printing methods, the elimination of the printed part is an extremely technical process that involves the precise attraction of print while it is still attached to the build plate and encased in the build material. These processes generally require procedures of strict removal and machine operators who are highly skilled, together with controlled environments and safety equipment.

(v) Post-processing
The procedures for post-processing depend on printing technology. Some technologies entail a component to cure before handling by keeping under UV, while other technologies allow components to be handled right away. Some technologies employ support, which is also eliminated at the post-processing stage.

1.3 Fringe Benefits of 3D Printing Technology

a. Low cost
In traditional metal casting and injection molding processes, for each part, there is a requirement of a specific mold. These customized tools require high costing. To recuperate these costs, similar parts have been manufactured in thousands. 3D printing does not require any specialized tooling so there are fundamentally no start-up costs. The cost of a 3D printed component is dependent on the type of material utilized, time taken to print, and post-processing required to attain the desired product.

b. Complexities of geometry with no additional cost
3D printing facilitates the effortless fabrication of intricate shapes which is usually arduous to produce by other manufacturing technologies. The additive characteristics of

technology mean that the complexity of geometries does not proffer higher prices. Organic geometries and complex parts are optimized for performance costs.

c. Diverse range of materials
Nowadays, the materials employed for 3D printing are plastic, whereas metal 3D printing discovers an escalating number of industrial applications. There is inclusion of specialty materials with characteristics for pallets of 3D printing tailored for discrete applications. The components obtained from 3D printing has excellent stiffness and strength, high heat resistance, and better biocompatibility. Fabrication of composites is generally common in 3D printing technology. The most common materials are carbon particles, carbon fibers, metals, plastics, wood, ceramics, etc. The reinforcement of these materials results in unique properties for specific applications.

d. Customization of components
By employing the traditional manufacturing process, it is unambiguously economical to create and sell uniform products to the consumers. Contrarily, 3D printing facilitates contented customization. The cost of starts-ups is very low, only one desideratum to meliorate the digital 3D model to form a custom object. As a result, every object can be tailored to encounter the specific needs of customers without influencing the cost of manufacturing.

e. Low-cost prototyping
One of the imperative employments of 3D printing today is prototyping—both for function and form. This is conducted at the fraction of cost of procedures and speeds that no other manufacturing processes can truffle with: the objects fabricated by 3D printer desktop which are ordinarily kept overnight and the orders placed through professional services with immense industrial machines followed by their delivery in 2 to 5 days. The prototyping speed greatly stimulates the cycle of design (design, analysis, improvement, re-design). The products which require a long time can be readily processed in 8 to 10 weeks.

1.4 Compendium of Materials Employed as Matrix

Metals, ceramics, conventional thermoplastics, and graphene-based materials, a combination of hybrids or composites, are the most commonly employed materials that can be printed by 3D printing technology. 3D printing requires extortionate quality materials that encounter congruous specifications to fabricate consistent devices of high quality. The materials employed for 3D printing technology is categorized in metals, polymers, and ceramics which are discussed in Figure 1.2 as follows:

a. Metals
   Due to numerous advantages proffered by the technology, 3D printing has soared countless recognition in biomedical, automotive, aerospace, and manufacturing industries [12]. 3D printing technology generally employs a powdered form of metals. Utilizing a variety of metallic powders, metal printing allows load-bearing, functional, and high-quality parts to be produced. Flowability, shape, and distribution of particle size are essential properties that regulate how to expropriate a metallic powder for 3D printing.

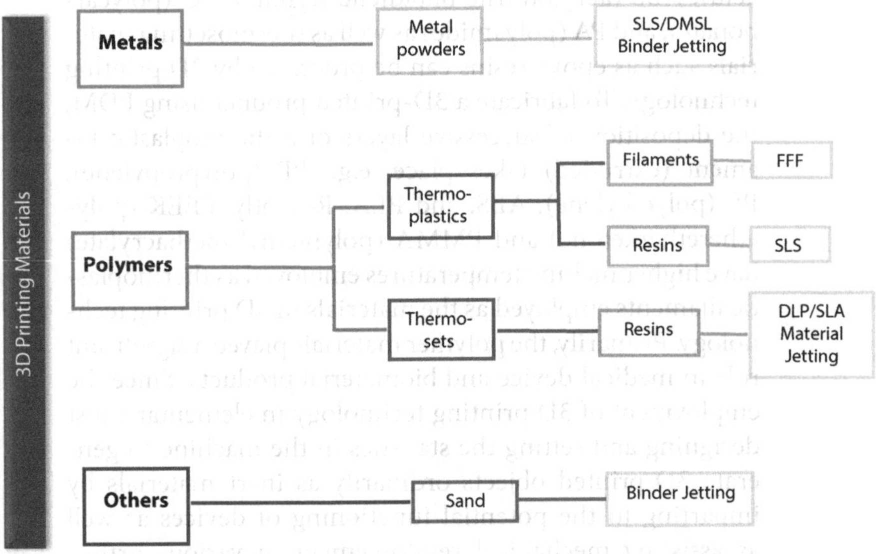

Figure 1.2 Flowchart of the materials employed in the classification of 3D printing technology.

Metal materials have remarkable physical properties, and the material is employed to fabricate complex manufacturers by printing aerospace parts to human organs—for instance, cobalt-based alloys [13], aluminum alloys [9], titanium alloys [14, 15], stainless steel [16], and nickel-based alloys [17]. Additionally, 3D printing technology can fabricate nickel-based alloys used in aerospace applications. The components produced by using nickel-based alloys can be employed in a hazardous environment because it has high resistance toward corrosion and resistant to temperatures up to 1,200°C. Cobalt-based alloy is competent to be employed in 3D printing dental applications. The objects fabricated by titanium-based alloys have exclusive properties such as low density, oxidation resistance, good corrosion, and ductility. It is utilized in a high stress and high operating temperatures—for instance, biomedical industry and aerospace components (see Table 1.1).

b. Polymers

3D printing technologies are extensively employed for the production of polymeric matrix components through prototypes to functional structures with onerous geometries [18]. Thermoplastics materials such as PLA (polylactic acid), ABS (acrylonitrile butadiene styrene), PC (polycarbonate), and PA (polyamide) as well as thermosetting materials such as epoxy resins can be processed by 3D printing technology. To fabricate a 3D-printed product using FDM, the deposition of successive layers of a thermoplastic filament (extruded) takes place, e.g., PP (polypropylene), PE (polyethylene), ABS, and PLA. Recently, PEEK (polyetheretherketone) and PMMA (polymethyl methacrylate) have higher melting temperatures employed as thermoplastic filaments employed as the materials in 3D printing technology. Primarily, the polymer materials played a significant role in medical device and biomaterial products. Since the employment of 3D printing technology in elementary, just designing and setting the statistics in the machine to generate 3D-printed objects ordinarily as inert materials by imparting to the potential functioning of devices as well as assisting mechanical reinforcement in various orthopedic implants. In the liquid state, 3D-printed polymers with a low melting point are enormously employed in the

Table 1.1 Depiction of characteristics of a range of materials employed for 3D printing technology.

	Metal filled	Wood filled	Carbon fiber filled	PP	PC	Nylon	ABS	PVA	PLA	HIPS	PETG	ASA	Flexible
Density (g/cm³)	2-4	1.15–1.25	1.3	0.9	1.2	1.06–1.14	1.04	1.23	1.24	1.03–1.04	1.23	1.07	1.19–1.23
Stiffness	10	8	10	4	6	5	5	3	7.5	10	5	5	26
Ultimate strength (MPa)	20–30	46	45–48	32	72	40–85	40	78	65	32	53	55	26–43
Coefficient of thermal expansion (μm/m°C)	33.75	30.5	57.5	150	69	95	90	85	68	80	60	98	57
Printability	7	8	8	4	6	8	8	5	9	6	9	7	6
Maximum service temperature (°C)	52	52	52	100	121	80–95	98	75	52	100	73	90	60–74
Extruder temperature (°C)	190–220	190–220	220–270	220–250	260–310	220–270	220–250	185–200	190–220	230–245	230–250	235–255	225–245
Bed temperature (°C)	45–60	45–60	45–60	85–100	80–120	70–90	95–110	45–60	45–60	100–115	75–90	45–60	90–110
Recommended build surfaces	Glue stick, PEI, painter's tape	Glue stick, PEI, painter's tape	Glue stick, PEI, painter's tape, glass plate	PP sheet, packing tape	PEI, glue stick, commercial adhesive	PEI, glue stick	ABS slurry, Kapton tape	Painter's tape, PEI	Glue stick, PEI, painter's tape, glass plate	Glue stick, glass plate, Kapton tape	Painter's tape, glue stick	PEI, glue stick	Painter's tape, PEI
Other hardware requirements	Part cooling fan, stainless steel or wear-resistant nozzle	Part cooling fan	Part cooling fan	Part cooling fan, enclosure recommended, heated bed	Enclosure recommended, heated bed	Enclosure recommended, heated bed	Enclosure recommended, heated bed	Heated bed, part cooling fan	Part cooling fan	Enclosure recommended, heated bed	Heated bed, part cooling fan	Heated bed	Part cooling fan
Price (kg) ($)	50–120	25–55	30–80	60–120	40–75	25–65	10–40	40–110	10–40	24–32	20–60	38–40	30–70

technology owing to their low weight, low cost, and flexibility. Despite being employed in structural components, most of the 3D-printed polymers have still been utilized as conceptual prototypes as pristine polymer products fabricated by 3D printing lack functionality and strength as fully load-bearing and functional parts. The SLA and material jetting phenomena utilize photopolymer thermosets which become hard on exposure to UV light or laser. Commonly, these thermoset products encompass high-temperature components and bowling balls such as knobs used on the stove. The progression of biocompatible materials recently has facilitated 3D bioprinting of functional living of tissues which can be implemented in regenerative medicines to inscribe the exigency for organ transplantation. Material selection is one of the imperative steps in the process of 3D printing. Usually, chitosan, alginate, PCL (polycaprolactone), acrylate-based polymers, sodium alginate, and modified diblock copolymers are the biocompatible polymers which can be employed for the 3D printing process.

c. Composites

3D-printed polymers have a lack of mechanical strength, complex geometry, and functionality, which is a considerable challenge for their broad area of applications. Particle reinforcements are broadly employed to ameliorate properties of the polymer matrix. Composite materials have low weight, tailorable properties, and exceptional properties which have become a revolution in high-performance industries. Glass fiber- and carbon fiber-reinforced polymer composites are the most studied examples of reinforced composites. Carbon fiber-based composites are widely employed in aerospace applications attributing to the excellent corrosion resistance, high specific stiffness, good fatigue performance, and stiffness. Concurrently, glass fiber-reinforced polymer composites have a low coefficient of thermal expansion and high thermal conductivity, which make it appropriate for applications in 3D printing technology as they are not affected by the curing temperature in the process of manufacturing [19, 20]. Particles can mix with polymers either in liquid form for SLA or powder form for SLS which can be extruded into printable filaments in the FDM process. An innovation in particle reinforces nanocomposites of 3D printing to fabricate objects for structural applications employed for real-world applications. Nanomaterials include graphite, graphene, and CNTs (carbon nanotubes); metal oxides often exhibit idiosyncratic thermal, electrical, and mechanical properties. Therefore, the inclusion of nanoparticles into polymers for 3D printing technology could validate the formation of high-performance functional composites. The inclusion of reinforcement particles into polymers assists in addressing some perplex difficulties in the printing technique. One of the prime stumbling blocks in the FDM printing process is the decomposition of the end product which is produced by thermal expansion of polymers. Reinforcing the metal particles into a polymer matrix could be an efficient riposte to this nuisance.

d. Ceramics

Importantly, ceramics and concrete are not appropriate for 3D printing as the pristine powder cannot be mixed by employing heat up to their melting point. Additionally, the high melting temperature of ceramics is the primary challenge in additive manufacturing technology. Presently, concrete and ceramics have been employed to fabricate 3D-printed objects without any cracks or artifacts that can be attained by optimizing parameters and enhancing mechanical properties [21]. Ceramics are durable, fire-resistant, and strong and can be implemented in any shape or geometry useful in building and constructions. Ceramic materials are also useful in aerospace and dental applications [22]—for instance, alumina, zirconia, and bioactive glasses [23–25]. Polymer materials filled with ceramics have enhanced wear resistance, which makes them potential material for tooling applications. STL printing propound ceramic powder filled with resin employed in the production of injection molds (high detail). Alumina has great curing complexity and formidable ceramic oxide with a broad myriad of applications comprising of chemicals, propulsion components, microelectronics, thermal protection systems, catalyst, and high technology industry [26]. STL machine was employed to fabricate bioactive and ceramic glass into dance props by improving the bending strength of these materials. The increased mechanical strength has the potential to be applied in pertinent clinical structures like bone and scaffold. Zirconia is the major constituent of construction materials used in nuclear power sectors applied in element tubing. The AM process which involves a single step is more pertinent to fabricate ceramic parts in a shortened time [27]. In a study, the low cost of raw material, manufacturing scalability, and powder-based technologies are the most economically viable AM process to create ceramic parts owing to the ease of parallel processing of multiple parts [28].

Box 2: Smart materials

Materials which have capabilities to alter the shape and geometry of the object are referred as smart materials [29]. These can be influenced by external conditions such as water and heat—for instance, 3D-printed components formed by employing smart materials are soft robotic systems and self-evolving structures. Smart materials can also be categorized as 4D printing materials—for instance, smart materials are shape memory polymers and shape memory alloys [30, 31]. Nevertheless, not all components produced by 3D printing processes, e.g., printed living hinges, are categorized as 4D printing as they do not have shape changing, self-actuating, and self-sensing properties [32]. Nickel-titanium, a category of shape memory alloy, is useful in bio-medical implants as micro-electromechanical device applications in which density, reproducibility of microstructures, and transformation temperatures are the essential parameters. By employing 3D printing technology, the complex shape of shape memory alloys could be favorable to produce. The assessment of the quality of material is performed based on part density, surface roughness, and dimensional accuracy.

With the advent of 3D printing technology, sustainable materials are of utmost significance in the present scenario as the 3D printing market is proliferating at a rapid rate. Recycling the unwanted or excess material into new-fangled feedstock or unraveling novel methods for composting and degeneration of materials into innocuous building blocks in the ecosystem is crucial. Biomaterials have created a quest in replacing traditional polymers which have a detrimental impact on the ecosystems due to their promising solutions to mitigate plastic waste. The selection of material specifically in tissue engineering applications is dependent on innumerable factors such as processability, degradability, surface characteristics, mechanical properties, and biocompatibility [33]. Both natural and synthetic polymers have been employed in 3D printing techniques. Natural polymers like chitosan, polysaccharides, collagen, chitin, PLA, etc., are widely employed for biomedical applications. Synthetic polymers like PGA, PCL, PVA, and PLGA are the most commonly used polymers that can bestow utmost flexibility and better physicochemical properties [34–36].

Box 3: Bioprinting: The exigency for definitions

Biological printing facilitates the precise cell deposition in viscous biomaterial in a discrete spatial arrangement utilizing a computer aided printer. The progression of this technology from conventional 3D printing has entailed complex printing processes to facilitate the role of biological inks like proliferation and cell adhesion to be assisted in addition to the mechanical characteristics required for traditional 3D printing ink. In bioprinting, multimaterial structures encompass cells which can be printed along with inorganic or organic materials to create a fully functional scaffold.

The pursuit of biocompatible materials is the paramount challenge in translational bioprinting. The stipulation to refine a biomaterial transformed into bioink places further behest on the criteria for biomaterials. The biomaterial must facilitate extrusion as a liquid bioink and yet be able to perpetuate its shape in the post-printing process to capitulate viable organs, tissues, and biological materials. Figure 1.3 delineates the classification of biomaterials utilized in 3D printing technology.

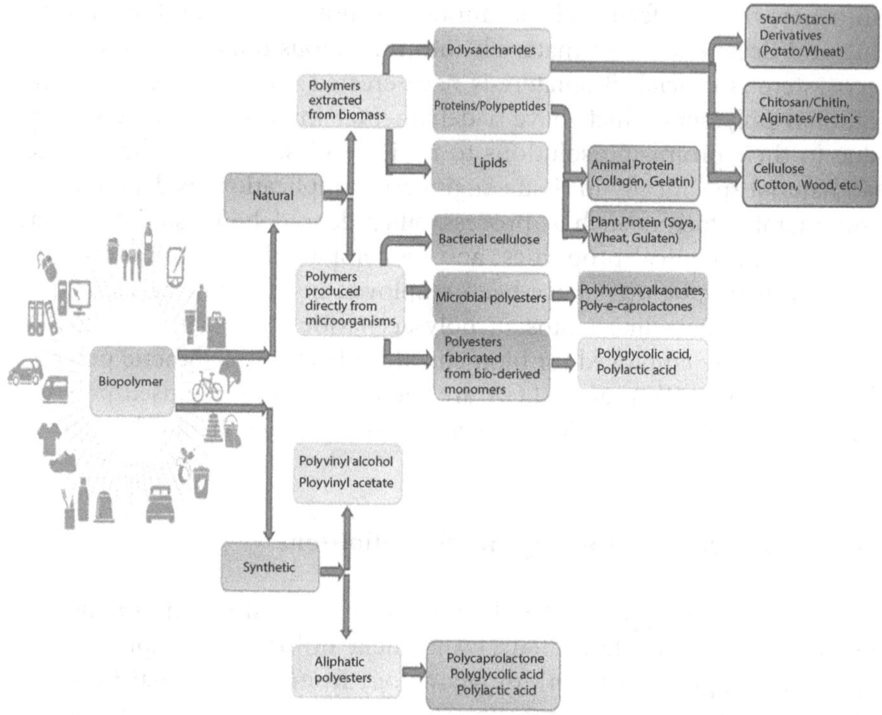

Figure 1.3 Outlining the classification of natural and synthetic biopolymers.

1.5 A Paradigm in 3D Printing Technology: Eminent Innovations for Mankind

- 3D printers have soared beyond plastics as there are new-fangled printers that can process recycled materials—for instance, one inventor came up with a strategy to print wet paper fibers. The paper pulp printer is the first innovation of a 3D printer that utilizes sustainable materials. Due to its durability and strong characteristics, it imparts a solution to circumvent plastics. Furthermore, it is contemplated as a closed-loop recycling system to recycle 3D-printed objects again in the near future.
- Ceramics is the material that 3D printing developers are presently forming waves with, i.e., woven ceramics which allow for objects and structures that have micro-pores employed of the circulation of air. Besides this, the aforementioned

3D-printed sculptures of ceramics are also dignitary of being allocated in the museum.

- Steel, an efficient material being leveraged in the process of 3D printing phenomena—for instance, a 3D-printed steel bridge was constructed to manifest the dormancy of multi-axis 3D printing technology. As 3D printing technology persists to proliferate and evolve, organizations may be able to manufacture city infrastructure faster and precisely.
- The healthcare industry is being revolutionized by the innovation of the recreation of the human heart by encapsulating fatty tissue employing genetic processes to transform them into heart cells. Ultimately, that biomaterial was converted into "bio-ink". It manifested that bio-printing could one day print biological structures and human organs such as teeth. The scientists who worked on printing the human heart analyzed that the printed organ was biological to the patient as it is from the similar genetic material, so the sufferer would not require to medicate with immune-suppressants. This will assist in circumventing the life-threatening organ repudiation that some patients suffer.
- Researchers are working on hair restoration procedures by developing "hair fam" which can be transplanted into a patient—for instance, strong and thick as bristles on a hairbrush and synthetic hair for a wig.
- Spritam is an innovation of 3D-printed medication that is approved by the FDA. The advantage of this anti-seizure drug is to dissolve rapidly on the tongue (something conventional medications could not do).
- Recently, Audi is adapting the 3D printer automotive parts efficiently which transforms the way their production and development are carried out. A prime advantage is to lessen the duration of the wait which customers experience while ordering intricate parts. By employing the SLS process, they can more affordably and efficiently create lighter parts with complex internal functions and geometry.
- In a project called ICON, researchers produced a 3D-printed home out of cement. The scientists of the University of Nantes have fabricated a 3D printing system that is capable of building a polyurethane foam (fully insulated housing structure) in just 20 to 30 min by the robotic arm.

- The scientists from the University of Minnesota created a prototype of an artificial type to overcome the barriers. This landmark research marks an imperative step towards developing a bionic eye which assists sighted or blind people to see better. Silver particles were printed on hemispherical glass dome, followed by the employment of semiconducting polymer to print photodiodes which transforms light into electrical signals that can be turned and processed into real images. Even though it may sound like science fiction, this 3D-printed bionic eye will soon become a reality, creating a new epoch in regenerative medicine.
- Relativity Space, a California-based startup, is designing to create a rocket in which all the components will be 3D-printed. The company has expanded "Stargate" to make it a reality. It is one of the leading arc-welding metal 3D printers powered with robotics and machine learning. The technology works on the principle of melting metal fire extruded through a nozzle onto a build platform.

Box 4: Metamaterials

3D printing innovation evinced metamaterials which are crucial myriad of materials science assisting technology. Specifically, functional metamaterials operate more like convoluted machines, which is made viable by 3D printing technology. They are implemented in door handles and hinges, etc. Auxetic metamaterials were formed by Delft University of Technology that facilitate better hip implants which reduced tear and wear on surrounding bone and distributed force more uniformly. The aforementioned materials have been designed with an internal geometry which permits them to expand on application of pressure. Scientists at Boston University have worked on innovative metamaterial which improvise the scan quality and speed of MRI (magnetic resonance imaging). This was conducted by 3D printing plastic coils employed in the material. They have also innovated a 3D-printed product which blocks 94% of sound while permitting airflow. USC Viterbi formed a material that manipulate sound as well which can be triggered by the magnetic field. With the inclusion of iron particles into the lattice structure, a magnetic field can be utilized to deform a structure into one by trapping sound instead of letting it pass through.

1.6 Limitations of 3D Printing Innovation

There are numerous disadvantages of 3D printing technology in the manufacturing industry. Exemplarily, the impact of employment of 3D printing technology will foreshorten the utilization of manufacturing labor; as a result, it will affect the countries' economy which relies on a gigantic number of low-skilled jobs. In the bargain, by adopting 3D printing technology, the end-user can print several types of objects such as guns, knives, and other dangerous products. Consequently, the employment of 3D printing technology should be restricted to specific people only to impede criminals and terrorists who bring guns without detection. Concurrently, people who get a grasp of the layout will be capable to counterfeit the products deftly. The 3D printers' mass flow rate which is based on hot melt or extrusion of gel is generally quite restricted, elucidating that mass food production utilizing 3D printers is a kind of a futuristic approach unless several printers are operating in parallel together. Developments in 3D printing of nanoscale regions are not precise in contrary to the fabrication of micro- to the macro-level. Additionally, nano-scale printing is arduous to purvey in most countries. Exemplarily, polymer-based 3D printing technology will not function for ceramics or metals and *vice versa*. This optimization of the process is very imperative for diverse materials which will be employed with a specific printer. The end-user must know which material should be employed and the resolution required for selecting a 3D printer. This stumbling block is the major reason why different organizations have several 3D printers for different areas of applications.

1.7 Conclusion and the Way Forward

The envisage of 3D printing technology has a rich landscape in the manufacturing industry. The 3D printing technology is progressively employed for the production or mass customization of diverse open-source designs in health, agriculture, automotive and electronic industry. The existence of layer-by-layer additive fabrication is evinced in that the technology imparts "complexity for free", which allows the physical realization of intricate structures enriched with complex external and internal architectures. To accomplish precise and accurate printing, the prime aspects should be contemplated like processing parameters, material properties, and post-processing parameters. Furthermore, the inclusion of feedstock material in 3D printing technology improvement could meliorate the final properties of the resultant printed material which enables large-scale applications. Recently, 3D printing in food has captivated the quest for a myriad of applications like space, military, special needs, etc. This technology facilitates the revolution in the food industry by forming a variety of meals in severe conditions, e.g., nutritional stability requirements, the safety of food, the acceptability of meals in space missions, etc. The major challenges concerned with 3D printing technology specifically in tissue engineering applications lie in attaining target mechanical properties, designed architectures, and durability to satiate both functional and mechanical requirements. Irrespective of some limitations, natural or biomaterials, renewable materials hold prominence for revolutionizing the bioprinting future. The risk associated with the adoption of 3D printing technology is yet to be comprehended and evaluated.

References

1. Standard AS, *Standard terminology for additive manufacturing technologies*, ASTM International F2792-12a, USA, 2012.
2. C.W. Hull, inventor. Apparatus for production of three-dimensional objects by stereolithography. United States patent US 45,753,301, 1986.
3. S.S. Crump, inventor; Stratasys Inc, assignee. Apparatus and method for creating three-dimensional objects. United States patent US 5,121,329, 1992.
4. E.M. Sachs, J.S. Haggerty, M.J. Cima, P.A. Williams, Three-dimensional printing techniques. U.S. Patent 5,204,055, issued April 20, 1993.
5. Jakus, A.E., An Introduction to 3D Printing—Past, Present, and Future Promise, in: *3D Printing in Orthopaedic Surgery*, pp. 1–15, Elsevier, Mexico, 2019.
6. DebRoy, T., Wei, H.L., Zuback, J.S., Mukherjee, T., Elmer, J.W., Milewski, J.O., Beese, A.M., Wilson-Heid, A., De, A., Zhang, W., Additive manufacturing of metallic components–process, structure and properties. *Prog. Mater Sci.*, 92, 112–224, 2018.
7. Xiong, R., Grant, A.M., Ma, R., Zhang, S., Tsukruk, V.V., Naturally-derived biopolymer nanocomposites: Interfacial design, properties and emerging applications. *Mater. Sci. Eng.: R: Rep.*, 125, 1–41, 2018.
8. MacDonald, E. and Wicker, R., Multiprocess 3D printing for increasing component functionality. *Science*, 353, 6307, 2016.
9. Shahrubudin, N., Lee, T.C., Ramlan, R., An overview on 3D printing technology: technological, materials, and applications. *Procedia Manuf.*, 35, 1286–1296, 2019.
10. Sniderman, B., Baum, P., Rajan, V., 3D Opportunity for Life: Additive Manufacturing Takes Humanitarian Action. *Deloitte Rev.*, 2016. Available

online:https://www2.deloitte.com/insights/us/en/deloitte-review/issue-19/3d-printing-for-humanitarian-action. html (accessed on 12 October; 2017).
11. Keleş, O., Blevins, C.W., Bowman, K.J., Effect of build orientation on the mechanical reliability of 3D printed ABS. *Rapid Prototyping J.*, 23, 2, 320–328, 2017.
12. Horst, D.J., Duvoisin, C.A., de Almeida Vieira, R., Additive manufacturing at Industry 4.0: a review. *Int. J. Eng. Tech. Res.*, 8, 8, 3–8, 2018.
13. Hitzler, L., Alifui-Segbaya, F., Williams, P., Heine, B., Heitzmann, M., Hall, W., Merkel, M., Öchsner, A., Additive manufacturing of cobalt-based dental alloys: analysis of microstructure and physicomechanical properties. *Adv. Mater. Sci. Eng.*, 2018, 2018.
14. Uhlmann, E., Kersting, R., Klein, T.B., Cruz, M.F., Borille, A.V., Additive manufacturing of titanium alloy for aircraft components. *Procedia CIRP*, 35, 55–60, 2015.
15. Trevisan, F., Calignano, F., Aversa, A., Marchese, G., Lombardi, M., Biamino, S., Ugues, D., Manfredi, D., Additive manufacturing of titanium alloys in the biomedical field: processes, properties and applications. *J. Appl. Biomater. Funct. Mater.*, 16, 2, 57–67, 2018.
16. DebRoy, T., Wei, H.L., Zuback, J.S., Mukherjee, T., Elmer, J.W., Milewski, J.O., Beese, A.M., A., Wilson-Heid, A., De, A., Zhang, W., Additive manufacturing of metallic components–process, structure and properties. *Prog. Mater Sci.*, 92, 112–224, 2018.
17. Murr, L.E., Frontiers of 3D printing/additive manufacturing: from human organs to aircraft fabrication. *J. Mater. Sci. Technol.*, 32, 10, 987–995, 2016.
18. Caminero, M.A., Chacón, J.M., García-Moreno, I., Rodríguez, G.P., Impact damage resistance of 3D printed continuous fibre reinforced thermoplastic composites using fused deposition modelling. *Composites, Part B*, 148, 93–103, 2018.
19. Amer, A.A.R., Abdullah, M.M.A.B., Ming, L.Y., Tahir, M.F.M., Performance and properties of glass fiber and its utilization in concrete-A review, in: *AIP Conference Proceedings*, vol. 2030, AIP Publishing LLC, p. 020296, 2018.
20. Liu, Z., Zhang, L., Yu, E., Ying, Z., Zhang, Y., Liu, X., Eli, W., Modification of glass fiber surface and glass fiber reinforced polymer composites challenges and opportunities: from organic chemistry perspective. *Curr. Org. Chem.*, 19, 11, 991–1010, 2015.
21. Baldassarre, F. and Ricciardi, F., The Additive Manufacturing in the Industry 4.0 Era: The Case of an Italian FabLab. *J. Emerg. Trends Market. Manage.*, 1, 1, 105–115, 2017.
22. Owen, D., Hickey, J., Cusson, A., Ayeni, O.I., Rhoades, J., Deng, Y., Zhang, Y. *et al.*, 3D printing of ceramic components using a customized 3D ceramic printer. *Prog. Addit. Manuf.*, 3, 1-2, 3–9, 2018.
23. Zocca, A., Lima, P., Günster, J., LSD-based 3D printing of alumina ceramics. *J. Ceram. Sci. Technol.*, 8, 1, 141–148, 2017.

24. Gmeiner, R., Deisinger, U., Schönherr, J., Lechner, B., Detsch, R., Boccaccini, A.R., Stampfl, J., Additive manufacturing of bioactive glasses and silicate bioceramics. *J. Ceram. Sci. Technol.*, 6, 2, 75–86, 2015.
25. Ianko, T., Panov, S., Sushchyns'ky, O., Pylypenko, M., Dmytrenko, O., Zirconium alloy powders for manufacture of 3d printed articles used in nuclear power industry. *Вопросы атомной науки и техники*, 113, 148–153, 2018.
26. Tang, X. and Y.Y., Electrospinning preparation and characterization of alumina nanofibers with high aspect ratio. *Ceram. Int.*, 41, 8, 9232–9238, 2015.
27. Deckers, J., Vleugels, J., Kruth, J.-P., Additive manufacturing of ceramics: a review. *J. Ceram. Sci. Technol.*, 5, 4, 245–260, 2014.
28. Zocca, A., Colombo, P., Gomes, C.M., Günster, J., Additive manufacturing of ceramics: issues, potentialities, and opportunities. *J. Am. Ceram. Soc.*, 98, 7, 1983–2001, 2015.
29. Lee, J.-Y., An, J., Chua, C.K., Fundamentals and applications of 3D printing for novel materials. *Appl. Mater. Today*, 7, 120–133, 2017.
30. Van Humbeeck, J., Additive manufacturing of shape memory alloys. *Shape Mem. Superelasticity*, 4, 2, 309–312, 2018.
31. Yang, Y., Chen, Y., Wei, Y., Li, Y., 3D printing of shape memory polymer for functional part fabrication. *Int. J. Adv. Manuf. Technol.*, 84, 9-12, 2079–2095, 2016.
32. Bogue, R., Smart materials: A review of capabilities and applications. *Assem. Autom.*, 34, 16–22, 2014.
33. Kim, B.-S., Park, I.-K., Hoshiba, T., Jiang, H.-L., Choi, Y.-J., Akaike, T., Cho, C.-S., Design of artificial extracellular matrices for tissue engineering. *Prog. Polym. Sci.*, 36, 2, 238–268, 2011.
34. Sousa, I., Mendes, A., Bártolo, P.J., PCL scaffolds with collagen bioactivator for applications in tissue engineering. *Procedia Eng.*, 59, 279–284, 2013.
35. Zhang, C., Sangaj, N., Hwang, Y., Phadke, A., Chang, C.-W., Varghese, S., Oligo (trimethylene carbonate)–poly (ethylene glycol)–oligo (trimethylene carbonate) triblock-based hydrogels for cartilage tissue engineering. *Acta Biomater.*, 7, 9, 3362–3369, 2011.
36. Islam, A., Yasin, T., ur Rehman, I., Synthesis of hybrid polymer networks of irradiated chitosan/poly (vinyl alcohol) for biomedical applications. *Radiat. Phys. Chem.*, 96, 115–119, 2014.

2

Delineating the Techniques Employed for the Fabrication of Sustainable Polymers *via* 3D Printing Phenomena

Reetu Sharma[1*], Nishi Verma[2], Ranjana Dewangan[2], Amit Kumar Sharma[3], Mamta Tripathi[4] and Balaram Pani[5]

[1]*Department of Chemistry, Netaji Subhas University of Technology, Dwarka, New Delhi, India*

[2]*Department of Chemistry, Bhilai Mahila Mahavidyalaya, Bharti Vishwavidyalaya, Durg (C.G.), India*

[3]*Department of Chemistry, Atma Ram Sanatan Dharma College, University of Delhi, Dhaula Kuan, New Delhi, India*

[4]*School of Studies in Chemistry, Pt. Ravishankar Shukla University, Raipur, India*

[5]*Department of Chemistry, Bhaskaracharya College of Applied Science, University of Delhi, Dwarka, New Delhi, India*

Abstract

Additive manufacturing, also recognized as 3D printing, converts virtual 3D designs generated through computer-aided design (CAD) into tangible objects. The field of 3D printing is swiftly advancing in both academic and industrial research contexts. The development of 3D printing technologies has led to new applications in various fields, including rapid prototyping, tooling, dentistry, microfluidics, biomedical devices, tissue engineering, drug delivery, and more. The progress in functional polymer production through 3D printing methods has been extensively investigated due to its cost-effectiveness, seamless integration into custom shapes, heightened efficiency, enhanced precision, and the ability to utilize materials without limitations, as opposed to conventional techniques such as injection molding and thermoforming. The range of polymers used in additive manufacturing encompasses thermoplastics, thermosets, elastomers, polymers combined with fillers, biopolymers, and blends of polymers with organic compounds. This chapter presents an overview of 3D printing techniques employed in

**Corresponding author*: ritus1066@gmail.com

Bhasha Sharma, Balaram Pani, Shashank Shekhar and Jude A. Okolie (eds.) Sustainable 3D Printing for Innovative Biopolymer Production and Applications, (25–48) © 2025 Scrivener Publishing LLC

the creation of polymer materials. Noteworthy 3D printing approaches, including fused deposition modeling, stereolithography, inkjet 3D printing, 3D plotting, and selective laser sintering, are introduced.

***Keywords*:** 3D printing, additive manufacturing, polymers, reinforcements, 3D printing techniques, advantage, disadvantage

2.1 Introduction

Additive manufacturing techniques, also referred to as 3D printing, rapid prototyping, or layered manufacturing, were initially introduced by Hull during the 1980s to cater to the specialized needs of model making and rapid prototyping (RP). Over time, additive manufacturing (AM), alias 3D printing, has evolved into a versatile technological platform for computer-assisted design (CAD) and rapid manufacturing [1–7]. This technology enables the production of intricate custom parts from metals, ceramics, and polymers without the requirement for molds or traditional machining methods [8].

This technological breakthrough has attracted considerable attention from polymer chemists, materials scientists, and engineers, owing to its significant contributions to innovations in polymer chemistry [9, 10]. In academic literature, terms like additive manufacturing, rapid prototyping, layered manufacturing, solid freeform fabrication, 3D fabbing, and 3D printing are frequently used interchangeably [11]. While the engineering community typically prefers the term "additive manufacturing," "3D printing" is more prevalent, especially in popular media.

In this context, both "additive manufacturing" (AM) and "3D printing" refer to the same fundamental manufacturing principle. Furthermore, this technology has the capability to fabricate objects over a wide range of sizes, from micrometers to meters. The cost of commercially available 3D printers has substantially decreased, enabling the production of customized products at home, such as bicycle components, jewelry, and electronic parts.

The benefits of 3D printing encompass ease of operation, flexible design, reliability, cost-effectiveness, and compatibility with a diverse array of materials, rendering it highly promising for various technological applications, including biotechnology, energy, environmental applications, robotics, and aerospace. Notably, one of the earliest commercial applications of 3D-printed organic materials, dating back to the early 1990s, involved manufacturing pharmaceutical capsules composed of a drug and a polymer binder [12]. While ongoing research and development persist in this domain [13], numerous additional medical applications have emerged, capitalizing on the geometric versatility of 3D printing. These applications encompass multilayered tubular structures, microfibril arrangements, and cellular structures as well as various internal and external anatomical configurations [14–18].

Computer-aided design (CAD) software plays an integral role in additive manufacturing (AM) technology, facilitating the creation of personalized digital designs and 3D models for prototype fabrication. More specifically, such software prepares 3D models in slice formats, representing layers in polymer printing. These digital files also contain information pertaining to material properties and printing parameters utilized in various printing methodologies [19].

At its core, AM represents a stark departure from traditional subtractive manufacturing methods. This approach operates within the "bottom-up" or "top-down" paradigms, where customized structures are meticulously crafted layer by layer, as opposed to the conventional casting or injection methods [20–23]. The world of AM offers a vast array of materials, including metals, ceramics, polymers, and various combinations thereof in the form of blends, hybrids, and composites.

The creation of intricate geometries, made possible by advanced computer-aided design software, has emerged as a thriving research focus within the realm of 3D printing. Limitations in the printability and mechanical properties of pure polymers have driven the scientific community to delve into the development of high-performance 3D-printable polymer blends and composites. These advancements yield numerous benefits, such as cost-efficiency, heightened precision, reduced polymer waste, minimized chemical usage, and the ability to produce customized intricate geometries [24]. Such advanced geometries find applications across diverse fields, spanning from packaging, medical, tissue engineering, automotive, and aerospace to electronics engineering [25–27].

Over the past three decades, 3D printing has captured substantial attention. While most previously published review articles have concentrated on processing techniques and the printing of pure polymer materials, the recent years have witnessed significant progress in the development of printable polymers with enhanced performance. Therefore, the aim of this chapter is to present and consolidate relevant information regarding the 3D printing of polymers. The core focus of this chapter revolves around the techniques utilized for fabricating polymers through the medium of 3D printing.

2.2 3D Printing Techniques

Numerous methodologies have emerged within the realm of 3D printing, encompassing a range of approaches such as fused deposition modeling (FDM), initially pioneered and commercialized by Stratasys, as well as stereolithography (SLA), digital light processing (DLP) introduced by Envision TEC, laminated object manufacturing (LOM) championed by Cubital and Helisys, and selective laser sintering (SLS) pioneered by DTM Corp., a subsidiary of 3D Systems [28]. Each of these techniques boasts its

Table 2.1 Overview of established rapid 3D prototyping techniques.

Technique	Typical polymer materials	State of starting materials	Working principle
FDM	Thermoplastics, such as PC, ABS, PLA, and nylon	Filament	Extrusion and deposition
SLA	Photocurable Resin (epoxy or acrylate-based resin)	Liquid photopolymer	Laser scanning and UV-induced curing
DLP	Liquid resin	Liquid photopolymer	Digital UV projector screen
SLS	PCL and polyamide powder	Powder	Drop-on-demand binder printing
LOM	ABS, TPU	Laminated sheet	Deposition of layers of laminated materials
PolyJet Printing	Photopolymers	Liquid or semi-liquid	UV-induced curing
3DP	Any materials can be supplied as powder, binder needed	Powder	Drop-on-demand binder printing
3D plotting	PCL, PLA, hydrogel	Liquid or paste	Pressurized syringe extrusion, and heat or UV-assisted curing

distinct merits and limitations in the realm of polymer product fabrication. The choice of a particular fabrication approach hinges upon factors such as the initial materials, processing speed and resolution requirements, cost considerations, and the performance specifications of the end products. A consolidated overview of established rapid 3D prototyping techniques is presented in Table 2.1.

2.2.1 Fused Deposition Modeling (FDM)

Among the diverse array of 3D printing techniques, fused deposition modeling (FDM) stands out as one of the most extensively utilized and readily accessible methods for polymer fabrication. FDM printers have gained widespread popularity due to their adaptability in working with thermoplastics characterized by relatively low melting temperatures. Materials like ABS [29–31], PLA [32], PVA [33, 34], TPU [35, 36], ULTEM (polyetherimide) [37, 38], nylon [39], ASA (acrylonitrile styrene acrylate) [40, 41], and PPSF (polyphenylsulfone) [42] are frequently employed in FDM-based printing.

FDM printers function by precisely extruding thermoplastic filaments under controlled conditions. Within the FDM process, these filaments experience controlled melting at the nozzle, followed by extrusion in a layer-by-layer manner onto the build platform. As the layers merge and solidify, the final components take shape, with this process continuing until the

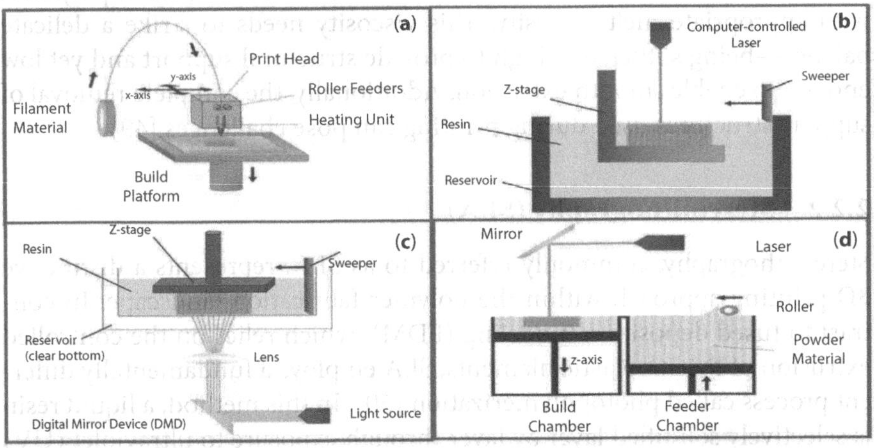

Figure 2.1 Schemes of (a) fused deposition modeling (FDM), (b) laser-based stereolithography, (c) digital light projection (DLP) system, and (d) selective laser sintering (SLS) techniques.

desired form is achieved [43]. Figure 2.1a illustrates the key components of the FDM process [44]. The quality of FDM printing can be adjusted by manipulating various parameters, including printing bed temperature, nozzle head temperature, nozzle size, layer thickness, printing orientation, printing speed, and raster angle [45]. Research indicates that precise control over bed temperature and convective heat transfer conditions can enhance the bonding strength between successive layers, thus improving the mechanical properties of the printed parts [46].

Advantages

FDM offers several notable advantages. Notably, it presents a cost-effective solution, especially for compact machines, and facilitates the printing of multi-material objects [47]. Additionally, FDM printers boast attributes such as cost-efficiency, rapid operation, and user-friendly interfaces. The utilization of multiple extruders further allows for the optimization of material properties at different print locations, resulting in distinct chemical and mechanical characteristics [48].

Disadvantages

FDM printing does come with its set of limitations. One prominent drawback relates to the requirement of polymer materials to be accessible in filament form to facilitate the extrusion process. Ensuring a consistent dispersion of reinforcements and the elimination of voids that may form during filament manufacturing can present challenges. Furthermore, FDM printers are restricted to the utilization of thermoplastic polymers with appropriate melt viscosity. This viscosity needs to strike a delicate balance—being sufficiently high to provide structural support and yet low enough to enable smooth extrusion. Additionally, the complete removal of support structures used during printing can pose challenges [49].

2.2.2 Stereolithography (SLA)

Stereolithography, commonly referred to as SLA, represents a distinctive 3D printing approach within the polymer fabrication landscape. In contrast to fused deposition modeling (FDM), which relies on the controlled extrusion of thermoplastic filaments, SLA employs a fundamentally different process called photopolymerization [50]. In this method, a liquid resin is selectively solidified layer by layer through exposure to ultraviolet (UV) light. This precise layer-by-layer solidification allows for the creation of intricate and highly detailed prints with a remarkably smooth surface finish. As for the bottom-up approach, a UV laser beam is directed upward, passing through a window at the resin tank's bottom (Figure 2.1b). SLA offers several advantages that make it an attractive choice for certain applications. Its ability to produce intricate geometries and intricate details with precision is particularly noteworthy. Moreover, SLA prints tend to exhibit minimal visible layer lines, contributing to their exceptional surface quality. Additionally, SLA technology can utilize a variety of specialized resin materials, each tailored to specific applications, such as dental, jewelry, and engineering prototypes [9].

However, SLA also presents its own set of challenges. One notable aspect is the necessity for post-processing, which typically involves the removal of excess, uncured resin from the printed object. This step may require additional time and effort compared to some other 3D printing techniques. Furthermore, the photopolymer resins used in SLA are sensitive to UV light exposure, and prolonged exposure to UV light can lead to undesirable effects on the printed object, such as yellowing or brittleness [51]. It is important to carefully handle and store photopolymer resins to maintain their integrity. In deciding between FDM and SLA for a given project, it is crucial to consider the specific requirements, including material characteristics, desired print quality, and the complexity of the design. Each technique has its strengths and limitations, making it essential for designers and engineers to choose the most suitable 3D printing method based on the unique demands of their application [53].

Advantages

SLA technology offers a principal advantage in the capability to produce high-resolution parts. Moreover, its nozzle-free nature eliminates the concern of nozzle clogging.

Disadvantages

However, the industrial application of SLA is hindered by its substantial cost. The potential cytotoxicity stemming from residual photo initiator and uncured resin presents another significant consideration [52].

2.2.3 Digital Light Processing (DLP)

Digital light processing (DLP) is a significant 3D printing technology that plays a crucial role in polymer fabrication. DLP shares commonalities with stereolithography (SLA) but operates differently (Figure 2.1c). In DLP 3D printing, a digital projector projects UV light or another light source onto a vat of photosensitive resin. This UV light selectively cures and solidifies the resin layer by layer, following the digital model's cross-sectional pattern. DLP offers notable advantages, such as speed and accuracy. It can produce objects quickly since entire layers are exposed to light simultaneously, making it efficient for complex designs. Additionally, DLP can achieve high levels of precision and resolution, resulting in smooth surface finishes with minimal visible layer lines, which is advantageous for applications requiring a polished appearance [54–56]. DLP has some considerations and limitations. Similar to SLA, DLP-printed objects typically require post-processing. This involves rinsing, curing, and potentially additional finishing steps to remove excess resin and achieve the desired properties. Proper material handling is crucial as photosensitive resin materials used in DLP can be sensitive to environmental factors like humidity and UV light, affecting the quality of the final prints. DLP printers may have restrictions on build volume, limiting the size of objects that can be printed. Additionally, DLP printers and materials can be relatively more expensive compared to some other 3D printing technologies, which should be factored into project costs and budgets.

Advantages

Digital Light Processing (DLP) 3D printing offers several key advantages. It is known for its speed, as it can produce objects quickly by exposing entire layers to light simultaneously, making it efficient for time-sensitive projects. DLP excels in achieving high-resolution prints with fine details, making it suitable for precision applications. The technology also delivers a smooth surface finish with minimal visible layer lines, reducing the need for extensive post-processing. DLP can work with a variety of photosensitive resin materials, offering versatility for different applications. It is particularly adept at handling complex geometries and internal structures without requiring support structures, which expands design possibilities. Additionally, DLP is widely used in industries like healthcare, aerospace, and art, highlighting its adaptability to diverse needs. However, users should be mindful of post-processing requirements, material handling considerations, and equipment costs when choosing DLP for specific projects [57].

Disadvantages

Digital light processing (DLP) 3D printing, while offering notable advantages, comes with some limitations. Post-processing is often required to remove excess resin and ensure the final object's desired properties, adding to the overall production time. Materials used in DLP can be sensitive to environmental factors like humidity and UV light, necessitating careful handling and storage. DLP printers may have restrictions on build volume, limiting the size of objects that can be produced. Additionally, DLP printers and materials can be relatively more expensive compared to some other 3D printing technologies, which can impact project budgets [58].

2.2.4 Selective Laser Sintering (SLS)

Selective laser sintering (SLS) is a highly regarded 3D printing technology with a notable presence in the realm of polymer fabrication. Unlike some other 3D printing methods, such as stereolithography (SLA) or digital light processing (DLP), SLS utilizes a powder bed fusion approach (Figure 2.1d). In SLS, a powerful laser selectively sinters or fuses powdered thermoplastic materials, layer by layer, following the precise digital 3D model. One distinctive feature of SLS is the use of the unfused powder surrounding the printed part as a natural support structure. This allows for the creation of intricate and complex geometries without the need for additional supports, which can be a significant advantage [59]. One of SLS's primary strengths is its material versatility. It can work with a wide range of thermoplastic materials, including nylon, polyamide, TPU, and even composites with additives like carbon fiber or glass fiber. This versatility enables the production of parts with specific mechanical, thermal, or chemical properties, making SLS suitable for various applications across industries [60, 61].

SLS excels at producing complex and intricate geometries, including internal cavities and moving parts. This capability is highly valuable in industries where design freedom and functional prototyping are critical, such as aerospace and automotive. Parts produced through SLS are known for their durability and good mechanical properties, making them suitable for functional prototypes and even some end-use components [62]. Additionally, SLS-printed parts often exhibit excellent heat resistance, which expands their applicability in situations where exposure to high temperatures is a concern. SLS is particularly well suited for small-batch or low-volume production runs, offering a cost-effective manufacturing solution for producing small quantities of parts [63–65]. However, there are several considerations when using SLS. Equipment costs for SLS printers are typically higher compared to some other 3D printing technologies, which can be a limiting factor for some users. Additionally, while SLS-produced parts are functional and durable, their surface finish may not be as smooth as those achieved with technologies like SLA or DLP. Post-processing steps may be necessary to improve surface aesthetics. Furthermore, handling powdered materials, a requirement in SLS, can be messy, and safety precautions are necessary to minimize exposure to fine particles. Lastly, SLS may have limitations in achieving the same level of fine detail and resolution as technologies like SLA or DLP, which are better suited for highly detailed models or small-scale objects [64].

Advantages
Selective laser sintering (SLS) boasts a unique set of advantages, including exceptional material versatility, the ability to produce complex and intricate geometries without support structures, excellent durability and mechanical properties, heat resistance, and suitability for small-batch production and functional prototyping. Its capacity to work with a wide range of thermoplastic materials and the freedom it offers in design make it an invaluable choice for diverse industries, despite considerations such as equipment costs and surface finish [63].

Disadvantages
Selective laser sintering (SLS) comes with certain drawbacks, including relatively high equipment costs, the potential for rougher surface finishes compared to some other 3D printing methods, the need for careful handling of powdered materials, and limitations in achieving fine levels of detail and resolution, which may make it less suitable for applications requiring intricate or highly detailed models [65].

2.2.5 Laminated Object Manufacturing (LOM)

Laminated object manufacturing (LOM) is a unique 3D printing method that distinguishes itself by its layering and bonding approach. It utilizes a variety of materials, including paper, plastic sheets, or metal-coated sheets, bonded together with adhesive. LOM's versatility allows users to choose materials that match the specific requirements of their applications, such as strength, flexibility, or aesthetics.

One of LOM's primary advantages is its cost-effectiveness, particularly for producing larger objects or prototypes. The availability and affordability of raw materials, especially paper, contribute to reduced production costs compared to some other 3D printing techniques (Figure 2.2). LOM's layering process involves precise cutting or scribing of each layer, guided by the digital 3D model [66]. The excess material between layers serves as natural support, eliminating the need for additional support structures during printing. This streamlined process simplifies production and minimizes material wastage. Furthermore, LOM is well suited for creating relatively large objects due to its inherent layering approach, making it a suitable choice for projects where size matters [67]. Depending on the chosen material, LOM-produced parts can exhibit durability and strength, meeting the

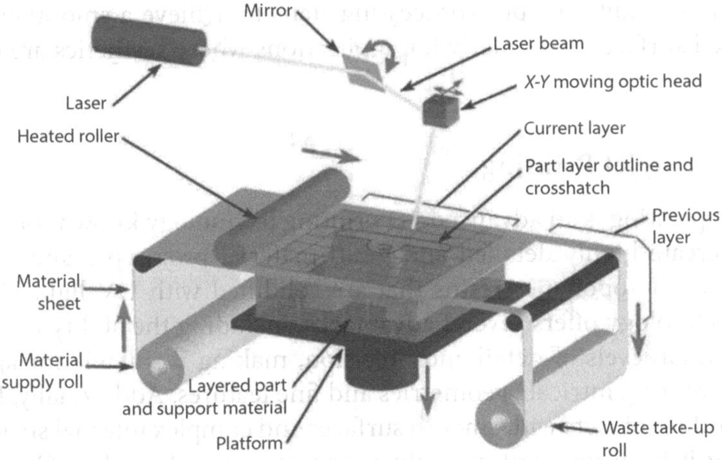

Figure 2.2 Scheme of laminated object manufacturing (LOM).

requirements of functional applications and prototypes [68]. However, it is essential to consider LOM's surface finish. While it can produce functional and durable parts, the surface may not be as smooth as those achieved with technologies like SLA or DLP. Therefore, additional post-processing steps may be necessary to enhance surface aesthetics, especially for applications where a polished finish is desired. Another consideration is LOM's limitations in achieving fine levels of detail and intricate geometries [69]. It may not excel in this regard compared to some other 3D printing methods like SLA or DLP. Thus, project requirements for high resolution or intricate designs should guide the selection of the appropriate printing technology.

Advantages
Laminated object manufacturing (LOM) excels in its versatility, accommodating a wide range of materials, including paper, plastic sheets, and metal-coated sheets. This adaptability allows for tailored material selection to meet specific application requirements, making LOM suitable for projects that demand diverse material properties, cost-effectiveness, and the production of larger objects [70].

Disadvantages
One limitation of laminated object manufacturing (LOM) lies in its surface finish, as LOM-printed objects may have a relatively rougher texture compared to certain other 3D printing methods. This roughness may

necessitate additional post-processing steps to achieve a smoother, more polished surface, particularly for applications where aesthetics are critical [80].

2.2.6 PolyJet Printing

PolyJet printing is an advanced 3D printing technology known for its ability to create highly detailed and multi-material prototypes and parts. It relies on photopolymer resins that are solidified with UV light [71–74]. This technology offers several advantages, including the ability to achieve exceptional levels of detail and precision, making it suitable for applications requiring intricate geometries and fine features. Additionally, PolyJet can produce objects with smooth surfaces and complex internal structures, making it ideal for creating realistic prototypes and models. One of the standout features of PolyJet is its capability to print with multiple materials simultaneously in a single job [75]. This allows for the creation of parts with varied material properties, colors, and transparencies within the same object (Figure 2.3). Full-color printing is another strength of PolyJet, making it valuable for applications where visual realism or color differentiation is crucial, such as product design and architectural modeling [76].

PolyJet printers can generate support structures using a different, dissolvable material, enabling the production of complex geometries and intricate overhangs. Support removal is often more straightforward compared to some other 3D printing methods. Moreover, PolyJet is well suited for rapid prototyping and design iterations, thanks to its speed and ability to create highly detailed, multi-material prototypes [77]. However, there

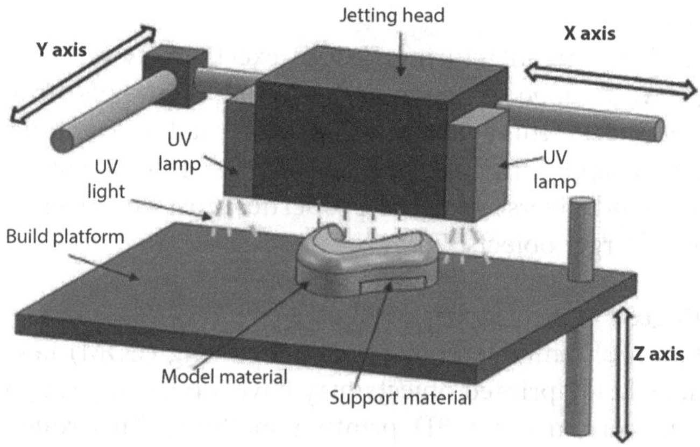

Figure 2.3 Scheme of PolyJet printing.

are considerations with PolyJet printing. Photopolymer resins used in PolyJet can be relatively expensive, impacting the overall production cost, especially for larger or multi-material prints. Some PolyJet-printed objects may require post-processing steps such as sanding or painting for applications requiring an even finer finish. Additionally, the durability and mechanical properties of PolyJet-produced parts may not be as robust as those produced with other 3D printing methods like fused deposition modeling (FDM) or selective laser sintering (SLS). Material selection is crucial, particularly for functional parts [78]. Lastly, PolyJet printers may have limitations regarding build volume, restricting the size of objects that can be printed.

Advantages

PolyJet printing stands out for its exceptional precision and versatility, capable of producing highly detailed and multi-material prototypes with smooth surfaces and intricate internal structures. It enables full-color printing, making it ideal for applications demanding visual realism, while its simultaneous support removal simplifies the production of complex geometries. PolyJet is also well suited for rapid prototyping, offering speed and design flexibility [78].

Disadvantages

A drawback of PolyJet printing is its relatively higher material costs, which can impact overall production expenses, especially for larger or multi-material prints. Additionally, some PolyJet-printed objects may require post-processing steps to achieve a smoother finish, and the durability and mechanical properties of parts may not be as robust as those produced using other 3D printing methods. Material selection is crucial for functional applications. Lastly, PolyJet printers may have limitations in terms of build volume, restricting the size of objects that can be printed [79].

2.2.7 Powder Bed and Inkjet Head 3D Printing (3DP)

Powder Bed and Inkjet Head 3D Printing (PBJ) is an additive manufacturing technique that utilizes a layer-by-layer approach to construct three-dimensional objects. The process begins by evenly spreading a fine powder material, often gypsum or plaster, across a build platform. A specialized print head, resembling those found in inkjet printers, selectively deposits liquid binding agents onto the powder bed. This precise application of binding agents solidifies the powder in specific areas, defining the cross-sectional shape of the object being printed. Emerging from the

Massachusetts Institute of Technology (MIT) in 1993 as a rapid prototyping innovation, Powder Bed and Inkjet Head 3D Printing (3DP) leverages powder-based processing, as depicted in Figure 2.4a (reproduced with permission from [80], 2017 @ ELSEVIER). After completing one layer, a new layer of powder is distributed, and the print head repeats the process until the entire object is formed. One of the distinctive features of PBJ is its ability to produce full-color, highly detailed prototypes and models. This is achieved by incorporating colored binders, allowing for intricate and vibrant 3D models with visual realism. PBJ technology finds extensive use in industries such as architecture, art, and design, where the ability to create visually appealing and colorful prototypes is critical. It excels in applications where precise color differentiation and detailed visualization are essential, making it a favored choice for architectural models, art installations, and conceptual design prototypes. However, it is worth noting that PBJ may have limitations in terms of material strength and durability compared to other 3D printing methods. Consequently, it is often preferred for producing visual and conceptual prototypes rather than functional parts or end-use products that require higher mechanical performance. As with any 3D printing technology, material selection should align with the specific requirements of the intended application.

Advantages

Powder Bed and Inkjet Head 3D Printing (PBJ) offers a unique advantage in its ability to produce highly detailed, full-color prototypes and models with visual realism. This technology's multi-material capabilities, utilizing colored binders, allow for the creation of intricate and vibrant 3D objects,

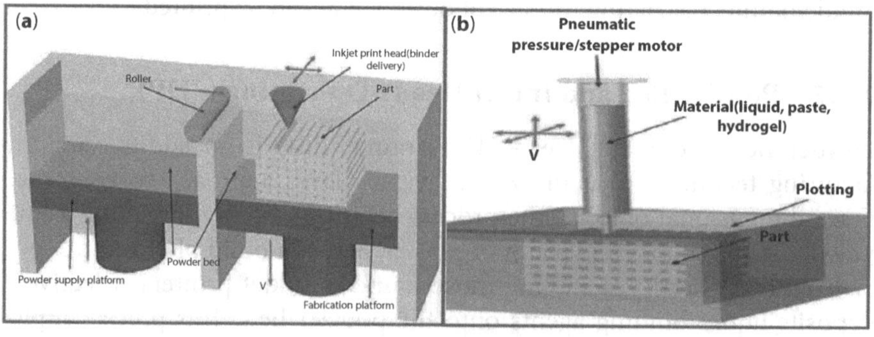

Figure 2.4 Schemes of (a) powder bed and inkjet head 3D printing (3DP) and (b) 3D plotting.

making it ideal for applications in architecture, art, and design where precise color differentiation and visual appeal are paramount.

Disadvantages

A limitation of Powder Bed and Inkjet Head 3D Printing (PBJ) is its potential weakness in terms of material strength and durability, which may not meet the requirements of functional parts or end-use products. While PBJ excels in producing visually appealing and highly detailed prototypes, its focus on aesthetics can compromise the mechanical properties of printed objects. Therefore, it may not be the best choice for applications demanding robust and mechanically sound components, necessitating careful consideration of material suitability for specific use cases.

2.2.8 3D Plotting/Direct-Write

3D Plotting, also referred to as Direct-Write 3D Printing, is an additive manufacturing technique that stands out for its precision and versatility in creating complex three-dimensional structures. Unlike some other 3D printing methods that rely on layer-by-layer deposition of material, 3D Plotting takes a more direct approach. In 3D Plotting, a nozzle or print head dispenses material with precision onto a substrate, effectively building the object layer by layer in a manner similar to 2D plotting or inkjet printing. This method allows for exceptional control over the placement and distribution of material, resulting in the creation of intricate geometries and fine features with high resolution. This level of precision makes 3D Plotting suitable for applications that demand detailed and complex designs [81].

One notable advantage of 3D Plotting is its material versatility. It can work with a wide range of materials, including various polymers, ceramics, metals, and even biological materials like living cells. This versatility enables diverse applications across industries, from creating customized medical implants to producing intricate microelectronics components. Additionally, 3D Plotting often eliminates the need for support structures, a common requirement in many other 3D printing methods. This feature can simplify the printing process and reduce post-processing requirements, as complex overhangs and structures can be created without additional support. However, there are considerations to bear in mind. While 3D Plotting offers high precision, it may not be the fastest 3D printing method, particularly for larger objects, which may result in longer print times. Material compatibility is another critical factor, as not all materials are suitable for this method. Users need to carefully evaluate material properties, viscosity, and compatibility with the specific 3D Plotting system. Additionally, there may be size limitations to consider, impacting the suitability of 3D Plotting for certain projects. Finally, equipment and material costs for 3D Plotting systems can vary, and some setups may be relatively expensive, influencing the overall cost of production. In summary, 3D Plotting/Direct-Write 3D Printing is a technology valued for its precision, material versatility, and suitability for applications requiring intricate details and customized structures. However, users should be mindful of factors like print speed, material compatibility, size constraints, and costs when assessing its suitability for specific projects [81].

Advantages

A significant advantage of 3D Plotting/Direct-Write 3D Printing is its exceptional precision, allowing for the creation of highly detailed and complex three-dimensional structures with intricate features. This precision makes it particularly well suited for applications where fine details and customized designs are essential. Additionally, its versatility in working with a wide range of materials, including polymers, ceramics, metals, and biological substances, makes it suitable for diverse applications across industries. Furthermore, the elimination of the need for support structures simplifies the printing process and reduces post-processing requirements, contributing to greater efficiency and cost-effectiveness in many projects.

Disadvantages

3D Plotting/Direct-Write 3D Printing offers exceptional precision and material versatility, making it ideal for detailed and customized structures. However, it may have slower print speeds, limiting its suitability for rapid production, and material compatibility and equipment costs should be carefully considered for specific applications.

2.3 Conclusion

Polymers constitute the foundational materials for the majority of AM processes. Ongoing technological advancements and explorations into polymers as 3D printing substrates reflect a swiftly progressing domain. The realm of 3D printing is becoming an essential tool for fabricating medical devices, electronics, and high-performance structural elements. Simultaneously, there remains significant uncharted territory in the chemistry realm for additive manufacturing, and the range of commercially available material properties remains constrained. Although AM holds the potential to substantially impact our daily lives, addressing current challenges and pursuing future objectives will require sustained efforts and progress.

References

1. Stansbury, J.W. and Idacavage, M.J., 3D printing with polymers: Challenges among expanding options and opportunities. *Dent. Mater.*, 32, 54, 2016.
2. MacDonald, E. and Wicker, R., Multiprocess 3D printing for increasing component functionality. *Science*, 353, aaf2093, 2016.
3. Jungst, T., Smolan, W., Schacht, K., Scheibel, T., Groll, J., Strategies and Molecular Design Criteria for 3D Printable Hydrogels. *Chem. Rev.*, 116, 1496, 2016.
4. Yang, Y., Song, X., Li, X., Chen, Z., Zhou, C., Zhou, Q., Chen, Y., Recent Progress in Biomimetic Additive Manufacturing Technology: From Materials to Functional Structures. *Adv. Mater.*, 30, 1706539, 2018.
5. Velasco-Hogan, A., Xu, J., Meyers, M.A., Additive Manufacturing as a Method to Design and Optimize Bioinspired Structures. *Adv. Mater.*, 30, 1800940, 2018.
6. Farahani, R.D., Dube, M., Therriault, D., 3D Printing of Multifunctional Nanocomposites:Manufacturing Techniques and Applications. *Adv. Mater.*, 28, 5794, 2016.
7. Truby, R.L. and Lewis, J.A., Printing soft matter in three dimensions. *Nature*, 540, 371, 2016.
8. Wendel, B., Rietzel, D., Kühnlein, F., Feulner, R., Hülder, G., Schmachtenberg, E., Additive Processing of Polymers. *Macromol. Mater. Eng.*, 293, 799–809, 2008.
9. Layani, M., Wang, X., Magdassi, S., Novel materials for 3D printing by photopolymerization. *Adv. Mater.*, 30, 1706344, 2018.
10. Zhang, J. and Xiao, P., 3D Printing of Photopolymers. *Polym. Chem.*, 9, 1530–1540, 2018.
11. ASTM F2792-12a, *Standard Terminology for Additive Manufacturing Technologies*, ASTM International, West Conshohocken, PA, 2012.
12. C.W. Rowe, W.E. Pryce Lewis, M.J. Cima, E.R.N. Bornancini, J.K. Sherwood, C.-C. Wang, C.M. Gaylo, J.A. Fairweather, . PCT Int. Appl. WO 2003041690, 2003.

13. Ursan, I., Chiu, L., Pierce, A., 3D drug printing: a structured review. *J. Am. Pharm. Assoc.*, 53, 136–144, 2013.
14. Pi, Q., Maharjan, S., Yan, X., Liu, X., Singh, B., van Genderen, A.M., Robledo-Padilla, F., Parra-Saldivar, R., Hu, N., Jia, W., Xu, C., Kang, J., Hassan, S., Cheng, H., Hou, X., Khademhosseini, A., Zhang, Y.S., Digitally Tunable Microfluidic Bioprinting of Multilayered Cannular Tissues. *Adv. Mater.*, 30, 1706913, 2018.
15. Kim, W., Kim, M., Kim, G.H., 3D-Printed Biomimetic Scaffold Simulating Microfibril Muscle Structure. *Adv. Funct. Mater.*, 28, 1800405, 2018.
16. Compton, B.G. and Lewis, J.A., 3D Printing:3D-Printing of Lightweight Cellular Composites. *Adv. Mater.*, 26, 5930, 2014.
17. Schaffner, M., Rühs, P.A., Coulter, F., Kilcher, S., Studart, A.R., 3D printing of bacteria into functional complex materials. *Sci. Adv.*, 3, eaao6804, 2017.
18. Hinton, T.J., Jallerat, Q., Palchesko, R.N., Park, J.H., Grodzicki, M.S., Shue, H.-J., Ramadan, M.H., Hudson, A.R., Feinberg, A.W., 3Dprinting of complex biological structures by freeform reversible embedding of suspended hydrogels. *Sci. Adv.*, 1, e1500758, 2015.
19. Somireddy, M., Czekanski, A., Singh, C.V., Computational modeling of constitutive behaviour of 3D printed composite structures. *Mater. Today Commun.*, 15, 143–152, 2018.
20. Daminabo, S.C., Goel, S., Grammatikos, S.A., Nezhad, H.Y., Thakur, V.K., Fused deposition modeling-based additive manufacturing(3Dprinting): techniques for polymer material systems. *Mater. Today Chem.*, 16, 100248, 2020.
21. Zhou, Y., Xia, X., Liu, X., Huang, B., Xiao, L., Qian, Q., Chen, Q., Rheology and 3D Printability of Percolated Graphene–Polyamide-6 Composites. *Macromol. Mater. Eng.*, 304, 1900021, 2019.
22. Melenka, G.W., Cheung, B.K.O., Schofield, J.S., Dawson, M.R., Carey, J.P., Computational modeling of constitutive behaviour of 3D printed composite structures. *Compos. Struct.*, 153, 866–875, 2016.
23. Rengier, F. *et al.*, 3D printing based on imaging data: review of medical applications. *Int. J. Comput. Assisted Radiol. Surg.*, 5, 4, 335–341, 2010.
24. Anderson, I., Mechanical Properties of Specimens 3D Printed with Virgin and Recycled Polylactic Acid. *3D Print. Addit. Manuf.*, 4, 110–115, 2017.
25. Jorfi, M. and Foster, E.J., Recent Advances in Nanocellulose for Biomedical Applications. *J. Appl. Polym. Sci.*, 132, 41719–41738, 2015.
26. Ligon, S.C., Liska, R., Stampfl, J., Gurr, M., M¨ulhaupt, R., Polymers for 3D Printing and Customized Additive Manufacturing. *Chem. Rev.*, 117, 10212–10290, 2017.
27. Wang, X., Jiang, M., Zhou, Z., Gou, J., Hui, D., 3D printing of polymer matrix composites:A review and prospective. *Compos. B Eng.*, 110, 442–458, 2017.

28. Wohlers, T. and Gornet, T., History of additive manufacturing. *Wohlers Rep.*, 24, 118, 2014.
29. Weng, Z., Wang, J., Senthil, T., Wu, L., Mechanical and thermal properties of ABS/montmorillonite nanocomposites for fused deposition modeling 3D printing. *Mater. Des.*, 102, 276–283, 2016.
30. Nuñez, P., Rivas, A., García-Plaza, E., Beamud, E., Sanz-Lobera, A., Dimensional and surface texture characterization in fused deposition modelling (FDM) with ABS plus. *Procedia Eng.*, 132, 856–863, 2015.
31. Samykano, M., Selvamani, S., Kadirgama, K., Ngui, W., Kanagaraj, G., Sudhakar, K., Mechanical property of FDM printed ABS: Influence of printing parameters. *Int. J. Adv. Manuf. Technol.*, 102, 2779–2796, 2019.
32. Cuiffo, M.A., Snyder, J., Elliott, A.M., Romero, N., Kannan, S., Halada, G.P., Impact of the fused deposition (FDM) printing process on polylactic acid (PLA) chemistry and structure. *Appl. Sci.*, 7, 579, 2017.
33. Ni, F., Wang, G., Zhao, H., Fabrication of water-soluble poly (vinyl alcohol)-based composites with improved thermal behavior for potential three-dimensional printing application. *J. Appl. Polym. Sci.*, 134 , 44966, 2017.
34. Goyanes, A., Kobayashi, M., Martínez-Pacheco, R., Gaisford, S., Basit, A.W., Fused-filament 3D printing of drug products: Microstructure analysis and drug release characteristics of PVA-based caplets. *Int. J. Pharm.*, 514, 290–295, 2016.
35. Xiao, J. and Gao, Y., The manufacture of 3D printing of medical grade TPU. *Prog. Addit. Manuf.*, 2, 117–123, 2017.
36. Han, Y. and Kim, J., A study on the mechanical properties of knit fabric using 3D printing-Focused on PLA, TPU Filament. *J. Fash. Bus.*, 22, 93–105, 2018.
37. Zaldivar, R., Witkin, D., McLouth, T., Patel, D., Schmitt, K., Nokes, J., Influence of processing and orientation print effects on the mechanical and thermal behavior of 3D-Printed ULTEM® 9085 Material. *Addit. Manuf.*, 13, 71–80, 2017.
38. Wu, H., Sulkis, M., Driver, J., Saade-Castillo, A., Thompson, A., Koo, J.H., Multi-functional ULTEMTM 1010 composite filaments for additive manufacturing using Fused Filament Fabrication (FFF). *Addit. Manuf.*, 24, 298–306, 2018.
39. Knoop, F., Schoeppner, V., Knoop, F., Schoeppner, V., Mechanical and thermal properties of FDM parts manufactured with polyamide 12, in: *Proceedings of the 26th Annual International Solid Freeform Fabrication Symposium—An Additive Manufacturing Conference*, Austin, TX, USA, pp. 10–12, August 10–12, 2015.
40. Ahn, S.H., Lee, C.S., Jeong, W., Development of translucent FDM parts by post-processing. *Rapid Prototyping J.*, 10, 218–224, 2004.

41. Cahyadi, W., *Mechanical Properties of 3D Printed Acrylonitrile Styrene Acrylate*, vol. 3187, Electronic Theses and Dissertations, 2019.
42. Novakova-Marcincinova, L., Novak-Marcincin, J., Barna, J., Torok, J., Special materials used in FDM rapid prototyping technology application, in: *Proceedings of the 2012 IEEE 16th International Conference on Intelligent Engineering Systems (INES)*, Lisbon, Portugal, pp. 73–76, June 13–15, 2012.
43. Hashemi Sanatgar, R., Campagne, C., Nierstrasz, V., Investigation of the adhesion properties of direct 3D printing of polymers and nanocomposites on textiles: Effect of FDM printing process parameters. *Appl. Surf. Sci.*, 403, 551–563, 2017.
44. Dilara, G.T., Tugba, D.U., Ayse, S.A., Deniz, Y., Nesrin, H., Vasif, H., 3D and 4D Printing of Polymers for Tissue Engineering Applications. *Front. Bioeng. Biotechnol.*, 7, 1–22, 2019, doi: 10.3389/fbioe.2019.00164.
45. Sheth, S., Taylor, R.M., Adluru, H., Numerical investigation of stiffness properties of fdm parts as a function of raster orientation, in: *Proceedings of the 28th Annual International Solid Freeform Fabrication Symposium—An Additive Manufacturing Conference*, Austin, TX, USA, pp. 1112–1120, August 7–9, 2017.
46. Sun, Q., Rizvi, G., Bellehumeur, C., Gu, P., Effect of processing conditions on the bonding quality of FDM polymer filaments. *Rapid Prototyp. J.*, 14, 72–80, 2008.
47. Espalin, D., Ramirez, J.A., Medina, F., Wicker, R., Multi-material, multi-technology FDM: Exploring build process variations. *Rapid Prototyp. J.*, 20, 236–244, 2014.
48. Chen, M.Y., Skewes, J., Woodruff, M.A., Dasgupta, P., Rukin, N.J., Multi-colour extrusion fused deposition modelling: A low-cost 3D printing method for anatomical prostate cancer models. *Sci. Rep.*, 10, 1–5, 2020.
49. Cicala, G., Ognibene, G., Portuesi, S., Blanco, I., Rapisarda, M., Pergolizzi, E., Recca, G., Comparison of Ultem 9085 used in fused deposition modelling (FDM) with polytherimide blends. *Materials*, 11, 285, 2018.
50. Melchels, F.P., Feijen, J., Grijpma, D.W., A review on stereolithography and its applications in biomedical engineering. *Biomaterials*, 31, 24, 6121–6130, 2010.
51. Santoliquido, O., Colombo, P., Ortona, A., Additive Manufacturing of ceramic components by Digital Light Processing: A comparison between the "bottom-up" and the "top-down" approaches. *J. Eur. Ceram. Soc.*, 39, 2140–2148, 2019.
52. Cho, Y., Lee, I., Cho, D.-W., Laser scanning path generation considering photopolymer solidification in micro-stereolithography. *Microsyst. Technol.*, 11, 2-3, 158–167, 2005.

53. Heller, C. *et al.*, Vinyl esters: low cytotoxicity monomers for the fabrication of biocompatible 3D scaffolds by lithography based additive manufacturing. *J. Polym. Sci., Part A: Polym. Chem.*, 47, 24, 6941–6954, 2009.
54. Wu, C., Yi, R., Liu, Y.-J., He, Y., Wang, C.C., Delta DLP 3D printing with large size, in: *Proceedings of the 2016 IEEE/RSJ International Conference on Intelligent Robots and Systems (IROS)*, Daejeon, Korea, pp. 2155–2160, October 9–14, 2016.
55. Yi, R., Wu, C., Liu, Y.-J., He, Y., Wang, C.C., Delta DLP 3-D printing of large models. *IEEE Trans. Autom. Sci. Eng.*, 15, 1193–1204, 2017.
56. de Camargo, E.J., Moreschi, E., Baseggio, W., Cury, J.A., Pascotto, R.C., Composite depth of cure using four polymerization techniques. *J. Appl. Oral Sci.*, 17, 446–450, 2009.
57. Shen, M., Zhao, W., Xing, B., Sing, Y., Gao, S., Wang, C., Zhao, Z., Effects of exposure time and printing angle on the curing characteristics and flexural strength of ceramic samples fabricated via digital light processing. *Ceram. Int.*, 46, 24379–24384, 2020.
58. Wu, D., Zhao, Z., Zhang, Q., Qi, H.J., Fang, D., Mechanics of shape distortion of DLP 3D printed structures during UV post-curing. *Soft Matter*, 15, 6151–6159, 2019.
59. Shirazi, S.F.S., Gharehkhani, S., Mehrali, M., Yarmand, H., Metselaar, H.S.C., Kadri, N.A., Osman, N.A.A., A review on powder-based additive manufacturing for tissue engineering: Selective laser sintering and inkjet 3D printing. *Sci. Technol. Adv. Mater.*, 16, 033502, 2015.
60. Eshraghi, S. and Das, S., Mechanical and micro structural properties of polycaprolactone scaffolds with one-dimensional, two-dimensional, and three-dimensional orthogonally oriented porous architectures produced by selective laser sintering. *Acta Biomater.*, 6, 2467–2476, 2010.
61. Vaezi, M., Seitz, H., Yang, S., A review on 3D micro-additive manufacturing technologies. *Int. J. Adv. Manuf. Technol.*, 67, 1721–1754, 2013.
62. Shuai, C., Zhuang, J., Peng, S., Wen, X., Inhibition of phase transformation from β to α -tricalcium phosphate with addition of poly (L-lactic acid) in selective laser sintering. *Rapid Prototyping J.*, 20, 369–376, 2014.
63. Schmid, M., Amado, A., Wegener, K., Polymer powders for selective laser sintering (SLS), in: *Proceedings of the AIP Conference Proceedings*, Cleveland, OH, USA, vol. 1664, p. 160009, June 6–12, 2014.
64. Gibson, I. and Shi, D., Material properties and fabrication parameters in selective laser sintering process. *Rapid Prototyp. J.*, 3, 4, 129–136, 1997.
65. Goodridge, R. *et al.*, Processing of a Polyamide-12/carbon nanofibre composite by laser sintering. *Polym. Test.*, 30, 1, 94–100, 2011.

66. Mueller, B. and Kochan, D., Laminated object manufacturing for rapid tooling and patternmaking in foundry industry. *Comput. Ind.*, 39, 47–53, 1999.
67. Subbiah, R., Arun, A., Lakshmi, A.A., Naga Sai Harika, A., Ram, N., Sateesh, N., Experimental Study of Wear Behaviour on Al-2014 Alloy Coated with Thermal Spray HVOF (High Velocity Oxy-Fuel) and Plasma Spray Process—A Review. *Mater. Today Proc.*, 18, 5151–5157, 2019.
68. Park, J., Tari, M.J., Hahn, H.T., Characterization of the laminated object manufacturing (LOM) process. *Rapid Prototyp. J.*, 6, 36–50, 2000.
69. Benjamin, D. and Nahum, T., Laminated Object Manufacturing of Ceramic-Based Materials. *Adv. Eng. Mater.*, 22, 9, 1–24, 2020, https://doi.org/10.1002/adem.202000256.
70. Paul, B.K. and Voorakarnam, V., Effect of layer thickness and orientation angle on surface roughness in laminated object manufacturing. *J. Manuf. Process.*, 3, 94–101, 2001.
71. Cazón, A., Morer, P., Matey, L., PolyJet technology for product prototyping: Tensile strength and surface roughness properties. *Proc. Inst. Mech. Eng., Part B: J. Eng. Manuf.*, 228, 1664–1675, 2014.
72. Miyanaji, H., Momenzadeh, N., Yang, L., Effect of printing speed on quality of printed parts in Binder Jetting Process. *Addit. Manuf.*, 20, 1–10, 2018.
73. Singh, R., Process capability study of polyjet printing for plastic components. *J. Mech. Sci. Technol.*, 25, 1011–1015, 2011.
74. Barclift, M.W. and Williams, C.B., Examining variability in the mechanical properties of parts manufactured via polyjet direct 3D printing, in: *Proceedings of the International Solid Freeform Fabrication Symposium*, pp. 6–8, Austin, TX, USA, August 6–8, 2012.
75. Razvan, U. and Ion, C.B., Polyjet technology applications for rapid tooling. *MATEC Web Conf.*, 112, 03011, 2017, doi: 10.1051/matecconf/201711203011.
76. Gaynor, A.T., Meisel, N.A., Williams, C.B., Guest, J.K., Multiple-material topology optimization of compliant mechanisms created via PolyJet three-dimensional printing. *J. Manuf. Sci. Eng.*, 136, 061015:1–061015:10, 2014.
77. Ibrahim, D., Broilo, T.L., Heitz, C., de Oliveira, M.G., de Oliveira, H.W., Nobre, S.M.W., dos Santos Filho, J.H.G., Silva, D.N., Dimensional error of selective laser sintering, three-dimensional printing and PolyJet. models in the reproduction of mandibular anatomy. *J. Craniomaxillofac. Surg.*, 37, 167–173, 2009.
78. Tan, W.S., Suwarno, S.R., An, J., Chua, C.K., Fane, A.G., Chong, T.H., Comparison of solid, liquid and powder forms of 3D printing techniques in membrane spacer fabrication. *J. Membr. Sci.*, 537, 283–296, 2017.
79. Jasiuk, I., Abueidda, D.W., Kozuch, C., Pang, S., Su, F.Y., McKittrick, J., An Overview on Additive Manufacturing of Polymers. *JOM*, 70, 275–283, 2018.

80. Xin, W., Man, J., Zuowan, Z., Jihua, G., David, H., 3D printing of polymer matrix composites: A review and prospective. *Composites, Part B*, 110, 442–458, 2017, doi: 10.1016/j.compositesb.2016.11.034.
81. Murphy, S.V. and Atala, A., 3D bioprinting of tissues and organs. *Nat. Biotechnol.*, 32, 8, 773–785, 2014.

3

Potential Roadmap of Synthetic Polymers in 3D Printing and Their Diverse Application

Toheeb Jimoh[1] and Jude A. Okolie[2]*

[1]*Department of Chemical Engineering, University of Ilorin, Ilorin, Nigeria*
[2]*Gallogly College of Engineering, University of Oklahoma, Norman, OK, United States*

Abstract
This chapter explores the evolving landscape of synthetic polymers in 3D printing, highlighting both their potential and the challenges they present. It starts with a discussion on common synthetic polymers used in 3D printing, along with current research in the field. This includes studies demonstrating the creation of stimulus-responsive capsules using plasmonic nanoparticles and 3D printing technology, the use of peptide-polymer conjugates for conditioned fabrication of scaffolds, and the enhancement of scaffold properties through the incorporation of cellulose nanofibrils for improved cellular responses in bone tissue regeneration. Furthermore, the chapter delves into how the mechanical properties of 3D-printed polycaprolactone (PCL) scaffolds can be tailored for specific tissue types. However, the use of synthetic polymers in 3D printing is not without challenges, such as material properties differing from traditional manufacturing methods, variations in print quality, the necessity of post-processing, higher costs, environmental impacts, slower production speeds, and limited material availability. Despite these challenges, continuous research and development are paving the way to overcome these obstacles through new materials, technologies, and techniques.

Keywords: Polymers, polycaprolactone, 3D printing, polyethylene glycol, biodegradeable, carbon fiber

**Corresponding author*: Jude.okolie@usask.ca

Bhasha Sharma, Balaram Pani, Shashank Shekhar and Jude A. Okolie (eds.) Sustainable 3D Printing for Innovative Biopolymer Production and Applications, (49–64) © 2025 Scrivener Publishing LLC

3.1 Introduction

The computer-aided additive manufacturing approach based on 3D printing has evolved as a ground-breaking manufacturing technique and is currently seen to be at the pinnacle of research in disciplines like biomedical engineering and material sciences [1]. 3D printing is believed to hold great promise for the emerging era of personalized medicine because it allows for the layer-by-layer creation of medical devices, tissues, and other bio-systems with heterogenous and diverse frameworks that can imitate their native equivalents [2]. It then becomes crucial in these biofabrication efforts to choose the right printing technology, design, and materials so as to maintain the connectivity between processing, structure, and property in the printed 3D structures [1].

Depending on the 3D printing applications, it may be necessary to create and optimize printable ink materials to attain the requisite physico-mechanical, rheological, and bifunctional qualities in order to produce functional products that are of great strength and self-supporting. Diversity in 3D printable ink materials is much needed despite the availability of numerous ink materials such as ceramics, composites, and polymers [3]. Hence, the advent of bioinks (3D printable ink materials impregnated with cells) has piqued the interest of researchers and is thought to be one of the most cutting-edge methods for opening up new opportunities in drug delivery, tissue engineering, regenerative medicine, etc.

Two types of polymer (natural and synthetic polymers) have been employed in producing ink with attractive biological and mechanical properties for the purpose of 3D printing [1]. Natural polymers are obtained from animals, plants, cells, etc. They are currently used extensively for producing bioinks because they are characterized with excellent cellular proliferation and biocompatibility rates. It can be difficult to use naturally produced polymers for biofabrication because they are frequently coupled with various drawbacks such as faster rate of biodegradability and batch-to-batch variations [4]. Synthetic polymers, on the other hand, can overcome these limitations and also offer the chance to be modified with certain biological, physical, and chemical properties which then pave the way for the creation of a vast range of novel materials for (bio)inks [1, 5]. By choosing the right monomer, controlling the architecture, and taking use of post-polymerization functionalization options, synthetic polymers also provide the chance to modify rheological and functional behaviors, such as the inks' mechanical integrity and ability to print [6].

In recent years, the usage of synthetic polymers in 3D printing has increased significantly. Several different synthetic polymer materials, such as PEG (polyethylene glycol), PLA (polylactic acid), PCL (polycaprolactone), and PETG (polyethylene terephthalate glycol), can be used in 3D printing [7]. These materials are excellent for a variety of applications because of their many qualities, including strength, flexibility, and thermal resistance—for instance, PLA is a biodegradable material that is frequently used for prototyping. A vast range of things, from tiny toys and figurines to substantial industrial parts and structural elements, have been produced using 3D printing and synthetic polymers [1]. The technique is employed across a wide range of sectors, including those of aircraft, automotive, healthcare, and architecture [8]. A lot of research is also being done to improve the sustainability, biodegradability, and environmental friendliness of these materials (synthetic polymers). Additionally, numerous novel synthetic polymer and composite varieties are being created especially for 3D printing. This entails the use of polymers reinforced with metals, glass, and carbon fiber to increase the strength and longevity of printed products [1]. Overall, synthetic polymers continue to be a frequently used and preferred material in 3D printing, but work is still being done to enhance their capabilities.

This chapter is then aimed at exploring the potential roadmap of synthetic polymers in 3D printing and other applications. To do this, this chapter has been prepared as follows; first, the common synthetic polymers employed in 3D printing coupled with the ongoing research in synthetic polymer will be discussed. Following this, the advancements in synthetic polymers for 3D printing will also be looked into. Lastly, this chapter will also present the diverse applications of synthetic polymers in 3D printing.

3.2 Types of Synthetic Polymers in 3D Printing

3.2.1 Overview of Common Synthetic Polymers Used in 3D Printing

Synthetic polymers also known as artificial polymer are produced via processes that can be fine-tuned in terms of their physical features and chemical frameworks. A vast amount of this type of polymer is characterized with super-mechanical properties such as fracture toughness, tensile strength, fatigue, and elastic modulus. While printing 3D, synthetic polymers with suitable mechanical properties can withstand internal and external stresses [1, 7]. They offer a tremendous deal of prospects to be employed as ink materials for 3D printing since they can be shaped into sophisticated and highly customizable designs for tissue (hard and soft tissues) scaffolds and medical devices. To preserve biocompatibility and biodegradation difficulties, a synthetic polymer needs to be carefully chosen [9]. The commonly used biodegradable synthetic polymers are, but not limited to, PU, PEG, PCL, and PLGA as detailed below:

(i) PCL (polycaprolactone)
The United States' Food and Drug Administration (FDA) has given their approval for the utilization of PCL, a thermoplastic polymer, in suture materials, drug delivery systems, and adhesion barriers in the human body [1]. This polymer is characterized with a melting point of about 60°C and a glass transition temperature (T_g) of about -60°C [10]. Owing to this polymer's biodegradability and its semicrystalline shape, it dissolves through hydrolysis of its ester bond under physiological circumstances. It has a slower rate of biodegradation when compared to natural polymers like gelatin, fibrin, and collagen [11]. PCL has long been utilized as a supplement for a variety of biomedical uses. As an illustration, PCL could be impregnated with starch to not only improve its biodegradability but also reduce its cost. PCL has no natural peptide patterns as with most other synthetic polymers. For PCL to be widely used, a typical strategy is to combine it with useful biomaterials or naturally occurring polymers to generate hybrid structures [7].

(ii) PU (polyurethane)
PU is a class of synthetic polymer that is made up of organic and oligodiol units which are connected by urethane linkages. Biodegradable or non-biodegradable polyurethanes have been extensively applied in biomedical engineering because of the outstanding biocompatibilities and exceptional mechanical qualities that they are known for. The physicochemical properties of polyurethanes are dependent on their chemical compositions [12]. To make waterborne PU, for instance, an ionic hydrophilic group is added to PU, which then causes it not to only disperse

in water but also become an ionomer. It has been found that oligodiols have a significant impact on the thermosensitivity of PU hydrogels [7]. Thermosensitive PUs can then be used for 3D printing of cells and tissues when an acrylate group is added to it.

(iii) PEG (polyethylene glycol)
PEG is also known as a synthetic polyether having a linear and branching structure that is hydrophilic, biocompatible, and non-immunogenic. It has been given FDA approval for use in biomedicine as a viable option in encapsulating cells. Their hydrogels are not biodegradable by nature; however, they can be modified to improve decomposition by adding degradable segments [13]. Undoubtedly, PEG and its variants appear to be the most studied synthetic polymer for soft tissue repair. PEG is frequently chemically altered to produce the photopolymerizable PEGDA, which allows for the rapid encapsulation of cells [14]. When PEGDA is polymerized by UV light, it becomes photopolymerized, which proves useful in extrusion-based 3D printing for the purpose of producing tissue engineering scaffolds with improved mechanical qualities [15].

(iv) PLGA [poly(lactic-co-glycolic acid)]
PLGA is formed from the copolymerization of lactic acid (LA) and glycolic acid (GA), while aluminum isopropoxide or tin (II) alkoxides is used as a catalyst. The successive units of PLGA's monomers are connected by ester linkages during its formation; hence, PLGA degrades by hydrolyzing its ester bonds in the presence of water [16]. The PLGA degradation duration is dependent on the proportions of the reactants (lactic acid and glycolic acid) used in the beginning materials; a higher quantity of glycolic acid will result in a faster degradation rate and then otherwise for lactic acid [13]. The United States' FDA has given its approval for the utilization of PLGA in biomedical applications (microspheres, porous scaffolds, and films). Glass transition temperatures for PLGAs commonly fall between 40°C and 60°C. PLGA is more suitable for load-bearing applications than natural polymers because of its better mechanical qualities [7].

Table 3.1 Properties of some synthetic polymers used in 3D printing.

Polymers	Properties	Benefits	References
Polycaprolactone (PCL)	1. It has a melting point of 58°C. 2. It requires more than 2 years to biodegrade. 3. It has a glass temperature of -72°C. 4. It has a tensile strength of 16 MPa. 5. It has a breaking strain of over 80%.	1. It is a non-toxic polymer, making it suitable for medical applications. 2. It is biodegradable; hence, it can be used in tissue engineering. 3. It can also be used in packaging foods.	[6, 17]
Polyurethane (PU)	1. It does not have a clear melting point. 2. Its glass temperature is dependent on its composition. PU foams' T_g ranges between -20°C and 20°C while that of PU elastomers ranges from -60°C to -40°C.	1. They can be used in making seals and gaskets because they are highly elastic. 2. Some PU foams are fire-resistant; hence, they can be used as insulation in buildings. 3. PUs are highly stable thermally, making them suitable for low- and high-temperature applications.	[18, 19]
Polyethylene glycol (PEG)	1. It has a glass temperature of about -73°C–20°C. 2. It has a breaking strain of over 60%. 3. It has a melting point of approximately 35°C–65°C.	1. It can be used as a lubricant because of its low friction coefficient. 2. It is soluble in water; hence, it can be used to dissolve drugs and other water-insoluble compounds. 3. It can be used as an emulsifying agent in food and cosmetic products.	[7, 17, 20]

(Continued)

Table 3.1 Properties of some synthetic polymers used in 3D printing. (*Continued*)

Polymers	Properties	Benefits	References
Poly(lactic-co-glycolic acid) (PLGA)	1. It is amorphous in shape. 2. It has a glass temperature of approximately -73°C–20°C. 3. It takes less than a year to biodegrade. 4. Its breaking strain is 20%.	1. It can be used in making scaffolds for tissue engineering because of its biodegradability. 2. It can also be used to create a drug delivery system.	[6, 7]

3.3 Comparison of Properties and Benefits of Type of Synthetic Polymers

Synthetic polymers possess exceptional mechanical properties such as thermal resistance, fracture toughness, tensile strength, fatigue, elastic modulus, etc. Table 3.1 provides a summary of the most popular synthetic polymers used in 3D printing.

A (bio)fabricated construction with a proper control over resolution, integrity, and size is produced when synthetic polymers and not conventional methods are employed as the ingredients for producing ink.

3.4 Discussion of Current Research and Development in Polymers for 3D Printing

In their study, [16] demonstrated how to combine useful plasmonic nanoparticles and 3D printing technology to produce stimulus-responsive capsules with programmable release from hydrogel matrices that are printed in a multi-plexed array and shelled in PLGA polymer. [21] reported a considerable improvement in NIH 3T3 fibroblast adhesion when they use multiple printer heads to examine RGES(azide)-PCL and RGDS(biotin)-PCL-based peptide-polymer conjugates for the conditioned fabrication of 3D printed scaffolds. [22] reported that when a PCL-based 3D-printed scaffold is doped with cellulose nanofibrils, it enhances the PCL scaffold surface's protein absorption and hydrophilicity. This then subsequently improves the mesenchymal stem cells' *in vitro* cellular response, which opens the possibilities of using the printed 3D PCL scaffolds in bone tissue regeneration.

The 3D-printed PCL scaffolds' mechanical properties could be fine-tuned by altering their structures' geometries and porosities so as to correspond with that of the host tissues in areas like trabecular bones, cartilage, and joints [23]. Peripheral nerve regeneration (PNG), human fibroblast proliferation (HFP), and human adipose-derived stem cell (hASC) accommodation were all made possible by the 3D-printed PCL scaffolds' customized porosities and geometries [24].

3.5 Advancements in Synthetic Polymer for 3D Printing

Challenges and Limitations of Synthetic Polymers in 3D Printing

The use of synthetic polymers in 3D printing is subject to a number of difficulties and restrictions, such as the following:

1) Material properties: The performance and durability of the finished product might be affected by the synthetic polymers used in 3D printing because they frequently have different properties from those used in conventional manufacturing processes, such as injection molding [25].
2) Print quality: Variations in print quality, including surface finish, layer alignment, and dimensional accuracy, can occur when using synthetic polymers for 3D printing [26].
3) Post-processing: In order to enhance the aesthetic and functionality of the synthetic polymers used in 3D printing, post-processing procedures like sanding, painting, or coating are frequently necessary [15].
4) Cost: The cost of the materials, machinery, and post-processing stages can make 3D printing using synthetic polymers more expensive than conventional production techniques [27].
5) Environmental impact: Although the effects of 3D printing with synthetic polymers on the environment are still being assessed, it is known that the process produces waste plastic that must be disposed of correctly [28].

6) Speed: The scalability of 3D printing with synthetic polymers is constrained by the fact that it is frequently slower than conventional manufacturing techniques, which might extend the lead time for production [29].
7) Material availability: Compared to more conventional production techniques, the selection of synthetic polymers available for 3D printing is currently fairly small [26].

Despite these difficulties and restrictions, research and development into 3D printing with synthetic polymers is still ongoing. Many of these difficulties are being resolved with new materials, technologies, and post-processing techniques.

3.6 Exploration of Potential Solutions and Future Development in Synthetic Polymers for 3D Printing

Even though polymers are known for their inexpensiveness, flexibility, and light density, their 3D-printed products are characterized with complicated geometric designs; besides that, their poor functionality and mechanical strength impedes their widespread application. These difficulties can be sorted by mixing different materials so that excellent functional and mechanical qualities can be obtained [25]. This then piqued the interest of researchers to develop composite materials that can work with the current printers. These composites could be (polymer + nanoparticles, polymer + particles, polymer + fibers) [25, 29]. These are discussed in detail below:

(i) Nanocomposites

A nanocomposite-reinforced polymer, also known as nanoengineered polymer, is formed when nanofillers such as carbon nanotubes (CNT), graphene, nanorods, nanofibers, quantum dots, and nanoclays are doped with polymer materials [30]. This product (nanoengineered polymer) has been found to demonstrate far better mechanical and electrical properties, rheological performances, and excellent functionalities than the ordinary polymer itself owing to the numerous benefits of the integrated nanomaterials [31], making the polymer nanocomposite a viable option for enabling additive manufacturing-based complex, useful, and highly personalized 3D constructs. When [32] produced a nanoengineered polymer (photopolymer + graphene oxide), they reported that the composite is of excellent ductility and strength. Not only that, they also revealed that 0.2% addition of graphene oxides causes 12.8% increase in elongation and a 62.2% increase in tensile strength, while the polymer's excellent ductility is due to the crystallinity of the graphene oxides.

(ii) Fiber-reinforced polymer
Incorporating fibers to polymer materials could also improve their properties. The common fiber materials that are frequently used in the 3D printing industry are standard short fibers, such as carbon fibers (CFs) and glass fibers [25]. The characteristics of finished composite products are significantly influenced by the void fraction and fiber orientation of composites. This incorporation could be done via FDM and the direct write technique. For the latter, fibers are first mixed with the polymer pellets in a blender before being supplied to an extruder to create filaments for FDM processing [31]. To ensure the uniform distribution of fibers, a second extrusion procedure could be used. [33] reported that a 3D-printed composite made of PLA and carbon fiber had a strength and tensile modulus of 185.2 ± 24.6 MPa and 19.5 ± 2.08 GPa, respectively, which is equivalent to 435% and 599% of the pristine PLA. The printed samples do, however, occasionally still have irregularities and discontinuities in the fiber. Despite a significant improvement over pure polymers in terms of mechanical properties, composites still fell short of the ruler of the mixture's calculated value [34].

(iii) Particle-reinforced polymer
Particle reinforcements are frequently utilized to enhance the qualities of polymer matrix because of their inexpensive cost. It is simple to combine particles with polymers via the SLA or FDM process whether they are in powder or liquid form. A fascinating development in the 3D printing of particle-reinforced polymer is its ability to produce structural components for potential practical uses [26]. Particles are included into polymers to help solve several printing-related issues. One difficulty with FDM printing technique is the distortion of the final printed parts, which

is brought on by the polymer's thermal expansion. This issue has been effectively solved by incorporating metal ions into polymers [35]. By adding magnetized alumina platelets to a polymer matrix, [36] were able to regulate the orientation of the printed particles using a unique magnetically assisted 3D printing platform.

3.7 Diverse Applications of Synthetic Polymers in 3D Printing

Overview of the Various Industries and Fields Utilizing Synthetic Polymers

(i) Automotive industry
Some aerospace and automotive parts are known to have complicated geometries, which makes them time-consuming and expensive to produce using traditional methods like injection molding [25]. Hence, 3D printing becomes a suitable alternative for their fabrication. The majority of their components such as turbine blade have so far been 3D-printed using metal materials [29] due to the metals' better strength and fire retardancy features compared to polymers. However, the use of metals has been found to normally result in higher fuel consumption in automobiles as a result of the higher density of the metals. This then stimulates researchers to look for materials that could help overcome this issue posed by metals. Polymer, which is known for its light weight, chemical resistance, and reduced fuel consumption, seems to be a viable option. Automotive and aerospace industries may use 3D-printed polymer composites because they have excellent reproducibility and consistency [25]—for instance, [37] reported excellent reproducibility when he used polymer composites (carbon fiber reinforced photopolymer + glass fiber) to print propeller and airfoil.

(ii) Electronic sector
By utilizing cutting-edge 3D printing technology, complicated-designed electrical prototypes may be created with more resolution and accuracy in less time [29]. When an electrically conductive material is doped with a polymer

to print 3D, the resulting composite formed functions like electrical device and hence has numerous applications [25]. [38] produces sensors ranging from the piezoresistive to capacitive sensors when he doped PCL with carbon black. The development of specialized 3D electrical structures has taken a lot of research work as a connector was made recently in a digital light processing (DLP) printer by employing silver particles and cross-linked photopolymer [29].

(iii) Biomedical sector

The idea of 3D printing in clinics has developed as a result of the rapid advancement of numerous 3D printing processes and their ability to create products in any shape or size that may be customized [25]. It is projected that 3D printing of any product at a clinic will become a viable service in the ensuing decades.

(iv) Tissue engineering

Synthetic polymers like polylactic acid (PLA) and polyethylene glycol (PEG) are frequently employed to build scaffolds for the growth of tissue. These polymers can be developed to imitate the extracellular matrix's (ECM) characteristics and be used to stimulate cell proliferation and differentiation [29]. To encourage tissue regeneration, bioactive chemicals like growth factors can be added to synthetic polymers [7]. Additionally, synthetic polymers are easily modified to alter the scaffold's mechanical characteristics and rate of disintegration, enabling fine-grained control of tissue growth and repairs [3].

(v) Drug delivery

Drugs can be delivered directly to the location of a disease by being enclosed and protected in nanoparticles made of synthetic polymers like polyethylene glycol (PEG), which is utilized in medication delivery. PEG can also be utilized to make hydrogels that deliver medications over a long period of time [39].

(vi) Dentistry

Fillings, sealants, and orthodontic devices are just a few of the dental applications where synthetic polymers are frequently used—for instance, to treat cavities in teeth, composite resin fillings, which are formed from a combination of synthetic resin and glass or ceramic particles, are employed [29]. They can be sculpted and polished to resemble the surrounding teeth's color and shape. Similar to this, by sealing off the grooves and fissures that are particularly vulnerable to decay, dental sealants, which are thin coatings of synthetic polymer placed to the surface of teeth, can be used to prevent tooth decay [40].

3.8 Conclusion

The exploration of synthetic polymers in 3D printing reveals a field rich in potential but faced with significant challenges. The advancements in creating functional and responsive materials through 3D printing technology demonstrate its capacity to revolutionize areas like tissue engineering and regenerative medicine. However, the discrepancies in material properties, print quality issues, post-processing requirements, cost implications, environmental concerns, production speed, and limited material availability present substantial hurdles. These challenges underscore the need for ongoing research and development in this area. As the field continues to evolve with new materials and technologies, it is anticipated that many of these challenges will be mitigated, further cementing the role of synthetic polymers in the future of 3D printing and manufacturing.

References

1. Kumar, S., Synthetic polymer-derived single-network inks/bioinks for extrusion-based 3D printing towards bioapplications. *Mater. Adv.*, 2, 21, 6928–6941, 2021.
2. Lee, A.R.H.A., Hudson, A.R., Shiwarski, D.J., Tashman, J.W., Hinton, T.J., Yerneni, S., ... & Feinberg, A.W., 3D bioprinting of collagen to rebuild components of the human heart. *Science*, 365, 6452, 482–487, 2019.
3. Pugliese, R., Beltrami, B., Regondi, S., Lunetta, C., Polymeric biomaterials for 3D printing in medicine: An overview. *Ann. 3D Print. Med.*, 2, 100011, 2021.
4. Raju, S.N., Polymers Used for 3D Printing. *IRJET*, 8, 9, 2021, Retrieved from https://www.irjet.net.
5. Guvendiren, M., Molde, J., Soares, R.M.D., Kohn, J., Designing biomaterials for 3D printing, 2016.
6. Gopakumar, D.A., Pasquini, D., Henrique, M.A., Morais, L.C., Grohens, Y., Thomas, S., Meldrum's acid modified cellulose nanofiber-based

polyvinylidene fluoride microfiltration membrane for dye water treatment and nanoparticle removal, 2017.

7. Hartley, E., Moon, H., Neves, A., Biodegradable synthetic polymers for tissue engineering: a mini-review. *Reinvention: An International Journal of Undergraduate Research*, 15, 1, 2022.
8. Fang, J., Liu, Z., Luan, H., Liu, F., Yuan, X., Long, S. *et al.*, Renewable energy, Elsevier Ltd, 2020, https://doi.org/10.1016/j.renene.2020.11.033.
9. Zhou, L. Y., Fu, J., He, Y., A review of 3D printing technologies for soft polymer materials. *Adv. Func. Mater.*, 30, 28, 2000187, 2020.
10. Liu, J., Sun, L., Xu, W., Wang, Q., Yu, S., Sun, J., Synthesis and characterization of polycaprolactone nanofibers by electrospinning. *Carbohydr. Polym.*, 203, 1–17, 2018, https://doi.org/10.1016/j.carbpol.2018.11.077.
11. Labet, M. and Thielemans, W., Synthesis of polycaprolactone: A review. *J. Polym. Sci. Part A: Polym. Chem.*, 48, 3, 405–427, 2009.
12. Hsiao, S.H. and Hsu, S.H., Synthesis and characterization of dual stimuli-sensitive biodegradable polyurethane soft hydrogels for 3D cell-laden bioprinting. *Acta Biomater.*, 84, 48–57, 2018.
13. Shi, J., Yu, L., Ding, J., PEG-based thermosensitive and biodegradable hydrogels. *Acta Biomater.*, 122, 168–178, 2021, https://doi.org/10.1016/j.actbio.2021.04.009.
14. Yeo, K. S., Wang, H., Zhao, H. Y., 3D printable thermosensitive bioink for hydrogel biomaterials. *J. Biomater. Appl.*, 33, 9, 1143–1156, 2019.
15. Yang, Y., Wang, X., Lin, X., Xie, L., Ivone, R., Shen, J., Wu, W., A tunable extruded 3D printing platform using thermo-sensitive pastes. *Int. J. Pharm.*, 587, 119360, 2020.
16. Gupta, M., Meng, F., Johnson, B., Kong, Y.L., Tian, L., Yeh, Y.W., 3D programmable release capsules. *Nat. Commun.*, 6, 10, 6245, 2015.
17. Poulopoulou, N., Smyrnioti, D., Nikolaidis, G.N., Tsitsimaka, I., Christodoulou, E., Bikiaris, D.N., Charitopoulou, M.A., Achilias, D.S., Kapnisti, M., Papageorgiou, G.Z., Sustainable plastics from biomass: Blends of polyesters based on 2,5-furandicarboxylic acid. *Polymers*, 12, 1, 225, 2020.
18. Leblon, C.E., Pai, R., Fodor, C.R., Golding, A.S., Coulter, J.P., Jedlicka, S.S., *In vitro* comparative biodegradation analysis of salt-leached porous polymer scaffolds. *Polym. Degrad. Stab.*, 98, 7, 127–136, 2012.
19. Hong, Y., Guan, J., Fujimoto, K.L., Hashizume, R., Pelinescu, A.L., Wagner, W.R., Biomaterials tailoring the degradation kinetics of poly (ester carbonate urethane) urea thermoplastic elastomers for tissue engineering scaffolds. *Biomaterials*, 31, 15, 4249–58, 2010, http://dx.doi.org/10.1016/j.biomaterials.2010.02.005.
20. Myung, D., Koh, W., Ko, J., Hu, Y., Carrasco, M., Noolandi, J. *et al.*, Biomimetic strain hardening in interpenetrating polymer network hydrogels. *Nat. Mater.*, 6, 5376–87, 2007.

21. Camacho, P., Fainor, M., Seims, K.B., Tolbert, J.W., Chow, L.W., Fabricating spatially functionalized 3D-printed scaffolds for osteochondral tissue engineering. *Tissue Eng. Part A*, 27, 1-2, 14–21, 2021.
22. Rashad, A., Mohamed-ahmed, S., Ojansivu, M., Berstad, K., Yassin, A., Kivijärvi, T. *et al.*, Coating 3D printed polycaprolactone scaffolds with nanocellulose promotes growth and differentiation of mesenchymal stem cells. *J. Biomed. Mater. Res. A*, 106, 7, 2219–30, 2018.
23. Hutmacher, D.W., Schantz, J.T., Xu, C., Lam, F., Tan, K.C., Lim, T.C., State of the art and future directions of scaffold-based bone engineering from a biomaterials perspective. *J. Tissue Eng. Regen. Med.*, 1, 4, 245–60, June 2007.
24. Kurtz, S.M. and Devine, J.N., PEEK biomaterials in trauma, orthopedic, and spinal implants. *Biomaterials*, 28, 32, 4845–69, 2007.
25. Wang, X., Jiang, M., Zhou, Z., Gou, J., Hui, D., 3D printing of polymer matrix composites: A review and prospective. *Compos. Part B*, 110, 442–58, 2017, http://dx.doi.org/10.1016/j.compositesb.2016.11.034.
26. Mandrycky, C., Wang, Z., Kim, K., Kim, D.H., 3D bioprinting for engineering complex tissues. *Biotechnol. Adv.*, 34, 4, 422–34, 2016, http://dx.doi.org/10.1016/j.biotechadv.2015.12.011.
27. Ivone, R., Yang, Y., Shen, J., Recent advances in 3D printing for parenteral applications. *J. Control. Release*, 330, 602–614, 2021.
28. Melchels, F.P.W., Feijen, J., Grijpma, D.W., Biomaterials A review on stereolithography and its applications in biomedical engineering. *Biomaterials*, 31, 24, 6121–30, 2010, http://dx.doi.org/10.1016/j.biomaterials.2010.04.050.
29. Misra, M., Additive manufacturing technology of polymeric materials for customized products: recent developments and future prospective. *J. Polym. Res.*, 28, 12, 36398–438, 2021.
30. Yan, X., Gu, J., Zheng, G., Guo, J., Galaska, A.M., Yu, J., 3D printing of graphene oxide reinforced polymer nanocomposites. *Polymer*, 112, 546–556, 2016.
31. Mobaraki, M., Ghaffari, M., Yazdanpanah, A., Luo, Y., Bioprinting bioinks and bioprinting: A focused review. *Bioprinting*, 18, e00080, January 2020, https://doi.org/10.1016/j.bprint.2020.e00080.
32. Lin, D., Jin, S., Zhang, F., Wang, C., Wang, Y., 3D stereolithography printing of graphene oxide reinforced complex architectures. *Nanotechnology*, 26, 43, 434003, 2015, http://dx.doi.org/10.1088/0957-4484/26/43/434003.
33. Matsuzaki, R., Ueda, M., Namiki, M., Jeong, T.K., Asahara, H., Three-dimensional printing of continuous-fiber composites by in-nozzle impregnation. *Nat. Publ. Gr*, 1–7, February 2016.
34. Tian, X., Liu, T., Yang, C., Wang, Q., Li, D., Composites: Part A interface and performance of 3D printed continuous carbon fiber reinforced PLA composites. *Compos. Part A: Appl. Sci. Manuf.*, 88, 198–205, 2016, http://dx.doi.org/10.1016/j.compositesa.2016.05.032.
35. Hwang, S., Reyes, E.I, Moon, K.S., Rumpf, R.C., Kim, N., Thermo-mechanical characterization of metal/polymer composite filaments and printing

parameter study for fused deposition modeling in the 3D printing process. *J. Electron. Mater.*, 44, 3, 771–777, 2014.

36. Kokkinis, D. and Schaffner, M., Multimaterial magnetically assisted 3D printing of composite materials. *J. Polym. Sci., Part B: Polym. Phys.*, 53, 13, 893–899, 2015.
37. Invernizzi, M., Natale, G., Levi, M., Turri, S., Griffini, G., UV-Assisted 3D printing of glass and carbon fiber-reinforced dual-cure polymer composites. *Polym. Compos.*, 39, 7, 1538–1549, 2018.
38. Leigh, S.J., Bradley, R.J., Purssell, C.P., Billson, D.R., Hutchins, D.A., A simple, low-cost conductive composite material for 3D printing of electronic sensors. *PLoS ONE*, 7, 11, 1–6, 2012.
39. Selvakumaran, S. and Muhamad, I.I., Designing polymeric nanoparticles for targeted drug delivery system. *J. Control. Release*, 173, 1–12, 2014.
40. Negrutiu, M.L. and Cam, C.A.D., State-of-the-art and future perspectives-A review. *Polym. Rev.*, 62, 2, 1–24, 2022.

4

Nanotechnology-Derived 3D-Printed Applications: Opportunities and Challenges in Biopolymers

Ali Mehboob[1], Muhammad Adeel Abid[2], Sonia Javed[3], Tanveer Hussain[4], Imad Barsoum[1,5,6] and Sharjeel Abid[7*]

[1]*Advanced Digital & Additive Manufacturing Center, Khalifa University, Abu Dhabi, United Arab Emirates*
[2]*Princess of Wales Hospital, Wales, UK*
[3]*Institute of Science and Technology for Ceramics (ISTEC), CNR, University of Parma, Parma, Emilia-Romagna, Italy*
[4]*Department of Textile Engineering, National Textile University, Faisalabad, Pakistan*
[5]*Department of Engineering Mechanics, Royal Institute of Technology – KTH, Stockholm, Sweden*
[6]*Department of Mechanical and Nuclear Engineering, Khalifa University, Abu Dhabi, United Arab Emirates*
[7]*Beacon Impex Pvt., Ltd., Faisalabad, Pakistan*

Abstract

3D printing has gained enormous attention from researchers due to its unique features. The flexibility of the design, minimum waste generation, optimum resource utilization, rapid prototyping, precision of print dimensions, ease of manufacturing, and print-on-demand are some of the key advantages of 3D printing. With the advancement and increasing role of nanotechnology in 3D printing, the outreach is expanding from common applications to biomedical products. This chapter highlights the recent advancements in 3D printing technologies and the emerging role of nanotechnology to further synergize the properties of 3D-printed materials across various end products. Furthermore, various materials used for 3D printing have been discussed with their advantages and challenges for specific applications.

Keywords: Nanotechnology, 3D printing, implants, hydrogels, biopolymers

**Corresponding author*: m.sharjeel.abid.gmail.com

Bhasha Sharma, Balaram Pani, Shashank Shekhar and Jude A. Okolie (eds.) Sustainable 3D Printing for Innovative Biopolymer Production and Applications, (65–94) © 2025 Scrivener Publishing LLC

4.1 Introduction

Additive manufacturing, also known as three-dimensional (3D) printing, has recently drawn keen interest from various industrial sectors due to massive applications in different fields, including textile, aerospace, biomedical, and biobased industries using biopolymers [1]. 3D printing in different fields also uses various bioactive powders and cells with base polymer materials to treat damaged physiological structures and make new structures [2]. Recent advancements in nanotechnology have made it an extensive, diverse, and adaptable field with the application of multiple blended materials. The contents of the materials can be manipulated at the nanoscale to tune the mechanical, chemical, and biological properties with controlled size and shape according to the target application [3, 4]. Different application fields call for different required properties—for example, in the biomedical field, the material should be biocompatible, printable, and degradable with good mechanical properties. On the other hand, textile products should be flexible, breathable, and comfortable to wear with proper strength [5]. Materials for food products should be edible and deformation resistant during food processing. The widespread adoption of 3D printing technology can produce material waste and landfill, with a high amount of energy consumption for recycling; thus, considering the sustainability goals, using bioplastics and biopolymers instead of artificial plastics is of great interest [6]. Incorporating nanoscale materials in metals, ceramics, and polymer substrate creates composites for various fields, including customized therapies. Nanotechnologies can be combined with 3D printing to prevent, diagnose, and treat various diseases, including applications like medical imaging, drug delivery, cancer therapy, tissue engineering, and immunotherapy [7].

Nerve injuries, either due to a disease like cancer or an accident, can become a source of weakness and disability, resulting in painful neuropathies. A major challenge is to perform reconstructive surgery using regenerative medicines or autografts. Autografts are picked from the host or donor with less important body sources, mainly when the injury is around 5 mm or more. Serious ethical issues with autografts involve donor compatibility, risk of disease transfer, and additional surgery. Nerve repair using synthetic and natural materials provides a favorable microenvironment altering the natural autografts. Functional materials are required for nerve repair, and several fabrication techniques are used by researchers, including molding, electrospinning, and dip coating using Schwann cells (SCs) or stem cells. However, these methods cannot control the architecture or sizes and have low accuracy. Three-dimensional printing can overcome these issues very easily, along with higher precision. A few studies have discussed the materials, methods, and application of 3D printing, but an in-depth review of the application of nanotechnology with 3D printing technology is not available to the best of our knowledge. A detailed review of the advancements in the application of nanotechnologies in 3D printing is the need of the hour, especially in the biomedical field, which is the focus of this chapter.

4.2 Principles of 3D Printing

3D printing, as advanced technology, uses multiple materials like nanoparticles, powders, ceramics, materials, and polymers to build layer-by-layer structures with different geometries. Instant prototyping in the medical area was possible because of imaging techniques, processing, and reconstruction. The steps in 3D printing include the following:

a. Preparing the 3D computer-assisted design (CAD) model
b. Getting the printer compatible
c. Preparing the printer bed, material loading, and nozzle cleaning
d. Printing the required model
e. Removal of excess material and cleaning

This is a fully computer-controlled production process with reduced time and developmental costs; even changes in the final product can be done by adjusting the CAD file instead of traditional manual production [8]. Different 3D printing processes include stereolithography, inkjet printing, selective laser sintering, and fused deposition modeling, which are discussed further. The detailed view is shown in Figure 4.1.

4.2.1 Stereolithography

Stereolithography (SLA or SL) is called vat photopolymerization, where layer-by-layer 3D printing is done using a photochemical process [9]. High-energy-emitting light causes the solidification of photopolymer resin within the tank, which causes monomers and oligomers to crosslink together to solidify. As soon as the resin layer solidifies according to the defined depth, the bed moves vertically down along the z-axis. The layer is

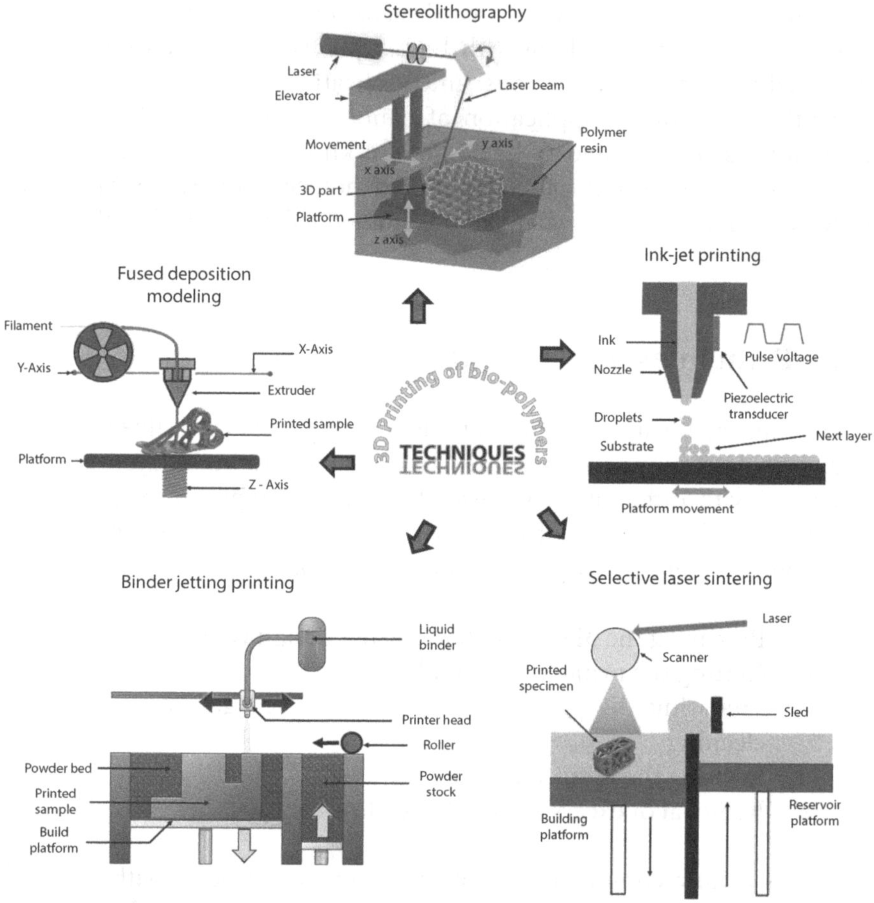

Figure 4.1 Types of 3D printing with their printing process.

again coated with resin, and the layer-by-layer process continues. Medical fields mostly use this technology for tissue engineering.

4.2.2 Inkjet Printing

It is a material jetting technique that follows the same working principle as the inkjet paper printers in houses and offices. However, the ink is replaced with the pharmaceutical liquid blended with the desired drugs on the substrate for biomedical applications. These are mostly oral dispersible films or aerogels.

4.2.3 Selective Laser Sintering

Selective laser sintering (SLS) 3D printing is primarily based on powder as the basic material. High-power laser is used on the powder bed to make the selective powder particles melt and bind into the layer. The feed rollers provide the powder for a new layer on the sintered material. Due to the high-power laser, the increased temperature may degrade the drug constituents, limiting its use in the medical field, while the precise laser can print with a high resolution of ~100 μm. On the other hand, this solvent-free printing does not require any post-fabrication drying time.

4.2.4 Fused Deposition Modeling

Fused deposition modeling (FDM) is an extrusion-based, most adaptive 3D printing technique. It is a versatile, practical, and cost-effective process that uses a heated nozzle to melt the loaded polymer-based filament that solidifies immediately after leaving the nozzle. According to the CAD model, it deposits the thin layer one over the other until the final product is formed. The traction gears feed the filament to the extrusion head for the layup process. The head is controlled by the motors in 2D directions x- and y-axis, while the build plate moves on the z-axis, allowing the model to develop in 3D motion. The filament has a predetermined diameter manufactured using melt extrusion. Depending on the wide requirement of plastic models, various filaments are available in the market, such as polylactic acid (PLA), acrylonitrile butadiene styrene (ABS), polyether ether ketone (PEEK), or high-impact polystyrene (HIPS). FDM can be used in various areas but requires new filaments with many properties such as color, bio-active, resistance, strength, flexibility, conductivity, and biocompatibility. Patient-specific implants with varying shapes and sizes can be printed easily with FDM.

4.2.5 Semisolid Extrusion

Semisolid extrusion is also based on material extrusion 3D printing. It consists of the syringe-based system, in which the gel or paste is extruded on the build bed layer by layer. Same as FDM, the material solidifies but through the evaporation process. Commonly used printers print through pastes, powders, dispersion gels, or solutions. This controlled process can use multiple materials like epoxy resins, edible materials, hydrogels, and biopolymers. Various studies have elaborated its scope in printing edible tablets with multiple-loaded drugs.

4.2.6 Binder Jetting

Binder jetting is a process that uses two materials to produce a 3D structure: a powder-based material and a binder. The binder acts as an adhesive between powder layers. The binder is usually in liquid form, and the build material is in powder form. A print head moves horizontally along the x- and y-axes of the machine and deposits alternating layers of the building material and the binding material. After each layer, the printed object is lowered on its build platform in the z-axis.

4.3 Advantages of 3D Printing

Improved 3D printing techniques using advanced functional materials can help solve genetic disorders and damaged tissue problems. Multiple medicines in a particular dose with target application, including patches, tablets releasing ions or nutrients, and supporting structures, can be prepared in a few minutes. Thus, a single dosage can cure several diseases at a time. Several healthcare facilities use primary and secondary treatment from pharmacies to hospital wards and emergencies. 3D-printed drug delivery systems can reduce medicine wastage, incorporate tunable ingredients, and enhance target-oriented drug release compared with traditional methods such as high-cost medication, changing densities, large production, and poor stability [10].

Moreover, the economic benefits of 3D printing enable its flexible use in hospitals or pharmacies for the on-demand preparation of instant medicines. Regardless of spending money on conventional methods, which may not treat patients, 3D printing can take it to the next level to quickly relieve patients' diseases by sharing digital data. However, software are required to ensure the data collection precision to secure a safe treatment. Thus, for the standardized procedure, the technical team, software developers, healthcare practitioners, and doctors should be on the same page. That is why there has been an increase in research studies in the last few decades, where 3D printing has been used in combination with nanotechnology.

A recent study categorized publications based on 3D printing for biomedical applications into various groups. Almost half (49%) of the studies focused on tissue repair merged with engineering technologies (see Figure 4.2). Furthermore, the research on biomedical and medicine-based 3D technologies dramatically increased by 2016, with record publications in 2019. This is clear evidence of the recent interest in sophisticated 3D printing methods.

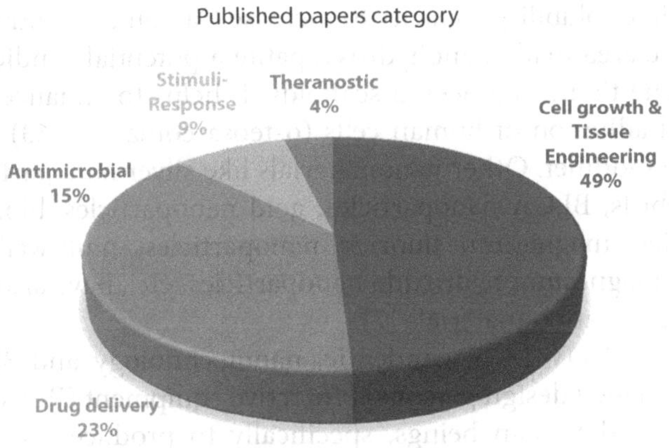

Figure 4.2 Papers published in biomedical fields.

4.4 Nanotechnology and Its Applications in 3D Printing

3D printing and nanotechnology are two branches in which progressive research has been done in the past few years. Research in these fields has made advancements in a wide range of applications in various fields like medicine, tissue engineering, surface science, semiconductor physics, nanoelectronics, molecular engineering, biomaterials, etc. Nanotechnology can be used to enhance the working ability of 3D-printed materials—for example, carbon-based filaments are used in fused deposition modeling (FDM) 3D printing to make them electrically conductible. FDM 3D-printed plastic objects can be strengthened by coating them with carbon nanotube ink [11].

Both synthetic and natural materials have been utilized for preparing 3D printing materials in bone tissue engineering. Polycaprolactone (PCL) is a biodegradable polyester used extensively, either individually or mixed with other materials, for various 3D printing applications [12]. Other materials like sodium alginate, poly (ethylene glycol), chitosan, and gelatin have also been used in this field. Some of these materials have comparatively greater biocompatibility; however, some showed a poor affinity with cellular structures, which can result in material rejection. To overcome these difficulties, 3D printing materials that improve cell adhesion and maintain the material's mechanical properties have been manufactured. Hydroxyapatite, known as bio-ceramic, is an inorganic bone component with excellent osteoinductivity, biocompatibility, and osteoconductivity.

Properties like solubility, capacity to increase Ca^{+} ion concentration, and a high specific area made nanohydroxyapatite a potential candidate for 3D printing bone tissue-engineered scaffolds. It helps to enhance the proliferation and adhesion of human cells (osteosarcoma MG-63) when used as a surface modifier. Other nanomaterials like silver nanoparticles, cellulose nanofibrils, BLGA nanoparticles, gold nanoparticles, bioactive glass nanoparticles, magnesium fluoride nanoparticles, multiwalled carbon nanotubes, magnesium hydroxide nanoparticles, etc., have also been used for bone regeneration materials [7].

During the COVID-19 pandemic, nanotechnology and 3D printing technology helped design personal protective equipment (PPEs) and other materials to aid human beings, specifically to produce the respiratory mask based on cellulose acetate and PLA nanofibers matrix containing graphene oxide and copper oxide nanoparticles. The polymeric network prevents airborne viral particles, and nanoparticles inactivate the viral particles. PLA is heated by the 3D printing technology to produce a fixed piece of face mask. A replica was prepared with functionalized graphene to improve the antimicrobial and antiviral features of a 3D-printed facial mask. It has been estimated that this facial mask can filter 98.2% of bacteria, and the graphene-coated layer can stop the complete transmission of the virus. The PPEs prepared were more cost-effective than commercially available equipment [13].

Nanotechnology is widely used for the synthesis of biomaterials. Nanophase biomaterials are used in 3D printing as bio-inks and for other applications such as orthopedic implant diagnostic and drug delivery applications. Bio-inks are known to be the most important part of bioprinting, consisting of printing cartridges containing liquid cell suspension. The ideal properties of a bio-ink for 3D printing have been highlighted in Figure 4.3. These suspensions help provide a favorable environment for cell development and growth. Moreover, bio-inks also protect the cells from damage during the printing process. With the advancement in research technology, bio-inks are prepared these days to have better functional properties that have changed the phase of the bioprinting business. As far as applicability is concerned, nanoscale bio-inks are useful in tissue engineering and skin repair for small injuries and burns. These days bio-inks have been developed and used in bone regeneration technology for cartilage and bone repair [14].

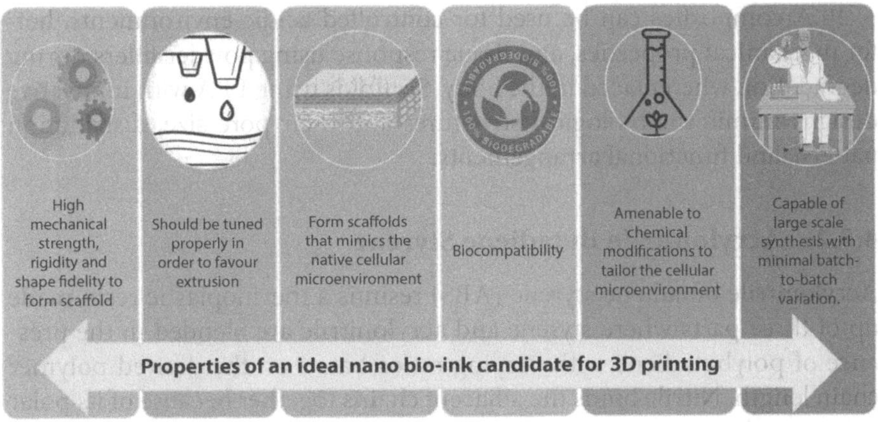

Figure 4.3 Properties of an ideal nano bio-ink candidate for 3D printing.

4.5 Polymers and Polymer Composites for 3D Printing

3D printing technology is a quick prototyping method that can prepare genuine products through layer-by-layer printing based on a three-dimensional digital model. Polymer is the most commonly used material for 3D printing [15–17]. Typical polymer materials used for 3D printing include ABS, PLA, PCL, and TPU photosensitive resin polymer hydrogel and their composites for applications in biomedical, electrical, aerospace, and other fields. Some of the most commonly used polymers are briefly discussed in the following subsections.

4.5.1 Polylactic Acid

Polylactic acid (PLA) is a thermoplastic polymer mostly generated from plant starch and is biocompatible and biodegradable. PLA may be used for 3D printing because of its low melting point, low viscosity, and great mechanical properties; however, it has high brittleness, transparency, and glossy nature. It has a glass transition of 60°C–65°C and a melting temperature of 130°C–180°C with Young's modulus of around 2 to 5 GPa. Its mechanical properties lie between polyethylene (PET) terephthalate and polystyrene. PLA structures can be easily fabricated using 3D printing, casting, extrusion, machining, injection molding, and solvent joining. PLA is used as base material during 3D printing in fused filament deposition.

PLA composites can be used for controlled acidic environments, better mechanical properties, and tissue response using powder fillers during degradation when placed in the body. Scaffolds using PLA with innovative designs mimic the biological structures and their pore size, distribution pattern, and functional arrangements.

4.5.2 Acrylonitrile Butadiene Styrene

Acrylonitrile butadiene styrene (ABS) resin is a thermoplastic resin made up of three parts where styrene and acrylonitrile are blended in the presence of polybutadiene with varying ratios based on the desired polymer chain length. Nitrile binds the adjacent chains together because of its polar nature, making it stronger than pure styrene. At the same time, the acrylonitrile compound is chemical- and heat-resistant with improved fatigue, rigidity, and hardness. Styrene's presence gives a shiny and impermeable surface and ease of processing. Polybutadiene acts as a rubbery material with better ductility and toughness even at minimal temperatures. Its mechanical properties vary along with changing temperatures between 20°C and 80°C. ABS material is appropriate for 3D printing. However, the unsaturation of butadiene causes ABS to have poor heat and ozone resistance as well as warping deformation and dimensional shrinkage when subjected to thermal stress, limiting the increase of pure ABS in the field of 3D printing technology.

4.5.3 Polycaprolactone

Polycaprolactone (PCL), which has strong biodegradability, biocompatibility, and non-toxicity, is extensively employed as a biodegradable medical material with a drug control release system. PCL can retain and recover distortion and keep up with the rapid development of 4D printing. It has a very low glass transition temperature of around 60°C. PCL is commonly used in polyurethane production to make them water-, oil-, and solvent-resistant. It is also added in resins to improve processing and biodegradable properties, thus widely used for 3D printing in biomedical applications.

4.6 Biopolymers for 3D Printing

Several biopolymers are used for 3D printing techniques, like chitosan, alginate, cellulose, collagen, gelatine, agarose, etc. Good biodegradability, biocompatibility, mechanical strength, non-toxicity, and sustainability are the key factors for supporting the use of biopolymers. Some biopolymers, their properties, and their usage are described below.

4.6.1 Lignin

Lignin is a biopolymer with a phenolic macromolecule and has a dense structure based on plant species [15]. Various studies reported high-performance 3D printing using lignin and its composites. In another study, lignin-coated cellulose nanocrystals (LCCNC) with methacrylate were used for 3D-SLS printing. Improved thermo-mechanical properties were achieved by adding 0.5% LCCNC [18]. The material should not lose its integrity during 3D printing; thus, it can withstand a temperature higher than 3.9°C without degrading. A 5% improvement in tensile strength was also observed.

4.6.2 Cellulose

Cellulose is the most abundantly available natural biopolymer. Scientists have recently investigated cellulose as 3D printing material. The high crystallinity in structure with stiff intra/intermolecular hydrogen bonds made its application difficult. However, chemical modifications can be done due to the rich hydroxyl groups on its surface. Fu *et al.* fabricated a complex three-dimensional structure using a desktop printer, where the microfluidic analytic device was fabricated using cellulose powder. Cellulose-based scaffold structures display a graded network, improved surface area, high porosity, and flexibility but show poor strength when wet [19, 20]. A drug delivery system was also developed using cellulose recently [21]. Furthermore, its derivatives in 3D printing inks are used as binders and viscosity changes [22]. In textiles, cellulose derivatives coat or 3D-print microstructures like 3D-printed cellulose on woven viscous and cotton fabric [23].

4.6.3 Chitosan

Chitosan is a naturally derived cationic polymer synthesized by chitin deacetylation. The chemical hydrolysis of chitin takes place using deacetylase enzymes or concentrated alkali water. It comprises deacetylated D-glucosamine units (GlcN) and N-acetyl-D-glucosamine (GlcNAc) connected by β-(1–4)-glycosidic bonds and scattered within the polymer. Chitosan is a non-toxic and biodegradable material used in many applications like bone regeneration, tissue engineering, biosensors, bio-inks, drug delivery, and water treatment [24]. Properties like suitable viscosity, solution stability under physiological conditions, cell proliferation, and differentiation ability made chitosan a suitable candidate for 3D printing. Moreover, fine-tuning chitosan hydrogels' properties can replace tissues' natural extracellular matrix [25].

The skin regeneration ability of 3D-printed chitosan–raffinose scaffolds was tested by comparing scaffolds with and without chitosan film. After 35 days, optimal cell growth on these 3D scaffolds was observed, where keratinocyte and fibroblast cells were grown only on chitosan film. It is due to the property of chitosan film to provide a boundary to capture the growing early skin layer cells. Moreover, chitosan contains step-forward cell regeneration ability when compared to other commercially available products [26]. Chitosan-based hydrogels are used as bio-inks in 3D printing technology for cell regeneration. The advantage of using chitosan-based ink is that it can be used for printing directly in the air, which causes solvent evaporation resulting in partial hardening, as shown in Figure 4.4. Physical gelation of chitosan scaffolds during the neutralization step greatly refines the structure of scaffolds. Chitosan bio-ink is prepared by dissolving 8% (w/v) chitosan in a solvent mixture containing citric acid/lactic acid/acetic acid. The amino group in chitosan is protonated by the acidic solution, which prevents chain expansion and forms semirigid rod structures, as charged amino groups repulse each other. The prepared chitosan solution was filled in a syringe and passed through the micronozzle; a certain pressure was applied for a smooth flow. As the solution passes through the micronozzle, the solvent evaporates and solidifies the chitosan filaments, providing layer-by-layer modeling of scaffolds. After drying of the remaining solvent, the filaments contracted and significantly flattened [27].

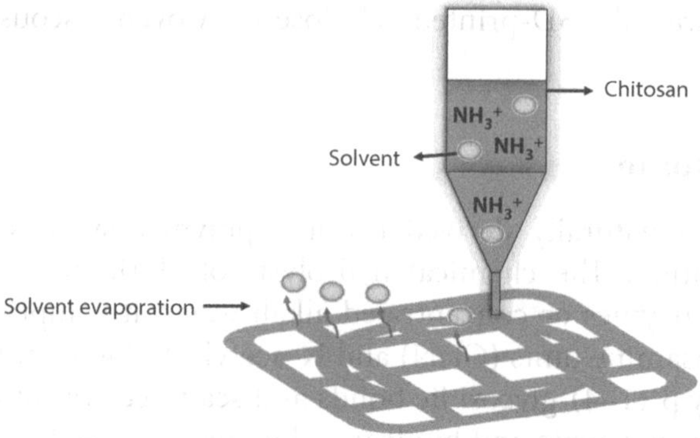

Figure 4.4 Chitosan-containing bio-ink fabrication for 3D printing.

4.6.4 Alginate

Alginate is a biological substance derived from brown algae. This biopolymer has properties like wound healing and regenerating tissues; thus, it is used in various biomedical applications, *i.e.*, drug delivery [28], bone tissue engineering, wound healing [29], etc. Alginate can transform into a gel by ionic crosslinking with a suitable divalent cation and is used in tissue engineering to prepare 3D bio-printed scaffolds. However, the mechanical strength of alginate is low compared to other biopolymers. Therefore, to enhance its rheological and mechanical properties, other polymers like chitosan, gelatin, collagen, polypyrrole, polysaccharide, hydroxyapatite (HA), and tetraethyl orthosilicate (TEOS) were added. Alginate composites mixed with nanocellulose, which provides rigidity and stiffness to the structure, were used as bio-inks for the bone regeneration technique. Adding nanocellulose provides the necessary strength to the alginate-printed scaffolds, which not only supports the proliferation of chondrocytes but also retains its properties. Thus, the bio-printed composites prevent chondrocyte dedifferentiation and improve cell adhesion [30].

A novel investigation was performed on 3D bio-printed scaffolds consisting of alginate–polyvinyl alcohol (PVA)–HA. Adding PVS and HA significantly contributed to the viscosity and printability of the hydrogel. The optimized ratio for 3D-printable hydrogel was 2.5% alginate and 2.5% hydroxyapatite, in addition to 0.15% Na_2HPO_4, 0.72% NaCl, and 0.20% $CaSO_4$. A 14-day-long degradation rate of 3D-printed alginate–HA–PVA scaffold was observed in α-MEM cell culture [31]. Bioactive glass is also helpful in increasing the mechanical strength of 3D-printed alginate-based scaffolds. Alginate scaffolds with and without 13-93 bioactive glass were studied for cell differentiation and proliferation. It was observed that an alginate scaffold containing 13-93 bioactive glass imparts more mechanical properties and cell proliferation activity on mesenchymal stem cells in a rat model. The optimized ratio of alginate/13-93 bioactive glass was 4:2 [32]. A new biometric approach was proposed to prepare 3D-printed alginate composite structures where liquid ink directly interacts with the polymer. Sodium alginate, ammonium hydrogen phosphate, phosphate source, calcium chloride, and calcium source were mixed to form an inorganic phase called (the calcium phosphate) CP phase. The interaction between the carboxylate group of alginate and HPO_4^{2-} resulted in the formation of homogenized and stable gel for composite material [33].

4.6.5 Silk Fibroin

Silk is a biopolymer consisting of large protein subunits produced by silkworms. *Antheraea mylitta* silkworms mainly produce silk fibroin comprising RGD peptide bonds that aid in cell adhesion; thus, they are used in several biomedical applications. Its remarkable mechanical properties, biocompatibility, versatile functionalization, facile processability, and controlled biodegradation have allowed its use in various applications [34]. Silk fibroin and hydroxyapatite nanocomposites (<100-nm width) were prepared by coprecipitation to form bio-inks for 3D-printed scaffolds. The mechanical strength of these scaffolds was determined to be suitable for bone regeneration applications. Sodium alginate/hydroxyapatite and silk fibroin/hydroxyapatite nanocomposite scaffolds showed higher mechanical strength and superior cell adhesion properties [35].

4.6.6 Collagen

Collagen is derived from natural materials, i.e., soft coral *Sarcophyton ehrenbergi*. It is the main component of the extracellular matrix (ECM). The bio-composite prepared from collagen showed specific hyper-elastic behavior, like native human tissues. It has excellent mechanical and biocompatible properties suitable for 3D printing [36]. Collagen is usually crosslinked via pH change, temperature change, or utilizing vitamin B2. Crosslinking of collagen increases its tensile strength and gives it greater viscoelastic properties. However, crosslinking is time-consuming and requires 30 mins at 37°C for collagen gelation. Therefore, its usage in 3D printing is challenging. To solve this problem, collagen is combined with other materials like alginate, sodium alginate, gelatine, and tannic acid in different proportions and combinations [37].

4.6.7 Gelatin

Gelatin is derived from collagen by partial hydrolysis with acid (gelatin type A) and alkali (gelatin type B), mainly from animal collagen. It is easily soluble in water at 37°C, biodegradable, and nonimmunogenic and imparts amphoteric behavior. These properties suit biomedical and 3D printing techniques [38]. Types A and B gelatin are used as a carrier for acidic proteins and prolong the discharge of basic molecules. However, its thermal and mechanical properties must be adjusted by crosslinking with other materials like genipin, glutaraldehyde, PVA, methacrylate, and hydroxyapatite to be used in 3D printing [39].

4.6.8 Agarose

Agarose is a polysaccharide derived from seaweed. Due to its exceptional gelation properties, it has been used in various biomedical applications. It consists of a linear polymer chain with repeating units of agarose which comprises 3,6-anhydrous-L-galactopyranose and D-galactose. Agarose is used in 3D printing technology for its properties like biodegradability, high gelation ability, and rheology [40]. Agarose forms 3D bio-inks blended with fibrinogen and collagen separately and supports fibroblast and endothelial cell growth. The agarose, collagen, and fibrinogen concentration were set at 0.5% and 1.25%, respectively. After 14 days of submerging both alginate blends in endothelial cell culture, it was observed that the alginate–fibrinogen sample lost its shape; however, the alginate–collagen fiber retained its initial shape and thus qualified for 3D printing. The printability of collagen was analyzed both separately and with agarose. It was concluded that adding agarose radically improved the sheer modulus and restricted the remodeling and deformation of the cellular matrix [41].

In another study, agarose was chemically modified to carboxylate form to synthesize 3D-printed structures with tunable mechanical properties. It was noted that modified agarose showed higher cell viability of 95% in human mesenchymal stem cells than the natural agarose gel. On tuning the carboxylation of agarose, the gel's mechanical properties can vary for different applications [42].

4.7 Applications of 3D-Printed Biopolymers

There are several applications of nanotechnology-derived 3D-printed biopolymers in pharmaceuticals, like drug delivery, tissue engineering, bone regeneration, stimuli-responsive materials, cancer treatment, antimicrobials, nerve repair, biosensors, and many other fields. Some of them are mentioned in Figure 4.5.

4.7.1 Medicine/Drug Delivery

A liposome is a spherical vesicle created by rupturing biological membranes and contains at least one lipid bilayer. It can be used as a vehicle for the delivery of nutritional supplements and other medications. To create an injectable liposome–hydrogel system for tissue regeneration, thin-film evaporation and the (TE-scCO_2) supercritical CO_2 technology were coupled. At 25°C, this thermo-responsive hydrogel was a colloidal

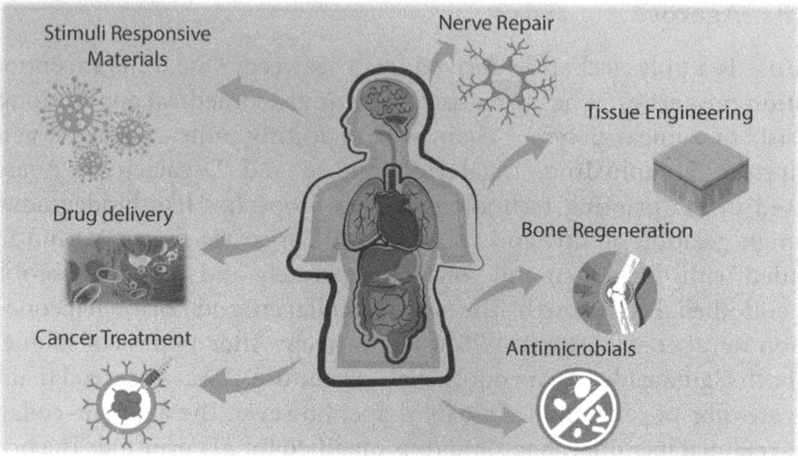

Figure 4.5 Applications of biopolymers in 3D printing technology.

sol; nevertheless, it became cured at physiological temperature. Injectable at room temperature, the liposome hydrogels turned into gel at 37°C. Curcumin liposome hydrogels (Cur-Lps-H) were developed and embedded into a permeable 3D glycerophosphate/chitosan. Fluorescent probes were used to embed curcumin in a liposome bilayer at a concentration corresponding to the molar ratios of curcumin to lecithin, i.e., 0.012. Compared to Cur-Lps-H generated by thin-film hydration, the TE-scCO_2-based Cur-Lps-H indicated greater stability and entrapment efficiency. The *in vitro* delayed release property of the Cur-Lps-H (over 12 days) rendered injectable liposome–hydrogels a potential drug delivery system for tissue repair [43].

Using direct ink writing (DIW), a complex-shaped 3D printing emulsion system made of an oil-in-water dispersion stabilized by chitosan containing silica nanoparticles was created. A water-soluble chitosan oligomer was adsorbed onto a 22-nm silica suspension to create an emulsion for hydrophobic encapsulation. Crosslinking and relatively stable emulsion droplets are required to obtain the viscoelastic qualities necessary for DIW. As a result, a hierarchical structure comprising droplets with a network of encapsulated functionalized silica particles was formed. Additionally, the agglomerated particle network gives emulsion stability during centrifugation and drying. The emulsion-based ink allows it to include hydrophobic elements in a hydrophilic substance. This stabilized chitosan–silica emulsion has applications in bioactive scaffolds and regulated drug release [44].

4.7.2 Tissue Engineering

In the body, the ducts of soft tissues, i.e., blood vessels, pancreatic duct, and weasand, are very important for transporting the metabolic waste, secretions, and nutrition. Lesions in these ducts need to be repaired as soon as possible. 3D-printed chitosan-containing ducts were prepared and analyzed for their performance in diverse tissues. Three different acidic solutions, lactic acid, acetic acid, and glycolic acid, were used to dissolve chitosan. The 30% (wt) glycolic acid solution was the best for 3D printing with sheer thinning properties. The Young's modulus, fracture strain, and tensile strength of manufactured chitosan rods were 12.38 ± 1.19 MPa, 146.03 ± 15.05%, and 10.98 ± 0.61 MPa. The prepared chitosan ducts by 3D printing showed great mechanical strength required for soft tissues, are biocompatible, and are excellent for restoring soft tissues [45]. Biopolymers have poor printability; therefore, a method to convert them into bio-inks was proposed to elevate their mechanical strength. Chitosan and collagen bio-inks were observed to have improved printability by hydrogen bond interaction and were used to form scaffolds for tissue engineering [46]. Agarose bio-inks are prepared with single-walled carbon nanotubes (SWCNT) via the excursion printing technique and are used in tissue engineering and organ printing [47]. Hydrogel ink comprises gelatin and alginate embedded with cellulose nanocrystals said to have better rheological properties for 3D printing than hydrogel inks without nanostructures. From the X-ray scattering images, it was found that cellulose nanocrystals were 80% in the direction of printing. Thus, the scaffolds prepared were more aligned, with controlled pore size and structure. The cellulose nanocrystals' inherent stiffness and mechanical strength were advantageous for composite preparation via stereolithography and 3D printing [48].

Nanocellulose incorporated in a polymer network of sodium alginate and gelatin produces 3D scaffolds with controlled gradient pore structure and pore size for optimal tissue engineering applications [49]. The direct ink write technique is also very useful in preparing 3D structures from cellulose nanocrystals (CNC)/aerogels. Fabrication of pure cellulosic aerogels is difficult, but combining the direct ink write technique and freeze-drying method solves that problem and produces 3D structures of CNC aerogels with reduced damage and shrinkage. An elevation in the printing quality of scaffolds was observed by increasing the printing resolution and CNC concentration. The designed scaffolds are suited for tissue engineering, drug delivery, and packaging [50].

The conductive hydrogels are produced by reinforcing a chitosan–lactic acid matrix with graphene oxide at various concentrations for tissue engineering applications. By using extrusion printing, these composites were quickly transformed into 3D scaffolds. A 200-μm-diameter nozzle was used to print scaffolds with 30 layers of each chitosan/graphene dispersion at a feed rate of 150 mm/min, orientation of 0°/90°, and 0.6 mm of strand spacing. Mechanical strength significantly increased because of the incorporation of graphene, and adding 3 wt.% graphene increased the tensile strength by over 200%. The prepared structures were subjected to fibroblast (L– 929) cells which proliferated on multiple layers of the 3D films. These scaffolds demonstrated superior conductivity as substrates for developing electro-responsive cells [51].

4.7.3 Bone Regeneration

The major research portion (around 70% publications) of the nanomaterial-based bio-active 3D printing structures is related to bone regeneration. Bone fractures could occur due to injury, diseased bones, and damaged tissues. Supporting structures are required to stabilize the impaired bones and improve tissue regeneration. Researchers use various terms to describe these fields, such as bone fracture repair, bone construction, repairing bone defects, and tissue repair. Various bio-active natural and synthetic materials and their composites are used for 3D printing applications. PLA, poly (lactic-co-glycolic acid) PLGA, and other bio-active polymers are used alone or in combination with reinforcement materials to improve the biomechanical properties. Polymers cannot tolerate the body load to support the damaged load-bearing bones alone. Reinforcement materials are required to improve the mechanical properties of the polymers.

In many cases, implant material shows poor interaction during cell adhesion, thus resulting in material rejection. Thus, researchers' efforts toward material modification for better cell adhesion and propagation showed a significant improvement. Ceramic powders like hydroxyapatite (HA) are used in PLA polymer, the major constituent of the natural bone material with excellent biocompatibility and osteoconductivity. It also has higher solubility, better adhesion with calcium ion concentration, and a unique nanostructure for 3D printing scaffolds. Moreover, other coating techniques are also used to improve the biomechanical properties of structures for enhancing the treatment by stimulating bone production.

4.7.4 Nerve Repair

Partial or total loss of nerves during injury disturbs the autonomic and sensory functions of nerves. Around 0.2 million procedures for nerve repair are done just in the United States, costing millions of dollars and ultimately a burden on the government. Common treatment practices involve the autografts method, which shows obvious disadvantages like additional surgery, donor compatibility, chances of disease transfer, costly procedure, and time consumption. 3D printing technology was considered to overcome traditional issues. In a study, 3D-printed silicone channel was printed and grafted in *in vivo* body to repair the damaged nerve. Hydrogels are also potential materials to be used for 3D printing. Various hydrogels include alginate, gelatin, ethylene glycol, polycaprolactone, and polyurethane. Higher water content in gels makes them ideal candidates for cell growth. Moreover, various drug powders or constituents can be incorporated into gels to facilitate growth. The increasing interest in using 3D prints using nanotechnology can be seen in Figure 4.6, which depicts the potential of nanotechnology in 3D printing.

4.7.5 Antimicrobials

Manufacturing different 3D-printed complex structures with additives is a new approach in tissue engineering and related fields. Bovine silk gelatin (BSG), clove essential oil (CEO), and zinc oxide nanocomposite scaffolds were fabricated for antimicrobial food packaging materials. A combination of 25% CEO and 2% ZnO improved the microbiological properties, thermal barrier, optical structure, and elongation of films. XRD and TGA thermograms confirmed the presence of ZnO and the thermal stability of

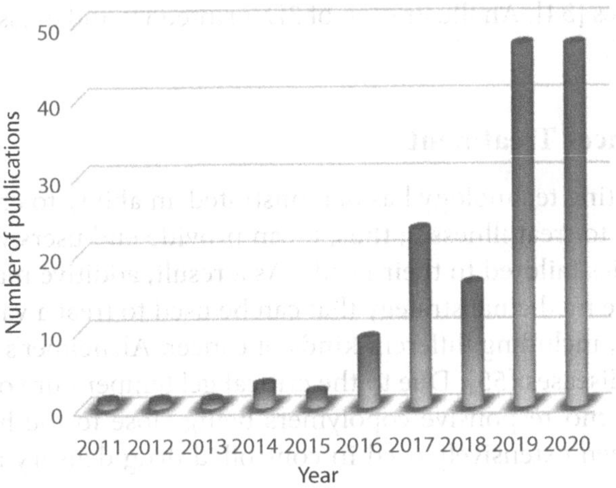

Figure 4.6 3D printing linked with nanotechnology-based publications per year.

gelatin composites. A twin screw extrusion process was adopted to develop a semisolid viscous mass of the BSG/ CEO/ZnO blend. This slurry was further used to produce 3D printable bio-inks by adding sodium alginate [52].

In another study, a combination of chitosan/tea polyphenol/halloysite (CS/TP/HNTs) nanotube nanocomposites was developed and then subjected to 3D printing for antibacterial and antioxidant food packaging materials. The optimized ratio for CS/HNTs was 6:4, and 10% TP content as the scanning electron microscopy (SEM) results showed the dense internal structure and smooth surface of nanocomposite films. Prepared nanocomposite films showed bacterial growth inhibition against *Staphylococcus aureus* 0.87 ± 0.51% and *Escherichia coli* 0.85 ± 0.26%. 3D printing of nanocomposite films was carried out at 25°C. The homogeneity between the crosslinked nanocomposite films was an important factor in printing. High-temperature extrusion rate increased even at the lowest possible pressure [53].

Using a photo-crosslinking approach, a novel 3D-printed dressing with biocompatibility and strong antibacterial activity was created using gelatin methacryloyl (GelMA). A matrix containing xanthan gum/carboxymethylcellulose sodium (CMC) was combined with antibacterial components of N-halamine hybrid nanoparticles (NPs)/CeO_2. The 3D-printed dressing showed great mechanical performance and shaped fidelity with significantly improved blood clotting ability and swelling efficiency. The prepared structures had antibacterial activity of 99.8% and 99.6% against *E. coli* and *S. aureus*, respectively. *In vivo* analysis resulted in excellent efficiency of dressings for wound closure by accelerating the epidermis and dermis tissues [54]. An illustration of 3D-printed wound dressing is shown in Figure 4.7.

4.7.6 Cancer Treatment

The 3D printing technology has demonstrated an ability to create personalized drugs to treat illnesses; thus, it can provide end users with targeted therapy tactics tailored to their needs. As a result, additive manufacturing is an effective medicinal strategy that can be used to treat a variety of clinical illnesses, including different kinds of cancer, Alzheimer's disease, and respiratory diseases [55]. Due to the critical gel temperature of *in situ*-gel-forming thermo-responsive copolymers being close to the human body, they have been extensively used in controlled drug delivery applications.

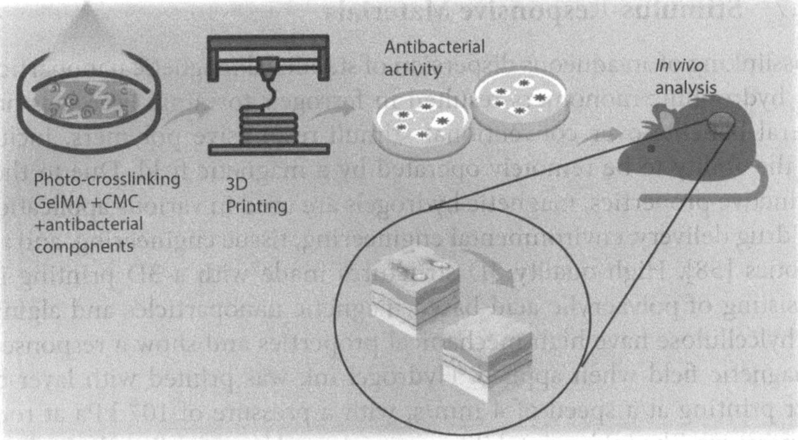

Figure 4.7 Photo-crosslinking approach for 3D-printed antibacterial wound dressing.

However, due to the inadequate networking and strengthening between the copolymer networks, it is not easy to regulate the release of drugs from a hydrogel network. 3D-printed injectable hydrogels solved this problem. Cellulose nanocrystals are incorporated in a complex copolymer to develop *in situ*-forming injectable 3D-printable hydrogel structures. The hydrogels coordinated with water-soluble lysosome and hydrophobic doxorubicin by a combination of hydrogen bonding and hydrophobic interaction and regulated the drug release. Up to 10 layers of 3D structures could be created with a high interlayer bonding among the printed layers. Doxorubicin (DOX)-incorporated hydrogel inhibited tumor growth, as observed by *in vivo* antitumor analysis on the MDA-MB-231 tumor xenograft model [56].

In another study, using an FDM printer, a scaffold containing chitosan, polycaprolactone, β-tricalcium phosphate, and chitosan-modified montmorillonite clay coated with DOX solution (DESCLAYMR-DOX) was fabricated to inhibit breast tumor growth in a mouse model. These drug-releasing implants demonstrated a large-burst release of DOX within the first 24 h, followed by a steady release over 4 weeks. Comparing the subcutaneous implantation of both INJECTION-DOX and DESCLAYMER-DOX in a mouse model resulted in prolonged release and stay of DOX at the respective site and showed greater inhibition of tumor cells. Local delivery of DOX via DESCLAYMER-DOX also reduced cytotoxicity and multiorgan metastasis compared to INJECTION-DOX [57].

4.7.7 Stimulus-Responsive Materials

Crosslinking of an aqueous dispersion of stabilized magnetic nanoparticles and hydrophilic monomers resulted in ferrogels forming. Ferrogels have several benefits over conventional stimuli-responsive polymers, including the ability to be remotely operated by a magnetic field. Due to these distinctive properties, magnetic hydrogels are used in various applications like drug delivery, environmental engineering, tissue engineering, and soft robotics [58]. High-quality 3D structures made with a 3D printing ink consisting of polyacrylic acid-based magnetic nanoparticles and alginate methylcellulose have high mechanical properties and show a response to a magnetic field when applied. Hydrogel ink was printed with layer-by-layer printing at a speed of 4 mm/s, with a pressure of 107 kPa at room temperature. The ink's printability was evaluated by extruding the hydrogel precursor using a 25G-type nozzle with a diameter of 250 m. The construct had the finest printing quality of any printout and kept its original shape for up to 30 min after deposition before calcium ions were used to crosslink it. The average diameter of the extruded layer was 268 ± 20 μm, indicating good printing resolution ($\varepsilon = 7.2\%$). The hydrogel with the highest PAA/MNPs ratio (1:2) produced 3D-printed structures of higher quality. The MNPs' inclusion within the hydrogel and its precursor (ink) affected their mechanical stability and rheological characteristics. MNPs have been discovered to have two distinct functions: first, as a function provider making a bio-nanocomposite magnetic, and second, as a nanofiller interacting with the polymer backbone and changing its physicochemical characteristics. The creation of multi-material structures like hydrogels with a gradient of magnetic nanoparticles is made possible by magnetic ink. Magnetic fields can remotely actuate 3D-printed items [59].

Based on chemically and physically crosslinked networks, the development of an electrically and mechanically self-healing hydrogel was reported. The process took two steps, and chitosan was double-bonded to polypyrrole nanostructures to produce polypyrrole-grafted chitosan, followed by chemical polymerization of acrylic acid in the presence of ferric ions. The double-network hydrogel showed ultra-stretchability of 1,500% with pressure sensitivity and 100% mechanical recovery in only 2 min in applicable strain and pressure sensing devices [60].

4.8 Challenges and Future Perspectives

3D printing has enabled many advancements in biopolymer research and development. The applications include rapidly fabricating complex, customized constructs for personalized treatments, despite some limitations in printing diverse materials. The biopolymer in 3D printing has good biocompatibility, biodegradability, and bioactivity but lacks the mechanical properties necessary for high load bearing and controlled bioactivity. Organic powders and nanomaterials as fillers allow much denser and higher-integrity constructs while maintaining biological attributes. Many organic powders either do not react to light or disintegrate in heat, making it impossible for biopolymers to be used in many thermal and light-based 3D printing techniques. A different technique for hardening the powdered material is used in the 3D printing method known as powder inkjet printing.

In some cases, the successful implementation of bio-polymer is restricted by a relatively narrow range of biocompatible materials. Despite the variable processability that 3D printing techniques offer, many issues must be addressed, including the printable ink's rigid and difficult flow behavior while maintaining the required functional properties. The development of printable biobased inks can be accelerated by rheological technique, which also provides valuable characterization tools, mathematical models, and a deep understanding of the printability of biopolymers. However, characterization methodologies suffer from transferability and comparability issues for biomaterials with unknown printability properties. Reproducible, comparable, and efficient characterization methods regarding the printing performance would be a key advantage, particularly for scientists entering the bio-fabrication field.

The functional qualities of many printable biobased materials are great for external applications; however, body performance and behavior need to be understood in detail. The term "rapid prototyping" is frequently used with the 3D printing technique, which is sometimes harmful in terms of construction speed. For material scientists, pharmacologists, food scientists, mechanical bioengineers, and others working in the 3D printing field, the transition from quick prototyping to fast building presents several challenges. Additionally, even if the 3D printing technology speeds up construction by reducing the time, it now only allows creation of working prototypes at the beginning stages of development. Comparing 3D printing techniques to mass production techniques, they are still lagging.

To prevent potential deformation when printing, the material should be easily extruded from the nozzle and have a sufficient elastic modulus. The functionalities of 3D-printed items are expected to be the same as those made using proven traditional techniques, as 3D printing is used for everything from (visual) prototyping to the creation of fully functional 3D structures. Despite several studies, the structural and mechanical strength of 3D structures created using 3D printing technology is below standard. The mechanical properties of printed objects also behave like an anisotropic network due to the layered manufacturing method, with the boundary between deposited layers denoting poor areas with strong residual forces and the main concerns being structural stability and geometrical structure alignment. Extrusion-based technologies, mechanical characteristics, and anisotropy depend highly on the biopolymers and the processing environment.

Furthermore, the fundamental issue with printing techniques and the components used is spatial resolution, which is very important in printing. Several researchers reported that the 3D object's precision was satisfactory; however, 3D constructs might be too thin for image-based analysis. Despite certain limitations of geometrical instability, 3D-printed products can still display high levels of reliability with simulated design.

Certainly, multidisciplinary research is already being done in nanotechnology-derived 3D printing technology. However, the existing approaches for the biopolymers obtained from natural sources need to be expanded by modification or combining them with additional components to introduce or enhance the material's workability and product functionality. It will take interdisciplinary teams from biology, engineering, computers, and statistics to advance further in organ transplants, enhanced *in vitro* models, and the creation and understanding of more complex systems. The possibilities are endless when a powerful technology becomes more available.

References

1. Ngo, T.D., Kashani, A., Imbalzano, G. *et al.*, Additive manufacturing (3D printing): A review of materials, methods, applications and challenges. *Compos. Part B Eng.*, 143, 172–196, 2018, https://doi.org/10.1016/J.COMPOSITESB.2018.02.012.
2. Rani, R., Dilbaghi, N., Dhingra, D., Kumar, S., Optimization and evaluation of bioactive drug-loaded polymeric nanoparticles for drug delivery. *Int. J. Biol. Macromol.*, 78, 173–179, 2015, https://doi.org/10.1016/J.IJBIOMAC.2015.03.070.

3. Ali, W., Mehboob, A., Han, M.G., Chang, S.H., Effect of fluoride coating on degradation behaviour of unidirectional Mg/PLA biodegradable composite for load-bearing bone implant application. *Compos. Part A Appl. Sci. Manuf.*, 124, 105464, 2019, https://doi.org/10.1016/J.COMPOSITESA.2019.05.032.
4. Agarwal, S., Curtin, J., Duffy, B., Jaiswal, S., Biodegradable magnesium alloys for orthopaedic applications: A review on corrosion, biocompatibility and surface modifications. *Mater. Sci. Eng., C*, 68, 948–963, 2016, https://doi.org/10.1016/J.MSEC.2016.06.020.
5. Chatterjee, K., Ghosh, T.K., Chatterjee, K., Ghosh, T.K., 3D Printing of Textiles: Potential Roadmap to Printing with Fibers. *Adv. Mater.*, 32, 1–24, 2019, https://doi.org/10.1002/adma.201902086.
6. Pradhan, R.A., Rahman, S.S., Qureshi, A., Ullah, A., Biopolymers: opportunities and challenges for 3D printing, in: *Biopolymers and their industrial applications*, pp. 281–303, Elsevier Inc, 2021. https://doi.org/10.1016/B978-0-12-819240-5.00012-2.
7. dos Santos, J., de Oliveira, R.S., de Oliveira, T.V. *et al.*, 3D Printing and Nanotechnology: A Multiscale Alliance in Personalized Medicine. *Adv. Funct. Mater.*, 31, 16, 1–31, 2021.
8. Madla, C.M., Trenfield, S.J., Goyanes, A. *et al.*, 3D printing technologies, implementation and regulation: An overview. *AAPS Adv. Pharm. Sci. Ser.*, 31, 21–40, 2018, https://doi.org/10.1007/978-3-319-90755-0_2.
9. Najmon, J.C., Raeisi, S., Tovar, A., Review of additive manufacturing technologies and applications in the aerospace industry, in: *Additive Manufacturing for the Aerospace Industry*, Chapter 2, pp. 7–31, Elsevier, 2019. https://doi.org/10.1016/B978-0-12-814062-8.00002-9.
10. Awad, A., Trenfield, S.J., Gaisford, S., Basit, A.W., 3D printed medicines: A new branch of digital healthcare. *Int. J. Pharm.*, 548, 586–596, 2018, https://doi.org/10.1016/J.IJPHARM.2018.07.024.
11. Tharayil, A., Rajakumari, R., Chirayil, C.J. *et al.*, A short review on nanotechnology interventions against COVID-19. *Emergent Mater.*, 4, 131–141, 2021, https://doi.org/10.1007/s42247-021-00163-z.
12. She, Y., Fan, Z., Wang, L. *et al.*, 3D Printed Biomimetic PCL Scaffold as Framework Interspersed With Collagen for Long Segment Tracheal Replacement. *Front. Cell Dev. Biol.*, 9, 33, 2021, https://doi.org/10.3389/FCELL.2021.629796/BIBTEX.
13. Mallakpour, S., Behranvand, V., Hussain, C.M., Worldwide fight against COVID-19 using nanotechnology, polymer science, and 3D printing technology. *Polym. Bull.*, 80, 165–183, 2022, https://doi.org/10.1007/s00289-021-04006-z.
14. Bhatt, A. and Anbarasu, A., Nanoscale Biomaterials for 3D printing. *IOSR J. Pharm. Biol. Sci.*, 12, 80–86, 2017, https://doi.org/10.9790/3008-1203068086.
15. Collins, M.N., Nechifor, M., Tanasă, F. *et al.*, Valorization of lignin in polymer and composite systems for advanced engineering applications-A review, 2019, https://doi.org/10.1016/j.ijbiomac.2019.03.069.

16. Lam, C.X.F., Mo, X.M., Teoh, S.H., Hutmacher, D.W., Scaffold development using 3D printing with a starch-based polymer. *Mater. Sci. Eng., C*, 20, 49–56, 2002, https://doi.org/10.1016/S0928-4931(02)00012-7.
17. Chaunier, L., Guessasma, S., Belhabib, S. *et al.*, Material extrusion of plant biopolymers: Opportunities & challenges for 3D printing, 2018, https://doi.org/10.1016/j.addma.2018.03.016.
18. Feng, X., Yang, Z., Chmely, S. *et al.*, Lignin-coated cellulose nanocrystal filled methacrylate composites prepared via 3D stereolithography printing: Mechanical reinforcement and thermal stabilization. *Carbohydr. Polym.*, 169, 272–281, 2017, https://doi.org/10.1016/j.carbpol.2017.04.001.
19. He, Y., Gao, Q., Wu, W.-B. *et al.*, 3D Printed Paper-Based Microfluidic Analytical Devices. *Micromachines*, 7, 1–12, 2016, https://doi.org/10.3390/mi7070108.
20. Dai, L., Cheng, T., Duan, C. *et al.*, 3D printing using plant-derived cellulose and its derivatives: A review, 2018, https://doi.org/10.1016/j.carbpol.2018.09.027.
21. Boetker, J., Water, J., Aho, J. *et al.*, Modifying release characteristics from 3D printed drug-eluting products, 2016, https://doi.org/10.1016/j.ejps.2016.03.013.
22. Tanwilaisiri, A., Zhang, R., Xu, Y. *et al.*, A manufacturing process for an energy storage device using 3D printing. *Proc. IEEE Int. Conf. Ind. Technol. 2016*, pp. 888–891, May 2016, https://doi.org/10.1109/ICIT.2016.7474869.
23. Tenhunen, T.M., Moslemian, O., Kammiovirta, K. *et al.*, Surface tailoring and design-driven prototyping of fabrics with 3D-printing: An all-cellulose approach. *Mater. Des.*, 140, 409–419, 2018, https://doi.org/10.1016/J.MATDES.2017.12.012.
24. Wu, Q., Therriault, D., Heuzey, M.C., Processing and Properties of Chitosan Inks for 3D Printing of Hydrogel Microstructures. *ACS Biomater. Sci. Eng.*, 4, 2643–2652, 2018, https://doi.org/10.1021/acsbiomaterials.8b00415.
25. Taghizadeh, M., Taghizadeh, A., Yazdi, M.K. *et al.*, Chitosan-based inks for 3D printing and bioprinting. *Green Chem.*, 24, 62–101, 2022, https://doi.org/10.1039/d1gc01799c.
26. Rajabi, M., McConnell, M., Cabral, J., Ali, M.A., Chitosan hydrogels in 3D printing for biomedical applications. *Carbohydr. Polym.*, 260, 2021, https://doi.org/10.1016/j.carbpol.2021.117768.
27. Wu, Q., Maire, M., Lerouge, S. *et al.*, 3D Printing of Microstructured and Stretchable Chitosan Hydrogel for Guided Cell Growth. *Adv. Biosyst.*, 1, 1–6, 2017, https://doi.org/10.1002/adbi.201700058.
28. Hasnain, M.S., Nayak, A.K., Kurakula, M., Hoda, M.N., Alginate nanoparticles in drug delivery. *Alginates Drug Deliv.*, 129–152, 2020, https://doi.org/10.1016/B978-0-12-817640-5.00006-6.
29. Tomić, S.L., Babić Radić, M.M., Vuković, J.S., Filipović, V.V., Nikodinovic-Runic, J., Vukomanović, M., Alginate-Based Hydrogels and Scaffolds for Biomedical Applications. *Mar. Drugs*, 21, 2023, https://doi.org/10.3390/MD21030177.

30. Ahmad Raus, R., Wan Nawawi, W.M.F., Nasaruddin, R.R., Alginate and alginate composites for biomedical applications. *Asian J. Pharm. Sci.*, 16, 280–306, 2021, https://doi.org/10.1016/j.ajps.2020.10.001.
31. Bendtsen, S.T., Quinnell, S.P., Wei, M., Development of a novel alginate-polyvinyl alcohol-hydroxyapatite hydrogel for 3D bioprinting bone tissue engineered scaffolds. *J. Biomed. Mater. Res. Part A*, 105, 1457–1468, 2017, https://doi.org/10.1002/jbm.a.36036.
32. Luo, G., Ma, Y., Cui, X. *et al.*, 13-93 bioactive glass/alginate composite scaffolds 3D printed under mild conditions for bone regeneration. *RSC Adv.*, 7, 11880–11889, 2017, https://doi.org/10.1039/c6ra27669e.
33. Egorov, A.A., Fedotov, A.Y., Mironov, A.V. *et al.*, 3D printing of mineral-polymer bone substitutes based on sodium alginate and calcium phosphate. *Beilstein J. Nanotechnol.*, 7, 1794–1799, 2016, https://doi.org/10.3762/BJNANO.7.172.
34. Nguyen, T.P., Nguyen, Q.V., Nguyen, V. *et al.*, Silk Fibroin-Based Biomaterials for Biomedical. *Polymers Basel*, 11, 1–25, 2019.
35. Huang, T., Fan, C., Zhu, M. *et al.*, 3D-printed scaffolds of biomineralized hydroxyapatite nanocomposite on silk fibroin for improving bone regeneration. *Appl. Surf. Sci.*, 467–468, 345–353, 2019, https://doi.org/10.1016/j.apsusc.2018.10.166.
36. Benayahu, D., Sharabi, M., Pomeraniec, L. *et al.*, Unique collagen fibers for biomedical applications. *Mar. Drugs*, 16, 1–11, 2018, https://doi.org/10.3390/md16040102.
37. Gopinathan, J. and Noh, I., review 5 2018 Recent trends in bioinks for 3D printing.pdf. *Biomater. Res.*, 22, 1–15, 2018.
38. Jaipan, P., Nguyen, A., Narayan, R.J., Gelatin-based hydrogels for biomedical applications. *MRS Commun.*, 7, 416–426, 2017, https://doi.org/10.1557/mrc.2017.92.
39. Suleman, A., Kondiah, P.P.D., Mabrouk, M., Choonara, Y.E., The Application of 3D-Printing and Nanotechnology for the Targeted Treatment of Osteosarcoma. *Front. Mater.*, 8, 1–23, 2021, https://doi.org/10.3389/fmats.2021.668834.
40. Salati, M.A., Khazai, J., Tahmuri, A.M. *et al.*, Agarose-Based biomaterials: Opportunities and challenges in cartilage tissue engineering. *Polymers Basel*, 12, 1–15, 2020, https://doi.org/10.3390/POLYM12051150.
41. Kreimendahl, F., Köpf, M., Thiebes, A.L. *et al.*, Three-Dimensional Printing and Angiogenesis: Tailored Agarose-Type i Collagen Blends Comprise Three-Dimensional Printability and Angiogenesis Potential for Tissue-Engineered Substitutes. *Tissue Eng. Part C Methods*, 23, 604–615, 2017, https://doi.org/10.1089/ten.tec.2017.0234.
42. Forget, A., Blaeser, A., Miessmer, F. *et al.*, Mechanically Tunable Bioink for 3D Bioprinting of Human Cells. *Adv. Healthc. Mater.*, 6, 1–7, 2017, https://doi.org/10.1002/adhm.201700255.

43. Abdollahiyan, P., Baradaran, B., de la Guardia, M. *et al.*, Cutting-edge progress and challenges in stimuli responsive hydrogel microenvironment for success in tissue engineering today. *J. Control. Release*, 328, 514–531, 2020, https://doi.org/10.1016/j.jconrel.2020.09.030.
44. Sommer, M.R., Alison, L., Minas, C. *et al.*, 3D printing of concentrated emulsions into multiphase biocompatible soft materials. *Soft Matter*, 13, 1794–1803, 2017, https://doi.org/10.1039/C6SM02682F.
45. Zhao, C.Q., Liu, W.G., Xu, Z.Y. *et al.*, Chitosan ducts fabricated by extrusion-based 3D printing for soft-tissue engineering. *Carbohydr. Polym.*, 236, 2020, https://doi.org/10.1016/j.carbpol.2020.116058.
46. Suo, H., Zhang, J., Xu, M., Wang, L., Low-temperature 3D printing of collagen and chitosan composite for tissue engineering. *Mater. Sci. Eng., C*, 123, 111963, 2021, https://doi.org/10.1016/j.msec.2021.111963.
47. Nadernezhad, A., Khani, N., Skvortsov, G.A. *et al.*, Multifunctional 3D printing of heterogeneous hydrogel structures. *Sci. Rep.*, 6, 1–12, 2016, https://doi.org/10.1038/srep33178.
48. Hassan, M., Dave, K., Chandrawati, R. *et al.*, 3D printing of biopolymer nanocomposites for tissue engineering: Nanomaterials, processing and structure-function relation. *Eur. Polym. J.*, 121, 109340, 2019, https://doi.org/10.1016/j.eurpolymj.2019.109340.
49. Sultan, S. and Mathew, A.P., 3D printed scaffolds with gradient porosity based on a cellulose nanocrystal hydrogel. *Nanoscale*, 10, 4421–4431, 2018, https://doi.org/10.1039/c7nr08966j.
50. Li, V.C.F., Dunn, C.K., Zhang, Z. *et al.*, Direct Ink Write (DIW) 3D Printed Cellulose Nanocrystal Aerogel Structures. *Sci. Rep.*, 7, 1–8, 2017, https://doi.org/10.1038/s41598-017-07771-y.
51. Silva, M., Pinho, I.S., Covas, J.A. *et al.*, 3D printing of graphene-based polymeric nanocomposites for biomedical applications. *Funct. Compos. Mater.*, 2, 2021, https://doi.org/10.1186/s42252-021-00020-6.
52. Ahmed, J., Mulla, M., Joseph, A. *et al.*, Zinc oxide/clove essential oil incorporated type B gelatin nanocomposite formulations: A proof-of-concept study for 3D printing applications. *Food Hydrocoll.*, 98, 105256, 2020, https://doi.org/10.1016/j.foodhyd.2019.105256.
53. Wang, Y., Yi, S., Lu, R. *et al.*, Preparation, characterization, and 3D printing verification of chitosan/halloysite nanotubes/tea polyphenol nanocomposite films. *Int. J. Biol. Macromol.*, 166, 32–44, 2021, https://doi.org/10.1016/j.ijbiomac.2020.09.253.
54. Yang, Z., Ren, X., Liu, Y., N-halamine modified ceria nanoparticles: Antibacterial response and accelerated wound healing application via a 3D printed scaffold. *Compos. Part B Eng.*, 227, 109390, 2021, https://doi.org/https://doi.org/10.1016/j.compositesb.2021.109390.
55. Shahbazi, M. and Jäger, H., Current Status in the Utilization of Biobased Polymers for 3D Printing Process: A Systematic Review of the Materials,

Processes, and Challenges. *ACS Appl. Bio Mater.*, 4, 325–369, 2021, https://doi.org/10.1021/acsabm.0c01379.

56. Phan, V.H.G., Murugesan, M., Huong, H. *et al.*, Cellulose Nanocrystals-Incorporated Thermosensitive Hydrogel for Controlled Release, 3D Printing, and Breast Cancer Treatment Applications. *ACS Appl. Mater. Interfaces*, 14, 42812–42826, 2022, https://doi.org/10.1021/acsami.2c05864.
57. Talebian, S., Foroughi, J., Wade, S.J. *et al.*, Biopolymers for Antitumor Implantable Drug Delivery Systems: Recent Advances and Future Outlook. *Adv. Mater.*, 30, 1–31, 2018, https://doi.org/10.1002/adma.201706665.
58. Xie, W., Gao, Q., Guo, Z. *et al.*, Injectable and self-healing thermosensitive magnetic hydrogel for asynchronous control release of doxorubicin and docetaxel to treat triple-negative breast cancer. *ACS Appl. Mater. Interfaces*, 9, 33660–33673, 2017, https://doi.org/10.1021/acsami.7b10699.
59. Podstawczyk, D., Nizioł, M., Szymczyk, P. *et al.*, 3D printed stimuli-responsive magnetic nanoparticle embedded alginate-methylcellulose hydrogel actuators. *Addit. Manuf.*, 34, 2020, https://doi.org/10.1016/j.addma.2020.101275.
60. Darabi, M.A., Khosrozadeh, A., Mbeleck, R. *et al.*, Skin-Inspired Multifunctional Autonomic-Intrinsic Conductive Self-Healing Hydrogels with Pressure Sensitivity, Stretchability, and 3D Printability. *Adv. Mater.*, 29, 1–8, 2017, https://doi.org/10.1002/adma.201700533.

5

Innovations in 3D Printing-Assisted Biopolymers for Biomedical Applications

Elhady, Sherifa[1], Ibrahim A. B. D. Ellatif[1], Kareem M. Abdelrahman[1,2], Alshaymaa S. Mostaf[2] and Irene Samy Fahim[1]*

[1]*School of Engineering and Applied Sciences, SESC Research Center, Nile University, Sheikh Zayed, Giza, Egypt*

[2]*Department of Materials Science, Institute of Graduate Studies and Research, Alexandria University, Alexandria, Egypt*

Abstract

Polylactic acid, also known as PLA, is one of the most widely utilized materials for desktop 3D printing. It is the filament of choice for most extrusion-based 3D printers because it can be printed at low temperatures and does not need a heated bed. PLA is an excellent material to use in 3D printing because it is simple to print and inexpensive and produces parts that can be used for a variety of purposes. It is also one of the most eco-friendly filaments currently available. PLA, which is derived from corn and sugarcane, is renewable and, more importantly, biodegradable. Unfortunately, the disadvantages of PLA polymer, such as high brittleness, poor heat bending temperature, poor thermal stability, and non-conductivity, restrict its use in a variety of industrial applications. The preparation of PLA composites with suitable additives is the most effective method for improving the properties of FDM-printed PLA components.

Keywords: Polylactic acid, biopolymer, FDM, biomedical applications

5.1 Introduction

The term 3D printing is widely used for additive manufacturing technologies which use three-dimensional printing technology to enable the

**Corresponding author*: ISamy@nu.edu.eg

Bhasha Sharma, Balaram Pani, Shashank Shekhar and Jude A. Okolie (eds.) Sustainable 3D Printing for Innovative Biopolymer Production and Applications, (95–116) © 2025 Scrivener Publishing LLC

production of components and assemblies without the usage of tools. Additive manufacturing is utilized for low-production part runs and lowers design risk while also improving manufacturing efficiency. Complex or customized constructions can be produced quickly and precisely with 3D printing; this is shown in Figure 5.1 [1].

There is a continuous change in the manufacturing landscape, and with the rising competition for sustainable designs and the possibility of using bio-materials, it makes it very difficult to achieve the highest material attributes and performance under the latest technologies. Due to the heightened competition for eco-friendly designs and continuous efforts to develop the highest attributes of matter, 3D printing appeared as a breakthrough technology with huge potential for environmental transformation. It is the most useful technology with low production volumes, regular design modifications, and extreme complexity of the necessary designs [2]. The most popular sectors that are moving toward using 3D printing are shown in Figure 5.2.

However, a major obstacle to the widespread use of 3D printing for the production of goods is the unavailability of bio-based materials with user-defined mechanical and biochemical properties. The creation of environmentally useful biopolymers of natural origin that are available for 3D printing and have the possibility of being used in a variety of fields is of great academic and environmental interest [3]. This chapter reviews the latest in 3D printing technology using naturally derived raw materials,

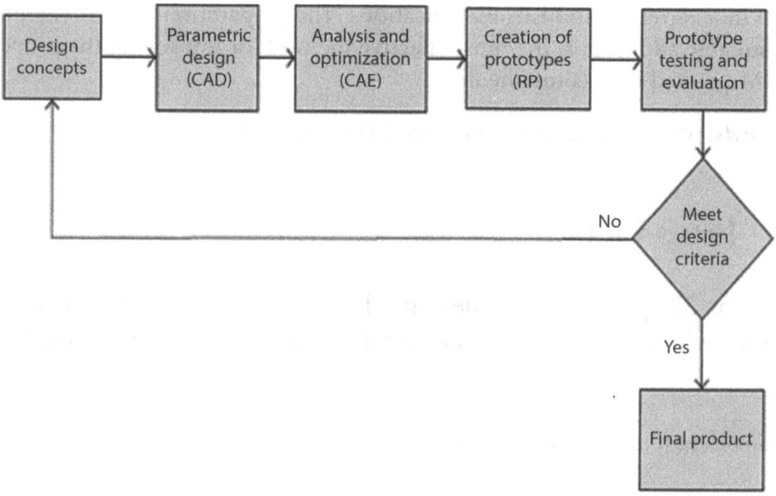

Figure 5.1 Product cycle by 3D printing.

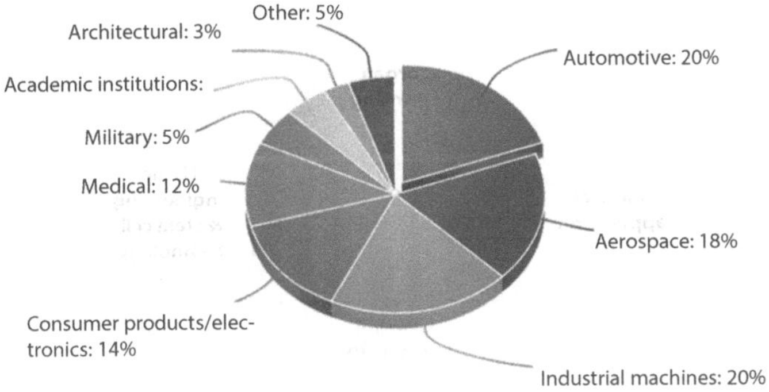

Figure 5.2 Industrial adoption of 3D printing.

including chitosan-based biopolymers, starch-based biopolymers, and polylactic acid (PLA). It acts as a guideline seeking to investigate biopolymers produced from natural sources as unique feedstocks for various 3D printing technologies that may be utilized in diverse fields.

5.2 Applications of 3D Printing Technology in Biomedicine

Recently, the use of 3D printing in biomedical applications has attracted the interest of numerous researchers, and many organ tissues printed with this technique continue to face significant hurdles, from cancer therapy to individualized prostheses. In several sectors of medicine, 3D-printed innovations are sought to improve patients' quality of life or save their lives. All of these applications can be classified as illustrated in Figure 5.3 [4]. 3D printers are now able to print with a variety of materials, including metals, wood products, and thermoplastics like polylactic acid (PLA) [5].

5.2.1 Surgical Application

Surgeons can better see the patient's anatomy with 3D printing's improved imaging technology. A surgical template guides the procedure in estimating angles and predicting bone orientation and size. Today, this technology guides operation planning and supervision to give accurate internal imaging. Technology can enhance surgical outcomes and eliminate medical errors, improving patient safety. This strategy may also help medical students and surgical assistants learn during and before surgery. 3D

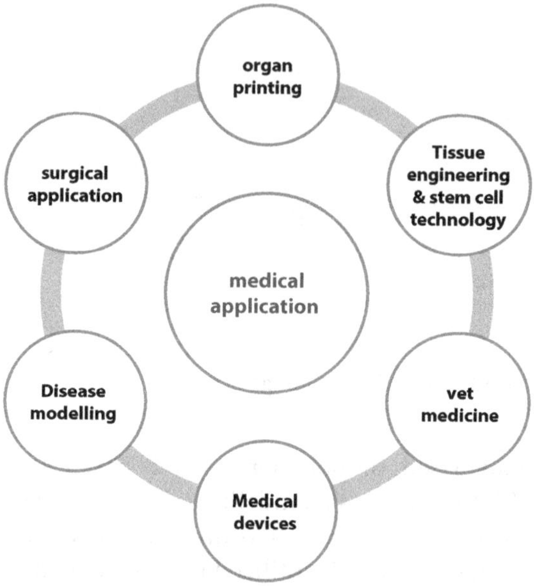

Figure 5.3 Medical application for 3D printing.

printers are used more in surgical applications every year, and research is being done to improve them [6]. Since PLA has been shown to be safe for surgical implantation, it was chosen as a cost-effective, safe, and environmentally friendly material for 3D printing a surgical instrument. The capacity to sterilize a 3D-printed equipment is essential for its use. PLA is extruded at temperatures considerably over the recommended 121°C for steam sterilization or 170°C for dry heat sterilization [7]. However, research indicates that autoclaving PLA impairs its structural integrity [8]. Even *in vitro*, when physiological circumstances are reproduced for days to weeks, PLA polymers degrade minimally [9, 10]. Even while lower-temperature techniques of sterilization, such as ethylene oxide "gas" sterilization have no effect on the strength of PLA, the presence of dangerous levels of ethylene oxide residue is a major worry. In contrast, glutaraldehyde, an excellent room temperature sterilant, has been demonstrated to preserve the most PLA strength when compared to other chemical sterilant [9].

5.2.2 Organ Printing

Using PLA in bioprinting is an innovative and significant area of research. Physical organ models are things that duplicate the patient's anatomical structure and have played crucial roles in modern illness diagnosis and treatment. As a robust, multifunctional manufacturing technology, 3D printing surpasses the constraints of conventional techniques and offers tremendous potential for the production of organ models. However, the therapeutic application of organ models is currently limited due to obstacles such as high cost, poor imitating performance, and inadequate precision. The most prevalent 3D printing technologies are described, and the existing production processes are separated into "directly printing" and "indirectly printing" with an emphasis on selecting the most appropriate techniques and materials [10].

5.2.3 Tissue Engineering and Stem Technology

Tissue engineering is a recently emerging biomedical technology and approach for assisting and accelerating the regeneration and repair of faulty and damaged tissues based on the patients' own natural healing potentials. For the novel treatment method, it is essential to provide cells with a local environment that promotes and regulates their proliferation and differentiation in order to regenerate cell-based tissue. The development of this cell environment involves a significant contribution from biomaterial technology [11]. The acellular approach comprises the use of natural or synthetic matrices, also known as scaffolds, to promote the body's natural capacity for self-repair and aid in determining the direction of new tissue growth. The cellular approach principle is to use donor cells that have been treated prior to implantation and either seeded into the scaffold (cell-seeded scaffold approach) or used alone—the stem cell approach—to promote the growth or regeneration of functional tissue [12].

5.2.4 Vet Medicine

Recent technological advancements in 3D printing have led to a rise in the usage of this technology in human medicine, while its lowering price makes it more affordable for veterinary applications [13]. Anatomical models utilized in learning and training facilities must be accurate at all levels of the veterinary education system, including undergraduate and graduate programs, clinical skills training, and client education [14]. It is utilized in a variety of veterinary applications, for example, in order to control and comprehend the anatomy of the animal and the veterinarian's practice prior to surgery. By planning beforehand, the veterinarian may anticipate any anatomical variances, decrease the margin of error, and ensure the animal's safety [15].

5.2.5 Medical Devices

In the creation of numerous medical equipment, the use of 3D printing technology is a prevalent preference and has become increasing. Polylactic acid (PLA) is one of the most widely utilized polymers in medical devices due to its biocompatibility, biodegradability, and absorbability [16]. With the introduction of fused deposition modeling (FDM), one of the most renowned 3D printing technologies in the biomedical field, interest in PLA has expanded significantly due to its advantageous thermoplastic properties. It can be heated to its melting point, cooled, and then reheated without degrading much. Together with FDM technology, this enables the rapid fabrication of patient-specific implants, surgical guides (cranial and maxillofacial surgery), and surgical tools for a vast array of research and clinical applications, including patient-specific implants, surgical guides (cranial and maxillofacial surgery), and surgical tools [17].

5.2.6 Disease Modeling

Using 3D printing technology, a duplicate of the patient's unique anatomy is created prior to difficult operations or to better comprehend the ailment. Thus, the copy of the organ may be evaluated considerably more clearly than with imaging technologies such as tomography and can be practiced prior to surgery, and the margin of error can be diminished. In addition, it can be used as a simple and rapid way for diagnosing and treating the disease—for example, high-fidelity kidney models allow for the planning and simulation of complex surgical procedures [18]. This strategy facilitates medical students' understanding of the disease through modeling and affords them the opportunity to investigate the complicated anatomy and pathology of the kidneys. In modern times, kidney tumors can be diagnosed using technical means such as CT and

MRI; yet, in some instances, it is extremely challenging to comprehend the relationship between kidney tumors and surrounding renal anatomical components. Modeling, therefore, plays a significant part in the identification of kidney tumors and the evaluation of the patient's status. Numerous studies argue that modeling will be of considerable value not only for kidney illness but also for cardiovascular disease and liver disease. Due to the limitations of traditional imaging techniques on a flat screen, they may not be adequate for sophisticated procedures such as cardiovascular surgery. Therefore, 3D modeling can provide a comprehensive depiction of the anatomy, prevent unanticipated results, and provide individualized treatment for numerous surgical procedures. Thus, it can shorten the duration of the procedure, reduce the risk of error, and ensure patient safety [19].

5.3 Biopolymers from Natural Sources

There are a variety of sources and techniques to obtain bio-based polymers, as shown in Figure 5.4.

5.3.1 Chemical Synthesis Biopolymers (PLA)

Overpopulation restricted nature from sheltering people's demands through the last decade; consequently, scientists and engineers had to find a suitable and eco-friendly replacement of natural raw materials. In addition, the fourth industrial revolution integrated technology with industry which led to gigantic constructive adjustments in the recycling industry.

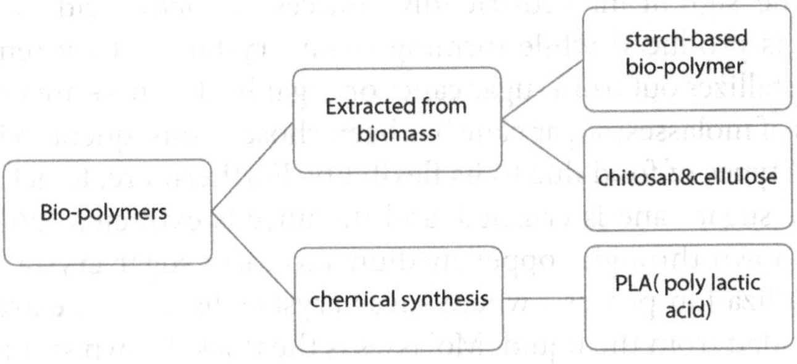

Figure 5.4 Bio-based polymers from biomass.

However, there are various prevalent applications' materials, and the concern of the scientific community rotated around an organic acid called lactic acid. It inspired scientists because of its widespread applications. Also, lactic acid is involved in the pharmaceutical, textile, chemical, cosmetic, and food production industrial sectors. The food production industry uses approximately 70% of produced lactic acid [20, 21]. Lactic acid (2-hydroxy propionic acid or 2-hydroxypropanoic acid) is an organic acid with the molecular formula $CH_3CHOHCOOH$.

The anhydrous racemic lactic acid is a white crystalline with a low melting point and two optical forms, D- and L+. Also, L+ is a biological isomer found naturally in human bodies, and the D- lactic acid microorganisms produce organic acids such as acetic acid [22]. The concern of scientists moved around lactic acid because it has widespread sources that start from unrenewable sources such as petrochemicals and renewable sources such as biological sources like cellulose and glucose. Industrial wastes include wastepaper, pulp, and wood, and agricultural wastes include sugarcane, beet, apples, pineapples, alfalfa, corn, wheat, rice, etc. [23]. Molasses produced out of sugarcane and beet are the most acceptable option for lactic acid production in Egypt due to the high production quantity in Egyptian land, up to 250,000 tons in 2006 [24]. Besides that, this is cheaper than other materials as one ton of molasses costs 4,000 EGP according to an interview with Egyptian sugar and integrated industry company, Cairo (ESIIC) manager—no pretreatment needed, with high sugar concentration which reaches 50.86%, and low level of contaminants, for instance, ash is 11.8% only [25].

5.3.1.1 PLA Production from Molasses

One of the significant eco-friendly sources of lactic acid is molasses. Molasses is produced while forming sugar crystals as they remain after sugar crystallizes out from sugar canes or sugar beets. There are various raw materials of molasses; sugar cane has been chosen consequent to its utilization in all types of food due to its flexibility. Furthermore, to get molasses, firstly, the sugar cane is crushed, and the juice is extracted [26]. The sap is boiled down through copper mediums to form sugar crystals through the crystallization process, where sugar crystals have been extracted and then excluded from the liquid. Molasses is the thick, brown syrup left after removing the sugar from the juice. This process is repeated several times, and each time a different type of molasses is produced. Moreover, there are three types of molasses: light molasses, which is formed after the first boiling and is the most golden in color; the second type is dark molasses which is being extracted after the second boiling, and it is more concentrated, which means thicker and darker with increasing impurities and different sources and decreasing sugar percentage; and the last type is blackstrap which is formed after the third boiling, and it is the thickest, darkest, and most concentrated one [27].

5.3.1.2 PLA Production by Using Bacteria

Various microorganisms produce lactic acid, including bacteria, fungi, yeast, algae, and cyanobacteria [28]. Lactic acid bacteria (LAB) are gram-positive organisms considered as one of the leading industrial-scale lactic acid producers [29]. The used type of lactic acid bacteria (LAB) is the factor of fermentation process selection; moreover, LAB can be homofermentative or heterofermentative. Hexose and pentose LAB metabolism produces glycolysis that produces LA under anaerobic conditions. In heterofermentative, there are identical molar amounts of lactic acid, acetic acid, ethanol, and carbon dioxide; on the other hand, in homofermentative, only lactic acid is formatted [30]. The yield and efficiency depend on the pH (3.5–9.6), temperature (5–45°C), nutrients' presence, and LAB strain producers [31]. However, LAB contains lactobacillus, *Lactococcus*, leuconostoc, streptococcus, and pediococcus; there are advantages of using commercial LAB in LA production, such as the high need for complex nutrients and low fermentation temperature, which lead to low-yield lactic acid [29]. The suitable alternative is laboratory-made bacteria with high yield, such as *Kosakonia cowanii* [32].

5.3.1.3 Lactic Acid Polymerization

Three processes for preparing polylactic acid are accessible: direct condensation, azeotropic dehydrative condensation, and ring-opening polymerization. The azeotropic dehydrative realized a breakthrough in enlarging PLA molecular weight. While the method takes a long reaction time, the azeotropic distillation technique occurs by using a solvent with a high boiling point to remove the dissociated water [33]. Recently, direct condensation and ring-opening polymerization have been the most commonly used production techniques [34]. Direct condensation is implemented in bulk by distilling condensed water with or without catalyst while the vacuum and temperature rise. As it is hard to remove water from the highly viscous reaction mixture, the obtained polymer has a lower molecular weight [35]. The ring-opening polymerization consists of two steps: the first one is switching of lactic acid feedstock to lactide, and the second is polymerization of lactide to give polylactic acid as a product. Besides this, a high-molecular-weight polylactic is obtained [36].

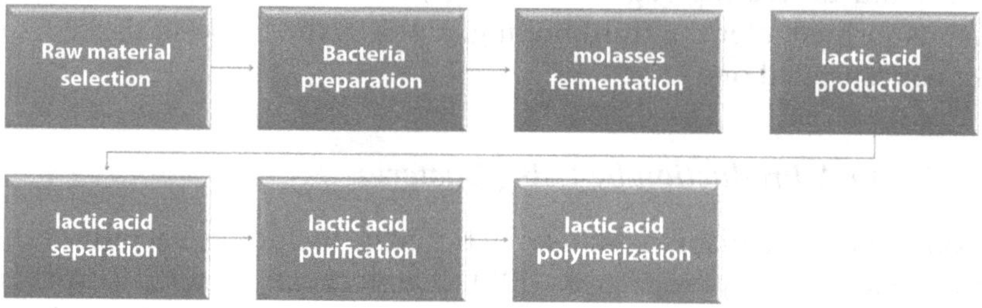

5.3.1.4 *Uses of PLA*

The lactic acid market demand increased in recent years, consequent to its usage as a monomer in polylactic acid preparation (PLA) [37]. Furthermore, PLA is considered as a renewable and biodegradable polymer with several properties as well as crystallinity and thermal stability [39], which has widespread applications such as in the medical field as in the case of prostheses, surgical sutures, and also screws and micro-titration palates to reduce the stress-shielding effect [40] and in packaging as in food and beverage containers, wrappings for sweets, single-use tray, and garbage bags as well as high-value films, rigid thermoforms, drinking cups, sundae and salad cups, lamination films, blister packages, water bottles, and coated papers [41]. Besides these, labeling, textile in non-woven textile for clothes, and pharmaceutical industries are also included.

5.3.1.5 *PLA Based-3D Printing*

3D printing is an additive manufacturing process in which materials are extruded layer by layer through a nozzle to form the printed part shape [41], as illustrated in Figure 5.5.

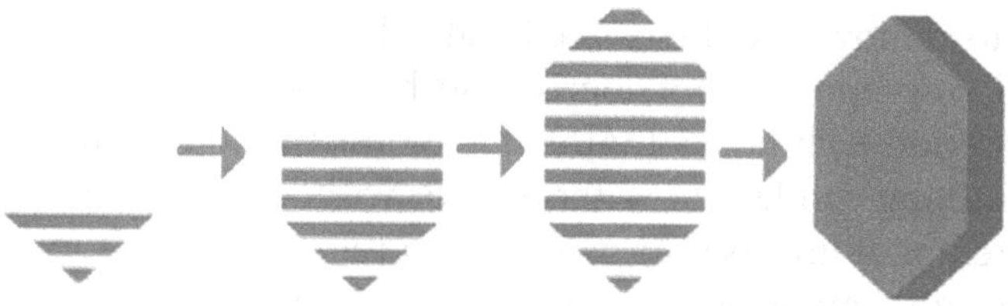

Figure 5.5 3D printing process.

3D printing is used widely for various applications like a product prototype due to its ability to create a complex shape more quickly than other processes. For prototyping aims, the mechanical properties for the printed part using 3D printing are very critical. Because of that, it is very important to understand the impact for different parameter values of the 3D printing process on the mechanical properties for the printed part. Polymers are widely used as materials for a prototype due to its having a lower density as well as higher strains during failure. Also, polymers have higher strength than metals. PLA (polylactic acid) is a polymer which is most widely used like in 3D printing material as it required inferior melting point, therefore less power to print. Also, PLA is considered a biodegradable polymer that does not spread toxic vapors. Finally, it sticks to the platform effectively and does not need a heated base [42]. The most common 3D printing methods are illustrated in Figure 5.6. FDM method is a simple and also more cost-effective technique of fabrication process in comparison with other 3D printing techniques. Unfortunately, there are shortcomings of the FDM technique, such as weakness of mechanical properties for FDM parts compared to those produced by compression molding and conventional injection methods. Thus, preparation of PLA composite materials with suitable additives is the most effective technique to enhance the qualities of the 3D-printed PLA parts accomplished by the FDM technique [43].

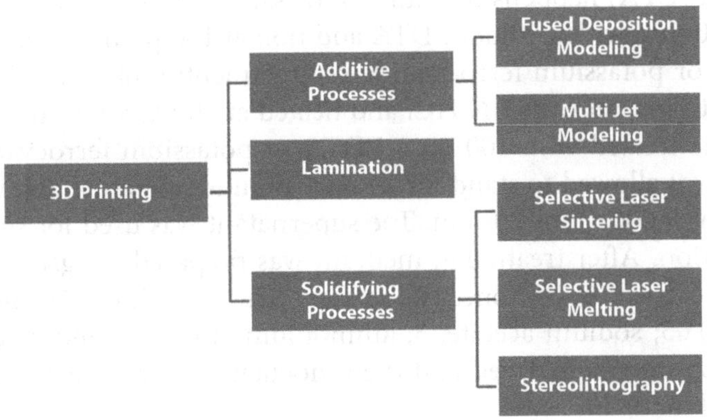

Figure 5.6 Common 3D printing methods.

5.3.1.6 *Methodology of Environmentally Friendly PLA Production for 3D Printing*

A. Isolation and Purification of Bacterial Isolates

Kosakonia cowanii is a predominant bacteria, and it can be isolated from the natural environment, animal hosts, clinical environments, industrial environments, and food [44]. To isolate bacteria, beet and sugarcane molasses were used independently. Both types of molasses were subjected to different pretreatments to depolymerize the existing sugar and remove fermentation inhibitors. These methods include the utilization of physico-chemical treatments by using heat, some acids such as sulfuric acid (SA) and orthophosphoric acid (OPA), besides some chemicals such as tricalcium phosphate (TCP), potassium ferrocyanide (KFCN), EDTA, and charcoal in combinations. Furthermore, five different microbial strains were evaluated for the efficiency of the lactic acid produced. Sulfuric acid pretreatment has several methods; all of them are by adding H_2SO_4 to molasses until pH 3.0. After that, it follows one of the following paths: heated at 100°C for 30 min, allowed to settle for 24 h at 25°C, centrifuged for 15 min at 5,000×*g*, allowed to settle for 24 h at 25°C, or then subjected to 3% (w/v) activated carbon pretreatment and filtration using Whatman filter paper. Another method is active carbon pretreatment; molasses solution was treated with 3% activated carbon and mixed for 24 h at 25°C and then filtered with a Whatman no. 1 filter. For tri-calcium phosphate treatment (TCP) method: pH was adjusted to 7.0 using 1 M NaOH, then treated with tricalcium phosphate (2%, w/v), autoclaved at 105°C for 5 min, and then cooled and centrifuged for 15 min at 3,000 rpm. The supernatants were used after adjusting the pH at 7.0. However, orthophosphoric acid treatment (OPA) happens by adding molasses until pH 3.0 and heating at 100 for 30 min with/without EDTA addition at 100 ppm. The last method is EDTA or potassium ferrocyanide pretreatment; molasses solution was adjusted to pH 5.5 with 5 N HCl and heated at 100°C for 15 min. The hot liquid was treated with 100 ppm EDTA or potassium ferrocyanide. The mixture was allowed to stand for 24 h at room temperature and then centrifuged at 5 000×*g* for 20 min. The supernatant was used for subsequent fermentation. After treatment, medium was prepared (in g/L): molasses, 20; yeast extract, 5; peptone, 10; beef extract, 8; K_2HPO_4, 2; $MgSO_4$, 0.1; $MnSO_4$, 0.05; sodium acetate, 5; ammonium citrate, 2; and Tween 80, 1 mL—these were autoclaved and then inoculated by microorganisms and incubated for 48 h.

B. Lactic Acid Separation Methods

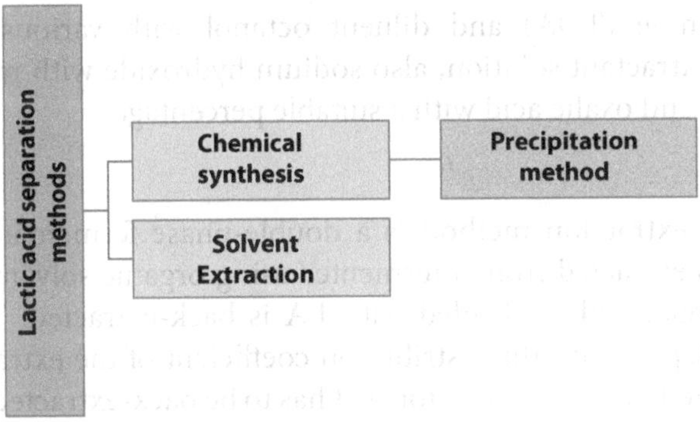

B.1 Chemical Synthesis (Precipitation Method)

Materials

A total of 10 g of calcium lactate $(CH_3CHOHCOO-)_2Ca_2$, 4 mm of sulfuric acid H_2SO_4, and water were needed for the experiment.

Method

The first process is producing calcium lactate from the fermented broth. Consequently, the fermentation substance is heated to dissolve all calcium lactate and treated with stoichiometric amounts of sulfuric acid. Then, 10 g of diluted calcium lactate reacted 10 wt% with 4 ml of diluted sulfuric acid 50 wt% to produce dilute crude lactic acid and calcium sulfate or gypsum. The reaction has arisen at 30°C and 300 rpm speed for 30 min. After that, the sulfate was separated using a centrifuge at a speed of $44*10^3$ rpm for 40 min. The method produced yellowish gypsum (calcium sulfate), as shown in Figure 5.9, containing free organic acid diluting crude lactic acid. According to the presence of impurities from the fermented process, the color starts to get darker and becomes black during the evaporation process, as shown in Figure 5.7.

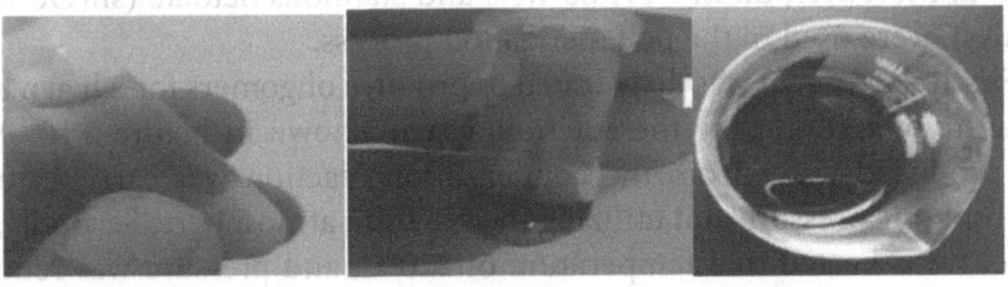

Figure 5.7 Lactic acid became darker.

B.2 Solvent Extraction Method

Materials

Tri-octyl amine (TOA) and diluent octanol with various percentage weights as extractant solution, also sodium hydroxide with multiple concentrations, and oxalic acid with a suitable percentage

Method

The solvent extraction method is a double-phase fermentation method where LA is extracted from a fermenter using organic solvent extractant; then, the organic phase loaded with LA is back-extracted. The reactive extraction depends on the distribution coefficient of the extractant. After LA is extracted into an organic form, it has to be back-extracted using head or chemical agents [23]. Tri-octyl amine was used as a solvent and diluent octanol with different percentage weights as extractant solution. Sodium hydroxide of various concentrations is used with TOA to free the lactic acid, which results in sodium lactate. After lactic acid tracking, oxalic acid is added with a suitable percentage to extract free lactic acid. The trials used 30:70 wt% of TOA and octanol. Besides this, there are two experiment procedures: static by separator funnel and mixing average in normal temperature. Both types combine the solvent at 1,500 rpm in a cold bath at 0°C and then centrifuged at 7,000 rpm. After many trials using different concentrations of TOA and octanol that represent a low concentration of free lactic acid, 50:50 wt% of TOA and octanol results in the best concentration of free lactic acid. Also, the results of both procedures or methods are shown in Figures 5.8 and 5.9 for solvent extraction by separator funnel and for solvent extraction by mixing separation.

C. Polymerization Setup

The dehydration of the monomer is the starting point of the polymerization process as it produces a prepolymer chain of oligomers and low molecular weight PLA. The first stage is committing LA in three distinct phases: polycondensation, obtaining lactide, and ring-opening polymerization. Moreover, diluted LA 85 wt% and stannous octoate (sn(Oct)2), a catalyst, were used in the polymerization process.

Firstly, lactic acid was dehydrated to produce oligomers for 2 h at a temperature of 160°C using the reaction system shown in Figure 5.10. Also, the produced water from the condensation reaction was carried out to atmospheric pressure and utilizing the N2 solid atmosphere. Secondly, the system was heated to a temperature of 220°C, and pressure was reduced to 200 mmHg, which led to the recovery of the produced lactide by distillation and collected in a condensate flask. After 4 h of reaction, the

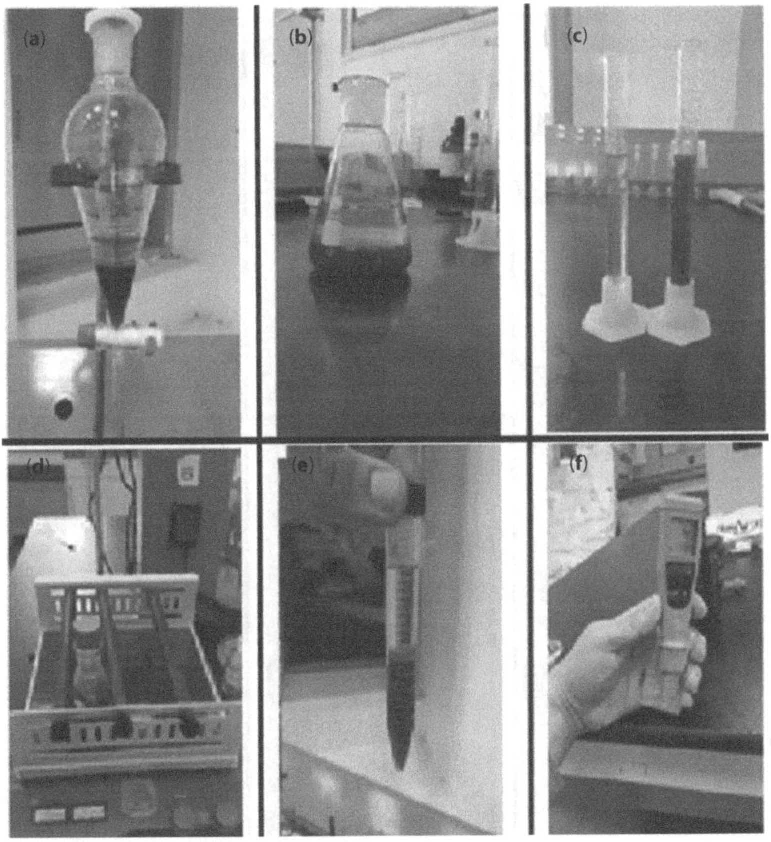

Figure 5.8 Solvent extraction by a separation funnel.

condenser was heated to 90°C to avoid solidification of the product. The lactide was then extracted by distillation, and the solid phase was recovered in a condensing flask. The crude product was washed with cold water, separated by filtration, and then dried overnight at a temperature of 40°C. Finally, the lactide produced was mixed with the catalyst (1 wt%) at a temperature of 140°C for 2 h, and PLA was made. Another method using the polycondensation method, lactic acid (99.6%), $Sncl_2$, $2H_2O$, and diphenyl ether has been brought up. Furthermore, the reaction was operated in a 250-mL, three-necked flask reactor equipped with a magnetic stirrer and a water-cooling condenser. Lactic acid and 0.05 wt % ($SnCl_{2.}2H_2O$) diphenyl ether have been added to the reactor. The temperature was gradually increased to 120°C in 3 h, and the reaction mixture was stirred continuously. During the reaction process from 8 to 24 h, the pressure decreased,

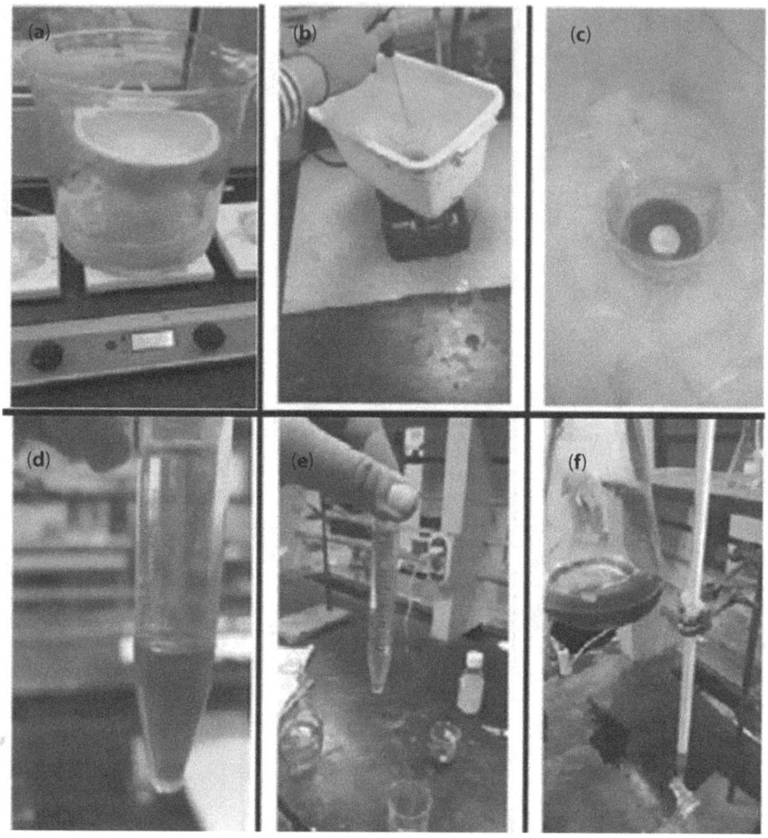

Figure 5.9 Solvent extraction by cold mixing separation.

and the temperature increased to 160–200°C. By the end of the reaction, it was cooled down to room temperature. A white powder appeared after precipitating in methanol. The product was dissolved in chloroform and purified in methanol. PLA was then filtrated and dried at 60°C under vacuum.

5.3.1.7 Challenges for PLA in 3D Printing

Although PLA is highly used as raw material in extrusion-based three-dimensional (3D) printing in different areas due to its biodegradability and being environmentally friendly, even so, its utilization is bounded due to several disadvantages such as high brittleness, low toughness, low melt strength, poor heat of bending temperature, low thermal stability,

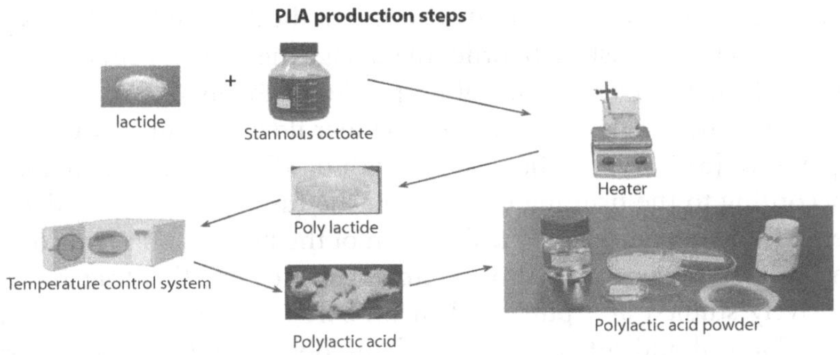

Figure 5.10 Polymerization process.

constricted processing window, and also non-conductivity. All of these reasons limit its use in multiple industrial applications [45, 46]. PLA has a weak mechanical strength compared to acrylonitrile butadiene styrene (ABS) and polycarbonate (PC) polymers. PLA has stronger mechanical properties due to its linear molecular chain structure that prevents chain tangling. Additionally, there is weakness of dimensional stability for PLA objects after printing due to the volume alteration occurring and stress formation during the 3D printing process, which is because of the variation of PLA crystallinity. Three steps are needed to achieve adhesion between two fused PLA filaments while 3D printing. Initially, two separate PLA filaments' surfaces come into contact with one another, and later a neck forms between them, and finally, the PLA chains diffuse at the interphase region between two adjoining filaments to eliminate the solid surface meniscus' presence [47, 48]. In general, the weakness of interfacial adhesion between the filaments leads to the poor mechanical behaviors of the 3D-printed PLA items. It was suggested that the brief contact duration of the extruded PLA filaments, which are held around their melting temperatures, is the cause of the weak interfacial adhesion since the filaments' temperature drops quickly to the melting point of PLA in a short time because of the quick heat transmission to the surroundings [43]. This occurs because of the interfacial region between two nearby filaments, the linear PLA macromolecules cannot completely diffuse into one another, and typically there are some cavities in the interphase which cause the decrease in the mechanical strength of the printed object [49]. When the cooling speed of the injected filaments is sluggish, areas of the object are disorganized, and the surface quality is poor. In contrast, when the cooling speed of the injected filaments is high, the PLA macromolecular diffusivity at the interface between neighboring filaments is poor, resulting in weak interfacial bonding [50]. This is the primarily because of the 3D-printed object's poor mechanical characteristics. In order to manage the state of PLA macromolecules and the related mechanical properties of the final item, it is essential to select the nozzle temperature, polymer flow rate, and local printing temperature [51]. The mechanical strength of 3D-printed PLA items varies according to the printing direction of the fused filaments, as depicted schematically in Figure 5.5. The direction of the filaments should coincide with the direction of the major stretching force, and this trait has been extensively studied and published in numerous studies [43]. Improving the mechanical and other qualities of homopolymer PLA-printed products using the FDM method is restricted. To boost the mechanical strength of the PLA homopolymer, it was recommended to employ PLA composites instead of pure PLA polymer by preparing the composite from two main components, the matrix and the reinforcing agent [52]. Principally, micro- or nano-particles of inorganic, carbon, and organic origin, short or long continuous fibers, and a few specialized metal particles are utilized to produce PLA composites with reinforcement [53]. Another way is by the inclusion of continuous fibers into PLA using specialized FDM-based 3D printers to create mechanically robust printed items.

References

1. Kalender, M., Kılıç, S.E., Ersoy, S., Bozkurt, Y., Salman, S., Additive manufacturing and 3D printer technology in aerospace industry, in: *2019 9th International Conference on Recent Advances in Space Technologies (RAST)*, IEEE, pp. 689–694, June 2019.
2. Vaneker, T., Bernard, A., Moroni, G., Gibson, I., Zhang, Y., Design for additive manufacturing: Framework and methodology. *CIRP Ann.*, 69, 2, 578–599, 2020.
3. Liu, J., Sun, L., Xu, W., Wang, Q., Yu, S., Sun, J., Current advances and future perspectives of 3D printing natural-derived biopolymers. *Carbohydr. Polym.*, 207, 297–316, 2019.
4. Weems, A.C., Pérez-Madrigal, M.M., Arno, M.C., Dove, A.P., 3D printing for the clinic: Examining contemporary polymeric biomaterials and their clinical utility. *Biomacromolecules*, 21, 3, 1037–1059, 2020.
5. Rankin, T.M., Giovinco, N.A., Cucher, D.J., Watts, G., Hurwitz, B., Armstrong, D.G., Three-dimensional printing surgical instruments: are we there yet? *J. Surg. Res.*, 189, 2, 193–197, 2014.
6. Weiss, R. and Read-Fuller, A., Cone beam computed tomography in oral and maxillofacial surgery: an evidence-based review. *Dent. J.*, 7, 2, 52, 2019.

7. Fedorovich, N.E., Alblas, J., Hennink, W.E., Öner, F.C., Dhert, W.J., Organ printing: the future of bone regeneration? *Trends Biotechnol.*, 29, 12, 601–606, 2011.
8. Meseguer-Olmo, L., Vicente-Ortega, V., Alcaraz-Baños, M., Calvo-Guirado, J.L., Vallet-Regí, M., Arcos, D., Baeza, A., *In-vivo* behavior of Si-hydroxyapatite/polycaprolactone/DMB scaffolds fabricated by 3D printing. *J. Biomed. Mater. Res. Part A*, 101, 7, 2038–2048, 2013.
9. Athanasiou, K.A., Niederauer, G.G., Agrawal, C.M., Sterilization, toxicity, biocompatibility and clinical applications of polylactic acid/polyglycolic acid copolymers. *Biomaterials*, 17, 2, 93–102, 1996.
10. Jin, Z., Li, Y., Yu, K., Liu, L., Fu, J., Yao, X., He, Y., 3D printing of physical organ models: recent developments and challenges. *Adv. Sci.*, 8, 17, 2101394, 2021.
11. Tabata, Y., Biomaterial technology for tissue engineering applications. *J. R. Soc. Interface*, 6, suppl_3, S311–S324, 2009.
12. Shokeir, A.A., Harraz, A.M., El-Din, A.B.S., Tissue engineering and stem cells: basic principles and applications in urology. *Int. J. Urol.*, 17, 12, 964–973, 2010.
13. Seoane-Viaño, I., Trenfield, S.J., Basit, A.W., Goyanes, A., Translating 3D printed pharmaceuticals: From hype to real-world clinical applications. *Adv. Drug Deliv. Rev.*, 174, 553–575, 2021.
14. Hespel, A.M., Wilhite, R., Hudson, J., Invited review-applications for 3d printers in veterinary medicine. *Vet. Radiol. Ultrasound*, 55, 4, 347–358, 2014.
15. Javaid, M. and Haleem, A., Additive manufacturing applications in orthopaedics: a review. *J. Clin. Orthop. Trauma*, 9, 3, 202–206, 2018.
16. Pérez-Davila, S., González-Rodríguez, L., Lama, R., López-Álvarez, M., Oliveira, A.L., Serra, J., González, P., 3D-Printed PLA Medical Devices: Physicochemical Changes and Biological Response after Sterilisation Treatments. *Polymers*, 14, 19, 4117, 2022.
17. Tyler, B., Gullotti, D., Mangraviti, A., Utsuki, T., Brem, H., Polylactic acid (PLA) controlled delivery carriers for biomedical applications. *Adv. Drug Deliv. Rev.*, 107, 163–175, 2016.
18. Giannopoulos, A.A., Mitsouras, D., Yoo, S.J., Liu, P.P., Chatzizisis, Y.S., Rybicki, F.J., Applications of 3D printing in cardiovascular diseases. *Nat. Rev. Cardiol.*, 13, 12, 701–718, 2016.
19. Mertz, L., New world of 3-D printing offers" completely new ways of thinking": Q&A with author, engineer, and 3-D printing expert Hod Lipson. *IEEE Pulse*, 4, 6, 12–14, 2013.
20. Eş, I., Mousavi Khaneghah, A., Barba, F.J., Saraiva, J.A., Sant'Ana, A.S., Hashemi, S.M.B., Recent advancements in lactic acid production-a review. *Food Res. Int.*, 107, 763–770, 2018, doi: 10.1016/j.foodres.2018.01.001.

21. Castro-Aguirre, E., Iniguez-Franco, F., Samsudin, H., Fang, X., & Auras, R., Poly (lactic acid)—Mass production, processing, industrial applications, and end of life. *Adv. Drug Deliv. Rev.*, 107, 333–366, 2016.
22. Alexandri, M., Schneider, R., Mehlmann, K., Venus, J., Recent advances in d-lactic acid production from renewable resources: Case studies on agro-industrial waste streams. *Food Technol. Biotechnol.*, 57, 3, 293–304, 2019, doi: 10.17113/ftb.57.03.19.6023.
23. Wee, Y.J., Kim, J.N., Ryu, H.W., Biotechnological production of lactic acid and its recent applications. *Food Technol. Biotechnol.*, 44, 2, 163–172, 2006.
24. Keshk, S.M.A.S., Razek, T.M.A., Sameshima, K., Bacterial cellulose production from beet molasses. *Afr. J. Biotechnol.*, 5, 17, 1519–1523, 2006, doi: 10.4314/ajb.v5i17.43149.
25. Zohri, A.A., "Ethanol production from Egyptian sugar cane molasses by six yeast strains using batch,". *J. Basic Appl. Mycol.*, vol, 43–49, 2014.
26. Sherif, N.J., *Characterization and Optimization of Lactic Acid Produced from Sugar Cane Molasses by Using Lactobacillus Plantarium Bacteria Isolated from "Kocho"*, Doctoral dissertation, Addis Ababa University Addis Ababa, Ethiopia, 2018.
27. Vidra, A., Tóth, A.J., Németh, A., Lactic acid production from cane molasses. *Waste Treat. Recovery*, 2, 1, 13–16, 2017, doi: 10.1515/lwr-2017-0003.
28. Rodrigues, C., Vandenberghe, L.P.S., Woiciechowski, A.L., de Oliveira, J., Letti, L.A.J., Soccol, C.R., Production and Application of Lactic Acid, in: *Current Developments in Biotechnology and Bioengineering: Production, Isolation and Purification of Industrial Products*, pp. 543–556, 2016.
29. Abedi, E. and Hashemi, S.M.B., Lactic acid production–producing microorganisms and substrates sources-state of art. *Heliyon*, 6, 10, e04974, 2020, doi: 10.1016/j.heliyon.2020.e04974.
30. Mazzoli, R., Bosco, F., Mizrahi, I., Bayer, E.A., Pessione, E., Towards lactic acid bacteria-based biorefineries. *Biotechnol. Adv.*, 32, 7, 1216–1236, 2014.
31. Miller, C., Fosmer, A., Rush, B., McMullin, T., Beacom, D., Suominen, P., Industrial production of lactic acid, in: *Comprehensive Biotechnology*, pp. 208–217, January 2019.
32. El-Sheshtawy, H.S., Fahim, I., Hosny, M., El-Badry, M.A., Optimization of lactic acid production from agro-industrial wastes produced by Kosakonia cowanii. *Curr. Res. Green Sustain. Chem.*, 5, 100228, 2022.
33. Ren, J., Synthesis and Manufacture of PLA, in: *Biodegradable Poly (Lactic Acid) Synthesis, Modification, Processing and Applications*, pp. 15–37, 2010.
34. Lopes, M.S., Jardini, A.L., Filho, R.M., Synthesis and characterizations of poly (lactic acid) by ring-opening polymerization for biomedical applications. *Chem. Eng. Trans.*, 38, 331–336, 2014, doi: 10.3303/CET1438056.
35. Södergård, A., Stolt, M., Inkinen, S., INDUSTRIAL PRODUCTION OF HIGH-MOLECULAR-WEIGHT POLY (LACTIC ACID), in: *Poly (Lactic Acid) Synthesis, Structures, Properties, Processing, Applications, and End of Life*, pp. 29–44, 2022.

36. Kim, S.H. and Kim, Y.H., Direct condensation polymerization of lactic acid. *Macromol. Symp.*, 144, 277–287, 1999, doi: 10.1002/masy.19991440125.
37. Komesu, A., de Oliveira, J.A.R., da S. Martins, L.H., Wolf Maciel, M.R., Maciel Filho, R., Lactic Acid Production to Purification: A Review. *BioResources*, 12, 2, 4364–4383, 2017, doi: 10.15376/biores.12.2.komesu.
38. Eiteman, M.A. and Ramalingam, S., Microbial production of lactic acid. *Biotechnol. Lett*, 37, 5, 955–972, 2015, doi: 10.1007/s10529-015-1769-5.
39. Avérous, L., Polylactic acid: Synthesis, properties and applications, in: *Monomers, Polymers and Composites from Renewable Resources*, pp. 433–450, 2008.
40. Pattinson, S.W. and Hart, A.J., Additive manufacturing of cellulosic materials with robust mechanics and antimicrobial functionality. *Adv. Mater. Technol.*, 2, 4, 1600084, 2017.
41. Suteja, T.J. and Soesanti, A., Mechanical properties of 3D printed Polylactic acid product for various infill design parameters: A review. *J. Phys. Conf. Ser.*, IOP Publishing, 1569, 4, 042010, July 2020.
42. Tümer, E.H. and Erbil, H.Y., Extrusion-Based 3D printing applications of PLA composites: a review. *Coatings*, 11, 4, 390, 2021.
43. Iversen, C., ELECTRICAL TECHNIQUES | Enterobacter, in: *Encyclopedia of Food Microbiology*, Second Ed, pp. 653–658, Jan. 2014.
44. Eynde, M.V.D. and Puyvelde, P.V., 3D Printing of Poly (lactic acid), in: *Industrial Applications of Poly (lactic acid)*, pp. 139–158, 2017.
45. Wickramasinghe, S., Do, T., Tran, P., FDM-based 3D printing of polymer and associated composite: A review on mechanical properties, defects and treatments. *Polymers*, 12, 7, 1529, 2020.
46. Sun, Q., Rizvi, G.M., Bellehumeur, C.T., Gu, P., Effect of processing conditions on the bonding quality of FDM polymer filaments. *Rapid Prototyping J.*, 14, 2, 72–80, 2008.
47. Tjong, S.C., Structural and mechanical properties of polymer nanocomposites. *Mater. Sci. Eng.: R: Rep.*, 53, 3-4, 73–197, 2006.
48. Yan-qing, W., Zhong-ming, W., Cheng-jin, S., You-liang, W., Research on enhancement of GFRP-anchor's torsional strength. *Sci. Eng. Compos. Mater.*, 19, 4, 423–429, 2012.
49. Wang, S., Daelemans, L., Fiorio, R., Gou, M., D'hooge, D.R., De Clerck, K., Cardon, L., Improving mechanical properties for extrusion-based additive manufacturing of poly (lactic acid) by annealing and blending with poly (3-hydroxybutyrate). *Polymers*, 11, 9, 1529, 2019.
50. Behzadnasab, M. and Yousefi, A.A., Effects of 3D printer nozzle head temperature on the physical and mechanical properties of PLA based product, in: *12th international seminar on polymer science and technology*, pp. 2–5, November 2016.
51. Rajpurohit, S.R. and Dave, H.K., Analysis of tensile strength of a fused filament fabricated PLA part using an open-source 3D printer. *Int. J. Adv. Manuf. Technol.*, 101, 5, 1525–1536, 2019.

52. Lebedev, S.M., Gefle, O.S., Amitov, E.T., Zhuravlev, D.V., Berchuk, D.Y., Mikutskiy, E.A., Mechanical properties of PLA-based composites for fused deposition modeling technology. *Int. J. Adv. Manuf. Technol.*, 97, 1, 511–518, 2018.
53. Ochi, S., Flexural properties of long bamboo fiber/PLA composites. *Open J. Compos. Mater.*, 5, 03, 70, 2015.

6

Innovations in 3D Printing-Assisted Biopolymer Composites for Aerospace Applications

Md Masiat Roushan Masrafee[1], Adib Bin Rashid[2] and Md Enamul Hoque[3*]

[1]Department of Aeronautical Engineering, Military Institute of Science and Technology (MIST), Dhaka, Bangladesh
[2]Department of Industrial and Production Engineering, Military Institute of Science and Technology (MIST), Dhaka, Bangladesh
[3]Department of Biomedical Engineering, Military Institute of Science and Technology (MIST), Dhaka, Bangladesh

Abstract
This century has seen significant progress in green technology and materials research, which have been made possible by a growing concern about environmental and sustainability issues. The assembly of 3D-printed biocomposites has numerous implementations in the rapidly expanding aerospace industry. Furthermore, in the air defense system, in-orbit manufacturing, ground preparation of space systems, or even *in situ* resource utilization on the surface of other planes have a broad prospect of 3D printing with biocompatible composites. Several years from now, technical demonstrations employing satellite platforms in Low Earth Orbits (LEOs) will follow current advancements in composites additive manufacturing (AM) technologies, which include indoor experiments on the International Space Station. This book chapter surveys the potential of biocomposites and their prospects in the additive manufacturing of structure, interior, and shielding facilities for spacecraft and aircraft to promote a cost-effective, eco-friendly, and sustainable future.

Keywords: 3D printing, biopolymer, biocomposites, composites, aerospace application

**Corresponding author*: enamul1973@gmail.com

Bhasha Sharma, Balaram Pani, Shashank Shekhar and Jude A. Okolie (eds.) Sustainable 3D Printing for Innovative Biopolymer Production and Applications, (117–148) © 2025 Scrivener Publishing LLC

6.1 Introduction

The Wright Brothers created their first "flying airplanes" more than a century ago using natural composite materials like wood. Even now, due to its lightness, durability, rigidity, and strength, the spars and ribs of a little aircraft's mainframe are made from wood [1]. In this era, the aerospace industry has expanded hugely. The aerospace industry has experienced rapid expansion in recent years, as evidenced by the exponential increase in air travelers and the number of satellites deployed into orbits. The fastest way of travel is by air; not only people but also goods are sent through cargo aircraft. In addition, the development of fighter aircraft, different guided weapons like ballistic missiles, and missile detection defense systems like radar has also become part of the aerospace industry [2].

Compared to other aircraft materials, wood has a lower density, Young's modulus, fracture toughness, and tensile strength. Additionally, because of its hygroscopic nature, it changes in density and mechanical characteristics when exposed to humidity in the air. As a result, it becomes vulnerable to attack from fungi, insects, and other microorganisms. Thus, different metals, metal alloys, and other aerospace materials have replaced wood to serve the requirements of today's demanding aerospace industry. However, the manufacturing processes of these aerospace materials take a lot of energy and keep a carbon footprint on the environment at an alarming rate. Therefore, the way metal and metal alloys have replaced wood, eco-friendly materials should replace these metal and metal alloys. Moreover, these new materials should serve the required purposes with recyclability and renewability capability [3].

In this case, biopolymers have a very prestigious prospect. Biocomposites are lightweight; so, that will reduce the overall fuel cost. Additionally, composites' high specific stiffness and strength allow for the more aerodynamic construction of complicated geometries, and their lack of corrosion concerns lowers maintenance costs. Furthermore, biocomposites release lesser or no toxic gases, and as they are biodegradable, they are eco-efficient. However, along with a bunch of advantages, biocomposites come with some disadvantages, namely [1]:

- Natural fiber-reinforced biocomposites have a low degree of thermal stability.
- Natural fibers' propensity to absorb moisture reduces their mechanical characteristics.

- Polymers are not well compatible with natural fibers because natural fibers are polar (hydrophilic), whereas most polymers are nonpolar (hydrophobic).
- Natural fibers' characteristics vary widely. Their characteristics vary according to the region harvested, the time harvested, the amount of sunlight and rain, and the kind of soil.

So, there is a vast void space for researching biocomposites to achieve specific criteria like flame redundancy, high moisture resistance, and improved interfacial bonding between polymeric matrix and natural fiber as well as high specific stiffness and strength to handle various stresses experienced in airplane flight to employ them in aerospace application. Furthermore, adopting additive manufacturing, like 3D printing in composites, has excellent potential—for example, big companies like SpaceX and NASA have already incorporated the 3D printing manufacturing process in developing heat shields, radiation shields, and spare parts for their spacecraft. Additionally, for the long-term objective of deep space exploration and human habitation on the moon or in Mars, transporting all the spare parts will not be feasible where on-spot 3D printing will be the solution. Moreover, for sustainable outer space and less debris, there is no alternative to using biodegradable and recyclable eco-friendly materials in 3D printing [4].

So, to protect the earth from being more affected by carbon pollution and to make a sustainable debris-free space, biodegradable and recyclable biocomposites have no other alternative. A sustainable modern ecosystem, both in earth and space, needs more research and inventions in biocomposites and their use in additive manufacturing/3D printing. Hence, this chapter aims to portray an overview of the various applications of 3D-printing-assisted biopolymer composites for aerospace industries.

6.2 Overview of Biocomposites

A composite material is developed by mixing two or more distinct types of materials to obtain a set of new characteristics that none of the individual components could have developed on their own. Biopolymer composites are composites that are biodegradable and created by fusing natural fibers sourced from plants and animals with synthetic or natural biopolymers [5]. In biopolymer composites, a reinforcing agent (natural fiber) in the discontinuous phase is introduced to the biopolymer matrix in the continuous phase to enhance the tensile strength as well as stiffness of the resulting composite. Figure 6.1 illustrates the detailed classifications of biocomposites.

A biopolymer composite's density, thermal properties, water sensitivity, electrical conductivity, gas permeability, morphological characteristics, shelf life, degradability, and crystallinity, along with production cost, depend on the type of natural fiber such as coir, banana, hemp, sisal rice straw, wood flour, clays, sisal rice straw, and natural biodegradable matrices including soy, coir shell, starch, polylactic acid (PLA), etc. [6–16], amount of fiber content, surface modification process and moisture absorption of fiber, interfacial adhesion between fiber and matrix, design and structure of the composite, presence of voids and incorporation of additives like nanofiller, compatibilizer, plasticizers, and binding agent [17, 18].

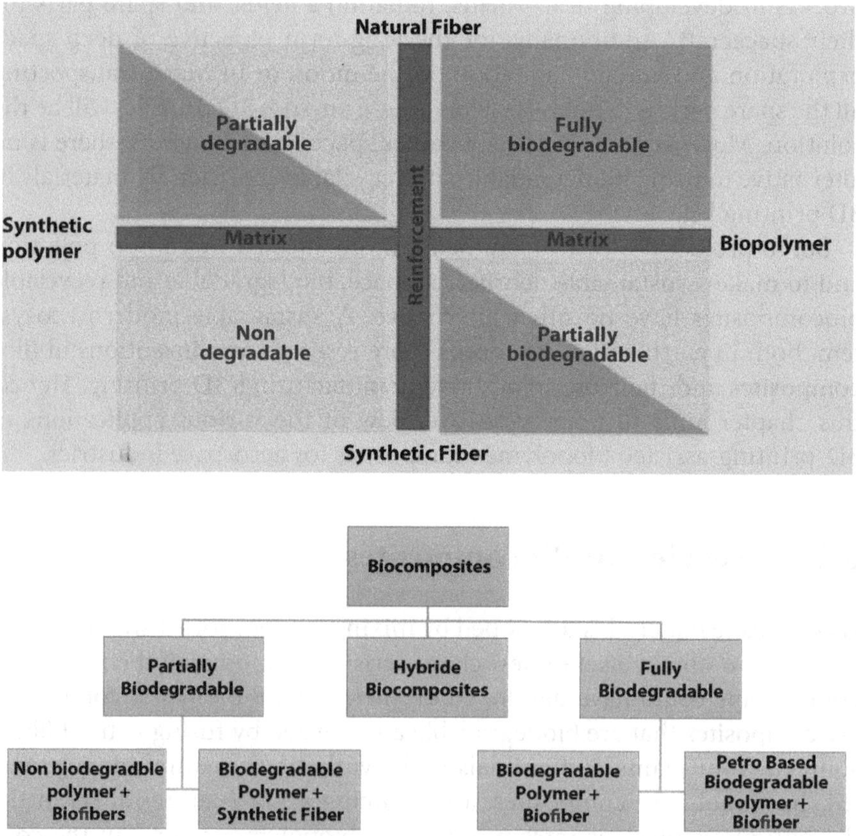

Figure 6.1 Classifications of biocomposites [1].

In biopolymer composite, biopolymer and natural fiber are utilized to produce a composite with superior mechanical properties as well as durability. Figure 6.2 includes different sources of natural fibers and Table 6.1 illustrates the basic properties of various natural fibers. Sustainability, appreciable specific strength, cost-effectiveness, lightweight properties, biocompatible, eco-friendliness of recyclable materials, and safety and health of consumers and manufacturers are some of the most important benefits of using biopolymer composites [21, 22]. Most biocomposites have a tensile strength of 20 to 200 MPa and a maximum stiffness of 1 to 4 GPa [23].

The aviation sector has strict requirements for aircraft components. Recently, the research of Gomez-Campos *et al.* [24] on the change in the properties of flax fiber composite over time showed that biocomposite materials are 14% heavier than regular aviation composite materials. So, the hybridization of natural and synthetic fibers is required to solve this

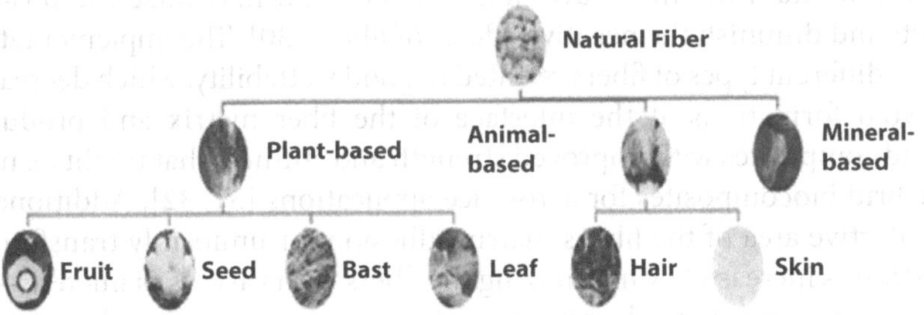

Figure 6.2 Different sources of natural fibers [1].

Table 6.1 Basic properties of different natural fibers [19, 20].

Fiber	Elongation at break (%)	Tensile strength (MPa)	Elastic modulus (GPa)	Microfibrillar angle (degrees)	Density (g/cm^3)
Bamboo	4.0–7.0	140–230	11–17	40–60	0.6–1.1
Hemp	1.5–4.0	690	30–70	50–90	1.5
Flax	2.0–4.0	350–1,040	28–70	6–10	1.5
Coir	30.0	180	4–6	30–49	1.2
Kenaf	1.6	930	53	7.1	1.4–1.5
Sisal	3.0–7.0	100–800	9–22	10–25	1.5
Jute	2.0–3.0	200–770	20–55	8	1.3–1.5

problem. In a hybrid composite, the combined effect of two different reinforcing fibers will be advantageous compared to using naturally sourced fiber alone in a composite.

6.2.1 Hybrid Biocomposites

Biocomposite with biodegradable matrix or reinforcing phase is called hybrid biocomposites [25–29]. Biocomposites containing two or more reinforcing elements in a single matrix are called hybrid biocomposites. The bolstering could take many different shapes physically. Examples of possible combinations include fibers with particle fillers, fibers with two different types of particle fillers, fibers with other fibers, layered fibrous mats, and more. A summary of hybrid biocomposites and other composites is shown in Figure 6.3.

Researchers have chosen the best composition of natural fibers for their studies to attain the most favorable outcome and maximize the positive effects and diminish the negative effects of fibers [30]. The implementation of two different types of fibers resulted in good wettability, which decreased the void formations at the interface of the fiber matrix and produced hybrid composites with improved strength and stiffness that might be used in hybrid biocomposites for aerospace applications [31, 32]. Additionally, the effective area of the fiber's matrix adhesion for uniformly transferring the stress is increased by hybridizing the fibers with various diameters—for example, the mechanical properties can be preserved even when a fiber with low elongation is used by transferring the stress to a composite fiber with an improved elongation [23].

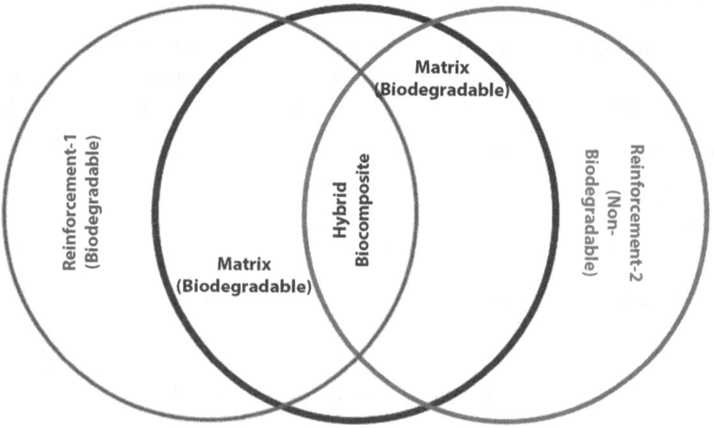

Figure 6.3 Representation of a hybrid biocomposite [1].

Bax and Müssig [33] found that the reinforcement of synthetic cellulose, jute fibers, and abaca significantly increased the PLA composite's rigidity and strength; however, the mechanical characteristics of the composites varied based on the type of fiber [34]. The nanofiber-reinforced composite film was created by solution-casting a mixture of 10% cellulose nanofiber (from different sources such as rutabaga, kraft pulp, flax, or hemp) and 90% PVA. Around five times higher enhancement in Young's modulus was observed compared to pure PVA film [35]. The characteristics of PLA composites made from hemp, kenaf, cotton, and lyocell were also seen. According to the study, PLA/cotton composite demonstrated good impact strength, while hemp/PLA and Kenaf composite displayed excellent Young's modulus and tensile strength [36].

However, until now, biopolymer composite's durability and biocompatibility are of great concern because no suitable solution has been developed to control these two factors fully.

6.3 Additive Manufacturing/3D Printing with Biocomposites

The advancement of 3D printers that can create composite components has created new opportunities for product development and prototyping in the aerospace industry. The manufactured parts can be adapted to take advantage of the anisotropy and strength of the material by making use of the high design freedom of the process. Contrary to previously used composite manufacturing methods like automated fiber placement, it is now possible to realize more complex geometries with higher tow path curvature and loading conditions. Additionally, better process automation and waste-free manufacturing are made possible with additive manufacturing of composites, which could result in significant cost savings [37].

Only pure plastic parts could be designed for additive manufacturing before the development of fused deposition modeling (FDM) machines with two extrusion nozzles (Figure 6.4(a)), whereas now, it is possible to support material (reinforcement) and extrude build material (matrix) simultaneously [38]. However, few studies have documented biocomposite structures' production for aerospace applications—for example, according to Figures 6.4(b and c), Fasel *et al.* investigated additively manufactured composite drones for morphing actuation [39].

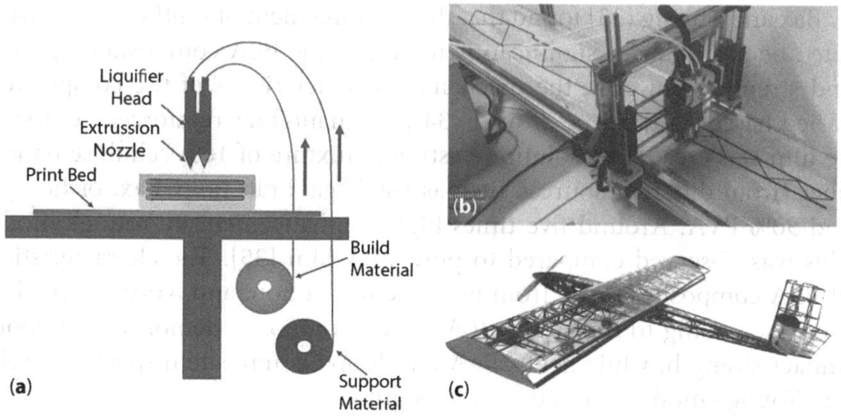

Figure 6.4 (a) Fused deposition modeling machine (FDM) with dual extrusion nozzles [38]. (b) Continuous carbon fiber (CF) 3D printer, printing the fuselage truss-structure. (c) Additively manufactured composite drone illustrating morphing actuation [39].

6.4 Process of Filament Fabrication with Biocomposites

A continuous thermoplastic filament is used in the 3D printing process [40]. A moving, heated printer extruder head feeds the filament from a large spool, which is then deposited on the growing work.

Making filament starts with the granulated raw resin with clear or white color. Then, additive is added to the resin for structural integrity, strength, impact resistance, and magnetic properties. Then, the drying process is incorporated at 60°C to 80°C. After the addition of additives, the extrusion process is introduced. Here, through mixing and heating, a solid monofilament is extruded. Following that, a hot water tank is introduced to the extruded material to avoid the oval shape of the filament. A perfectly round filament can be achieved if the hot water, nozzle, and resin temperature are set at the right temperature through a uniform gentle cooling process [41].

Biobased filaments using biobased resin and biocomposite fiber for the 3D printer can reduce the 3D printing cost. In addition, the printed component will be biodegradable, potentially reducing the carbon footprints significantly by replacing non-ecofriendly materials and structures. Thus, research on biocomposites has drawn noticeable concern for a better ecofriendly future—for example, in 2021, Chuanchom *et al.* gave a synoptic demonstration of the biocompatible filament fabrication (Figure 6.5) process with Kenaf fiber [42]. In addition, an American industrial 3D printing

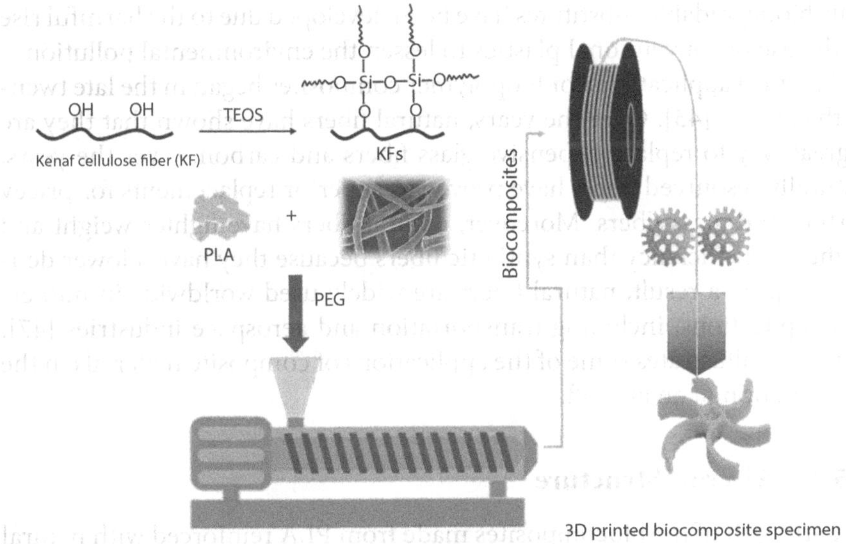

Figure 6.5 3D printer filament fabrication with biocomposite [42].

Figure 6.6 Biobased 3D printer filament using ULTM [43].

company, 3DGence, gave an example of their fabricated biocompatible filament with ULTEM (Figure 6.6) for 3D printing [43].

6.5 Use of Biocomposites for Aerospace Application

The recent years have seen a noticeable increase in concern over the noise produced by aerospace vehicles, environmental damage brought on by manufacturing materials, and post-consumer disposal [44]. As a result, fully biodegradable substitutes have been developed due to the harmful rise in the use of conventional plastics to lessen the environmental pollution.

Practical applications for biopolymer composites began in the late twentieth century [45]. Over the years, natural fibers have shown that they are a great way to replace expensive glass fibers and carbon. Over the years, naturally resourced fibers have proven as superior replacements for pricey carbon and glass fibers. Moreover, natural fibers have lighter weight and higher fuel efficiency than synthetic fibers because they have a lower density [46]. As a result, natural fibers are widely used worldwide in numerous applications, including transportation and aerospace industries [47]. Figure 6.7 illustrates some of the applications of composite materials in the construction of an aircraft.

6.5.1 Aircraft Structure

The strength of the biocomposites made from PLA reinforced with natural fibers like bamboo, jute, hemp, and kenaf is essential for performance in interior components like floor mats, luggage compartments, and package selves. Therefore, these composites have significant applications in these areas.

Natural fiber and dicumyl peroxide (DCP) addition improves the tensile and flexural strength of the composites and significantly lowers the

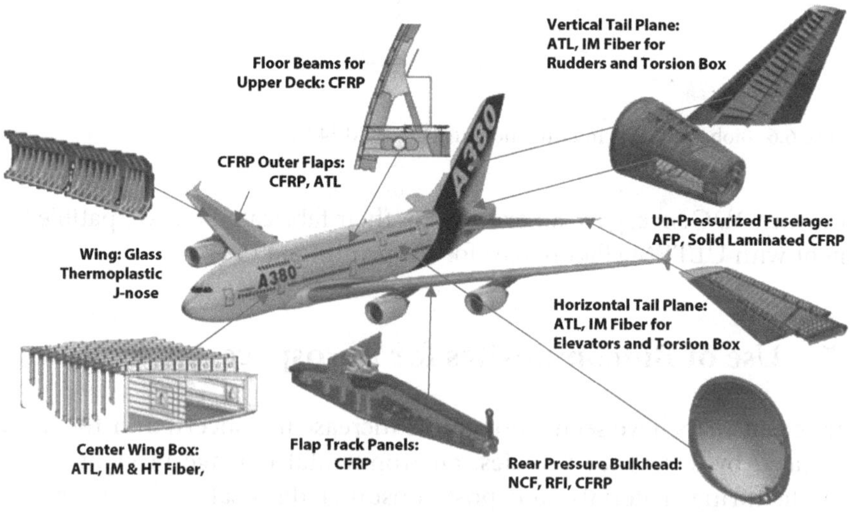

Figure 6.7 Use of composite materials in aircraft construction [1].

onset degradation temperature. The thermal stability of miscanthus fiber-reinforced PBAT composites showed similar results [48].

Improvements in thermal stability have been seen when acrylonitrile grafting and acetylation are used to modify natural fibers [23]—for example, in the aerospace industry, insulation materials are made from cotton and coconut fibers. In addition, components like headliners, package trays, dashboards, cabin lining, and padded seats are produced using natural fiber-reinforced biopolymer composites [49].

Mitsubishi Motors Corporation has developed PBS and PLA composites reinforced by bamboo fiber for ceiling materials and car floor mats, both of which can be used inside aircraft [46, 50]. The mechanical properties (tensile modulus, tensile strength, flexural modulus, and flexural strength) and the vibration properties of sisal, banana, and coconut sheath fiber-based polyester hybrid composites were studied (natural frequency and damping). It was discovered that the stack sequence impacted the vibration and mechanical characteristics.

Sisal-glass fiber polypropylene (PP) biocomposites' volumetric coefficient of thermal expansion decreased from +30°C to -30°C and their moisture resistance increased, demonstrating that hybrid biocomposites are less susceptible to dimensional change when exposed to varying atmospheric conditions [51]. The best hybrid composite was made from sisal, coconut sheath, and bananas that had been alkali-treated, and it could be used for interior aerospace application components [52].

Compared to neat polypropylene, non-woven kenaf and hybrid non-woven/woven kenaf fabric were found to be more lightweight and have better flexural and tensile strength, according to a research by Yahaya *et al.* [53]. The composite is thus appropriate for map pockets on doors. Atiqah *et al.* [54] manufactured thermoset kenaf fiber using the hand lay-up technique and used a pre-gel coat during the molding method to create a good surface finish.

Phenolic and epoxy resins are primarily used for aircraft interior panels and secondary structures [55]. However, due to their aliphatic-cyclic design, thermosetting epoxies made from gallic acid, itaconic, and rosin have been observed as promising bio-based resin alternatives and can outperform current solutions in terms of mechanical properties and chemical resistance [56, 58]. Furthermore, adding additives like graphene, graphene oxide [59], CNT [60], and silicon carbide nanoparticles (also known as nanotubes or nanowhiskers) may be very effective in thermal conductivity enhancement [61–63], along with electric conductivity [64] and improving mechanical properties [65, 66]. On the other hand, Amiri *et al.* [67]

discovered that resins reinforced with flax fiber and made from soybean oil had potential uses in the structural support of aircraft.

An acoustic panel for airplane propellers made of laminated hybrid glass/ramie reinforced was created by Liu *et al.* [68]. Strong specific stiffness and high strength characteristics can be seen in the ramie. Additionally, they noted that the fire safety performance of the created biocomposite was reasonable and could be utilized for interior applications of the aircraft fuselage. Incorporating sheets of ceramic into a polypropylene (PP) biocomposite reinforced by Kenaf fiber could enhance its mechanical and flammability properties, leading to the use of hybrid biocomposites in applications of the aerospace sector [69]. With benefits of being lightweight, cost-effective, self-extinguishing, and biodegradable, kenaf fiber-reinforced acrylonitrile butadiene styrene (ABS) with nano-clay hybrid biocomposite can self-extinguish and has the capability of efficient fire extinction. Therefore, it can be used in the secondary structure of the airplane. Furthermore, for confined spaces in aircraft, it was determined that a formulation of ABS/kenaf fiber (50/50) and 1% nano-clay was appropriate [70].

In 2012, plasma-treated PLA composites reinforced by coconut fiber were created by Jang *et al.* [71] using the commingled yarn method. The flammability performance was examined using the LOI (loss on ignition) method. Despite that, there was no discernible difference in the LOI values for the PLA/coconut fiber composites; they reported that it had more than 20 vol.%, which indicates a non-flammability trait. Analyzing the flame resistance of thermoplastic starch (TPS) composites with ammonium polyphosphate (APP) added revealed that APP improved the flame resistance of the biopolymer composites. The same kind of pattern was also seen where 20%–30% of the APP was replaced with keratin fibers [72]. By introducing hot hydrolysis of PBS and phosphorylation of natural fibers before fire exposure, the APP enhanced the fire retardancy property and created a barrier layer on top of the composite. Dorez *et al.* [73] prepared PBS-based composites for designing interior body parts by incorporating natural fibers and ammonium polyphosphate (APP) as a fire retardant.

As a substitute for alloys of aluminum and epoxy/E-glass composites, biodegradable bamboo fiber-reinforced composite boards and panels of PLA with high mechanical strength have been made [74]. Because of the strong interfacial adhesion between natural fiber and biopolymers, the PBAT/PLA composites were created by reinforcing surface-modified kenaf fiber with 2 wt% (3-aminopropyl) trimethoxysilane (APTMS) as a coupling agent with the combination of 90:10 weight percent ratio [75]. As a result, these composites are significantly improved in toughness, tensile, and flexural strength. Zin *et al.* [76] used an automated spray-up process to investigate the fabrication process of a hybrid biocomposite made from pineapple leaf fiber (PALF). They discovered that the hybrid biocomposite's low variation led to an increase in tensile strength. Compared to the single reinforcement of PALF composite, the synergistic impact of strong glass fiber characteristics and high-strength PALF enhances the tensile strength attributes up to 71.86 MPa. This hybrid PALF/glass biocomposite was suggested for aircraft interior applications as an alternative.

Karthigairajan *et al.* [77] have looked into developing rice husk-based high-performance epoxy biocomposite for coating air ducts. However, with just 2 vol.% of silane-treated rice husk added to epoxy, it was discovered that the thermal water uptake and conductivity were significantly reduced. To prevent the initiation of welding cracks, it was suggested that epoxy composite coating reinforced by silane-treated rice husk would be a promising coating solution. AIRBUS, a multinational European aerospace company, and the Council of Scientific and Industrial Research (CSIR) researchers collaborated on research to develop thermoset panels reinforced by natural fibers, especially for aeronautical applications. They created aqueous-based fire-retardant treatments for flax fibers to ensure that the composite panels meet Federal Aviation Administration (FAA) airworthiness regulations. On flax fiber-reinforced epoxy composites, the zirconia nanoparticles (ZrO_2) graft increases tensile strength without affecting the tensile modulus and inhibits the growth of fungal colonies [78].

Components made using 3D printing are about 55% lighter than those made using more conventional techniques. Additionally, they permit a 90% reduction in material consumption. One of the few 3D printing polymers authorized for use in aviation components is ULTEM resin. Its main differentiating traits are that it meets the UL94 standard's flammability class V0 requirements and has excellent FST (flame, smoke, and toxicity) properties. As a result, heat-curing plastics and ULTEM have been replacing metals more frequently. This material is largely used in the manufacture of electric and lighting systems, including headlights, engine or ignition components, frames, electric switches, and bulb sockets [79]. Figure 6.8 provides a picture of a 3D-printed gearbox cover developed by 3DGence, an American industrial 3D printing company, using ULTM filament.

6.5.2 Ultraviolet (UV) Protection of Aircraft Parts

In 1801, Johann Wilhelm Ritter, a German physicist, made the discovery of UV radiation. Ultraviolet radiation is a form of electromagnetic radiation emitted by the sun and is transmitted as waves or particles with various wavelengths and frequencies [80]. Numerous natural and synthetic

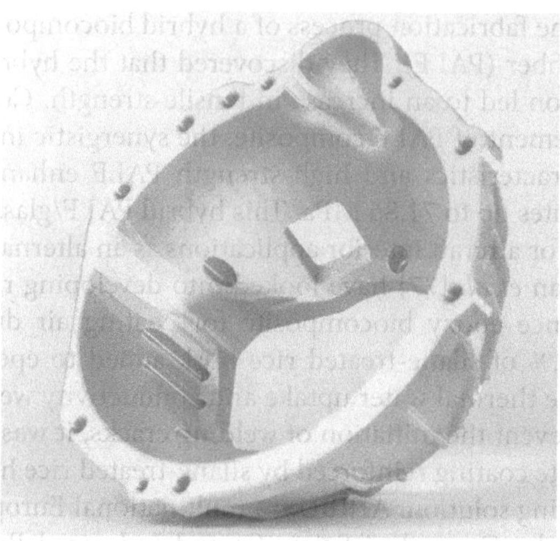

Figure 6.8 3D-printed gearbox cover using ULTM filament [79].

polymers, such as certain rubbers, neoprene, and polyvinyl chloride (PVC), lose strength, become less flexible, crack, disintegrate, and fade in color due to exposure to UV radiation. Textile materials can receive UV protection from a hybrid biopolymer made of multiwalled carbon nanotubes (MWCNT) and zinc oxide (ZnO) [81, 82]. The use of this biocomposite for UV protection of aircraft structures may be made possible by broad research, especially for the windscreen which will also protect the pilot's eyes from UV radiation.

6.5.3 Anechoic Chambers

An anechoic chamber is a protected area that has been built to withstand invasive electromagnetic radiation and inhibit the emissions of electromagnetic waves to the outside environment [83]. In the aerospace industry, anechoic chambers are frequently used for in-field testing of aircraft and rocket nozzle noise levels, as demonstrated in Figure 6.9. Anechoic chambers can be constructed using natural materials with good sound-absorbing qualities, such as wood and coconut coir, according to some earlier studies [84, 85].

Figure 6.9 An F-16 fighting falcon in eglin air force Base's anechoic chamber [83].

6.5.4 Radome Application

A radome (a combination of the words "radar" and "dome") is a structural, weatherproof casing that shields a radar antenna. The radome conceals the electronic equipment of the antenna and shields the weather radar housed on the nose of the aircraft [86]. It also provides an aerodynamic shape for the nose of the aircraft (Figure 6.10). In radome applications, plastic reinforced by glass fiber was employed because of its radio-frequency (RF) transparency, which allows the RF field to pass through without heat generation. The possibility of organically sourced fiber for radome applications has been discussed by Ilyas *et al.* [87]. They also looked at how the reinforcement of hybrid glass/kenaf fiber in the epoxy composite can lighten the material and improve the low dielectric constant of the radome.

6.5.5 Heat Shields for Suborbital Spacecraft

Heat shields are used in spacecraft to prevent the spacecraft from exploding and crashing at entry and re-entry into a planet's atmosphere since the forces during re-entry are far greater than those during launch and are strong enough to shatter an unprotected spacecraft [89]. Table 6.2 compares mixtures containing varying filler weight percent of multiwalled

Figure 6.10 Radome as a cover of the weather radar at the nose section of an aircraft [88].

Table 6.2 Thermal conductivity of composites made of polylactic acid with various filler contents (W/mK) [90].

Filler (wt%)	PLA + MWCNTs	PLA + GNPs PLA + GNPs + MWCNTs	PLA + GNPs + MWCNTs
3.0	0.231	0.323	0.270
6.0	0.232	0.448	0.352
9.0	0.268	0.550	-
12.0	0.365	0.664	0.533

carbon nanotubes (MWCNT) and graphene nanoplatelets with thermoplastic polylactic acid (PLA, [-CHCO-]$_n$) in its pure composition to show the linear increase in thermal conductivity following different filler contents (GNPs).

The outcome demonstrates that, in comparison to neat PLA, adding any fillers increased the thermal conductivity, but the magnitude of increment varied among the compositions of the fillers—for example, GNP-doping PLA achieved comparatively improved thermal conductivity values for a given filler proportion (wt%); MWCNT-doped PLA measured only about half as much thermal conductivity as GNP-doped PLA. The development of biocomposite thermal protection surfaces (TSP) for heat shields for suborbital spacecraft is now possible.

6.5.6 Aircraft Electronics

In the prospect of developing complicated multi-layered electronics by 3D printing, research conducted by Podsiadły *et al.* revealed that the tensile strength of substrate–component junctions was highest for acrylonitrile butadiene styrene (ABS), followed closely by polystyrene (PS, $[C8H8]_n$), and lowest for polylactic acid (PLA) [91]. In a research project, European Space Agency examined polyetheretherketone (PEEK) filaments doped with carbon nanotubes (CNT) or/and graphite nanoplates. The conductive PEEK had a conductivity of 10 S/m, a 100-fold improvement over pure PEEK [92]. Data transmission of 9,600 bits per second (bps or baud) between two computers is possible using electrically conductive pathways, USB cables, and USB-RS232 converters made of polyetheretherketone (PEEK) [93].

Researchers from the Air Force Research Laboratory (AFRL) developed a relatively straightforward microcontroller system-on-chip (SoC) prototype in 2017 using conductive traces made of environmentally friendly thermoplastic polyurethane that has been infused with silver [94, 95]. This stretchable and flexible SoC will be used in various Internet of Things (IoT) devices that users can wear, with potential uses in space exploration. Because the PC copolymer improves ULTEM 9085's flow property and results in a PEI/PC denotation, the material is adaptable to 3D printing [96]. The experimental Slosh avionics box was printed with this material.

6.5.7 Electromagnetic Interference Shielding

A straightforward, inventive method created PLA scaffolds based on TPMS and constrained conductive CNT networks for 3D-printed structured composites. The EMI shielding property and mechanical performance of 3D-CNT/PLA were improved by the novel segregated structure created by 3D printing. In-depth research was done on electrical, microstructural, and mechanical performances and EMI shielding characteristics. After compression molding, the highly conductive CNT-coated 3D-printed PLA scaffolds were transformed into linked networks. The composite EMI SE value rose to 67.0 dB when 5.0 wt% CNT was added.

In addition, the mechanical robustness of 3D-CNT/PLA was enhanced by the continuous PLA scaffold, exhibiting Young's modulus and bending strength of 4.43 and 87.8 GPa, respectively, which are 101% and 43%

higher, respectively, than those of 43.7 MPa and 3.08 GPa for CNT/PLA composites with randomly agglomerated CNTs. Furthermore, by adjusting the periodic unit size of the TPMS-based PLA scaffold by 3D printing, we were able to generate EMI SE that was fully controllable for CNT/PLA composites. Overall, CNT/PLA composites produced through 3D printing processes have good mechanical qualities and tunable EMI SE, making them applicable in a wide range of electrical devices and radiation source applications [97].

6.5.8 Space Habitation and Space Operation

In space, we do not have easy access to medical care. It is crucial to provide a spacesuit, space vehicle, or outpost with health monitoring capabilities for space habitation and space missions. For adapting to the environment of space habitation, environment monitoring nano-sensors, life support systems, wearable sensors with the Internet of Medical Things (IoMT), and health-monitoring-system-embedded space suits can be developed with eco-friendly ULTEM [4].

On the other hand, medical and other life-supporting components for the astronauts make long-duration mission planning more difficult because these should be maintained between the shortage of the supply and over-stocking, and if the essential supplies run out during a deep space operation, it will be very difficult to replenish them from Earth [98]. So, 3D printing of necessary medical and life-supporting essential supplies in space can be the solution here and the use of biocomposite materials in this additive manufacture will keep the mission cost down and benefit the long-term viability of human space operations [99, 100]. The European Space Agency (ESA) and its industrial partners are working on a PEEK 3D printer that can be used in space. A previous project of International Space Station (ISS) called "3D Printing in Zero G" showed that printing parts in microgravity did not result in any noticeable engineering flaws compared to printing them on Earth [98]. For manufacturing in zero gravity, the additive manufacturing facility originally employed ABS and green polyethylene (PE), a resin made from sugarcane [4].

Boeing's CST Starliner, a space shuttle for transferring people and goods to and from the International Space Station, uses polyetheretherketone (PEKK), which has high stiffness along with high thermal resistance, and it is also possible that SpaceX utilized it to make helmets to be used in space [101]. On the other hand, PEEK complaints with the AMF on the ISS and is being evaluated by businesses for manufacturing in space or off the surface of the earth in the future [4]. Membranes made of polyarylether

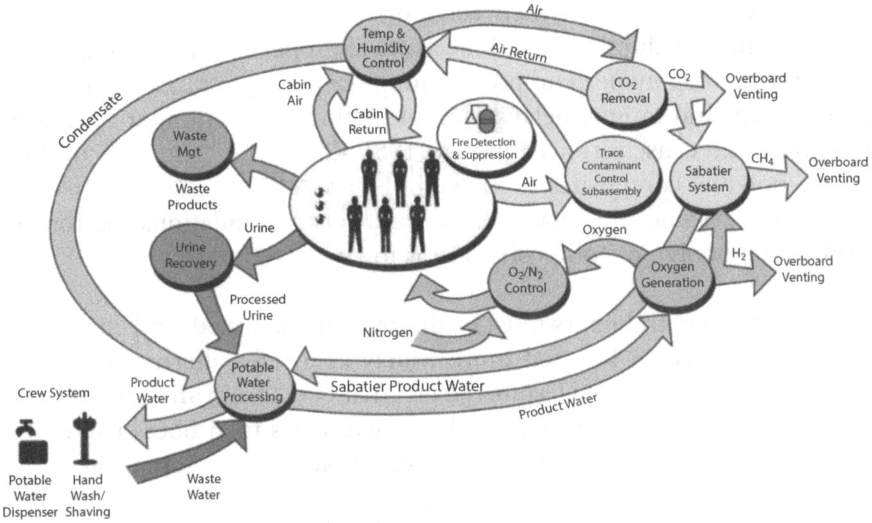

Figure 6.11 Environmental control and life support systems of the International Space Station [103].

sulphones (PAES) are produced for use in hemofiltration and dialysis [102]. PAES membranes were investigated for their potential use for various fuel cell configurations (alkaline, proton exchange, etc.) to be used in the Environmental Control and Life Support System (ECLSS) (Figure 6.11) which is a regenerative life-supporting hardware system to deliver clean air and water to the crews of ISS. Commercialization of PAES filaments and printers will accelerate their application in space [103].

6.6 Limitations and Prospects

Biopolymer composites have some apparent drawbacks in addition to their benefits. Compared to conventional plastics, biopolymer composites perform poorly in terms of dimensional stability, thermal decomposition temperature, moisture resistance, ultraviolet resistance, fire resistance, and biological resistance [23, 104]. The use of fibers from natural resources enhances the biopolymer composite's impact strength, and this property grows with the amount of fiber content. However, the composite's ductility, propensity to absorb water, and offensive odor all rise as the amount of fiber does [105]. The limited compatibility of natural fibers and biopolymers is another significant barrier to using biopolymer composites in interior, non-structural, and heavy load-bearing applications [45].

The degree of compatibility between biopolymer and natural fibers affects the performance, durability, and mechanical properties of the composites. In addition, biopolymer composites' long-term properties, such as durability, creep, and fatigue behaviors and lifetime prediction, are precisely unknown and cause serious concern [106]. Effect of moisture absorption on bio-polymer composites is shown in Figure 6.12.

Therefore, some important factors and recommendations for further studies of biopolymer composites are as follows [107]:

- New natural fibers must be discovered, identified, and characterized to be used more widely as a safer alternative to synthetic fibers. Despite the fact that fibers from animals are known to have mechanical characteristics than fibers from plants, there are still few studies on biopolymer composites reinforced with animal fibers.
- Understanding the durability of bio-based materials is one of their main drawbacks. The use of biopolymer composites in the aerospace industry is severely constrained by their unpredictable durability and low mechanical strength. New experimental methodologies should be investigated to assess, monitor, and regulate the durability of biopolymer composites—such as moisture absorption, fire/flame retardancy, thermal stability, UV resistance, and biodegradability.

To adopt more economical and environmentally friendly practices, it is necessary to explore the characteristics and modification of biocomposites

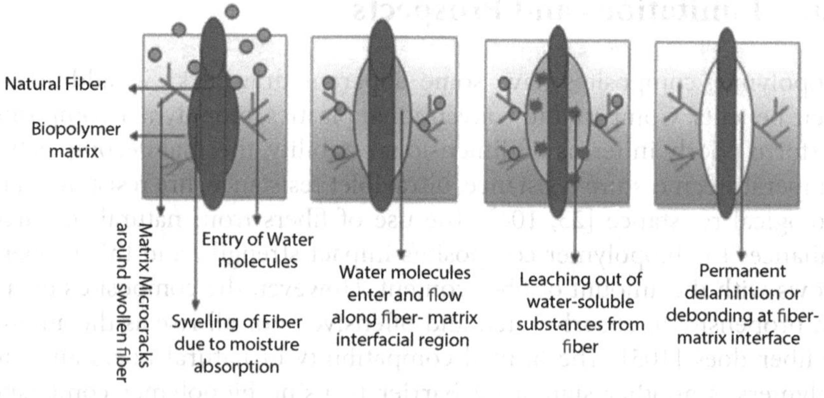

Figure 6.12 Effect of moisture absorption on biopolymer composites [107].

as recent innovations in biocomposites have created a huge prospect for the application of biocomposites in numerous areas.

- The application of nanotechnology has revealed a variety of possibilities for improving the properties of biopolymer composite products—for instance, research has shown that nanocrystalline cellulose is more durable than steel and stiffer than aluminum. So, using biopolymer composites structurally, the synthesis of nanofibers from natural fibers has a promising prospect.
- Thermal protection system (TPS) printing materials are much less demanding on the AM settings for temperature and pressure, functionally graded nanocomposites of phenolic resin and carbon nanofiber can be created, and the AM implementation of FGM would soon be possible [108, 109].
- An integrated circuit (IC) with hundreds of thousands of gates constructed using 28 nm or 14 nm process technologies is known as a field programmable gate array (FPGA) [4]. That is why research on 3D fabrication on a nanometer scale for producing programmable devices and aircraft sensors has a prominent prospect.

6.7 Conclusions

Industrial interest in biopolymer composites has gradually grown over the past few decades. When modified, natural fibers have demonstrated their superiority over synthetic fibers due to their wide availability, low cost, and eco-friendly features. Kenaf fiber has the most potent mechanical properties among the natural fibers, and jute fiber is the strongest and most compatible with biopolymers. Using natural fibers in composites and other applications benefits agriculture and the sericulture industry while lowering greenhouse gas emissions and enhancing soil fertility. Biopolymer composites can be tailored to meet different performance requirements because they are non-corrosive, renewable, lightweight, biodegradable, environmentally friendly, and simple to manufacture. The performance of biopolymer composites is influenced by a variety of elements, including the physical characteristics of the fiber and biopolymer, fiber modification methods, fiber volume fraction, chemical composition, the quantity of plasticizers and additives, the process of reinforcement, the orientation and geometry of the fiber in the matrix, the interaction between the biopolymer and fiber, structural flaws, and the environment in which the composites are used. Another widely used method for creating high-valued hybrid biopolymer composites is by combining various fiber types into a single matrix. Research showed that biopolymer composites have a wide range of uses and are suitable for the food, biomedical, construction, automotive, and aerospace industries. Future research on biopolymer composites may enable them to replace traditional petroleum-based plastics. New markets could open up when these biopolymer composites get better at performing and getting more durable. When a fiber from a natural resource is added as reinforcement to a biopolymer matrix, a few characteristics must be considered to understand the material's intended use, intended appearance, and environmental exposure. To determine the applicability of biopolymer composites, performance and life cycle assessment exploration must be ongoing.

References

1. R.A., Sapuan, S.M., Mazlan, N., *Advanced Composites in Aerospace Engineering Applications*, Springer Nature Switzerland AG, 2022.
2. Rashid, A. and Hoque, M.E., *Polymer Nanocomposites for Defense Application*, Woodhead Publishing in Materials, 2022.
3. Aranno, T.M. *et al.*, Fabrication, Experimental Investigation of Jute Fiber Reinforced Epoxy Composites and Hybrid Composites. *IOP Conference Series: Materials Science and Engineering*, 628, 1, Nov. 2019.
4. Paek, S.W., Balasubramanian, S., Stupples, D., Composites Additive Manufacturing for Space Applications: A Review. *Materials*, 6, 15, 13, MDPI, Jul. 01, 2022, doi: 10.3390/ma15134709.
5. Noor Azammi, A.M. *et al.*, Characterization studies of biopolymeric matrix and cellulose fibres based composites related to functionalized fibre-matrix interface, in: *Interfaces in Particle and Fibre Reinforced Composites: Current Perspectives on Polymer, Ceramic, Metal and Extracellular Matrices*, pp. 29–93, Elsevier, Woodhead Publishing, United Kingdom, 2019, doi: 10.1016/B978-0-08-102665-6.00003-0.
6. Punia Bangar, S. *et al.*, Development and characterization of physical modified pearl millet starch-based films. *Foods*, 2, 10, 7, Jul. 2021, doi: 10.3390/foods10071609.
7. Ilyas, R.A. *et al.*, Polylactic acid (Pla) biocomposite: Processing, additive manufacturing and advanced applications. *Polymers*, 2, 13, 8, MDPI AG, Apr. 02, 2021, doi: 10.3390/polym13081326.
8. Jumaidin, R., Diah, N.A., Ilyas, R.A., Alamjuri, R.H., Yusof, F.A.M., Processing and characterisation of banana leaf fibre reinforced thermoplastic

cassava starch composites. *Polym. (Basel)*, 13, 9, 1–2, 2021, doi: 10.3390/polym13091420.
9. Ilyas, R.A. *et al.*, Thermal, Biodegradability and Water Barrier Properties of Bio-Nanocomposites Based on Plasticised Sugar Palm Starch and Nanofibrillated Celluloses from Sugar Palm Fibres. *J. Biobased Mater. Bioenergy*, 14, 2, 234–248, Dec. 2019, doi: 10.1166/jbmb.2020.1951.
10. Syafiq, R. *et al.*, Antimicrobial activities of starch-based biopolymers and biocomposites incorporated with plant essential oils: A review. *Polymers*, 12, 10, MDPI AG, 1–26, Oct. 01, 2020, doi: 10.3390/polym12102403.
11. Ramesh, M., Palanikumar, K., Reddy, K.H., Comparative evaluation on properties of hybrid glass fiber-sisal/jute reinforced epoxy composites. *Procedia Eng.*, 51, 745–750, 2013, doi: 10.1016/j.proeng.2013.01.106.
12. Sathishkumar, T.P., Naveen, J., Navaneethakrishnan, P., Satheeshkumar, S., Rajini, N., Characterization of sisal/cotton fibre woven mat reinforced polymer hybrid composites. *J. Ind. Text.*, 47, 4, 429–452, Nov. 2017, doi: 10.1177/1528083716648764.
13. Ilyas, R.A. *et al.*, Sugar palm (Arenga pinnata [Wurmb.] Merr) starch films containing sugar palm nanofibrillated cellulose as reinforcement: Water barrier properties. *Polym. Compos.*, 41, 2, 459–467, Feb. 2020, doi: 10.1002/pc.25379.
14. Ilyas, R.A. *et al.*, Effect of sugar palm nanofibrillated cellulose concentrations on morphological, mechanical and physical properties of biodegradable films based on agro-waste sugar palm (Arenga pinnata (Wurmb.) Merr) starch. *J. Mater. Res. Technol.*, 8, 5, 4819–4830, Sep. 2019, doi: 10.1016/j.jmrt.2019.08.028.
15. Saxena, M. and Gupta, M.K., Mechanical, thermal, and water absorption properties of hybrid wood composites. *Proc. Inst. Mech. Eng., Part L: J. Mater.: Des. Appl.*, 233, 9, 1914–1922, Sep. 2019, doi: 10.1177/1464420718798661.
16. Jacob, M., Varughese, K.T., Thomas, S., Dielectric characteristics of sisal-oil palm hybrid biofibre reinforced natural rubber biocomposites. *J. Mater. Sci.*, 41, 17, 5538–5547, Sep. 2006, doi: 10.1007/s10853-006-0298-y.
17. Kabir, M.M., Wang, H., Lau, K.T., Cardona, F., Chemical treatments on plant-based natural fibre reinforced polymer composites: An overview. *Compos. B Eng.*, 43, 7, 2883–2892, Oct. 2012, doi: 10.1016/j.compositesb.2012.04.053.
18. Getme, A.S. and Patel, B., A review: Bio-fiber's as reinforcement in composites of polylactic acid (PLA). *Mater. Today: Proc.*, 26, 2116–2122, 2019, doi: 10.1016/j.matpr.2020.02.457.
19. da Silva, T.T. *et al.*, Thermal and chemical characterization of kenaf fiber (Hibiscus cannabinus) reinforced epoxy matrix composites. *Polym. (Basel)*, 4, 13, 12, Jun. 2021, doi: 10.3390/polym13122016.
20. Väisänen, T., Haapala, A., Lappalainen, R., Tomppo, L., Utilization of agricultural and forest industry waste and residues in natural fiber-polymer composites: A review. *Waste Manage. (Oxford)*, 54, Elsevier Ltd, 62–73, Aug. 01, 2016, doi: 10.1016/j.wasman.2016.04.037.

21. Mohamed, S.A.N., Zainudin, E.S., Sapuan, S.M., Azaman, M.D., Arifin, A.M.T., Introduction to Natural Fiber Reinforced Vinyl Ester and Vinyl Polymer Composites, in: *Natural Fiber Reinforced Vinyl Ester and Vinyl Polymer Composites: Development, Characterization and Applications*, pp. 1–25, Elsevier, Woodhead Publishing, United Kingdom, 2018, doi: 10.1016/B978-0-08-102160-6.00001-9.
22. Sharif A. and Hoque, M.E., Renewable Resource-Based Polymers, in: *Bio-based Polymers and Nanocomposites: Preparation, Processing, Properties & Performance*, M. Sanyang Muhammed Lamin and Jawaid (Eds.), pp. 1–28, Springer International Publishing, Cham, 2019, doi: 10.1007/978-3-030-05825-8_1.
23. Christian, S.J., Natural fibre-reinforced noncementitious composites (bio-composites), in: *Nonconventional and Vernacular Construction Materials*, pp. 111–126, Elsevier, Woodhead Publishing, United Kingdom, 2016, doi: 10.1016/b978-0-08-100038-0.00005-6.
24. Gomez-Campos, A. *et al.*, Natural Fibre Polymer Composites - A game changer for the aviation sector? *J. Cleaner Prod.*, 286, 15–18, Mar. 2021, doi: 10.1016/j.jclepro.2020.124986.
25. Aisyah, H.A. *et al.*, A comprehensive review on advanced sustainable woven natural fibre polymer composites. *Polymers*, 13, 3, MDPI AG, 1–45, Feb. 01, 2021, doi: 10.3390/polym13030471.
26. Alsubari, S., Zuhri, M.Y.M., Sapuan, S.M., Ishak, M.R., Ilyas, R.A., Asyraf, M.R.M., Potential of natural fiber reinforced polymer composites in sandwich structures: A review on its mechanical properties. *Polymers*, 13, 3, MDPI AG, 1–20, Feb. 01, 2021, doi: 10.3390/polym13030423.
27. Nurazzi, N.M. *et al.*, A review on natural fiber reinforced polymer composite for bullet proof and ballistic applications. *Polym. (Basel)*, 13, 4, 1–42, Feb. 2021, doi: 10.3390/polym13040646.
28. Venkatalaxmi, A., Padmavathi, B.S., Amaranath, T., A general solution of unsteady Stokes equations. *Dyn. Res.*, 35, 229–236, 2004, doi: 10.1016/j.uiddyn.2004.06.001.
29. Rajesh, M., Singh, S.P., Pitchaimani, J., Mechanical behavior of woven natural fiber fabric composites: Effect of weaving architecture, intra-ply hybridization and stacking sequence of fabrics. *J. Ind. Text.*, 47, 5, 938–959, Jan. 2018, doi: 10.1177/1528083716679157.
30. Asim, M., Saba, N., Jawaid, M., Nasir, M., Potential of natural fiber/biomass filler-reinforced polymer composites in aerospace applications, in: *Sustainable Composites for Aerospace Applications*, pp. 253–268, Elsevier, Woodhead Publishing, United Kingdom, 2018, doi: 10.1016/B978-0-08-102131-6.00012-8.
31. Islam, M.S. *et al.*, Effect of kenaf and EFB fiber hybridization on physical and thermo-mechanical properties of PLA biocomposites. *Fibers Polym.*, 18, 1, 116–121, Jan. 2017, doi: 10.1007/s12221-017-6208-x.

32. Ramamoorthy, S.K., Åkesson, D., Rajan, R., Periyasamy, A.P., Skrifvars, M., Mechanical performance of biofibers and their corresponding composites, in: *Mechanical and Physical Testing of Biocomposites, Fibre-Reinforced Composites and Hybrid Composites*, pp. 259–292, Elsevier, Woodhead Publishing, United Kingdom, 2018, doi: 10.1016/B978-0-08-102292-4.00014-X.
33. Bax, B. and Müssig, J., Impact and tensile properties of PLA/Cordenka and PLA/flax composites. *Compos. Sci. Technol.*, 68, 7–8, 1601–1607, Jun. 2008, doi: 10.1016/j.compscitech.2008.01.004.
34. Asha, A.B., Sharif, A., Hoque, M.E., Interface Interaction of Jute Fiber Reinforced PLA Biocomposites for Potential Applications, in: *Green Biocomposites. Green Energy and Technology*, M. Jawaid, M. Salit, O. Alothman (Eds.), Springer, Cham, 2017, https://doi.org/10.1007/978-3-319-49382-4_13.
35. Bhatnagar, A. and Sain, M., Processing of cellulose nanofiber-reinforced composites. *J. Reinf. Plast. Compos.*, 24, 12, 1259–1268, 2005, doi: 10.1177/0731684405049864.
36. Graupner, N., Herrmann, A.S., Müssig, J., Natural and man-made cellulose fibre-reinforced poly(lactic acid) (PLA) composites: An overview about mechanical characteristics and application areas. *Compos. Part A Appl. Sci. Manuf.*, 40, 6–7, 810–821, Jul. 2009, doi: 10.1016/j.compositesa.2009.04.003.
37. Eichenhofer, M., Wong, J.C.H., Ermanni, P., Continuous lattice fabrication of ultra-lightweight composite structures. *Addit. Manuf.*, 18, 48–57, Dec. 2017, doi: 10.1016/j.addma.2017.08.013.
38. Ning, F., Cong, W., Qiu, J., Wei, J., Wang, S., Additive manufacturing of carbon fiber reinforced thermoplastic composites using fused deposition modeling. *Compos. B Eng.*, 80, 369–378, Oct. 2015, doi: 10.1016/j.compositesb.2015.06.013.
39. Fasel, U., Keidel, D., Baumann, L., Cavolina, G., Eichenhofer, M., Ermanni, P., Composite additive manufacturing of morphing aerospace structures. *Manuf. Lett.*, 23, 85–88, Jan. 2020, doi: 10.1016/j.mfglet.2019.12.004.
40. Hamzah, H.H., Shafiee, S.A., Abdalla, A., Patel, B.A., 3D printable conductive materials for the fabrication of electrochemical sensors: A mini review. *Electrochem. Commun.*, 96, 27–31, 2018, doi: 10.1016/j.elecom.2018.09.006.
41. Kraft, C., How It Is Made: 3D Printing Filament, https://makezine.com/article/digital-fabrication/3d-printing-workshop/how-it-is-made-3d-printing-filament/ (accessed Nov. 08, 2022).
42. Aumnate, C., Soatthiyanon, N., Makmoon, T., Potiyaraj, P., Polylactic acid/kenaf cellulose biocomposite filaments for melt extrusion based-3D printing. *Cellulose*, 28, 13, 8509–8525, Sep. 2021, doi: 10.1007/s10570-021-04069-1.
43. 3DGence, ULTEM™ (PEI), [Online]. Available: https://3dgence.com/filaments/ultem/.
44. Scheff, S., Friedman-Berg, F., Shively, J., Carter, A., Human Factors Challenges in Urban Air Mobility. *Proceedings of the Human Factors and Ergonomics Society Annual Meeting*, vol. 64, no. 1, pp. 179–182, Dec. 2020, doi: 10.1177/1071181320641044.

45. Vinod, A., Sanjay, M.R., Suchart, S., Jyotishkumar, P., Renewable and sustainable biobased materials: An assessment on biofibers, biofilms, biopolymers and biocomposites. *J. Cleaner Prod.*, 258, 68, Elsevier Ltd, Jun. 10, 2020, doi: 10.1016/j.jclepro.2020.120978.
46. Bhagabati, P., Biopolymers and biocomposites-mediated sustainable high-performance materials for automobile applications, in: *Sustainable Nanocellulose and Nanohydrogels from Natural Sources*, pp. 197–216, Elsevier, Amsterdam, Netherlands, 2020, doi: 10.1016/B978-0-12-816789-2.00009-2.
47. Abidin, N.I.Z., Sabri, M.F.M., Kalantari, K., Afifi, A.M., Ahmad, R., Corrosion detection for natural/synthetic/textiles fiber polymer composites, in: *Structural Health Monitoring of Biocomposites, Fibre-Reinforced Composites and Hybrid Composites*, pp. 93–112, Elsevier, Woodhead Publishing, United Kingdom, 2018, doi: 10.1016/B978-0-08-102291-7.00006-X.
48. Muthuraj, R., Misra, M., Mohanty, A.K., Biodegradable biocomposites from poly(butylene adipate-co-terephthalate) and miscanthus: Preparation, compatibilization, and performance evaluation. *J. Appl. Polym. Sci.*, 134, 43, 2, Nov. 2017, doi: 10.1002/app.45448.
49. Mohamed, S.A.N., Ismai, H., Zainudin, E.S., Introduction to natural fiber reinforced vinyl ester and vinyl polymer composites, in: *Natural fibre reinforced vinyl ester and vinyl polymer composites*, Woodhead Publishing, United Kingdom, 2018, doi: https://doi.org/10.1016/C2016-0-03362-4.
50. Sarasini, F., Thermoplastic biopolymer matrices for biocomposites, in: *Biocomposites for High-Performance Applications: Current Barriers and Future Needs Towards Industrial Development*, pp. 81–123, Elsevier Inc., Woodhead Publishing, United Kingdom, 2017, doi: 10.1016/B978-0-08-100793-8.00004-1.
51. Kc, B., Tjong, J., Jaffer, S.A., Sain, M., Thermal and Dimensional Stability of Injection-Molded Sisal-Glass Fiber Hybrid PP Biocomposites. *J. Polym. Environ.*, 26, 3, 1279–1289, Mar. 2018, doi: 10.1007/s10924-017-1033-2.
52. Senthilkumar, K., Siva, I., Rajini, N., Jappes, J.T.W., Siengchin, S., Mechanical characteristics of tri-layer eco-friendly polymer composites for interior parts of aerospace application, in: *Sustainable Composites for Aerospace Applications*, pp. 35–53, Elsevier, Woodhead Publishing, United Kingdom, 2018, doi: 10.1016/B978-0-08-102131-6.00003-7.
53. Yahaya, R., Sapuan, S.M., Jawaid, M., Leman, Z., Zainudin, E.S., Review of kenaf reinforced hybrid biocomposites: Potential in defence applications. *Curr. Anal. Chem.*, 13, 110–112, Nov. 2017, doi: 10.2174/1573411013666171113150225.
54. Atiqah, A., Chandrasekar, M., Senthil Muthu Kumar, T., Senthilkumar, K., Ansari, M.N.M., Characterization and Interface of Natural and Synthetic Hybrid Composites, in: *Encyclopedia of Renewable and Sustainable Materials*, pp. 389–400, Elsevier, Elsevier, Amsterdam, Netherlands, 2020, doi: 10.1016/b978-0-12-803581-8.10805-7.

55. Pickering, S.J., Recycling technologies for thermoset composite materials-current status. *Compos. Part A Appl. Sci. Manuf.*, 37, 8, 1206–1215, Aug. 2006, doi: 10.1016/j.compositesa.2005.05.030.
56. Li, C., Liu, X., Zhu, J., Zhang, C., Guo, J., Synthesis, characterization of a rosin-based epoxy monomer and its comparison with a petroleum-based counterpart. *J. Macromol. Sci., Part A: Pure Appl. Chem.*, 50, 3, 321–329, Jan. 2013, doi: 10.1080/10601325.2013.755879.
57. Auvergne, R., Caillol, S., David, G., Boutevin, B., Pascault, J.P., Biobased thermosetting epoxy: Present and future. *Chem. Rev.*, 114, 2, 1082–1115, Jan. 22, 2014, doi: 10.1021/cr3001274.
58. Liu, X.Q., Huang, W., Jiang, Y.H., Zhu, J., Zhang, C.Z., Preparation of a bio-based epoxy with comparable properties to those of petroleum-based counterparts. *eXPRESS Polym. Lett.*, 6, 4, 293–298, 2012, doi: 10.3144/expresspolymlett.2012.32.
59. Tang, L.C. *et al.*, The effect of graphene dispersion on the mechanical properties of graphene/epoxy composites. *Carbon N Y*, 60, 16–27, Aug. 2013, doi: 10.1016/j.carbon.2013.03.050.
60. Sánchez, M., Campo, M., Jiménez-Suárez, A., Ureña, A., Effect of the carbon nanotube functionalization on flexural properties of multiscale carbon fiber/epoxy composites manufactured by VARIM. *Compos. B Eng.*, 45, 1, 1613–1619, Feb. 2013, doi: 10.1016/j.compositesb.2012.09.063.
61. Kavitha, N., Balasubramanian, M., Kennedy, A.X., Investigation of impact behavior of epoxy reinforced with nanometer- and micrometer-sized silicon carbide particles. *J. Compos. Mater.*, 47, 15, 1877–1884, Jul. 2013, doi: 10.1177/0021998312451920.
62. Naeimirad, M., Zadhoush, A., Neisiany, R.E., Fabrication and characterization of silicon carbide/epoxy nanocomposite using silicon carbide nanowhisker and nanoparticle reinforcements. *J. Compos. Mater.*, 50, 4, 435–446, Feb. 2016, doi: 10.1177/0021998315576378.
63. Yang, S.Y. *et al.*, Effect of functionalized carbon nanotubes on the thermal conductivity of epoxy composites. *Carbon N Y*, 48, 3, 592–603, Mar. 2010, doi: 10.1016/j.carbon.2009.08.047.
64. Choi, E.S. *et al.*, Enhancement of thermal and electrical properties of carbon nanotube polymer composites by magnetic field processing. *J. Appl. Phys.*, 94, 9, 6034–6039, Nov. 2003, doi: 10.1063/1.1616638.
65. Li, W., Dichiara, A., Bai, J., Carbon nanotube-graphene nanoplatelet hybrids as high-performance multifunctional reinforcements in epoxy composites. *Compos. Sci. Technol.*, 74, 221–227, Jan. 2013, doi: 10.1016/j.compscitech.2012.11.015.
66. Wang, X. *et al.*, Ultrastrong, Stiff and Multifunctional Carbon Nanotube Composites. *Mater. Res. Lett.*, 1, 1, 19–25, 2013, doi: 10.1080/21663831.2012.686586.
67. Amiri, A., Burkart, V., Yu, A., Webster, D., Ulven, C., The potential of natural composite materials in structural design, in: *Sustainable Composites for*

Aerospace Applications, pp. 269–291, Elsevier, Woodhead Publishing, United Kingdom, 2018, doi: 10.1016/B978-0-08-102131-6.00013-X.

68. Liu, X., Yi, X., Zhu, J., Bio-based epoxies and composites as environmentally friendly alternative materials, in: *Thermosets: Structure, Properties, and Applications: Second Edition*, pp. 621–637, Elsevier, Elsevier, Amsterdam, Netherlands, 2018, doi: 10.1016/B978-0-08-101021-1.00019-8.
69. Sim, I.N., Han, S.O., Jang, Y.H., Yoo, Y.J., Ceramic sheet hybrid kenaf reinforced polypropylene biocomposites. *J. Appl. Polym. Sci.*, 130, 3, 1917–1922, Nov. 2013, doi: 10.1002/app.39367.
70. Karunakaran, S., Majid, D.L., Tawil, M.L.M., Flammability of self-extinguishing kenaf/ABS nanoclays composite for aircraft secondary structure. *IOP Conf. Ser.: Mater. Sci. Eng.*, 152, 1, 10, Oct. 2016, doi: 10.1088/1757-899X/152/1/012068.
71. Jang, J.Y., Jeong, T.K., Oh, H.J., Youn, J.R., Song, Y.S., Thermal stability and flammability of coconut fiber reinforced poly(lactic acid) composites. *Compos. B Eng.*, 43, 5, 2434–2438, Jul. 2012, doi: 10.1016/j.compositesb.2011.11.003.
72. Rabe, S., Sanchez-Olivares, G., Pérez-Chávez, R., Schartel, B., Natural keratin and coconut fibres from industrial wastes in flame retarded thermoplastic starch biocomposites. *Materials*, 12, 3, 4, Jan. 2019, doi: 10.3390/ma12030344.
73. Dorez, G., Taguet, A., Ferry, L., Lopez-Cuesta, J.M., Thermal and fire behavior of natural fibers/PBS biocomposites. *Polym. Degrad. Stab.*, 98, 1, 87–95, Jan. 2013, doi: 10.1016/j.polymdegradstab.2012.10.026.
74. Morales, A.P., Güemes, A., Fernandez-Lopez, A., Valero, V.C., de La Rosa Llano, S., Bamboo-polylactic acid (PLA) composite material for structural applications. *Materials*, 10, 11, 19, Nov. 2017, doi: 10.3390/ma10111286.
75. Sis, A.L.M., Ibrahim, N.A., Yunus, W.M.Z.W., Effect of (3-aminopropyl) trimethoxysilane on mechanical properties of PLA/PBAT blend reinforced kenaf fiber. *Iran. Polym. J. (Engl. Ed.)*, 22, 2, 101–108, Feb. 2013, doi: 10.1007/s13726-012-0108-0.
76. Zin, M.H., Abdan, K., Mazlan, N., Zainudin, E.S., Liew, K.E., Norizan, M.N., Automated spray up process for Pineapple Leaf Fibre hybrid biocomposites. *Compos. B Eng.*, 177, 4, 5, Nov. 2019, doi: 10.1016/j.compositesb.2019.107306.
77. Karthigairajan, M. *et al.*, Effect of Silane-Treated Rice Husk Derived Biosilica on Visco-Elastic, Thermal Conductivity and Hydrophobicity Behavior of Epoxy Biocomposite Coating for Air-Duct Application. *Silicon*, 13, 12, 4421–4430, Dec. 2021, doi: 10.1007/s12633-020-00772-z.
78. Bachmann, J. *et al.*, Outlook on ecologically improved composites for aviation interior and secondary structures. *CEAS Aeronaut. J.*, 9, 3, 533–543, Sep. 2018, doi: 10.1007/s13272-018-0298-z.
79. 3D printed custom made gearbox cover, https://3dgence.com/case-studies/prototyping-3d-printed-with-ultem-gearbox-cover/.

80. Knezevic, J., Impact of High-Altitude Ultraviolet Radiation on Functionability of Flight Crews. *Arch. Biomed. Eng. Biotechnol.*, 2, 2, 1, Jul. 2019, doi: 10.33552/abeb.2019.02.000533.
81. Rajeswari, A., Stobel Christy, E.J., Pius, A., Biopolymer blends and composites, in: *Biopolymers and their Industrial Applications*, pp. 105–147, Elsevier, Elsevier, Amsterdam, Netherlands, 2021, doi: 10.1016/b978-0-12-819240-5.00005-5.
82. Nasouri, K., Fabrication of lightweight and flexible cellulose acetate composite nanofibers for high-performance ultra violet protective materials. *Polym. Compos.*, 40, 8, 3325–3332, Aug. 2019, doi: 10.1002/pc.25191.
83. Anechoic chamber, https://en.wikipedia.org/wiki/Anechoic_chamber (accessed Nov. 08, 2022).
84. Birat, K.C., Panthapulakkal, S., Kronka, A., Agnelli, J.A.M., Tjong, J., Sain, M., Hybrid biocomposites with enhanced thermal and mechanical properties for structural applications. *J. Appl. Polym. Sci.*, 132, 34, 2, Sep. 2015, doi: 10.1002/app.42452.
85. Ward, K.A. *et al.*, Safety systems, culture, and willingness to fly in autonomous air taxis: A multi-study and mediation analysis. *J. Air Transp. Manage.*, 91, 1–9, Mar. 2021, doi: 10.1016/j.jairtraman.2020.101975.
86. Radome, https://en.wikipedia.org/wiki/Radome#:~:text=A%20radome%20(a%20portmanteau%20of,antenna%20electronic%20equipment%20from%20view. (accessed Nov. 27, 2022).
87. Sapuan, S.M., Nurazzi Norizan, M., Mahamud, A., POTENTIAL OF NATURAL FIBRE COMPOSITES FOR TRANSPORT INDUSTRY: A REVIEW Characterization of Corn/Sugar Palm Fiber Reinforced Cornstarch Hybrid Composites View project Synthesis, Characterization and DC Electrical Properties of Polyamide and Polyester/Multi-Walled Carbon Nanotube Composites View project, 2019, [Online]. Available: https://www.researchgate.net/publication/332168594.
88. Freepik, https://www.freepik.com/premium-photo/white-passenger-airplane-maintenance-hangar-checking-mechanical-systems-flight-operations-aircraft-has-opened-weather-radar_20506435.htm (accessed Nov. 08, 2022).
89. Wu, R., Roberts, P.C.E., Soutis, C., Diver, C., Flexible heat shields deployed by centrifugal force. *Acta Astronaut.*, 152, 78–87, 2018, doi: https://doi.org/10.1016/j.actaastro.2018.06.021.
90. Spinelli, G. *et al.*, Nanocarbon/poly(lactic) acid for 3D printing: Effect of fillers content on electromagnetic and thermal properties. *Materials*, 12, 15, 12, Aug. 2019, doi: 10.3390/ma12152369.
91. Podsiadły, B., Skalski, A., Słoma, M., Soldering of Electronics Components on 3D-Printed Conductive Substrates. *Materials*, 14, 14, 10–11, 2021, doi: 10.3390/ma14143850.
92. Gonçalves, J. *et al.*, Electrically Conductive Polyetheretherketone Nanocomposite Filaments: From Production to Fused Deposition Modeling. *Polym. (Basel)*, 10, 8, 11, 2018, doi: 10.3390/polym10080925.

93. Saunders, S., ESA and Zortrax 3D Print PEEK Composites for Built-in Electronics, 2020, https://3dprint.com/274117/supported-by-esa-zortrax-3d-prints-composite-parts-using-two-peek-polymer-blends/ (accessed Nov. 09, 2022).
94. Alia-Novobilski, M., AFRL, American Semiconductor create flexible system-on-chip for 'internet-of-things', 2017, https://www.wpafb.af.mil/News/Article-Display/Article/1402631/afrl-american-semiconductor-create-flexible-system-on-chip-for-internet-of-thin/.
95. Alia-Novobilski, M., AFRL, Harvard researchers invent new hybrid 3-D printing method for flexible electronics, 2017, https://www.af.mil/News/Article-Display/Article/1345387/afrl-harvard-researchers-invent-new-hybrid-3-d-printing-method-for-flexible-ele/.
96. Zaldivar, R.J., Witkin, D.B., McLouth, T., Patel, D.N., Schmitt, K., Nokes, J.P., Influence of processing and orientation print effects on the mechanical and thermal behavior of 3D-Printed ULTEM ® 9085 Material. *Addit. Manuf.*, 13, 71–80, Jan. 2017, doi: 10.1016/j.addma.2016.11.007.
97. Wang, Y. *et al.*, 3D-printing of segregated carbon nanotube/polylactic acid composite with enhanced electromagnetic interference shielding and mechanical performance. *Mater. Des.*, 197, 2, Jan. 2021, doi: 10.1016/j.matdes.2020.109222.
98. Rajput, S., A review of space surgery - What have we achieved, current challenges, and future prospects. *Acta Astronaut.*, 188, 18–24, 2021, doi: https://doi.org/10.1016/j.actaastro.2021.07.012.
99. Chowdhury, S.R., Lokanathan, Y., Xian, L.J., Mh Busra, F., Yazid, M.D., Sulaiman, N., Lahiry, G., Hoque, M.E., 3D Printed Bioscaffolds for Developing Tissue-Engineered Constructs, in: *Design Engineering and Manufacturing*, E. Yasa (Ed.), IntechOpen, UK, Published on 28 May 2020, DOI: 10.5772/intechopen.92418.
100. Hoque, M.E., Rayhan, A.M., Shaily, S.I., Natural fiber-based green composites: Processing, properties and biomedical applications. *Appl. Sci. Eng. Prog.*, 14, 4, 689–718, 2021, doi: 10.14416/j.asep.2021.09.005.
101. Scott, A., Boeing's space taxis to use more than 600 3D-printed parts, https://www.reuters.com/article/us-boeing-space-exclusive-idUSKBN15I1HW (accessed Nov. 09, 2022).
102. Yumoto, M. *et al.*, *In Vitro* Evaluation of High Mobility Group Box 1 Protein Removal with Various Membranes for Continuous Hemofiltration. *Ther. Apher. Dial.*, 15, 4, 385–393, Aug. 2011, doi: 10.1111/j.1744-9987.2011.00971.x.
103. NASA Facts, 2004.
104. Faruk, O. and Ain, M.S., Biofiber reinforced polymer composites for structural applications, in: *Developments in Fiber-Reinforced Polymer (FRP) Composites for Civil Engineering*, pp. 18–53, Elsevier Inc., Woodhead Publishing, United Kingdom, 2013, doi: 10.1533/9780857098955.1.18.

105. Faruk, O., Bledzki, A.K., Fink, H.P., Sain, M., Biocomposites reinforced with natural fibers: 2000-2010. *Prog. Polym. Sci.*, 37, 11, 1552–1596, Nov. 2012, doi: 10.1016/j.progpolymsci.2012.04.003.
106. Dittenber, D.B. and Gangarao, H.V.S., Critical review of recent publications on use of natural composites in infrastructure. *Compos., Part A*, 43, 8, 1419–1429, Aug. 2012, doi: 10.1016/j.compositesa.2011.11.019.
107. Aaliya, B., Sunooj, K.V., Lackner, M., Biopolymer composites: a review. *Int. J. Biobased Plast.*, 3, 1, 40–84, Jan. 2021, doi: 10.1080/24759651.2021.1881214.
108. Bafekrpour, E., Yang, C., Natali, M., Fox, B., Functionally graded carbon nanofiber/phenolic nanocomposites and their mechanical properties. *Compos. Part A Appl. Sci. Manuf.*, 54, 124–134, 2013, doi: 10.1016/j.compositesa.2013.07.009.
109. Dhakate, S.R., Mathur, R.B., Dhami, T.L., Development of vapor grown carbon fibers (VGCF) reinforced carbon/carbon composites. *J. Mater. Sci.*, 41, 13, 4123–4131, Jul. 2006, doi: 10.1007/s10853-006-6320-6.

7

Innovations in 3D-Printing-Assisted Biopolymers for Electronic Applications

Md. Sohel Rana[1], M.M. Maruf Hasan[1], K. M. Anis-Ul-Haque[2] and Md. Wasikur Rahman[1]*

[1]Department of Chemical Engineering, Jashore University of Science and Technology, Jashore, Bangladesh
[2]Department of Chemistry, Jashore University of Science and Technology, Jashore, Bangladesh

Abstract
The integration of 3D printing technology with biopolymers has opened up exciting new possibilities in the field of electronic applications. Biopolymers, which are derived from natural sources, offer advantages over traditional synthetic polymers in terms of biocompatibility, biodegradability, and sustainability. In recent years, 3D printing has also found its way into the field of electronic applications, where it is being used to create novel devices such as sensors, batteries, and electronic circuits. The use of biopolymers in 3D printing for electronic applications is a growing area of research, as biopolymers offer several advantages over traditional materials such as plastics and metals. In this book chapter, we will discuss the innovations in 3D-printing-assisted biopolymers for electronic applications. It covers various aspects of the field, including the development of new biopolymer-based materials, printing techniques, and applications in the fields of biomedicine and wearable electronics. The potential benefits of using biopolymers in electronic devices are discussed, including improved biocompatibility, reduced environmental impact, and lower cost.

Keywords: 3D printing, renewable materials, biopolymers, electronic applications

**Corresponding author*: w.rahman@just.edu.bd; mwrahman.ump@gmail.com

Bhasha Sharma, Balaram Pani, Shashank Shekhar and Jude A. Okolie (eds.) Sustainable 3D Printing for Innovative Biopolymer Production and Applications, (149–176) © 2025 Scrivener Publishing LLC

7.1 Introduction

Sustainable 3D printing using biopolymers is an innovative approach that has the potential to revolutionize the production of materials and products in a sustainable manner. Biopolymers are derived from renewable resources, such as plant-based materials, and are biodegradable, which makes them an attractive alternative to traditional petroleum-based plastics.

3D printing, also known as additive manufacturing, is a technology that allows the creation of complex 3D structures by adding layer upon layer of material. This technology is being increasingly used in a wide range of industries, including aerospace, automotive, and medical devices. When combined, sustainable 3D printing using biopolymers can provide a range of benefits, such as reducing the carbon footprint of production, improving the recyclability of materials, and creating novel biodegradable products. Here are some ways in which sustainable 3D printing using biopolymers can be used for innovative biopolymer production and applications: Biopolymer filaments can be used as the raw material for 3D printing, replacing traditional petroleum-based plastic filaments. These filaments can be made from a range of biopolymers, including polylactic acid (PLA), cellulose, and chitosan [1]. Using biopolymer filaments for 3D printing can significantly reduce the carbon footprint of production.

3D printing using biopolymers can be used to create a range of biodegradable products, such as packaging materials, disposable cutlery, and medical implants [2]. These products can be designed to break down in a matter of weeks or months, rather than years, reducing the impact on the environment. Biopolymers can be recycled through a range of methods, such as composting or enzymatic degradation [3]. 3D-printed products made from biopolymers can be designed to be easily disassembled and recycled, reducing the amount of waste sent to landfill.

3D printing is a manufacturing technology that allows the creation of complex geometries with great precision and accuracy. It is a widely used technique in various fields, including engineering, architecture, and medicine. Sustainable 3D printing using biopolymers is an innovative approach that has the potential to provide a range of benefits, including reducing the carbon footprint of production, improving the recyclability of materials, and creating novel biodegradable products. In recent years, 3D printing has also found its way into the field of electronic applications, where it is being used to create novel devices such as sensors, batteries, and electronic circuits. The use of biopolymers in 3D printing for electronic applications is a growing area of research, as biopolymers offer several advantages over traditional materials such as plastics and metals. However, there are still challenges to be addressed in order to fully realize the potential of 3D-printing-assisted biopolymers for electronic applications. These include optimizing printing parameters to improve mechanical properties, developing new biopolymers with enhanced properties, and addressing issues related to scalability and reproducibility. With further research and development, these materials could lead to the creation of new devices that are more biocompatible, sustainable, and affordable.

7.2 Biopolymers for Electronic Applications

Biopolymers are natural polymers that are derived from biological sources such as plants, animals, and microorganisms. They are increasingly being used as a sustainable alternative to traditional polymers as they are biodegradable, renewable, and non-toxic. Biopolymers have unique mechanical, thermal, and electrical properties that make them suitable for use in electronic applications. Some of the commonly used biopolymers for electronic applications are cellulose, chitosan, gelatin, and silk.

7.2.1 Innovations in 3D Printing of Biopolymers for Electronic Applications

Innovations in 3D-printing-assisted biopolymers for electronic applications have led to the development of new and improved electronic devices with enhanced properties. The use of 3D printing for the fabrication of electronic devices using biopolymers has several advantages over traditional manufacturing techniques. 3D printing allows for the creation of complex geometries and the incorporation of multiple materials in a single device. The following subsections present some of the recent innovations in 3D-printing-assisted biopolymers for electronic applications.

7.2.1.1 3D-Printed Biopolymer Sensors

Researchers have developed 3D-printed biopolymer sensors that can detect various gases, humidity, and temperature changes. These sensors are made using cellulose and chitosan, and they have shown high sensitivity and stability.

There have been several research studies on the development of 3D-printed biopolymer sensors that can detect various gases, humidity, and temperature changes. The use of cellulose and chitosan in the fabrication of these sensors has been found to be particularly promising due to their biocompatibility, low cost, and availability—for example, researchers developed 3D-printed flexible sensors based on cellulose and chitosan for the detection of humidity, temperature, and ammonia gas [4]. The sensors showed high sensitivity and stability, with the ability to detect changes in relative humidity between 11% and 95%, temperature changes between 25°C and 70°C, and ammonia gas concentrations as low as 2.5 ppm.

In another study, researchers developed a 3D-printed cellulose–chitosan composite sensor for the detection of hydrogen gas [5]. The sensor showed high sensitivity and selectivity for hydrogen gas, with a response time of less than 5 s.

The procedure for 3D-printed biopolymer sensors that can detect various gases, humidity, and temperature changes, as described in the article by [6], involves the following steps: acetic acid is mixed with cellulose and chitosan in a solvent to make a biopolymer solution. The sensing characteristics of a biopolymer solution can be improved by using various sensing components, such as graphene oxide, carbon black, and silver nanoparticles. Once the desired sensor shape has been designed, the biopolymer solution containing the sensing components is fed into a 3D printer. Following printing, a glutaraldehyde solution is used to crosslink the sensor, boosting its mechanical strength and durability.

The printed sensor is next subjected to an electrical resistance test, during which its sensitivity to a range of gases, humidity levels, and temperature adjustments is evaluated. The 3D-printed biopolymer sensors made using cellulose and chitosan have shown high sensitivity and stability toward various gases, humidity, and temperature changes, making them suitable for various applications in the fields of healthcare, environmental monitoring, and food quality control.

7.2.1.2 3D-Printed Biopolymer Batteries

3D-printed biopolymer batteries have been developed using materials such as cellulose and gelatin. These batteries have shown good electrical performance and biodegradability.

Researchers have been exploring the use of biopolymers in the development of 3D-printed batteries 10% alginate and 8% gelatin [1]. Cellulose is a natural polymer that is found in plants and is the most abundant organic compound on Earth. Cellulose has excellent mechanical properties and is a good electrical insulator, making it an ideal material for battery applications. Researchers have used cellulose to create 3D-printed batteries that are flexible and lightweight, making them suitable for use in wearable electronics and medical implants [7]. Another biopolymer that has been used in 3D-printed batteries is gelatin, a protein-based material that is derived from collagen [8]. Gelatin has several advantages as a battery material, including its biocompatibility and ability to form complex structures. Researchers have been able to 3D-print gelatin-based batteries that have high power density and can be used in biomedical application [9]. Figure 7.1 shows the 3D printing sodium alginate-based biopolymer for batteries application [10].

The first step is to select the biopolymer materials that can be used for making the battery. Some of the commonly used biopolymer materials for battery fabrication include cellulose, gelatin, chitosan, and alginate.

Once the biopolymer material is selected, it needs to be converted into a printable ink. The ink can be prepared by dissolving the biopolymer in a solvent, such as water or ethanol. Other components, such as conductive nanoparticles or electrolytes, can be added to the ink to enhance its

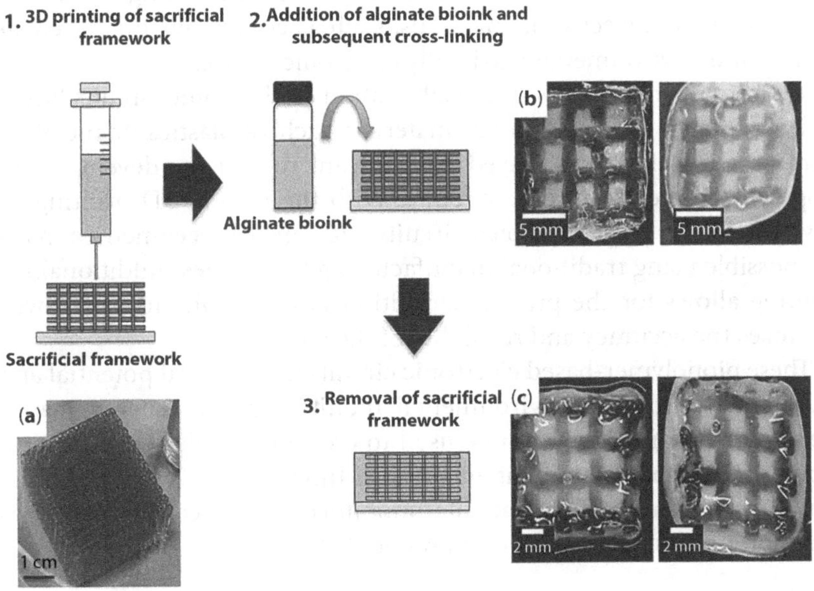

Figure 7.1 3D printing sodium alginate-based biopolymer for batteries application. (a) Gelatin 3D-printed sacrificial framework before (b) and after removal of the sacrificial framework (c). Sodium alginate 1.5% wt/vol (right) and 3.0% wt/vol (left) [10].

performance. The battery ink is then loaded into a 3D printer and deposited layer by layer to form the battery structure. The printer can be programmed to create different shapes and sizes of batteries, depending on the application. After printing, the battery components, such as the anode, cathode, and electrolyte, are assembled to form a complete battery [11]. This process involves the deposition of different ink formulations in a specific pattern to create the necessary components. The final step is to test the battery for its performance and optimize the ink formulation and printing parameters to improve the battery's efficiency and durability.

Although, the development of 3D-printed biopolymer batteries is an exciting area of research that has the potential to revolutionize the field of energy storage. By using sustainable and biodegradable materials, these batteries could have a much lower environmental impact than traditional batteries. Additionally, the ability to 3D-print complex structures could lead to new battery designs and applications.

7.2.1.3 3D-Printed Biopolymer Electronic Circuits

Biopolymer-based electronic circuits are a novel technology that combines the properties of biopolymers with electronic components to create 3D-printed circuits. These circuits are constructed by using a mixture of biopolymers and electronic materials, such as conductive ink, to create a flexible and environmentally friendly electronic device.

The use of biopolymers in the fabrication of electronic circuits has several advantages over traditional materials such as plastics or metals. 3D printing technology has played a significant role in the development of biopolymer-based electronic circuits. With the help of 3D printing, it is now possible to create complex circuits with intricate geometries that are not possible using traditional manufacturing techniques. Additionally, 3D printing allows for the precise deposition of electronic materials, which enhances the accuracy and reliability of the circuits.

These biopolymer-based electronic circuits have several potential applications in healthcare, environmental monitoring, and energy management—for instance, they can be used to create wearable sensors that can monitor the vital signs of patients in real time (Figure 7.2). Additionally, they can be used in environmental monitoring to detect pollution levels and monitor the quality of air and water [12].

The first step is to select the appropriate biopolymer material for the electronic circuit. Biopolymers such as cellulose, chitosan, and alginate are commonly used due to their biocompatibility, flexibility, and high electrical conductivity.

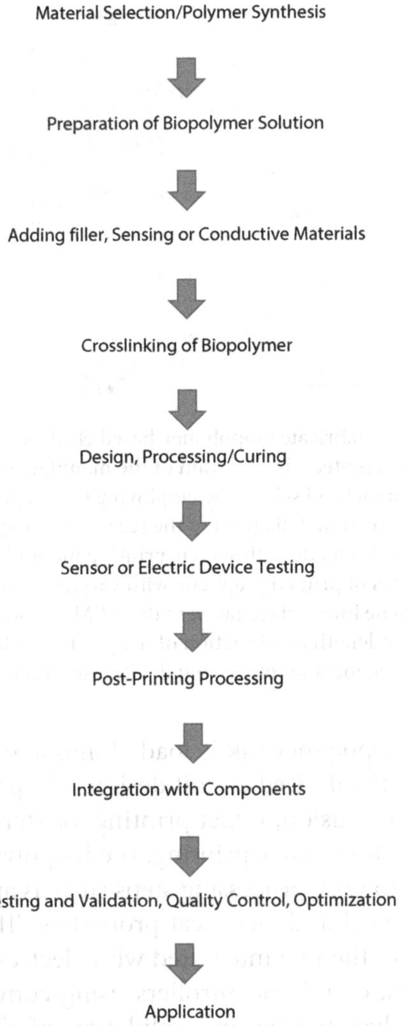

The biopolymer ink is prepared by dissolving the biopolymer in a solvent such as water or an organic solvent. The ink is then mixed with conductive materials such as carbon nanotubes or graphene to enhance electrical

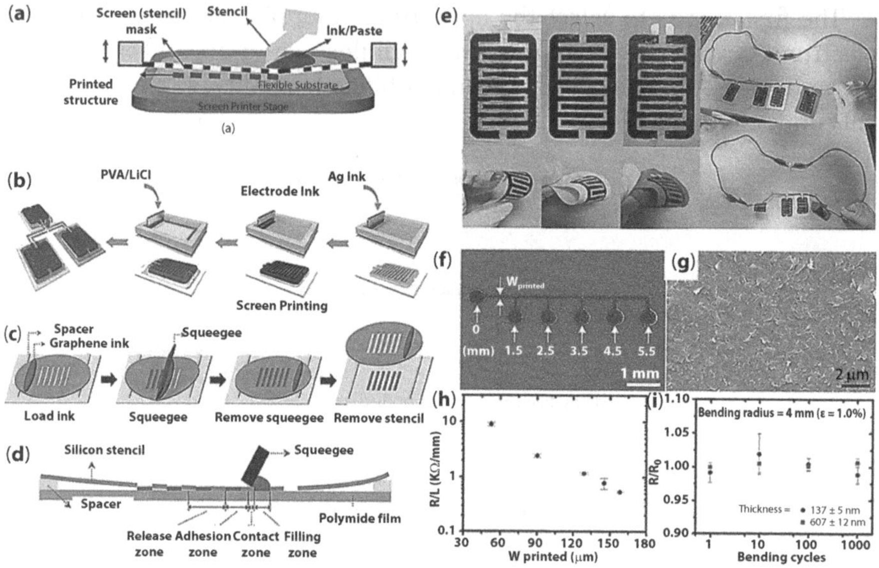

Figure 7.2 3D printing to fabricate biopolymer-based electronic circuits (a) Screen printing technology is presented. (b) Diagram of the manufacture of the transistor. (c) Electrode fabrication method schematic employing screen printing technology and graphene ink. (d) Cross-sectional diagram of the screen printing technique. (e) Optical images of printed electrodes made with screen-printing method are displayed. (f) Optical micrograph showing lines of printed graphene with varying degrees of printed width. (g) Images of the graphene line surface taken with a SEM. (h) Resistance of the graphene measured in units of unit length as a function of the printed width. (i) Changes in the relative resistance of graphene lines of varying thicknesses under bending conditions [13].

conductivity. The biopolymer ink is loaded into a 3D printer and printed layer by layer using the desired circuit design. The printer can use various techniques such as extrusion, inkjet printing, or stereolithography to create the desired structure. After printing, the biopolymer-based electronic circuit is subjected to post-processing steps such as annealing or curing to enhance its mechanical and electrical properties. The biopolymer-based electronic circuit can then be integrated with electronic components such as sensors, batteries, or microcontrollers using conventional techniques. The final step involves testing and validation of the biopolymer-based electronic circuit to ensure its functionality and reliability. Testing can be done using various methods such as electrical characterization, thermal analysis, and mechanical testing. After testing and validation, the biopolymer-based electronic circuit can be optimized for performance and scalability, if necessary. The optimized circuit can then be used for various applications such as wearable electronics, implantable medical devices, and environmental monitoring. The development of biopolymer-based electronic circuits through 3D printing is a significant breakthrough in the field of electronics. It offers several advantages such as biodegradability, low cost, and eco-friendliness, making it a sustainable option for future electronic devices. This technology could revolutionize the way we produce and dispose of electronic devices while reducing their environmental impact. It opens up new possibilities for developing biocompatible and wearable electronic devices. However, further research is necessary to improve the mechanical and electrical properties of the biopolymers used in 3D printing to make them more efficient and robust for commercial applications. Overall, the use of biopolymers in electronic circuits has immense potential in creating sustainable and innovative electronic devices.

This study demonstrates the use of 3D printing to fabricate biopolymer-based electronic circuits, which are environmentally friendly and biodegradable. Figure 7.2 demonstrates the 3D printing to fabricate biopolymer-based electronic circuits [13]. The researchers used a biopolymer ink made from cellulose nanofibrils and graphene oxide to print conductive tracks and electrodes. The resulting circuits were flexible and transparent and had high electrical conductivity [14]. This study reports the development of a biopolymer-based electronic circuit for biosensing applications. The circuit was 3D-printed using a biopolymer ink made from chitosan and graphene oxide. The resulting circuit showed good sensitivity and selectivity for glucose sensing. The biopolymer-based circuit is a promising platform for the development of biosensors and other electronic devices [15]. This study demonstrates the use of 3D printing to fabricate biopolymer-based electronic circuits on bioinspired surfaces. The researchers used a biopolymer ink made from gelatin and graphene oxide to print conductive tracks and electrodes on a bioinspired surface. The resulting circuits showed good adhesion and stability on the surface. The biopolymer-based circuits on bioinspired surfaces could be used in a variety of applications, including sensors and actuators [7].

7.2.1.4 3D-Printed Biopolymer Antennas

Biopolymer-based antennas have been developed using 3D printing. These antennas can be made from biopolymers such as cellulose, chitosan, and silk, and they have shown good performance in transmitting and receiving signals. The biopolymers used in the production of these antennas have unique properties that make them well suited for use in antennas—for example, cellulose is a strong, flexible material that can be easily molded into various shapes, while chitosan is known for its high biocompatibility and antimicrobial properties. Silk, on the other hand, is a strong, lightweight material that can be easily woven into complex structures.

One of the key benefits of using biopolymers in the production of antennas is that they are more environmentally friendly than traditional materials such as metal or plastic. Biopolymers are biodegradable and can be easily recycled, making them a sustainable choice for the production of electronic devices. In addition to their environmental benefits, biopolymer-based antennas have shown promising performance in laboratory tests. They have been shown to have good signal strength, low noise levels, and high efficiency, making them a viable alternative to traditional antennas. With further development and refinement, these antennas could find applications in a wide range of fields, including telecommunications, aerospace, and medical devices.

Figure 7.3 shows 3D-printed biopolymer antenna made from copper and graphene. 3D-printed biopolymer antenna made from polylactic acid

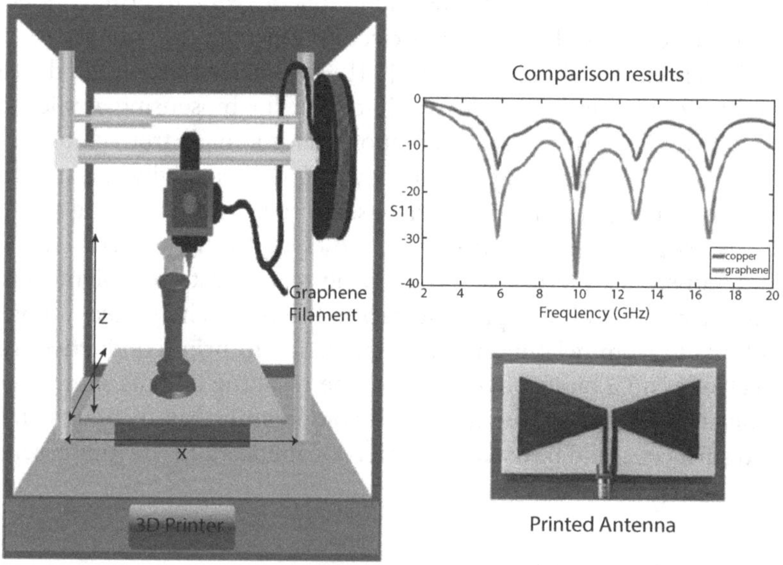

Figure 7.3 3D-printed biopolymer antenna made from copper and graphene. [7, 16] present a study of a 3D-printed biopolymer antenna made from polylactic acid (PLA) and silk fibroin. The antenna showed good performance in the frequency range of 2–6 GHz. Vittorio presents a review of the state-of-the-art in 3D printing of biopolymer-based antennas. The authors discuss the potential of biopolymers such as cellulose, chitosan, and silk for use in antennas, and they present some examples of 3D-printed biopolymer antennas [17]; another study provides a review of 3D printing of biopolymer-based structures, including antennas. The authors discuss the advantages and challenges of using biopolymers in 3D printing and present some examples of biopolymer-based antennas [18].

(PLA) and silk fibroin reported elsewhere [7, 16]. The antenna showed good performance in the frequency range of 2–6 GHz. Vittorio presents a review of the state-of-the-art in 3D printing of biopolymer-based antennas. The authors discuss the potential of biopolymers such as cellulose, chitosan, and silk for use in antennas, and they present some examples of 3D-printed biopolymer antennas [17]; another study provides a review of 3D printing of biopolymer-based structures, including antennas. The authors discuss the advantages and challenges of using biopolymers in 3D printing and present some examples of biopolymer-based antennas [18].

The use of biopolymers in 3D printing can provide a sustainable and eco-friendly alternative to traditional antenna manufacturing methods. The resulting antennas can have a range of applications, including in wireless communication, satellite technology, and more.

7.2.1.5 3D-Printed Biopolymer LEDs

Biopolymer-based LEDs are a new development in the field of lighting technology, where 3D printing technology is used to create lighting devices made from natural materials such as gelatin and silk. These LEDs are considered environmentally friendly as they do not require the use of synthetic polymers or heavy metals that are typically used in conventional LEDs. The resulting LED is flexible and can be easily shaped into any desired form, making it suitable for use in a wide range of applications. Figure 7.4 shows biopolymer-based 3D printing LED materials.

One of the major advantages of biopolymer-based LEDs is their good light emission properties, which have been demonstrated in lab tests. These LEDs emit light with high efficiency and color rendering, making them ideal for use in lighting fixtures, displays, and other applications that require high-quality illumination. These biopolymer-based LEDs also have the potential to be used in a range of applications such as in smart textiles and wearable technology, where their flexibility and biocompatibility would be highly advantageous. Additionally, the use of biopolymers may also lead to the development of new and innovative materials for other applications in fields such as biomedical engineering. Another benefit of biopolymer-based LEDs is their biodegradability, which makes them a sustainable alternative to traditional lighting materials. Once the LED reaches the end of its useful life, it can be easily disposed of without harming the environment.

In summary, biopolymer-based LEDs created using 3D printing offer a sustainable and innovative alternative to traditional lighting technology. They are environmentally friendly, have good light emission properties, and can be easily shaped into any desired form, making them suitable for a wide range of applications.

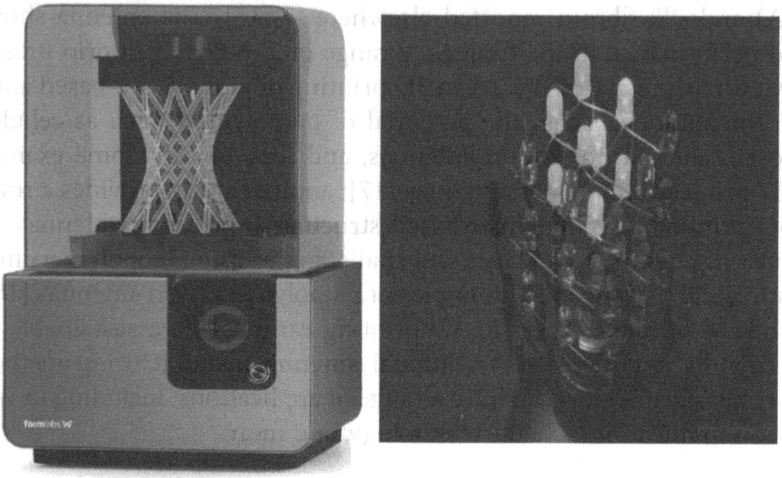

Figure 7.4 Biopolymer-based 3D printing LED materials.

The first step in making biopolymer-based LEDs is to prepare the materials. Gelatin and silk are two biopolymers that can be used for this purpose. The gelatin is extracted from animal bones, skin, and connective tissues, while silk is obtained from the cocoons of silkworm. Once the materials have been prepared, they need to be synthesized into polymers. This is done by adding chemical compounds to the materials and then heating them to a specific temperature. The resulting polymer is then cooled and can be used to create the LED [19]. The next step is to use 3D printing to create the LED. The biopolymer-based ink is loaded into a 3D printer, which then prints the LED layer by layer. The printer can be set to produce any shape or design. Once the LED has been printed, it needs to be assembled. This involves attaching the LED to a circuit board, which provides the power source. The LED is then connected to the power source and tested to ensure that it is working correctly. The final step in making biopolymer-based LEDs is to test them. The LED is placed in a dark room, and the light emission properties are measured. The testing will reveal whether the LED is producing enough light and whether it is working correctly. If the LED passes the test, it can be used in a range of applications, including lighting and displays [20].

The use of biopolymers for creating LEDs is a significant development in the field of sustainable materials and technology. The creation of such LEDs using 3D printing is expected to have a significant impact on the manufacturing industry. The use of biopolymers such as gelatin and silk as a base material for creating LEDs also has implications for the biodegradability and sustainability of electronic devices [21].

7.2.1.6 *Wearable Technology*

Wearable technology such as smart watches and fitness trackers have become increasingly popular in recent years. These devices use sensors and other technologies to monitor various aspects of a person's health and fitness, such as heart rate, sleep patterns, and activity levels. The use of biopolymers in 3D printing also allows for more customization and flexibility in the design of wearable technology. With 3D printing, it is possible to create unique and personalized products that fit the individual needs of each user. Biopolymer-based 3D printing has the potential to revolutionize the way we create wearable technology, making it more sustainable, customizable, and eco-friendly. Figure 7.5 demonstrates biopolymer-based 3D-printed wearable materials [12].

The first step in making biopolymer-based 3D printing is selecting a suitable biopolymer. Biopolymers are naturally occurring polymers that can be extracted from various sources such as plants, animals, and

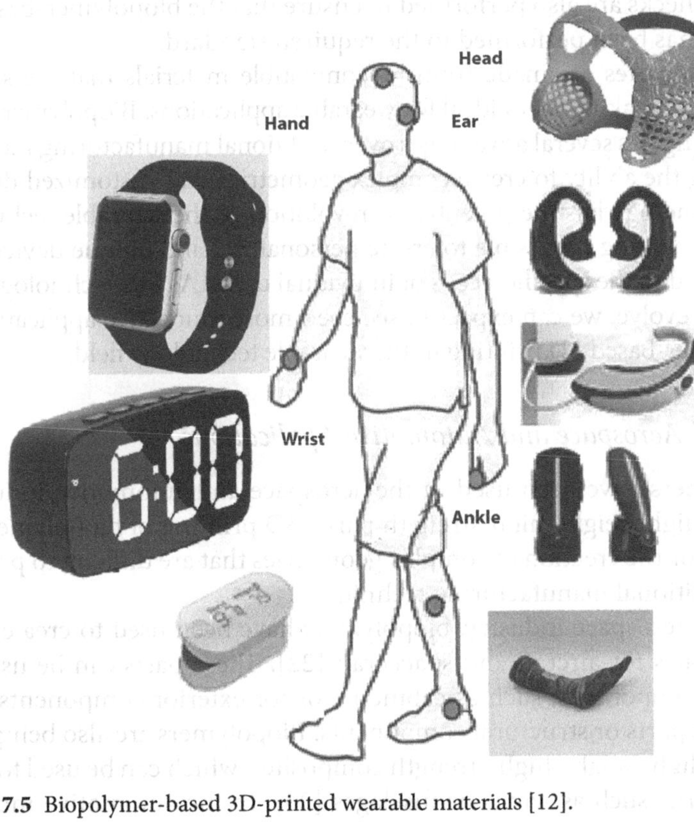

Figure 7.5 Biopolymer-based 3D-printed wearable materials [12].

microorganisms. Some commonly used biopolymers for 3D printing include polylactic acid (PLA), cellulose, chitosan, and starch.

Once the biopolymer is selected, the 3D printing process can begin. The process involves creating a 3D digital model of the wearable technology using specialized software. The model is then sliced into multiple layers, and the 3D printer deposits the biopolymer layer by layer to form the final product. Once the 3D printing is complete, the wearable technology may require post-processing. This can involve removing any support structures, sanding and polishing the surface, and adding any required finishing touches. After the post-processing is complete, any necessary electronics and sensors are integrated into the wearable technology. This may involve soldering, wiring, and programming. After the post-processing is complete, any necessary electronics and sensors are integrated into the wearable technology. This may involve soldering, wiring, and programming. The final step is testing the wearable technology to ensure that it functions as intended. This includes checking the fit and comfort of the product and verifying that the sensors and electronics are working correctly. Quality control checks are also performed to ensure that the biopolymer-based 3D printing has been performed to the required standard.

These devices are made from biocompatible materials that are safe for human use, making them ideal for wearable applications. Biopolymer-based 3D printing has several advantages over traditional manufacturing methods, including the ability to create complex geometries and customized designs. This technology has the potential to revolutionize the wearable technology industry, making it possible to create personalized and unique devices that are tailored to the specific needs of individual users. As this technology continues to evolve, we can expect to see even more innovative applications of biopolymer-based 3D printing in the wearable technology field.

7.2.1.7 Aerospace and Automotive Applications

Biopolymers have been used in the aerospace and automotive industries to create lightweight, high-strength parts. 3D printing of biopolymers has allowed for the creation of complex geometries that are difficult to produce with traditional manufacturing techniques.

In the aerospace industry, biopolymers have been used to create lightweight parts for aircraft and spacecraft [22]. These parts can be used for interior components, such as cabinetry, or for exterior components, such as engine parts or structural components. Biopolymers are also being used to create lightweight, high-strength composites, which can be used to build larger parts, such as wings or fuselages [23]. In the automotive industry,

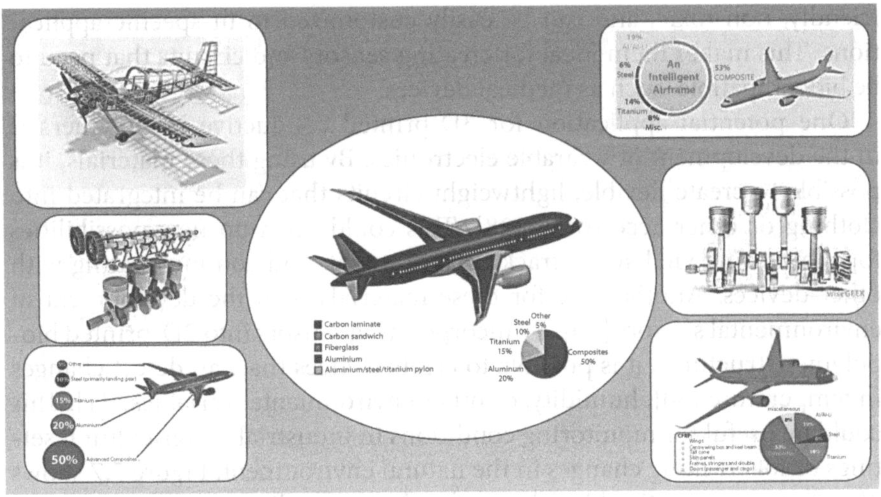

Figure 7.6 3D-printed biopolymer application in aerospace and automotive.

biopolymers are being used to create lightweight parts for cars and trucks. These parts can be used for interior components, such as dashboards and seat covers [24], or for exterior components, such as body panels or wheels [25]. Biopolymers are also being used to create composites that can be used to make entire car bodies [26]. Figure 7.6 shows 3D-printed biopolymer application in aerospace and automotive.

By using renewable materials to create lightweight, high-strength parts, manufacturers can reduce the environmental impact of their products while also improving performance. As 3D printing technology continues to evolve, we can expect to see even more innovative uses of biopolymers in the future.

7.2.1.8 Conductive Biopolymers

3D-printed conductive biopolymers are a relatively new technology that has the potential to revolutionize the field of electronics and sensors. These materials are made from biodegradable, renewable, and biocompatible polymers that can be 3D-printed into complex structures with integrated conductive pathways. One of the primary benefits of using biopolymers for 3D printing is their ability to conduct electricity [27]. By incorporating conductive additives, such as graphene or carbon nanotubes, into the biopolymer matrix, it is possible to create materials that can transmit electrical signals.

In addition to their conductivity, these materials also offer several advantages over traditional synthetic polymers. They are environmentally friendly, non-toxic, and can be easily customized to fit specific applications. This makes them ideal for creating sensors and circuits that need to be biocompatible, such as medical devices.

One potential application for 3D-printed conductive biopolymers is in the development of wearable electronics. By using these materials, it is possible to create flexible, lightweight circuits that can be integrated into clothing or other accessories [28]. This could open up new possibilities for monitoring vital signs, tracking movement, and communicating with other devices. Another use for these materials is in the development of environmental sensors [29]. By incorporating sensors into 3D-printed biopolymer structures, it is possible to create devices that can detect changes in temperature [30], humidity, or other environmental variables [31]. This could be useful for monitoring conditions in industrial or agricultural settings or for tracking changes in the natural environment. Figure 7.7 shows 3D-printed conductive biopolymer for electric application.

Overall, 3D-printed conductive biopolymers offer a promising new avenue for creating circuits and sensors. As the technology continues to develop, it is likely that we will see even more innovative applications emerge in the future.

Creating 3D-printed conductive biopolymers for circuits and sensors involves the following steps: Choose a biopolymer that can be 3D-printed and conductive. Biopolymers such as chitosan, alginate, and gelatin have been shown to be conductive.

Add a conductive filler such as carbon nanotubes, graphene, or silver nanoparticles to the biopolymer. The amount of the filler added determines the conductivity of the material. The biopolymer and conductive filler are mixed together to form a homogenous mixture. The conductive biopolymer mixture is 3D-printed using a 3D printer. The printer can print the conductive material in the desired shape and size for the circuit or sensor [32, 46]. Once the 3D printing is complete, the conductive biopolymer is post-processed to improve its conductivity. This can involve curing the material with heat or UV light. The 3D-printed conductive biopolymer can now be used to create circuits and sensors. This involves designing the circuit or sensor and integrating the conductive material into it [3]. The circuit or sensor is tested to ensure that it is functioning properly. With the above-mentioned steps, it is possible to create 3D-printed conductive biopolymers for circuits and sensors [17]. These materials can be used in a variety of applications such as wearable electronics, prosthetics, and biomedical device.

The development of 3D-printed conductive biopolymers has the potential to revolutionize the production of circuits and sensors. These

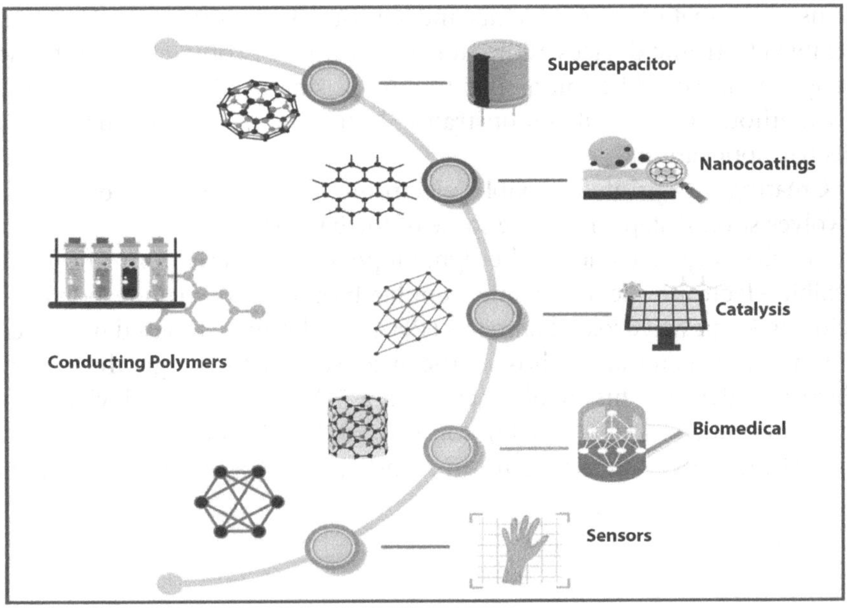

Figure 7.7 3D-printed conductive biopolymer for electric application.

biopolymers have unique properties that make them ideal for creating flexible and lightweight devices. The ability to 3D-print these materials allows for precise control over their design and the ability to create complex shapes and geometries [33]. This technology has a wide range of applications, from medical devices to consumer electronics [34]. However, further research and development are needed to optimize the performance of these materials and to scale up production for commercial use. With continued innovation, 3D-printed conductive biopolymers have the potential to transform the way we create and use electronic devices.

7.2.1.9 3D Biopolymer-Based Flexible Electronics

Flexible electronic devices are electronic components that can be bent, twisted, and stretched without losing their functionality. They are made using materials that are flexible and durable, such as biopolymers [35]. Flexible electronic devices can be used in a wide range of applications, such as wearable technology [36], medical devices [37], and sensors [38]. Printing of biopolymer flexible electronic devices is a revolutionary technology that is gaining popularity in the field of advanced manufacturing. Biopolymers such as polylactic acid (PLA) and cellulose acetate can be used to create flexible circuits and sensors. 3D printing of biopolymer flexible electronic devices allows for the creation of complex and intricate designs that would be impossible to create using traditional manufacturing methods. Figure 7.8 demonstrates 3D-printed flexible biopolymer for electric application.

Creating 3D-printed flexible electronic devices using biopolymers involves several steps. Here are some of the basic steps.

The first step is to choose the right biopolymer material for printing the flexible electronic device. Some popular biopolymers include polylactic acid (PLA), polyhydroxyalkanoates (PHA), and cellulose-based materials.

Once the material is chosen, the next step is to design the flexible electronic device. This involves creating a 3D model of the device using computer-aided design (CAD) software. After the design is complete, the 3D printer can be set up to print the device. The printer will use the

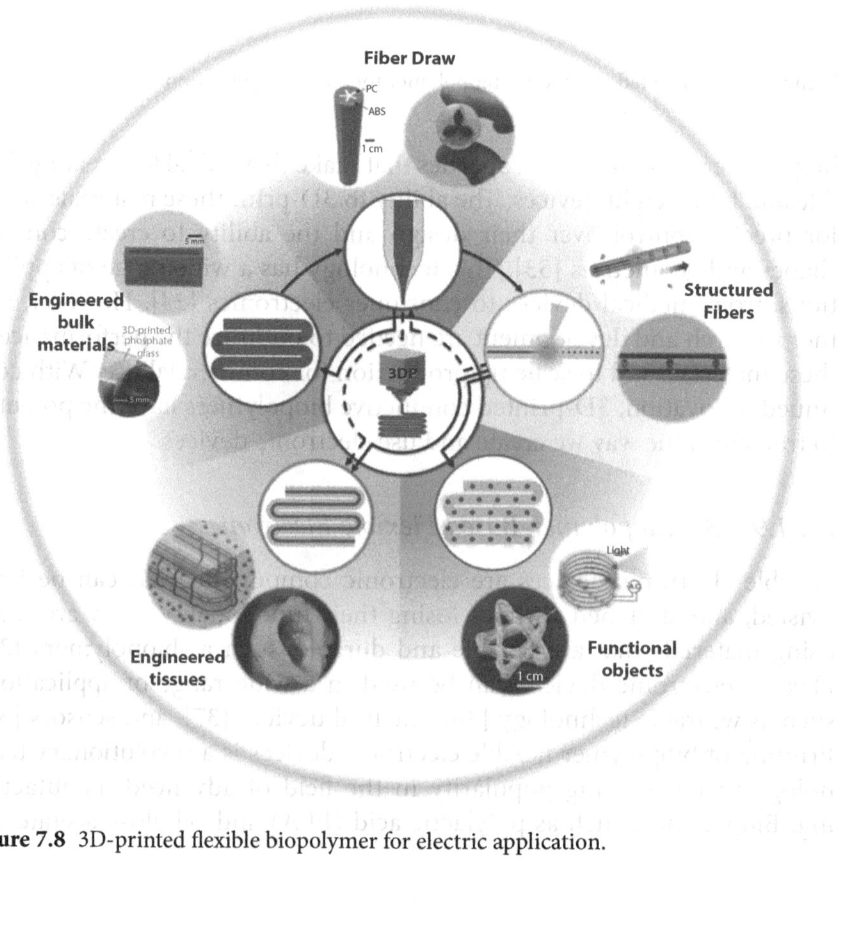

Figure 7.8 3D-printed flexible biopolymer for electric application.

biopolymer material to create the device layer by layer. Once the device is printed, electronic components such as sensors, transistors, and conductive tracks can be added. These components can be embedded directly into the biopolymer material or attached to the surface of the device. Finally, the printed device can be tested to ensure that it functions correctly. This may involve testing its mechanical flexibility, electrical conductivity, and other parameters. Keep in mind that creating 3D-printed flexible electronic devices using biopolymers is a complex process that requires specialized equipment and expertise. However, with the right tools and knowledge, it is possible to create functional and innovative devices that have many applications in fields such as healthcare, wearables, and environmental monitoring

3D printing allows for the creation of complex geometries and customized designs, which is important for the development of flexible electronic devices that can conform to the shape of the human body. Additionally, the use of biopolymers in 3D printing can lead to the development of environmentally sustainable electronics. While there are still challenges to overcome in the field of 3D printing of biopolymers for flexible electronic devices, such as achieving high resolution and mechanical strength, advancements in technology and materials are continuously being made. With further research and development, 3D printing of biopolymers has the potential to revolutionize the production of flexible electronic devices, leading to the creation of more comfortable, durable, and sustainable devices for a wide range of applications, including medical devices and wearables.

7.2.1.10 Bio-Inspired 3D Printing

Researchers are using bio-inspired 3D printing techniques to create electronic devices using biopolymers. This involves using natural structures, such as spider webs and leaf veins [39], as templates for 3D printing. The resulting electronic devices have improved properties, such as increased strength and flexibility. One of the techniques used is the printing of electronic devices using nanocellulose [40], which is a biopolymer derived from plant matter. The nanocellulose is mixed with conductive materials such as graphene or silver nanoparticles to create conductive inks. The ink is then printed onto a substrate using a 3D printer, resulting in a biopolymer-based electronic device. Another technique involves using biopolymer templates found in nature to create electronic devices—for example, the structure of a butterfly wing can be used as a template to print electronic circuits [41, 42]. The wing's structure has a complex network of veins and pores that can be mimicked using biopolymers to create conductive pathways for electronic devices. Bio-inspired 3D printing techniques using biopolymers have the potential to revolutionize the way electronic devices are produced [43]. The use of natural materials and structures can lead to the development of sustainable and biodegradable electronics, which are less harmful to the environment and compatible with living systems. Following is the procedure of bio-inspired 3D printing techniques to create electronic devices using biopolymer. Figure 7.9 shows bio-inspired 3D printing biopolymer using electronic devices.

The first step is to select a suitable biopolymer material for the 3D printing process. The biopolymer should have the required properties, such as mechanical strength, thermal stability, and biocompatibility, to support the electronic components that will be integrated. The second step is to design the electronic circuit that will be printed onto the biopolymer material. The circuit can be designed using computer-aided design (CAD) software. The third step is to prepare the biopolymer material for 3D printing. The biopolymer material can be dissolved in a suitable solvent and then extruded through a 3D printer nozzle. The fourth step is to print the biopolymer material layer by layer using the 3D printer. The electronic circuit is printed onto the biopolymer material using conductive inks. The fifth step is to cure the biopolymer material using UV light or heat. Post-processing techniques such as cleaning and surface treatment can also be performed to enhance the properties of the printed device. The final step is to integrate

Figure 7.9 Bio-inspired 3D printing biopolymer using electronic devices.

the electronic components such as sensors, LEDs, and batteries into the printed device. This can be done using traditional assembly techniques or by 3D printing the components directly onto the biopolymer material.

The bio-inspired 3D printing technique involves mimicking natural biological systems to create electronic devices using biopolymer materials. This approach has the potential to produce complex and functional electronic devices with high precision and accuracy.

7.2.1.11 Biopolymer-Based Energy Harvesting

Biopolymer-based energy harvesting devices, such as piezoelectric sensors, have been developed using 3D printing. These devices can convert mechanical energy into electrical energy, which can be used to power other electronic devices [44].

Piezoelectric sensors are commonly used in applications such as vibration monitoring, pressure sensing, and motion sensing [45]. They work by converting mechanical energy into electrical energy and can be used to power small electronic devices, such as sensors or wireless transmitters [46, 47].

One example of a biopolymer-based energy harvesting device with a piezoelectric sensor is a wearable energy harvester that can generate electricity from the movements of the wearer [48]. The device is made of a biopolymer material that is both lightweight and flexible, allowing it to be worn comfortably on the body. The use of biopolymer-based energy harvesting devices with piezoelectric sensors has the potential to revolutionize the field of renewable energy by providing a sustainable and efficient way to generate electricity from natural movements and vibrations.

The first step in developing a biopolymer-based energy harvesting device is to select the appropriate biopolymer material. Biopolymers are natural, sustainable, and biodegradable materials that are derived from renewable resources.

The next step is to design the piezoelectric sensor using computer-aided design (CAD) software. The sensor can be designed in different shapes and sizes, depending on the intended application. The designed sensor is then 3D-printed using a biopolymer-based filament. 3D printing allows for the fabrication of complex shapes and structures with high precision. Once the sensor is 3D-printed, it needs to be poled. Poling is the process of applying an electric field to the piezoelectric material to orient its internal dipoles in a specific direction. This process is crucial to enhance the piezoelectric properties of the sensor. After poling, the sensor is tested to ensure its functionality. The sensor is subjected to mechanical vibrations, and its output voltage is measured using an oscilloscope.

If the sensor does not perform as expected, its design can be optimized, and the printing parameters can be adjusted to improve its performance. Once the sensor is optimized and tested, it can be integrated into a larger system, such as a wearable device, to harvest energy from mechanical vibrations. Biopolymer-based energy harvesting devices can be used in a variety of applications, such as self-powered sensors, medical devices, and Internet of Things (IoT) devices. Figure 7.10 shows 3D-printing-derived protein energy harvesting device [49-51].

In the future, these biopolymer-based energy harvesting devices may play a significant role in creating sustainable energy solutions, especially in the field of wearable technology and medical implants. As research and development continue, it is possible that we may see more advanced and innovative biopolymer-based energy harvesting devices in the market, which will further reduce our dependence on non-renewable energy sources.

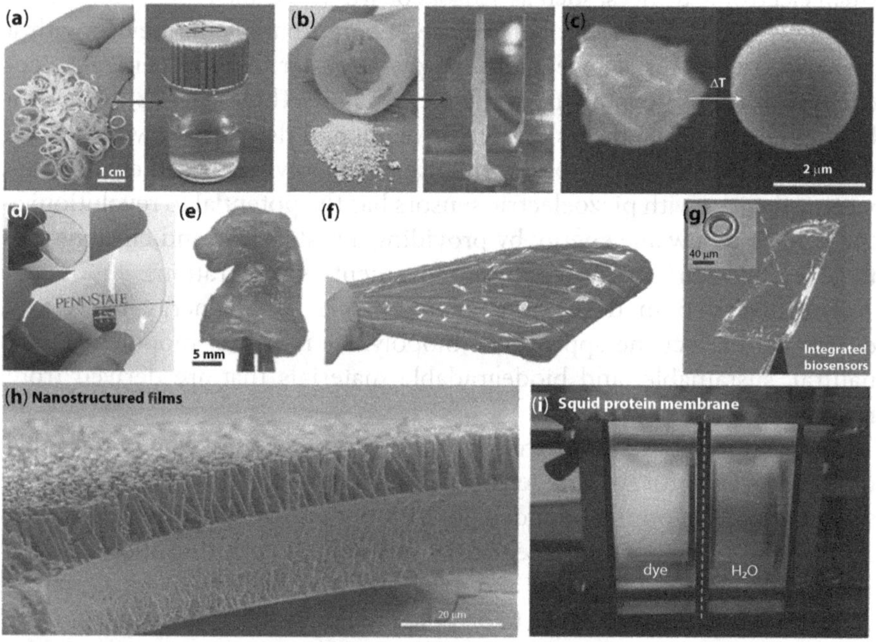

Figure 7.10 3D-printing-derived protein energy harvesting device [49, 51]: (a) solution-based or (b) thermal-based methods into a variety of materials. (c) Colloids. (d) Free-standing transparent flexible films. Reproduced with permission. (e) Complex 3D geometries. (f) Biomimetic materials. (g) Optical microresonators. (h) Nanostructured surfaces. Reproduced with permission. (i) SRT-based membranes.

Milad Abyaneh reported that their devices exhibited high energy conversion efficiency and could be potentially used for powering low-power. Another researcher found good sensitivity and stability that could be used for monitoring human activities and detecting environmental stimuli [50].

7.2.1.12 Biopolymer-Based 3D Printing Inks

Researchers have developed 3D printing inks made from biopolymers that can be used to create electronic devices. These inks can be used to create circuits, sensors, and other electronic components.

Biopolymer-based 3D printing inks can be used to print a variety of electronic devices, such as sensors, batteries, and even circuits. The ink is made from a mixture of biopolymers and conductive materials, such as graphene, carbon nanotubes, or silver nanoparticles [17].

The biopolymers provide the structure for the ink, while the conductive materials provide the electrical conductivity needed to create electronic devices [52]. These inks can be printed onto a variety of substrates, such as paper, plastic, or even textiles.

One of the advantages of biopolymer-based 3D printing inks is their sustainability [53]. Another advantage of these inks is their versatility. They can be used to create electronic devices of various shapes and sizes, making them ideal for a range of applications. They can also be customized to suit specific requirements, such as flexibility, durability, or conductivity [54].

Furthermore, biopolymer-based 3D printing inks are an exciting new technology that has the potential to revolutionize the field of electronics. By using sustainable materials, these inks offer a greener alternative to traditional polymers while also providing a versatile and customizable solution for creating electronic devices. Figure 7.11 shows alginate-based 3D printing inks in electric application, for example, 3D objects printed from water and culture medium solution [55].

The following is a general procedure of how biopolymer-based 3D printing inks can be used to create electronic devices: The first step is to select a biopolymer material that can serve as a base material for the 3D printing ink. The biopolymer material should be able to form a stable and conductive ink.

The biopolymer material is then mixed with other conductive materials, such as silver or graphene, to create a conductive ink. The ink should be prepared in such a way that it has a suitable viscosity for 3D printing. The prepared ink is loaded into a 3D printer and used to print the desired electronic device. The printer should be set to the appropriate temperature and printing speed to ensure that the ink is deposited correctly. Once the device has been printed, it is then cured to solidify the ink and create a conductive path between the different components of the device. Curing can be done using heat, UV light, or other methods, depending on the biopolymer material used. The printed device is then tested to ensure that it functions correctly. This involves checking the conductivity of the ink and verifying that the device performs the desired function. Biopolymer-based 3D printing inks offer a promising approach to creating electronic devices that are more sustainable and environmentally friendly. By using biodegradable and renewable materials, it is possible to create electronic devices that have a lower impact on the environment while still maintaining high performance and functionality.

Biopolymer-based 3D printing inks have immense potential in the creation of electronic devices. These inks offer a sustainable and eco-friendly solution to traditional polymer-based inks that can have harmful environmental impacts. Biopolymer-based inks are also biocompatible, making them suitable for use in biomedical applications. The properties of biopolymers, such as their mechanical and electrical properties, can be tailored to meet specific requirements. Additionally, the use of 3D printing technology enables the precise and accurate deposition of the ink, allowing

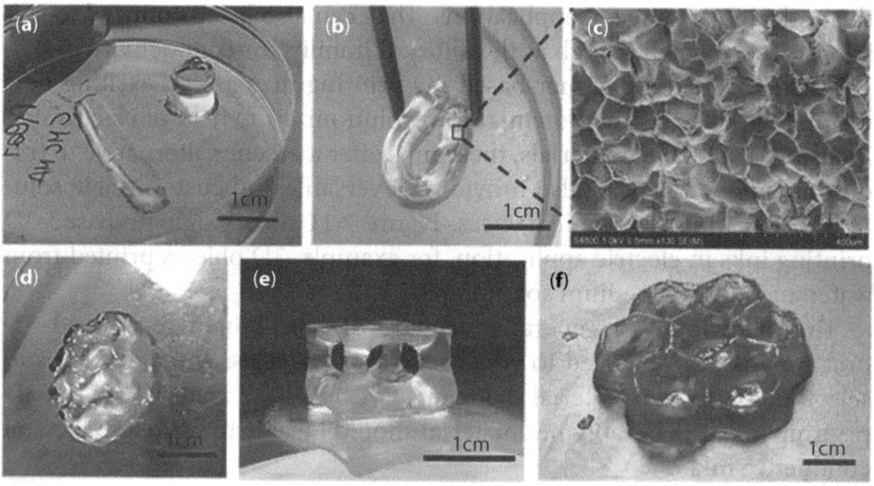

Figure 7.11 Alginate-based 3D printing inks in electric application. (a) Simple cylinders and parallelepipeds (solvent/water). (b) The hydrogel exhibited good flexibility and handleability. (c) SEM analysis performed on the freeze-dried hydrogel. (d–f) 3D objects printed from water (d) and from culture medium solution (e, f) [55].

for the creation of complex and intricate electronic devices. Overall, biopolymer-based 3D printing inks provide a promising avenue for the development of sustainable and functional electronic devices.

7.3 Conclusion

In conclusion, the use of biopolymers in 3D printing for electronic applications is a growing area of research. Biopolymers offer several advantages over traditional materials such as plastics and metals, including biodegradability, renewability, and non-toxicity. The recent innovations in 3D-printing-assisted biopolymers for electronic applications such as sensors, batteries, and electronic circuits have shown promising results, and it is expected that more developments will be made in this field in the future.

Acknowledgments

The authors acknowledge the financial support from Ministry of Science and Technology (MoST) under the NST Grants and technical support from Laboratories of Materials, Energy, and Nanotechnology MENTECH Labs, Jashore University of Science and Technology (JUST).

References

1. Finny, A.S., Jiang, C., Andreescu, S., 3D printed hydrogel-based sensors for quantifying UV exposure. *ACS Appl. Mater. Interfaces*, *12*, 39, 43911–43920, 2020.
2. Concas, A. and Cao, G., 3D Printing of Biopolymer-Based Composites for Applications in Tissue Engineering and Biomedical Devices. *Polymers*, 11, 11, 1731, 2019.
3. Díez-Pascual, A.M., Biopolymer Composites: Synthesis, Properties, and Applications. *Int. J. Mol. Sci.*, *23*, 4, 2257, 2022.
4. Li, L., Li, Y., Li, L., Zhang, C., Li, Y., A 3D-printed biopolymer-based temperature and humidity sensor with a graphene/PEDOT:PSS composite electrode. *J. Mater. Sci.*, 55, 24, 10935–10945, 2020, doi: 10.1007/s10853-020-04929-1.
5. Li, L., Li, Y., Li, L., Zhang, C., Li, Y., 3D-printed flexible sensors based on cellulose and chitosan for simultaneous detection of humidity and temperature. *Sens. Actuators, B*, 296, 126652, 2019, doi: 10.1016/j.snb.2019.126652.

6. Li, X., Guo, Z., Tang, J., Biopolymer-Based Light-Emitting Diodes: From Fabrication to Application. *Adv. Mater. Technol.*, 5, 11, 2000445, 2020, https://doi.org/10.1002/admt.202000445.
7. Liu, Y. *et al.*, 3D Printing of Biopolymer-based Electrolytes for Solid-State Batteries. *ACS Appl. Mater. Interfaces*, 10, 25, 21543–21550, 2018, doi: 10.1021/acsami.8b05554.
8. Wang, Z. *et al.*, 3D Printing of Biopolymer-based Electrodes for Energy Storage Devices. *ACS Appl. Mater. Interfaces*, 9, 11, 9795–9802, 2017, doi: 10.1021/acsami.6b16739.
9. Cheng, M., Jiang, Y., Yao, W., Yuan, Y., Deivanayagam, R., Foroozan, T., Huang, Z., Song, B., Rojaee, R., Shokuhfar, T., Pan, Y., Elevated-temperature 3D printing of hybrid solid-state electrolyte for Li-ion batteries. *Adv. Mater.*, *30*, 39, 1800615, 2018.
10. Piras, C.C. and Smith, D.K., Multicomponent polysaccharide alginate-based bioinks. *J. Mater. Chem. B*, *8*, 36, 8171–8188, 2020.
11. Wang, W., Zhu, C., Li, Y., Yang, J., Li, S., Luo, D., Biopolymer-based LEDs: A review of recent progress. *J. Mater. Sci.*, 55, 21, 8920–8940, 2020, https://doi.org/10.1007/s10853-020-04960-9.
12. Kumar, P. and Misra, M., Biopolymer-based 3D printing for next generation wearable technology. *Polym. Rev.*, 59, 4, 623–655, 2019, doi: 10.1080/15583724.2019.1581742.
13. Li, Q., Zhang, J., Li, Q., Li, G., Tian, X., Luo, Z., Qiao, F., Wu, X., Zhang, J., Review of printed electrodes for flexible devices. *Front. Mater.*, *5*, 77, 2019.
14. Kim, S. and Kim, S., Recent advances in 3D printing of functional materials for flexible and wearable electronics. *Chem. Eng. J.*, 371, 443–458, 2019.
15. Lyu, Z., Lim, G.J., Koh, J.J., Li, Y., Ma, Y., Ding, J., Wang, J., Hu, Z., Wang, J., Chen, W., Chen, Y., Design and manufacture of 3D-printed batteries. *Joule*, 5, 1, 89–114, 2021.
16. Avşar Aydın, E., 3D-Printed Graphene-Based Bow-Tie Microstrip Antenna Design and Analysis for Ultra-Wideband Applications. *Polymers*, *13*, 21, 3724, 2021.
17. Ajiteru, O., Sultan, M.T., Lee, Y.J., Seo, Y.B., Hong, H., Lee, J.S., Lee, H., Suh, Y.J., Ju, H.W., Lee, O.J., Park, H.S., A 3D printable electroconductive biocomposite bioink based on silk fibroin-conjugated graphene oxide. *Nano Lett.*, *20*, 9, 6873–6883, 2020.
18. Zhang, Y., Cai, X., Yang, H., Wang, J., Lu, J., Huang, W., Biopolymer-based light-emitting diodes: a review. *Mater. Sci. Eng.: R: Rep.*, 128, 1–27, 2018, https://doi.org/10.1016/j.mser.2018.03.002.
19. Wang, Y., Xia, Y., Rogers, J.A., 3D printing of functional materials for flexible and wearable electronics. *Adv. Mater.*, 31, 45, 1904391, 2019.
20. Zhang, M., Yang, Y., Xu, F., 3D printing of flexible and stretchable electronics using biopolymers: A review. *Mater. Today Bio*, 10, 100124, 2021.
21. Li, W., Yao, C., Li, Z., 3D printing of biopolymers for flexible electronics. *Mater. Today Chem.*, 20, 100452, 2021.

22. Ebnesajjad, S. ed., Handbook of biopolymers and biodegradable plastics: properties, processing and applications., Wiley-VCH, 2012.
23. Ngo, T.D., Kashani, A., Imbalzano, G., Nguyen, K.T.Q., Hui, D., Additive manufacturing (AM) and 3D printing in aerospace and defence industry. *Addit. Manuf.*, 19, 1–15, 2018.
24. Yu, C., Schimelman, J., Wang, P., Miller, K.L., Ma, X., You, S., Guan, J., Sun, B., Zhu, W., Chen, S., Photopolymerizable biomaterials and light-based 3D printing strategies for biomedical applications. *Chem. Rev.*, 120, 19, 10695–10743, 2020.
25. Chen, Y., Tan, J., Li, Y., Li, J., Biopolymers and their Composites for Additive Manufacturing: A Review. *Polymers*, 10, 8, 839, 2018.
26. Birla, S., Singh, N., Shukla, N.K. eds., Nanotechnology: device design and applications. CRC Press, 2022.
27. Lee, J., Lee, Y.J., Kim, J., Shin, H., 3D printing of conductive polymer nanocomposites for electronic applications. *Polymers*, 11, 2, 306, 2019.
28. Sathishkumar, S. and Mohan, R., Recent progress in 3D printing of biopolymers and their composites: A review. *Composites, Part B*, 183, 107663, 2020.
29. Park, J.H., Han, S.I., Kwon, H., 3D printing of bio-based conductive polymer composites for electronics. *J. Ind. Eng. Chem.*, 79, 169–180, 2019.
30. Kim, K.H. and Kim, M., Conductive 3D printing of biopolymer-based composites for wearable and implantable electronic devices. *Adv. Healthc. Mater.*, 9, 19, 2000864, 2020.
31. Kim, J., Lee, J., Park, J., Shin, H., Conductive 3D printing of biopolymers for flexible electronics: A review. *Composites, Part B*, 211, 108662, 2021.
32. Zhang, X., Yu, J., He, F., 3D printing of conductive biopolymers for biomedical applications. *Mater. Today Chem.*, 3, 41–50, 2017.
33. Awadhiya, P. and Srivastava, R.K., 3D Printing of Biopolymers: A Review. *Mater. Today: Proc.*, 4, 8, 9194–9202, 2017.
34. Colaco, R. and Mota, M., Biopolymer-based 3D printing: a comprehensive review. *Int. J. Biol. Macromol.*, 124, 965–975, 2019.
35. Bhardwaj, N. and Kundu, S.C., 3D printing of biomaterials for tissue engineering. *J. Tissue Eng. Regener. Med.*, 12, 5, e2028, 2018.
36. Zhao, Y., Liu, Y., Wang, Z., Sun, J., Biopolymer-based 3D printing for flexible electronics: A review. *J. Mater. Sci. Technol.*, 58, 105–118, 2020.
37. Kumar, L.J., Pandey, P.M., Wimpenny, D.I. eds., 3D printing and additive manufacturing technologies, vol. 311, Singapore, Springer, 2019.
38. Zheng, X., Zhou, Y., Wei, W., 3D printing of biopolymers for flexible electronic devices. *J. Mater. Sci. Technol.*, 69, 95–105, 2021.
39. Liu, Y., Wang, H., Wang, J., 3D printing of biopolymers for flexible and wearable electronics: A review. *J. Mater. Sci. Technol.*, 47, 144–153, 2020.
40. Yu, C., Schimelman, J., Wang, P., Miller, K.L., Ma, X., You, S., Guan, J., Sun, B., Zhu, W., Chen, S., Photopolymerizable biomaterials and light-based 3D printing strategies for biomedical applications. *Chem. Rev., 120*, 19, 10695–10743, 2020.

41. Zuev, D.M., Nguyen, A.K., Putlyaev, V.I., Narayan, R.J., 3D printing and bioprinting using multiphoton lithography. *Bioprinting, 20*, e00090, 2020.
42. Mishra, P.K. and Jagadesh, T., Applications and Challenges of 3D Printed Polymer Composites in the Emerging Domain of Automotive and Aerospace: A Converged Review. *J. Inst. Eng. India Ser. D*, 104, 2, 1–18, 2022.
43. Hassan, M., Dave, K., Chandrawati, R., Dehghani, F., Gomes, V.G., 3D printing of biopolymer nanocomposites for tissue engineering: Nanomaterials, processing and structure-function relation. *Eur. Polym. J., 121*, 109340, 2019.
44. Wang, C., Yokota, T., Someya, T., Natural biopolymer-based biocompatible conductors for stretchable bioelectronics. *Chem. Rev., 121*, 4, 2109–2146, 2021.
45. Ahmed, A., Azam, A., Wang, Y., Zhang, Z., Li, N., Jia, C., Mushtaq, R.T., Rehman, M., Gueye, T., Shahid, M.B., Wajid, B.A., Additively manufactured nano-mechanical energy harvesting systems: advancements, potential applications, challenges and future perspectives. *Nano Convergence, 8*, 1, 1–42, 2021.
46. Huang, X., Tan, L.P., Chua, C.K., 3D printing of polymer matrix composites: A review and prospective. *Composites, Part B*, 77, 18–36, 2015.
47. Mukherjee, I., Recent Development of Polysaccharide-Derived Hydrogel: Properties, Stimuli-Responsiveness and Bioapplications, 2022.
48. Kalas, D., Pretl, S., Reboun, J., Soukup, R., Hamacek, A., Towards hand model with integrated multichannel sensor system for thermal testing of protective gloves. *Period. Polytech. Electr. Eng. Comput. Sci.*, 62, 165–171, 2018.
49. Pena-Francesch, A. and Demirel, M.C., Squid-inspired tandem repeat proteins: Functional fibers and films. *Front. Chem., 7*, 69, 2019.
50. Lui, H., Lu, B., Lin, S., Qu, K., Xu, J., Luo, J., Zhao, X., 3D printing of conducting polymers. *Nat. Commun., 11*, 1, 1604, 2020.
51. Kamyshny, A. and Magdassi, S., Conductive nanomaterials for 2D and 3D printed flexible electronics. *Chem. Soc. Rev.*, 48, 1712–1740, 2019, doi: 10.1039/c8cs00595b.
52. Koç, B. and Kuo, C.C., 3D printing of biopolymers and biocomposites for tissue engineering applications. *Prog. Biomater.*, 7, 1, 1–18, 2018.
53. Lee, J. *et al.*, 3D Printing of Biopolymer-based Energy Storage Devices. *ACS Appl. Mater. Interfaces*, 9, 9, 8048–8055, 2017, doi: 10.1021/acsami.6b14332.
54. J.S. Choi *et al.*, Biopolymer-based 3D printing ink for fabricating electrical devices. US Patent Application 20200215260A1, 2020.
55. Melilli, G., Carmagnola, I., Tonda-Turo, C., Pirri, F., Ciardelli, G., Sangermano, M., Hakkarainen, M., Chiappone, A., DLP 3D printing meets lignocellulosic biopolymers: Carboxymethyl cellulose inks for 3D biocompatible hydrogels. *Polymers, 12*, 8, 1655, 2020.

8

Recent Trends in 3D-Printed Biopolymers for Structural Applications

Prakash Chander Thapliyal

Advanced Concrete Steel and Composites Group, CSIR-Central Building Research Institute, Roorkee, Uttarakhand, India

Abstract

Renewable materials have been fruitfully applied in 3D printing. Applications such as packaging, an eco-friendly answer depending on the re-use or upcycling of waste material, can be created. Local sourcing of renewable materials in 3D printing can be the upcoming stride so as to trim down the transportation cost. Some researchers have started experiments to enhance the build magnitude and pace of printers. Newer applications like in temporary furniture need to be further explored. This chapter deals with ways of utilizing the 3D manufacturing process making use of renewable bio-based materials, with spotlight on structural applications, and provides a smart situation for sustainable fabrication and recycling or upcycling of waste material, e.g., rice husk, wood flour, or *Miscanthus* fiber. To use renewable raw materials, 3D printing process can be modified and may need a unique binder. In addition, a software device can be developed to generate tailored component designs by individual part geometry. Like this and in arrangement with 3D manufacturing processes, sustainable as well as biodegradable complex components can be generated within a short time.

Keywords: 3D printing, renewable materials, biopolymers, structural applications

8.1 Introduction

Following three decades of growth, 3D printing is all the time more popular now for structural applications in civil engineering area.

Email: pct866@yahoo.com

Bhasha Sharma, Balaram Pani, Shashank Shekhar and Jude A. Okolie (eds.) Sustainable 3D Printing for Innovative Biopolymer Production and Applications, (177–190) © 2025 Scrivener Publishing LLC

Nevertheless, feeble interlayer performance along with need of reinforcement considerably confines its relevance. These days, structural functions that incorporated energy-efficient building resources, hypothetically requiring balance between mechanical functions with thermal resistance, are up-and-coming interdisciplinary investigation topics in the civil engineering area.

The 3D method of printing makes use of some software to facilitate slicing of 3D form into <0.01-mm-thick layers. Every stratum is then traced onto build plate by 3D printer; once the outline is done, a build plate is lowered, and a subsequent coat is added on top of the preceding one. The 3D method of printing is also branded as "additive manufacturing" by reason of information that as a replacement for removing material to produce a part, the method adds material in succeeding patterns to produce the preferred form. 3D printing is mainly used in (a) prototyping, (b) specific parts—military, aerospace, dental, biomedical, etc., (c) hobbies along with house use, etc., and (d) forthcoming applications—buildings, medical (body parts), cars etc.

3D printing uses the theory of additive manufacturing (AM) and hence is also known for rapid prototyping (RP) as well as free-form manufacturing by a layer-by-layer progression to build up parts or things of tiny dimension, inadequate quantities, and adapted plan [1]. Overall, 3D printing (3DP) is an enormously flexible production method and can be imposed on a range of materials like ceramics, metals, and polymers as well as other materials [2].

Additive manufacturing is the procedure that applies successive substance layers to outline a 3D replica requiring the exploitation of a computer-based 3D model requiring 3D printer as well as a course to build up the physical model. Over time, additive manufacturing (AM) has been regularly utilized as a helpful means of prototyping along with generating parts offering capacity to generate supplies at a superior speed and with dimensional accuracy than most modeling approaches [3, 4].

Traditional mechanized techniques are branded "subtractive manufacturing" for the reason that the practice is of removing material from preformed mass. Processes, for instance, milling along with cutting, are subtractive manufacturing techniques. This kind of development creates a lot of waste because the material that is peeled usually cannot be used for anything else and also is sent as fragment. 3D printing removes such waste given that the material is positioned in the site only where it is required; the remaining will be missing out as a blank gap.

8.2 3D-Printed Materials

As soon as it comes to 3D printing, materials are time and again one of the main significant preferences. Previously, AM technologies were using materials which were non-resistant as much as necessary and degrading rapidly. Gradually, the understanding about 3D printing has been extending and has been thinning out all over humankind and the initiative that this technology has is prospectively showing the way to pioneer ways of manufacturing; more studies as well as analysis are conducted, and new materials are produced. At present, the 3D printing market offers a range of materials, beginning from composites and ceramics to metals and polymers; many are created with individual advantage in addition to disadvantages (Figure 8.1).

Natural resources display budding mechanical properties due to their nanostructures with optimized pecking order, anisotropy, and nanoporosity. The production of such intricate systems is challenging presently as high-value three-dimensional (3D) printing is restricted to homogeneous materials. To augment the properties of two-photon three-dimensional (3D) printed polymeric structures, photo-initiators are introduced to enhance the generation of radicals along with getting better polymerization. Researchers observed a progress in two-photon 3D printing where atomic metal clusters are used as two-photon absorbers to augment the mechanical properties of the printed structures [5]. There are different

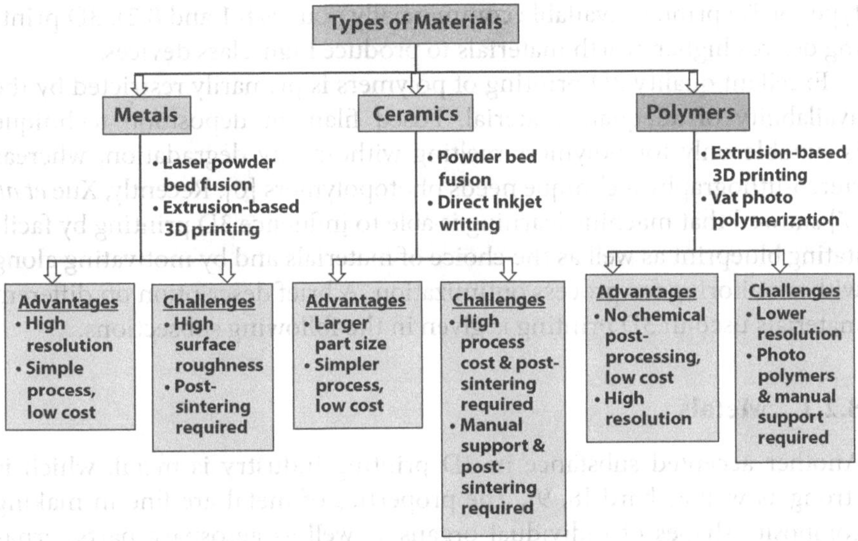

Figure 8.1 Types of materials used in 3D printing.

Table 8.1 Different types of 3D printers available based on the printing process for metals.

S. no.	Technology	Medium
1.	Digital light processing (DLP)	UV light
2.	Laminated object manufacturing (LOM)	Laser
3.	Selective laser melting (SLM)	Laser
4.	Direct metal laser sintering (DMLS)	Laser
5.	Electronic beam melting (EBM)	Electron beam
6.	Electron beam additive manufacturing (EBAM)	Electron beam

Table 8.2 Different types of 3D printers available based on the printing process for polymers.

S. no.	Technology	Medium
1.	Fused deposition modeling (FDM)	Nozzle
2.	Material jetting (MJ)	Inkjet
3.	Selective laser sintering (SLS)	Laser

types of 3D printers available commercially (Tables 8.1 and 8.2). 3D printing desires higher-worth materials to produce high-class devices.

Excellent quality 3D printing of polymers is primarily restricted by the availability of adequate material. Fused filament deposition technique is suitable only for polymers melting without any degradation, whereas stereo-lithography technique needs photopolymers [6]. Recently, Xue *et al.* [7] showed that machine learning is able to influence 3D printing by facilitating blueprint as well as the choice of materials and by motivating along with monitoring for process optimization. A brief description on different materials used in 3D printing is given in the following subsections.

8.2.1 Metals

Another accepted substance in 3D printing industry is metal, which is strong as well as hard [8, 9]. The properties of metal are fine in making composite shapes of individual organs as well as aerospace parts, ornaments, etc. A variety of metals used are as follows:

1) Stainless steel—Decorating utensils, cookware, and additional stuff that could, in the end, get in touch with liquids
2) Bronze—Used for vases and other similar goods, etc.
3) Gold—Decorated for ornaments like rings, earrings, etc.
4) Nickel—Used for printing of coins
5) Aluminum—Applied for decorating slim metal stuff
6) Titanium—Finest choice for well-built hard fittings

8.2.2 Powders

3D printers use powdered materials nowadays and melted within the printer, also disseminated in layer units with chosen breadth as well as patterns that are finished [9, 10]. There are lots of powders used in 3D printers, and the most frequent are as follows:

1) Polyamide (nylon)—Nylon's power as well as flexibility are excellent and are used for the unification of pieces as well as interlocking of parts in 3D models.
2) Alumide—This powder makes the strongest goods and is used in support of creating industrialized prototypes.

8.2.3 Polymers

Polymers are the most frequent material used for 3D printing, such as 3D-printed toys as well as domestic goods, and accessible in clear as well as in colors like green, red, yellow, etc. [10, 11]. The types of polymers used are finished by commencing with one of the resources given below:

1) Polylactic acid (PLA)—Environment-friendly PLA is prepared from sugarcane in addition to corn starch and, for that reason, are biodegradable and accessible in flexible as well as hard forms.
2) Acrylonitrile butadiene styrene (ABS)—ABS is the best option for 3D printers and available in various colors suitable for products like stickers, toys, and ornaments along with vases.
3) Polyvinyl alcohol (PVA)—It is used in low-end 3D printers, has a low price tag, and is used for provisional objects.
4) Polycarbonate (PC)—It is just used on 3D printers operating at a high temperature.

8.3 3D-Printed Polymers

Thermoplastics are melt-processable plastics having the capacity to repetitively soften on heating along with hardening on cooling and, during the procedure, retaining their built-in properties. Conventionally, processes such as material extrusion or injection molding are used to make thermoplastic components, but 3D printing technologies like FDM, SLS, etc., have made it achievable to produce complex thermoplastic parts. Some of the regular thermoplastics applied are ABS, acrylic, nylon, PC, PLA, PE, etc. Gladman *et al.* [12].

Sanchez *et al.* [13] used carbon fiber in acrylonitrile styrene acrylate (ASA) and found that 20 wt% of carbon fiber improved Young's modulus by 350% along with thermal conductivity by 500%. The addition of glass fiber makes it apt for load carrying functions and results in stiffness higher than that of aluminum [14]. Carneiro *et al.* [15] prepared glass-reinforced polymer composite using FDM to produce small parts and components.

Different AM processes are used to build up a range of components that find applications in a variety of engineering areas. Earlier, Gkartzou *et al.* [16] applied 3D printing for handing out bio-based blends of polylactic acid (PLA) by means of FDM and studied factors contributing to stress force on melt, shear rate, printing speed, extrusion temperature, etc. A brief description about the different applications of 3D printed polymers is given in the discussion below.

8.3.1 Aerospace Industry

In the USA, engineers drew on originality and sophisticated technology by making the parts of the Rover digitally, straight from computer designs, in Stratasys 3D Printer. This fused deposition modeling (FDM) method creates intricate shapes that are robust for the Martian topography [17]. For their 3D-printed portions, NASA scientists used PCABS and ABS as well as polycarbonate supplies. FDM is just a 3D printing technique which ropes lightweight durable thermoplastics for harsh end-use parts [18].

8.3.2 Architectural Work

Poly-Jet 3D printing know-how generates amazingly soft, thorough architectural models in a range of resources. For load-bearing models, FDM know-how builds sturdy parts in manufacture-grade thermoplastics [8].

8.3.3 Vehicle Industry

One of the main challenges is to lessen the time to the market place for fresh goods by sinking the design phase. To assist rally on this test, full design development is approved using fused deposition modeling (FDM) prototyping assemblies from Fortus. FDM facilitates Ducati to construct mutually conceptualized models as well as functional prototypes using polycarbonate and ABS as well as poly-phenylsulfone [8].

8.3.4 End-User Products

8.3.4.1 Consumer Electronics

Poly-Jet expertise can create models by way of an unusually slim parapet (0.6 mm or fewer), best for tiny devices amid an even finish as well as pragmatic colors such that it is impossible to differentiate from the final product.

8.3.4.2 Sporting Goods

Prototypes frequently necessitate a blend of stiff and elastic materials. Poly-Jet know-how can create prototypes by way of many materials as well as colors in sole, mechanized build such as a helmet's firm casing along with a padded core, or a couple of ski safety glasses amid colored lenses as well as rubber over molding. On the other hand, FDM works with production-grade thermoplastics to make higher-impact-strength and greater-durability parts just right for components required to withstand harsh and recurring well-designed testing.

8.3.4.3 Toys

To confine the appearance of potential products, Poly-Jet can bring excellent niceties, soft surfaces, good-natured textures, and diverse resources along with vibrant colors in solitary, automated build development [19].

8.3.4.4 Dental Industry

As an alternative of outsourcing production, "APEX Dental Milling Center" diverted to in-house 3D printing. The printed models delivered remarkably superior particulars and an exceptional surface finish [8].

8.3.4.5 *Jigs and Fixtures*

3D printing jigs as well as fittings with inkjet or FDM can assist lessen log jam for in-house machine shop and potentially useful as bridge-to-tooling resolution [19].

8.3.4.6 *Medical Industry*

The UK-based company "Design Reality" uses multi-material 3D printer to sample gas mask respirators for the Ministry of Defense as well as US Fire Services and drops the 56-day prototyping practice to hours now [19].

8.4 Assortment and Adaptation of 3D-Printed Polymers

To work in the direction of circular financial system along with small environmental footprint targets, the reuse of actual throw-away resources and their upcycling is the springboard. At this time, effort is at the use of waste material having no further distant application from energetic/burning or to employ locally accessible materials amid modest or zero shipping need as well as lower price tags, e.g., *Miscanthus* particles, wood flour, seashell powder, fruit stone flour, etc. *Miscanthus* is a quick-budding lightweight plant having first-class damping properties used for fiber casting such as to create egg cartons. For this, it has to be ragged, pulverized, and sieved to working particle allocation, and the waste product of the fiber is used for 3DP. Wood flour is the remains of mills and wood-working plants, and the short fibers are suitable for 3DP. The wood types assessed included beech, willow, maple, oak, larch, spruce, lime, etc. Seashells are abundant and used in 3DP; seashells are milled to an appropriate dimension. For 3DP, hard parts of apricot stone are crushed and sieved.

One of the most promising approaches is use of natural polymers (polysaccharides and proteins) as reinforcement [20]. These natural polymers can be extracted from plants or else obtained from biomasses or agricultural/industrial wastes. Using these natural materials as fillers for 3D printing has dominance linked to their accessibility jointly with likelihood of producing printed specimens with minor environmental impact as well as greater biodegradability. For that reason, they symbolize a "green option" for 3D printing processing.

Tanase and others [21] established that PLA on compounding with lignin gave biocomposites which performed fine in 3D printing functions with no phase partition. An X-ray analysis made known that lignin improved the crystallinity, but the mechanical properties showed downgrading.

8.5 Applications of 3D-Printed Biopolymers

Other than biomedical and related practical uses, polymer-based 3D printing processes are established in intricate prototype structures. Additionally, polymers are pooled with different reinforcements to get better mechanical properties [22].

Wang *et al.* [23] used epoxy resin and chloroprene latex-based polymer-modified 3D printed mortars as interlayer interface reinforcement materials to beat the declining outcome of water flow on interlaminar contact performances. Better interfaces are more than 200% higher than those of non-binding samples for tensile bonding strength, along with more than 180% superior to non-treated samples for shear bonding strength. The molecular dynamics computation demonstrated that the tensile strength as well as the shear strength of the system was augmented by 220 % and 200 %, respectively, due to epoxy resin.

Biomass-initiated natural polymers as strengthening agents in 3D printing processes cut the costs of the finished printed stuff, together with offering a green alternative with regards to making additional biodegradable and environment-friendly materials. Besides that, this scientific advance improves the thermal and mechanical properties of printed stuff, encouraging the re-use and offering a "second life" to numerous agro-industrial wastes [20].

A new digital cement-based manufacturing know-how by integrating 3D printing know-how by way of casting technology is the via incorporation of tension-resistant polymeric truss network (PTN), 3D fcc motivated cement-based hybrid composites (FCHCs). Experimental outcomes demonstrated that the 3D-printed PTN is able to successfully make better the tension-resistant capacity of cement mortar displaying multiple cracking damage as well as small thermal conductivity in comparison to the usual cement-based materials. Further on evaluating with simulation program, 34% heating energy saving is predicted for FCHCs building [24].

The 3D printing skill is normally extra cost-efficient and more rapid in comparison to long-established formative techniques. Advancements in 3D printing make it possible for companies to expand, commencing from designing 3D-printed prototypes to speedy manufacturing of end-products and providing special features counting minimized decentralized manufacturing with inventories, wherein routine designed objects are developed when they are required.

Intricate and sometimes contradictory necessities of biobased polymers have resulted in a rigorous move forward to develop novel techniques. The biobased polymers provide supramolecular well-designed properties to 3D ink dispersion such as better rheological characteristics and printing performance than synthetic polymers [25, 26]. As a result, the tuning properties of biopolymers provide approach for the most wanted ink flow as well as superior printability. Near the beginning, 3D-printed goods were compromised amid mechanical properties, rheological parameters, and cytocompatibility to build up biobased polymeric inks [27, 28].

To accomplish the preferred properties, supramolecular biopolymers are tuned with various methods, together with biopolymers joining together, nanoparticle amalgamation, as well as cross-linking. According to Shahbazi *et al.* [29], cross-linking is an efficient procedure to get better flow performance of ink-containing biopolymers, which can be photo-cross-linked, chemical, enzymatic, and ionic. It allows post-printing alterations of biomaterials, making enrichment of flow behavior, thermal, and mechanical parameters as well as upgrading of declaration with performance.

Utilization of lignocellulosic fibers as reinforcement in 3D printed biocomposite is based on regular fiber arrangement wherein there is sufficient stress shift amid fibers and resin. Precautions are needed so as not to deteriorate or break the reinforcing fibers throughout modification and processing [30].

Wang *et al.* [31] showed the 3D printing of cellulose nanocrystal reinforced nanocomposites with methacrylates via stereolithography, having superior thermal stability as well as mechanical strength with fine particles distribution. Yang *et al.* [32] established biomimetic anisotropic reinforcement architectures by means of 3D printing and made up artificial meniscus with multi-wall CNTs.

Blind persons cannot access graphics and images effectively, which complicates learning as well as data sharing amid normal and blind persons. As blind persons employ tactile senses to think about data, a sole data setup that can be easily visualized is required. Koone *et al.* [33] discovered that graphics can be three-dimensionally printed into tactile graphics which glow with similar video resolution via lithophane effect. Lithophane forms of micrographs, gel electropherograms, and electronic and mass spectra, in addition to textbook illustrations, could be interpreted by contact or vision at ≥79% precision. The lithophane data setup enables complete revelation of data regardless of eyesight stage.

8.6 Pros and Cons of 3D Printing

Layer-by-layer fabrication permits for better elasticity as well as ingenuity in design process. Parts can be entirely re-designed so that they are stronger in the selected areas where needed in addition to the whole becoming lighter. 3D printing considerably fastens the design and prototyping, bringing the design cycle to days or weeks compared to months. The main limitations of 3D printing comprise costly hardware along with pricey biomaterials. 3D printing is not possible for all construction schemes.

8.7 Conclusions

3D-printed biopolymers encompass numerous capable areas of probable upcoming applications. The latest 3D printing processes contain abridged time for designers along with engineers to conceptualize, generate, and exam prototypes. Some of the most potential fields consist of health applications, customized parts substitution, and customized consumer products as resources get better and expenses slide downward. Possibly the greatest part of probable development for 3D printing is in the health field, where researchers are experimenting by way of creating synthetic bones using 3D printers. Research is still ongoing, and further studies are necessary to define well again the functionality of every biopolymer as well as its part of applicability.

Acknowledgments

The author acknowledges the constant support from the director and staff of CSIR-Central Building Research Institute, Roorkee.

References

1. Gu, S.H., Nicolas, V., Lalis, A., Sathirapongsasuti, N., Yanagihara, R., Complete genome sequence and molecular phylogeny of a newfound hantavirus harbored by the Doucet's musk shrew (Crocidura douceti) in Guinea Infection. *Genet. Evol.*, 20, 118–23, 2013.
2. Valvez, S., Santos, P., Parente, J., Silva, M., Reis, P., 3D printed continuous carbon fiber reinforced PLA composites: A short review. *Procedia Struct. Integrity*, 25, 3949, 2020.

3. Dhinakaran, V., Kumar, K.P.M., Ram, P.M.B., Ravichandran, M., Vinayagamoorthy, M., A review on recent advancements in fused deposition modeling. *Mater. Today: Proc.*, 27, 752–6, 2020.
4. Sandanamsamy, L., Mogan, J., Halim, N.A., Harun, W.S.W., Kadirgama, K., Ramasamy, D., A review on 3D printing bio-based polymer composite. *IOP Conf. Ser.: Mater. Sci. Eng.*, 1078, 012031, 2021.
5. Li, Q., Kulikowski, J., Doan, D., Tertuliano, O.A., Zeman, C.J., Wang, M.M., Schatz, G.C., Gu, X.W., Mechanical nanolattices printed using nanocluster-based photoresists. *Science*, 378, 6621, 768–773, 2022.
6. Henríquez, C.M.G., Vallejos, M.A.S., Hernandez, J.R., Polymers for additive manufacturing and 4D-printing: materials, methodologies, and biomedical applications. *Prog. Polym. Sci.*, 94, 57116, 2019.
7. Xue, K., Wang, F., Suwardi, A., Han, M.Y., Teo, P., Wang, P., Ye, E., Li, Z., Loh, X.J., Biomaterials by design: Harnessing data for future development. *Mater. Today Bio*, 12, 100165, 2021.
8. Ramya, A. and Vanapalli, S.L., 3D Printing Technologies in Various Applications. *Int. J. Mech. Eng. Technol.*, 7, 3, 396–409, 2016.
9. Shinde, A.A., Patil, R.D., Dandekar, A.R., Dhawale, N.M., 3D Printing Technology, Material Used For Printing and its Applications. *Int. J. Sci. Eng. Res.*, 11, 7, 105–108, 2020.
10. Mpfou, T.P., Mawere, C., Mukosera, M., The impact and Application of 3-D Printing Technology. *Int. J. Sci. Res.*, 3, 6, 2148–2152, 2014.
11. Shahrubudin, N., Lee, C., Ramlan, R., An overview of 3D Printing Technology: Technological, Materials, and Applications, published in second international conference on sustainable materials processing and manufacturing. *Procedia Manuf.*, 35, 1286–1296, 2019.
12. Gladman, A.S., Garcia-leiner, M., Sauer-budge, A.F., Emerging polymeric materials in additive manufacturing for use in biomedical applications. *AIMS Bioeng.*, 6, 1–20, 2019.
13. Sanchez, D.M., de la Mata, M., Delgado, F.J., Casal, V., Molina, S.I., Development of carbon fiber acrylonitrile styrene acrylate composite for large format additive manufacturing. *Mater. Des.*, 191, 108577, 2020.
14. Sathishkumar, T.P., Satheeshkumar, S., Naveen, J., Glass fiber-reinforced polymer composites-A review. *J. Reinf. Plast. Compos.*, 33, 1258–1275, 2014.
15. Carneiro, O.S., Silva, A.F., Gomes, R., Materials & Design Fused deposition modeling with polypropylene. *Mater. Des.*, 83, 768–776, 2015.
16. Gkartzou, E., Koumoulos, E.P., Charitidis, C.A., Production and 3D printing processing of bio-based thermoplastic filament. *Manuf. Rev.*, 4, 1, 2017.
17. Liu, W., Li, Y., Liu, J., Niu, X., Wang, Y., Li, D., Application and performance of 3D printing. *J. Nanomater.*, 1, 1–7, 2013.
18. Yagnik, D., Fused Deposition Modeling - A Rapid Prototyping technique for Product Cycle Time Reduction cost effectively in Aerospace Applications. *IOSR J. Mech. Civ. Eng. (IOSR-JMCE)*, 1, 62–68, 2012.
19. Madhav, C.V., Kesav, R.S.H., Narayan, Y.S., Importance and Utilization of 3D Printing in Various Applications. *Int. J. Mod. Eng. Res.*, 6, 24–29, 2016.

20. Sabbatini, B., Cambriani, A., Cespi, M., Palmieri, G.F., Perinelli, D.R., Bonacucina, G., An Overview of Natural Polymers as Reinforcing Agents for 3D Printing. *Chem. Eng.*, 5, 4, 78–100, 2021.
21. Tanase, M.O., Victor, E.E., Rodriguez, A., Carrasco, G.C., Lignin: a biopolymer from forestry biomass for bio-composites and 3D Printing. *Materials*, 12, 18, 3006, 2019.
22. Alghamdi, S.S., John, S., Choudhury, N.R., Dutta, N.K., Additive Manufacturing of Polymer Materials: Progress, Promise and Challenges. *Polymers*, 13, 753, 1–39, 2021.
23. Wang, L., Tian, Z., Ma, G., Zhang, M., Interlayer bonding improvement of 3D printed concrete with polymer modified mortar: Experiments and molecular dynamics studies. *Cem. Concr. Compos.*, 110, 103571, 2020.
24. Li, L., Chen, J., Tan, X., Zhou, Y., Zhang, Y., Zhou, W., Xiao, R., Li, X., Lu, Y., Su, M., Ding, Z., Long, W., Song, J., 3D face-centered-cubic cement-based hybrid composites reinforced by tension-resistant polymeric truss network. *Autom. Constr.*, 120, 103380, 2020.
25. Hennink, W.E. and Nostrum, C.F., Novel crosslinking methods to design hydrogels. *Adv. Drug Deliv. Rev.*, 64, 223–236, 2012.
26. Shahbazi, M., Jager, H., Ahmadi, S.J., Lacroix, M., Electron beam crosslinking of alginate/nanoclay ink to improve functional properties of 3D printed hydrogel for removing heavy metal ions. *Carbohydr. Polym.*, 240, 116211, 2020.
27. Chimene, D., Kaunas, R., Gaharwar, A.K., Hydrogel bioink reinforcement for additive manufacturing: a focused review of emerging strategies. *Adv. Mater.*, 32, 1, 1902026, 2020.
28. Zeidler, H., Klemm, D., Hiller, F.B., Fritsch, S., Guen, M.J.L., Singamneni, S., 3D printing of biodegradable parts using renewable biobased materials. *Procedia Manuf.*, 21, 117–124, 2018.
29. Shahbazi, M. and Jager, H., Current Status in the Utilization of Biobased Polymers for 3D Printing Process: A Systematic Review of the Materials, Processes, and Challenges. *ACS Appl. Bio Mater.*, 4, 325–369, 2021.
30. Joffre, T., Miettinen, A., Berthold, F., Gamstedt, E.K., X-ray micro-computed tomography investigation of fiber length degradation during the processing steps of short-fiber composites. *Compos. Sci. Technol.*, 105, 127–133, 2014.
31. Wang, B., Liu, J., Chen, K., Wang, Y., Shao, Z., Three-dimensional printing of methacrylic grafted cellulose nanocrystal-reinforced nanocomposites with improved properties. *Polym. Eng. Sci.*, 60, 782–792, 2020.
32. Yang, Y., Chen, Z., Song, X., Zhang, Z., Zhang, J., Shung, K.K., Zhou, Q., Chen, Y., Biomimetic anisotropic reinforcement architectures by electrically assisted nanocomposite 3d printing. *Adv. Mater.*, 29, 1605750, 2017.
33. Koone, J.C., Dashnaw, C.M., Alonzo, E.A., Iglesias, M.A., Patero, K.S., Lopez, J.J., Zhang, A.Y., Zechmann, B., Cook, N.E., Minkara, M.S., Supalo, C.A., Wedler, H.B., Pfeffer, M.J.G., Shaw, B.F., Data for all: Tactile graphics that light up with picture-perfect resolution. *Sci. Adv.*, 8, eabq2640, 1–10, 2022.

9

Implementation of 3D-Printed Materials for Water Purification Applications

Mahrukh Aslam[1], Tanvir Shahzad[2], Sabir Hussain[2], Muhammad Waseem[2], Saima Muzammil[3], Muhammad Afzal[1] and Muhammad Hussnain Siddique[1*]

[1]*Department of Bioinformatics and Biotechnology, Government College University Faisalabad, Faisalabad, Pakistan*

[2]*Department of Environmental Sciences, Government College University Faisalabad, Faisalabad, Pakistan*

[3]*Institute of Microbiology, Government College University Faisalabad, Faisalabad, Pakistan*

Abstract

3D printing, also referred as additive manufacturing, is propelling development across a wide range of fields of study, including that of groundwater sources. The advantages of manufacturing techniques over traditional fabrication processes include its capacity to produce diverse products quickly and affordably. Science employing 3D printing for separation process, salinity, and water purification activities has exploded recently, modifying this industry. This study consists of new discoveries in 3D-printed materials and processes for water-related tasks, covering breakthroughs in module spacers, innovative purification and salinity layers, adsorbents, water restoration, solar steam supplies, catalysis, etc. The implementation of smart 3D printers with improved printing quality, reliability, pace, and material application has generated more attention and may have a significant impact on future exploration and advancements in the area. This chapter covers the area of classification of 3D printing techniques that are used in the treatment of wastewater and various applications of 3D printed materials for the removal of effluents from the polluted water.

***Keywords*:** Additive manufacturing, water purification, polymers, fabrication, membrane, spacers

**Corresponding author*: mhs1049@gmail.com

Bhasha Sharma, Balaram Pani, Shashank Shekhar and Jude A. Okolie (eds.) Sustainable 3D Printing for Innovative Biopolymer Production and Applications, (191–214) © 2025 Scrivener Publishing LLC

9.1 Introduction

Water is a basic need to sustain the life on this planet Earth. It is also a constrained and susceptible useful resource. Therefore, it is much needed to save it so that it can be available to all living things [1]. The availability of freshwater is a serious threat for the present world, and it is all because of human activities. Furthermore, 1.6 billion population of the world are facing inadequacy of water, which is approximately two-thirds of the total population [2]. There are many methods like adsorption, filtration, coagulation, and sedimentation that are being used these days for the treatment of wastewater, but these techniques have limitations as they are not cost-effective and a laborious and time taking process. To address these problems, 3D printing technology, generally known as additive manufacturing, has been used for the removal of effluents from polluted water.

Even complex designs can be made and constructed using 3D printing, which allows for the fabrication of items with seemingly infinite physical limits [3]. 3D printing technique has been classified into seven different types, which are binder jetting, sheet lamination, material jetting, material extrusion, direct energy deposition, power bed fusion, and vat photopolymerization. Prefabrication for a variety of technical and non-engineering projects using a wide spectrum of substances (e.g., polymeric, earthenware, alloys, etc.) has been made possible by the development of 3D printing technology [4]. Researchers have recently begun examining 3D-printed materials for separation processes, water recycling, and purification process applications. This is a result of the worldwide problem of scarce water resources, which calls for cooperation to find long-term remedies [5].

Applications of membrane technology (spacers, panels, and matrix manufacture) are among these alternatives, as are elements for photovoltaic absorbers/steam production, reducing agents for oil/water segregation, pigment destruction, and catalysis [6]. Different types of spacers with 3D fabrication have been used in water purification processes such as forward osmosis, membrane distillation, reverse osmosis and ultrafiltration. Researchers have developed some approaches in which they do direct printing on the porcelain and alloys, but these methods are not effective in the term of smooth fabrication.

The overall rise of scientific articles and writings in the literature clearly illustrates that curiosity in 3D printing for water treatment/purification and membrane filtration activities has been rising lately. The number of papers relating to 3D-printed materials for water-related applications has increased from fewer than 20 articles in 2010 to an average of 300 papers per year over the last 3.5 years alone. In the current chapter, we place a special emphasis on the most recent advancements and evolutions in 3D-printed components (polymeric and porcelain), with a particular emphasis on their applications in membrane separation, wastewater treatment, desalination, and water purification, particularly those that have been reported in the last three years. It is anticipated that the thorough presentation and in-depth discussion will be useful for materials scientists in two ways: first, to create innovative methods and functional materials for applications involving water purification, and second, to understand the pollutant adsorption and desorption mechanisms of 3D-based materials toward a variety of pollutants in water systems.

9.2 Classification of 3D Printing Techniques

3D printing automation is playing a crucial role not only in prototyping but also in many other industrial sectors to manufacture products on a large scale [7]. There are many methods of 3D printing depending on the nature of the primal matter and its various modeling techniques that provide layered construction [8]. The American Society for Testing and Materials (ASTM F2792) discovered about 50 various automations for 3D printing and distributed these automations to seven other major groups as shown in Figure 9.1 [4]. It includes binder jetting, directed energy deposition,

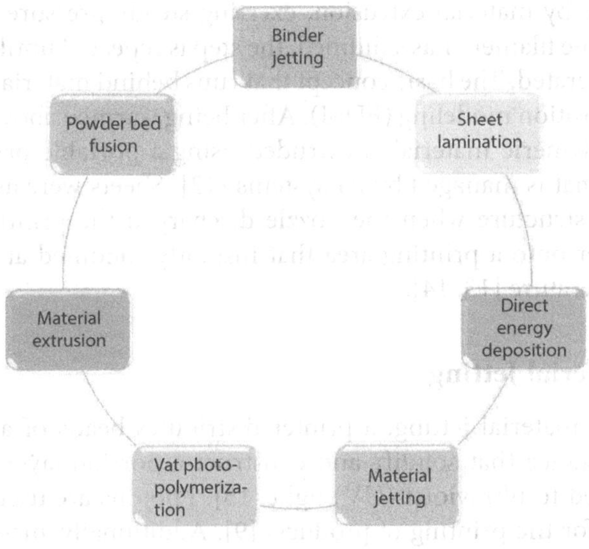

Figure 9.1 Classification of 3D printing techniques.

material extrusion, material jetting, powder bed fusion, and sheet lamination and vat photopolymerization.

9.2.1 Binder Jetting

Binder jet printing is a cheap and the easiest form of 3D printing. This technique includes two important materials for the formation of the product, which are the binder and the powder bed [9]. The binder material is normally a liquid agent which makes bonding with the powder present on the bed. Once the liquid binding material binds with the powder, the new layer or sheet of the powder is formed on the bed so that more liquid binders can attach with it. After that, the final product is jolted to separate it out from the bed and to rule out impurities. The fine quality of the product is mainly dependent on the consistent distribution of the powder on the bed to which a liquid binder can attach more stably. This method is widely used in making ceramics, metals, polymers, and hybrid. Injection casting by using stainless steel powder is the prominent example of this method [10].

9.2.2 Material Extrusions

3D printing based on material extrusion is affordable and the most commonly used method. It has a wide range of applications in food, living cell and polymer casting. Additionally, this method can create fully operational parts and components [11]. The material is driven out through an ink cartridge by material extrusion, exerting steady pressure to the base plate. Once the filament has solidified, the step is repeated until the desired model is generated. The basic concept that runs behind material extrusions is fused deposition modeling (FDM). After being warmed above glass temperature, polymeric material is extruded using a portable print (suction hose) head that is managed by AI systems [12]. Sheets were assembled to create a 3D structure when the nozzle discharged the printing product layer by layer onto a printing area that instantly solidified at a predetermined temperature [13, 14].

9.2.3 Material Jetting

When using material jetting, a printer distributes beads of a photoconductive substance that solidify and construct a portion layer upon layer while exposed to ultraviolet (UV) light [15]. Photons are used instead of normal ink for the printing of products [9]. Additionally, material jetting produces products with excellent geometric precision and an exceptionally nonporous texture. Material jetting brings luxurious printing and a large range of materials, including plastics, porcelain, composites, cell lines, and alloys.

9.2.4 Powder Bed Fusion

3D printing based on powder bed fusion requires a heating source for the melting, binding and molding of the powder material into different layers to attain the desirable geometry. The pore size of the geometry is dependent on the nature of the powder material, either it is solid-based or liquid-based. Three approaches are used for this process, which are as follows:

i. The first one is electron beam melting (EBM) in which improved power source like electron beam from an electron gun is used to warm and liquefy the powder material. Once the primary layer is formed, another sheet of powder is formed on the bed, and the cycle continues until the full structure is formed [16].
ii. The second one is selective laser sintering (SLS); in this method, a high laser beam is used on the fine layer of the powder material present on the printing surface until the full 3D structure is formed. The powder material could either be of porcelain, alloys, and metals [17].
iii. The last one is selective laser melting (SLM); this method is similar to SLS but with one difference, that is SLM uses a high laser beam or energy source that melts down the powder material on the printing area. This method is suitable to produce tough and rigid parts [18].

9.2.5 Sheet Lamination

It is the method in which the layers of the objects are joined together to form a part of a substance. There are two examples that utilize this approach, which are laminated object manufacturing (LOM) and ultrasound additive manufacturing (UAM).

i. In LOM, the separated printed materials on the bed are bound with each other by exerting pressure with the help of rollers. This method is very easy to handle, cheap, and less time consuming and develops more stable parts and structures [19].

ii. UAM method is used to generate 3D metals parts. Instead of using rollers or photons, high-range spectroscopy is employed for the binding of one layer of the metal sheet with another.

9.2.6 Vat Photopolymerization

The primary method of additive manufacturing processes that is widely employed is photopolymerization, which relates directly to fixing photo-reactive monomers using a beam, flashlight, or UV irradiation. The techniques that have been used in vat photopolymerization are stereolithography (SLA), digital light processing (DLP), and photopolymer jetting. Vat photopolymerization's critical variables include irradiation period, intensity, and source of energy.

SLA was affected by the laser activator, the specific irradiate contact circumstances, or any extra UV absorbent such as colors or tints. The working principle of SLA is that the liquid light monomers exudate on the bed and when the beam of light bombarded on monomers they become solidified, and this cycle is repeated until the whole model is fabricated [20].

DLP has a liquid crystal display panel attached to an electric lamp; the digital light method utilizes a more traditional light beam [21]. It may extend to the entire area of the photosensitive monomer vat in a quick pass, making it significantly quicker than stereolithography.

9.2.7 Direct Energy Deposition

A more complicated printing method frequently used to fix or add more stuff to current parts is known as directed energy deposition. This method is kind of similar to material extrusion, but there are some differences such as it has multidirectional portable nozzle and it is solely used for metallic blends [22]. Laser-engineered net shaping (LENS) method is used in this process which utilizes high energy source like laser to convert the metallic sheets into the molten state and then transform them into a very refined 3D model. The further correlation between these technologies are given in Table 9.1.

9.3 Applications of 3D Material for Water Purification

Additive manufacturing technique is being used by many industries like automobile, food, textile, and many more for the production of a variety

Table 9.1 Correlation between various 3D printing automations [3, 23, 24].

3D printing automation	Examples	Light source	Printing resources	Platform portability	Model shape	Limitations
Binder jetting	-	No	Powder of porcelain, alloys	Flexible	Any shape	Rough surface of 3D parts
Material extrusion	Fused deposition modeling	Beam for assembly	Liquid and powder	Flexible	Any shape	Printing of large models
Material jetting	-	Photons	Powder	-	Fixed	-
Powder bed fusion	EBM SLS SLM	Electron rays Laser Laser	Liquid Powder Powder	Flexible Flexible Flexible	Any shape Any shape Any shape	Post-processing of parts required and bad quality of model
Sheet lamination	SDL UAM	No No	Paper sheet Metallic sheet	Flexible Fixed	Flat sheet Flat sheet	Toxic fumes and waste formation
Vat photopolymerization	SLA DLP	Laser for assembly UV light	Laser Laser	Flexible Flexible	Any shape Any shape	Chances of ear damage due to ultrasonic waves
Direct energy deposition	DLW	Laser for assembly	Powder (metal)	Flexible	Any shape	Time consuming

of useful products. In recent years, 3D-printed materials are being used to treat impurities from contaminated water by various means as shown in Figure 9.2, like upgrading membrane materials, polymer membranes, progress to remove oil from water and to treat polluted water. Table 9.2 shows the various applications of 3D-printed materials for the treatment of water-related problems. This table indicated that fused deposition modeling (FDM), selective laser sintering (SLS), and stereolithography (SLA) are the major 3D printing techniques for the treatment of water.

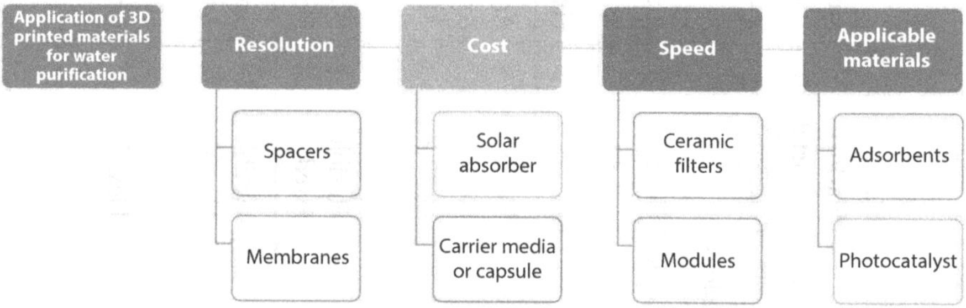

Figure 9.2 Application of 3D-printed material for water purification.

Table 9.2 Applications, techniques, and 3D materials used for water purification.

Applications	3D printing technique	3D-printed part	Material	Reference
a) **Oil water separation**				
Oil water separation	ISA 3D printing	Egg mixer; super hydrophobic geometry	E-glass with carbon nanotubes	[25]
Oil water separation	SLA	Oil skimmer mesh	-	[26]
Oil water separation	DLP	Porcelain lattice	Alumina	[27]

(*Continued*)

Table 9.2 Applications, techniques, and 3D materials used for water purification. (*Continued*)

Applications	3D printing technique	3D-printed part	Material	Reference
Oil water separation	DLP	Permeable layer	PDMS	[28]
b) **Membrane separation**				
Spacer				
UF	DLP	Spacer	Molten resin	[29]
MF	SLS	Shaking spacer	Solid polyamide	[30]
Filtration	SLS	Spacer	-	[31]
MD	SLS	Spacer	Polyamide 12	[32]
RO, UF	FDM	Spacer	Acrylonitrile butadiene styrene (ABS)	[33]
UF	SLA	Spacer	-	[34]
UF	Polyjet	Spacer	-	[35]
UF	DLP	Spacer	Acrylate monomer	[36]
Membrane				
Anion exchange	-	Anion exchange layer	Quaternized poly (DUDA-co-PEGDA-co-VBC)	[37]
RO	Electrospraying	Flat film composite layer	Metaphenylene diamine (MPD)	[38]
Gas–liquid contactors	DLP	Casted layer	Polydimethylsiloxane	[39]
RO, NF	Inkjet printing	Flat film composite layer	Fluorinated diamine	[15]

(*Continued*)

Table 9.2 Applications, techniques, and 3D materials used for water purification. (*Continued*)

Applications	3D printing technique	3D-printed part	Material	Reference
c) **Wastewater treatment**				
Moving bed biofilm reactor (MBBR)	Polyjet	Biofilter media conveyor	Molten acrylate-based monomer resin	[40]
Biofilm activator	SLS	Fullerene-based bio- carrier media	Nylon	[41]
Deterioration of drugs	Printers	Ferrate-loaded capsules	PVA	[42]
d) **Pigment adsorption**				
Rhodamine B deterioration	SLA	Monomer-coated material on 3D printers	-	[43]
MB expulsion	FDM	MOF complex	Cu-BTC-plated ABS	[44]
e) **Solar steam**				
Salt purification	Upright 3D fabrication Direct writing	Thermal evaporator like jelly fish Hybrid aerogel sheet	Porous carbon Carbon nitride (2D)	[45] [46]
Salt purification	FDM	Thermal evaporator with hollow geometry	Carbon nanotubes with graphene oxide	[47]

9.3.1 Sorbents for the Removal of Heavy Metals

The volatility of toxic substances like mercury, lead, cadmium, and copper need to be removed from freshwater. By accumulating within living organisms, these dangerous elements have a serious negative impact on human health [48]. Therefore, an effective and suitable approach has to be applied for the eradication of such harmful elements. Adsorption is the most reliable technique for this purpose as it is cheap, stable, efficient, and easy to handle. Metal ions removal through a permeable medium has been considered for use with biosorbents such biochar, carbon nanofibers, nanoparticle materials, and other permeable carbon materials. Due to its inherent benefits of affordability, efficiency, and renewability, the usage of bio-inspired substances is ubiquitous [49]. Many conventional materials like chitosan have been used to remove toxic elements from freshwater, but these materials are not advantageous because they are not easy to handle and non-recyclable.

3D-printed materials have solved these problems as they are more cost-effective, have high porosity, have more surface availability, and are renewable. 3D hydrogels and hexagonal geometry have shown great potential to eliminate metal substances from wastewater. The results of [50] showed that hexagonal pores have actively removed Cu(II) ions from the solution due to the availability of large space and holes, and its reusability is tested with ethylenediaminetetraacetic acid (EDTA) solution. Absorptive capacity near 92% was attained.

[51] reported that graphene-based 3D adsorptive materials like graphene oxide/DNA-based hydrogels are being employed for the sorption of lethal metals and toxic pigments present in untreated water. The result of [52] indicated that spongy graphene (SG) produced by hydrothermal method not only adsorbed pigments and metals from the polluted water but was also able to eliminate oil base products from it. [53] produced 3D Fe_3O_4 graphene-based nanocomposite by mixing the hydrogel of graphene with Fe_3O_4 for the sorption of arsenic like As^{+3} and As^{+5} from wastewater.

Nanofillers can also be used as sorbent material, but there is a drawback of using it—the use of polymers and high temperature deformed the 3D geometry of the functional parts of the model [54].

9.3.2 Sorbents for Dye Removal

Dyes are used by many industries including food, textile, pharmaceuticals, printing, leather, cosmetics, and plastics. They are carcinogenic and have very lethal effect on the health of living life. About 5×10^5 tons are used and about 200 to 300 m^3 colored water is released into the streams [55]. A hazardous amount of naturally bonded chlorine is present in around 40% of the pigments, and 70% of textile dyes are using it globally. Pigments are chemical by nature and have a unique color and aroma usually made from petroleum products and coal tar, and these dyes are xenobiotic in nature [56]. Adsorption is the most well-known method for dealing with dye-laden wastewater, and it has practical value in a number of ways. It is a more efficient and promising technique because of its low cost, low energy demand for the treatment process, easy process control, and versatility in the physical properties of adsorbents [57].

Adsorbents made of carbon are frequently employed to remove organic dyes; however, the segregation and regeneration process is inherently difficult. This is as a result of the elements' lack of mobility as powder [58, 59]. The 3D printing technology overcome this problem by producing conductive polymers with permeable and accessible architectures that may effectively and immediately adsorb organic colors; these materials may also be recyclable.

[44] reported adsorption of methylene blue dye by using 3D material commonly known as metal organic framework (MOF). They put the layer of powdery sportive material Cu–benzene tricarboxylic acid onto 3D-printed acrylonitrile butadiene styrene (ABS) to increase the elasticity of MOF for the better elimination of MB dye. Absorptive capacity of 93.3% to 98.3% was achieved. [60] reported that 3D GA-ZnO, which was hydrothermally produced, has the ability to degrade methylene blue dye (MB) from wastewater by up to 97.6% when a source of light is applied on it. The activity of 3D graphene-based ZnO composite to remove MB is much more than simple ZnO.

[61] developed a nanocomposite by coating graphene oxide on chitin nanofibrils substratum and established the spongy 3D structure as GO-CNF that was capable to remove 98.5% of methylene blue colorant from the wastewater and was also able to remove heavy alloys. It can be reused for several cycles without any functional discontinuity. Future work should concentrate on increasing the bond between fillers and the matrix material and upgrading the product of 3D printers (such as FDM) so that minimal or no pre-processing or blending is required.

9.3.3 Solar Absorbent for Solar Vapor Evaporation

Past developments have seen a lot of interest in solar-driven water evaporation as a reasonable alternative to the clear water scarcity. This method involves using sun energy beamed on photocatalytic elements [62]. SSE requires the following properties in its photothermal elements: spongy geometry having water repellent upper layer, minimum heat transfer, and ability for maximum adsorption in the presence of sunlight [63]. In the last 10 years, SSEs with metal compounds (like gold or anodized aluminum oxide) and activated carbons (like carbon nanotubes and graphene oxide, for example) have received a lot of attention. The key hurdle for the research is to develop scalable, easily manufacturable methods that can efficiently transform solar energy into usable heat energy. This problem is solved out by using the 3D technology, by coating photonic elements in such a manner that gives 3D printing parts with maximum efficiency.

3D printing automations like material extrusion and inkjet printing are widely used in SSE [45]. [46] used 3D printing for the formation of good light absorbing membrane. They created ink by combining gold with g-C_3N_4 (Au-CNNS) and also added sodium alginate to maintain the consistency of the media so that it can easily come out from the nozzle. The result showed that these printed parts are 2.5 times better in terms of efficiency and activity compared to normal samples. With regard graphene, due to its extremely high conductivity, stretchability, and chemical inertness, its inks are the subject of intense research towards manufactured wearable electronics [64, 65]. [66] developed a 3D-printed model-based water heating system that has the ability to warm up 30 kg of water at 50°C for a long period of time, but there is one drawback of this model—it is not cost-effective. Therefore, cheaper carbon sources are taken into consideration.

9.3.4 Crude Water Segregation

Water contamination, which contributes to climate change and ecosystems around the globe, is now a widespread problem that can be ascribed to the rise in industrial filthy wastewater discharges and spillages. Reliable crude isolation has become a top priority for preserving the environment and promoting social progress [67]. Permeable models with maximum numbers of pores are actively used for the removal of oils from water. These pores let oil pass out and clogged water molecules. 3D models mostly made up of porcelain and plastics are being used for the formation of a water resistance membrane.

[68] established a 3D-printed container inspired by a cactus plant to develop bio-inspired polydimethylsiloxane (PDMS) sponge. The porosity of this model allows PDMS sponge to absorb oil four times better than a conventional sponge. [69] developed LA/F/rGO-based hydrogel to isolate oil from water. The results demonstrated that the particles of toluene were not able to pass through the hydrogel 3D structure and, on the other side, chloroform was able to pass through it instead of water. This behavior of hydrogels was due to its amphiphobic nature. [70] developed an extreme water repellant polysulfone membrane using 3D printing sintering technique. The result showed that about 99% of its activity was achieved even after 10 repeating cycles.

9.3.5 Bio-Materials for Water Purification

Polluted water contains large numbers of effluents which are insanely toxic, and a huge amount is required for its treatment. Chemical oxidation is a general method that effectively eliminates these lethal effluents from contaminated water. The oxidation of capsulated ferrate VI has a great potential to eradicate these pollutants [71, 72]. Ferrate VI itself is not a highly stable element for the removal of effluents, so [42] did capsulation of ferrate VI with polyvinyl alcohol (PVA) by using 3D printing automation. The results showed 80% elimination of micro living creatures. The multiple steps involved in the 3D printing of the container and subsequent loading of the ferrate into the capsule are one drawback of the current strategy. This limitation is addressed by using the Polyjet approach that prints capsules with ferrate VI in just one go. Moving bed biofilm reactors (MBBRs), which enable linked and detached growth systems, are extensively applied for waste management throughout the world [73].

[40] used wastewater coming from the fish farm to remove NH_3 from it by using circular supporting media manufactured by 3D Polyjet automation and applied it in MBBR. The precise process by which the 3D-printed material drives microbial consortia and metabolism to impact reactor performance, however, was not clarified and needs more research. [74] developed an approach to get benefit from the possible high bio-carrier filling ratio in the bioreactor while simultaneously allowing a constrained range of motion for the carriers; a semi-suspended spindle-shaped bio-carrier media was created using a 3D printing procedure. Spindle-shaped bio-carriers are difficult to manufacture using traditional molding techniques because of their intricate design and structure, which is why 3D printing is used. The 3D-printed semi-suspended bio-carriers showed good development of a diversified microbial population, according to the results.

9.3.6 Photocatalytic Component

When a photonic substance is subjected to a light source with a certain wavelength, it reacts and becomes activated, oxidizing the pollutants. This process is known as heterogeneous photocatalysis [75]. Titanium dioxide (TiO_2) is an efficient photocatalytic material for the removal of effluents from polluted water. The working principle of TiO_2 is that when the beam of the light bombard on the catalyst, it generates pores; as a result, this catalyst converts the molecule of water into hydroxyl ion (OH^-) and oxide ions (O_2^-) [76]. The limitation of using standard photocatalytic substance is the available area for reaction, and this problem is addressed by 3D-printed automation in which the catalytic material is incorporated on the substrate more efficiently.

[77] developed a titanium-based photocatalytic layered structure with a density of 25 μm by using SLA 3D printing approach. The holes created by the photocatalytic reaction allowed the active removal of effluents without the need of more filters. The main advantage of using this model is that it can work in inert conditions and even at 1,000°C. [78] developed two approaches to incorporate TiO_2 with low-density polyethylene (LDPE), either (a) by mixing TiO_2 with LDPE in a warm shaker or (b) by using anionic surfactant. FDL 3D method was employed for the production of hover photocatalyst.

3D thermoplastic polymers like ABS are not easy to produce because of several problems like poor durability, not tolerant to high temperature, and less surface area, and these problems can be solved by integration of organic materials in plastic polymers but still are not applicable for photocatalytic functions. Therefore, better 3D approaches still require to overcome these issues [79].

9.3.7 Filters for Wastewater Treatment

The great cost-effectiveness and efficacy of ultrafiltration have led to the replacement of many traditional approaches to water treatment [80, 81]. Although ceramic materials are occasionally used, polymer membranes are the most common type. Phase transition, hollow fiber spinning, bending, and projection are common methods for producing polymeric membranes [82, 83]. Electrospinning is the effective method used in many research work to discharge effluents form contaminated water and to remove salt [84]. It is encouraging that membrane production will soon be possible because of 3D printing's capacity to reliably create complex structure and scale. Unfortunately, owing to quality and material limitations, 3D printing has not yet been extensively researched for the construction of continuous polymeric membranes. The substances that 3D printers can use are confined.

[85] produced water repellent and submerged superoleophobic ZIF-L adorned polyamide (PA) membrane base that was 3D-printed using SLS technology. In order to prepare the nanocomposite membrane, two stages were necessary. The first phase involved creating two different kinds of ZIF-L particles with distinctive shapes, and the subsequent phase involved depositing the ZIF-Ls on the 3D-printed PA surface. The results showed that about 99% of the membrane activity was achieved, and 24,000 LMH of oil was removed.

[86] used a continuous liquid interface production (CLIP) technique with polymer compositions, which is built on a DLP system, to create flat

membranes. This technique can print items constantly rather than layer upon layer, which accelerates printing. The researchers discussed the possibility of adopting this technique to create flat supported membrane structures for water and gas isolation by utilizing a broad range of polymeric materials. Additionally, they discussed the possibility of examining the impact of process variables on the absorption and segregation properties.

9.3.8 Spacers for Polluted Water

Feed channel spacers play a crucial role in providing steady flow, recirculation, and solvent blending for many membrane filtration processes, particularly those using spiral wound membranes (SWM) [87]. Effective spacer design is crucial because sinkholes can contribute to particle deposition, clogging, and a reduction in the mass flow rate. To reduce dead zones and assess its impact on mass transport, pressure loss, and clogging, numerous spacer arrangements with different parameters have been tried. Triply periodic minimum surfaces, multi-layered spacer structures, herringbone and helices, twisted tapes, ladders, etc., are a few examples of spacer layouts.

[36] employed DLP to produce their spacer with column patterns. They wanted to achieve the same flow channel diameter while increasing the space between the spindle and the barrier using their design. In order to serve as swirling shading structures, the row kind connectors were introduced. Their 3D-printed spacer had reduced pressure gradient and less empty space, which was supported by their research observations as well as quantitative simulation. The 3D-printed spacer with column design had a specific energy use that was two orders of magnitude smaller than the standard commercial spacer.

[88] formed a polypropylene (PP) net-type input spacer using the SLS process. When 3D printing the PP spacer, they looked into the ideal building temperature and system variables. Evaluations were made of the tensile modulus and manufacturing quality of the printed part. The Young's modulus of the PP substance was discovered to be correlated with the precision of the parameters. According to published research, feed spacer elements have recently emerged as one of the most extensively studied materials for 3D printing applications, particularly for water treatment and desalination.

9.4 Conclusion

The use of 3D printing applied in different testing and assembly procedures, particularly those involving water, has a promising possibility. It can produce practically any structure and configuration and is quick, adaptable, and effective, permitting an advanced manufacturing model. This chapter provided a summary of the most recent and innovative 3D-printed products in the field of water, including feed channel spacers, membranes, solar absorbers, bio-carriers for wastewater treatment, adsorbents for oil–water separation and heavy metal treatment, and desalination, among others. Generally speaking, the innovative domain of 3D printing as a fabrication technique is a new way of doing things in novel element design and fabrication that has enormous potential in water-related activities. Although 3D printing has made enormous strides in recent years, numerous issues related to its component, technique, affordability, post-processing requirements, and even its effects on the environment and human health still need to be resolved in order to fully realize its limitless possibilities.

References

1. Muhammad, N. *et al.*, Assessment of industrial wastewater for potentially toxic elements, human health (dermal) risks, and pollution sources: A case study of Gadoon Amazai industrial estate, Swabi, Pakistan. *J. Hazard. Mater.*, 419, 126450, Oct. 2021, doi: 10.1016/j.jhazmat.2021.126450.
2. Mekonnen, M.M. and Hoekstra, A.Y., Four billion people facing severe water scarcity. *Sci. Adv.*, 2, 2, e1500323, Feb. 2016, doi: 10.1126/sciadv.1500323.
3. Lee, J.-Y. *et al.*, The potential to enhance membrane module design with 3D printing technology. *J. Membr. Sci.*, 499, 480–490, Feb. 2016, doi: 10.1016/j.memsci.2015.11.008.
4. Lee, J.-Y., An, J., Chua, C.K., Fundamentals and applications of 3D printing for novel materials. *Appl. Mater. Today*, 7, 120–133, Jun. 2017, doi: 10.1016/j.apmt.2017.02.004.
5. Perrotta, M.L. *et al.*, An ultrathin suspended hydrophobic porous membrane for high-efficiency water desalination. *Appl. Mater. Today*, 9, 1–9, Dec. 2017, doi: 10.1016/j.apmt.2017.04.009.
6. Hu, L. and Jiang, G., 3D Printing Techniques in Environmental Science and Engineering Will Bring New Innovation. *Environ. Sci. Technol.*, 51, 7, 3597–3599, Apr. 2017, doi: 10.1021/acs.est.7b00302.
7. Wang, Y., Blache, R., Xu, X., Selection of additive manufacturing processes. *Rapid Prototyp. J.*, 23, 2, 434–447, Mar. 2017, doi: 10.1108/RPJ-09-2015-0123.
8. Zhakeyev, A., Wang, P., Zhang, L., Shu, W., Wang, H., Xuan, J., Additive Manufacturing: Unlocking the Evolution of Energy Materials. *Adv. Sci.*, 4, 10, 1700187, Oct. 2017, doi: 10.1002/advs.201700187.
9. Low, Z.-X., Chua, Y.T., Ray, B.M., Mattia, D., Metcalfe, I.S., Patterson, D.A., Perspective on 3D printing of separation membranes and comparison to

related unconventional fabrication techniques. *J. Membr. Sci.*, 523, 596–613, Feb. 2017, doi: 10.1016/j.memsci.2016.10.006.
10. Calvert, P., Inkjet Printing for Materials and Devices. *Chem. Mater.*, 13, 10, 3299–3305, Oct. 2001, doi: 10.1021/cm0101632.
11. Tofail, S.A.M., Koumoulos, E.P., Bandyopadhyay, A., Bose, S., O'Donoghue, L., Charitidis, C., Additive manufacturing: scientific and technological challenges, market uptake and opportunities. *Mater. Today*, 21, 1, 22–37, Jan. 2018, doi: 10.1016/j.mattod.2017.07.001.
12. Gnanasekaran, K. *et al.*, 3D printing of CNT- and graphene-based conductive polymer nanocomposites by fused deposition modeling. *Appl. Mater. Today*, 9, 21–28, Dec. 2017, doi: 10.1016/j.apmt.2017.04.003.
13. Alaimo, G., Marconi, S., Costato, L., Auricchio, F., Influence of meso-structure and chemical composition on FDM 3D-printed parts. *Compos. Part B Eng.*, 113, 371–380, Mar. 2017, doi: 10.1016/j.compositesb.2017.01.019.
14. Melenka, G.W., Cheung, B.K.O., Schofield, J.S., Dawson, M.R., Carey, J.P., Evaluation and prediction of the tensile properties of continuous fiber-reinforced 3D printed structures. *Compos. Struct.*, 153, 866–875, Oct. 2016, doi: 10.1016/j.compstruct.2016.07.018.
15. Badalov, S., Oren, Y., Arnusch, C.J., Ink-jet printing assisted fabrication of patterned thin film composite membranes. *J. Membr. Sci.*, 493, 508–514, Nov. 2015, doi: 10.1016/j.memsci.2015.06.051.
16. Murr, L.E. *et al.*, Metal Fabrication by Additive Manufacturing Using Laser and Electron Beam Melting Technologies. *J. Mater. Sci. Technol.*, 28, 1, 1–14, Jan. 2012, doi: 10.1016/S1005-0302(12)60016-4.
17. Hyde, J., MacNicol, M., Odle, A., Garcia-Rill, E., The use of three-dimensional printing to produce *in vitro* slice chambers. *J. Neurosci. Methods*, 238, 82–87, Dec. 2014, doi: 10.1016/j.jneumeth.2014.09.012.
18. Hwa, L.C., Rajoo, S., Noor, A.M., Ahmad, N., Uday, M.B., Recent advances in 3D printing of porous ceramics: A review. *Curr. Opin. Solid State Mater. Sci.*, 21, 6, 323–347, Dec. 2017, doi: 10.1016/j.cossms.2017.08.002.
19. Vijayavenkataraman, S., Fuh, J., Lu, W., 3D Printing and 3D Bioprinting in Pediatrics. *Bioengineering*, 4, 3, 63, Jul. 2017, doi: 10.3390/bioengineering4030063.
20. Leong, K.F., Chua, C.K., Chua, G.S., Tan, C.H., Abrasive jet deburring of jewellery models built by stereolithography apparatus (SLA). *J. Mater. Process. Technol.*, 83, 1–3, 36–47, Nov. 1998, doi: 10.1016/S0924-0136(98)00041-7.
21. Philamore, H., Rossiter, J., Walters, P., Winfield, J., Ieropoulos, I., Cast and 3D printed ion exchange membranes for monolithic microbial fuel cell fabrication. *J. Power Sources*, 289, 91–99, Sep. 2015, doi: 10.1016/j.jpowsour.2015.04.113.
22. Dilberoglu, U.M., Gharehpapagh, B., Yaman, U., Dolen, M., The Role of Additive Manufacturing in the Era of Industry 4.0. *Procedia Manuf.*, 11, 545–554, 2017, doi: 10.1016/j.promfg.2017.07.148.

23. Hajash, K., Sparrman, B., Guberan, C., Laucks, J., Tibbits, S., Large-Scale Rapid Liquid Printing. *3D Print. Addit. Manuf.*, 4, 3, 123–132, Sep. 2017, doi: 10.1089/3dp.2017.0037.
24. Tan, W.S., Suwarno, S.R., An, J., Chua, C.K., Fane, A.G., Chong, T.H., Comparison of solid, liquid and powder forms of 3D printing techniques in membrane spacer fabrication. *J. Membr. Sci.*, 537, 283–296, Sep. 2017, doi: 10.1016/j.memsci.2017.05.037.
25. Yang, Y., Li, X., Zheng, X., Chen, Z., Zhou, Q., Chen, Y., 3D-Printed Biomimetic Super-Hydrophobic Structure for Microdroplet Manipulation and Oil/Water Separation. *Adv. Mater.*, 30, 9, 1704912, Mar. 2018, doi: 10.1002/adma.201704912.
26. Yan, C., Ji, Z., Ma, S., Wang, X., Zhou, F., 3D Printing as Feasible Platform for On-Site Building Oil-Skimmer for Oil Collection from Spills. *Adv. Mater. Interfaces*, 3, 13, 1600015, Jul. 2016, doi: 10.1002/admi.201600015.
27. Chen, Z., Zhang, D., Peng, E., Ding, J., 3D-printed ceramic structures with *in situ* grown whiskers for effective oil/water separation. *Chem. Eng. J.*, 373, 1223–1232, Oct. 2019, doi: 10.1016/j.cej.2019.05.150.
28. Lv, J. *et al.*, 3D printing of a mechanically durable superhydrophobic porous membrane for oil–water separation. *J. Mater. Chem. A*, 5, 24, 12435–12444, 2017, doi: 10.1039/C7TA02202F.
29. Kerdi, S., Qamar, A., Vrouwenvelder, J.S., Ghaffour, N., Fouling resilient perforated feed spacers for membrane filtration. *Water Res.*, 140, 211–219, Sep. 2018, doi: 10.1016/j.watres.2018.04.049.
30. Tan, Y.Z. *et al.*, Enhancing fouling mitigation of submerged flat-sheet membranes by vibrating 3D-spacers. *Sep. Purif. Technol.*, 215, 70–80, May 2019, doi: 10.1016/j.seppur.2018.12.085.
31. Balster, J., Punt, I., Stamatialis, D., Wessling, M., Multi-layer spacer geometries with improved mass transport. *J. Membr. Sci.*, 282, 1–2, 351–361, Oct. 2006, doi: 10.1016/j.memsci.2006.05.039.
32. Thomas, N., Sreedhar, N., Al-Ketan, O., Rowshan, R., Abu Al-Rub, R.K., Arafat, H., 3D printed spacers based on TPMS architectures for scaling control in membrane distillation. *J. Membr. Sci.*, 581, 38–49, Jul. 2019, doi: 10.1016/j.memsci.2019.03.039.
33. Shrivastava, A., Kumar, S., Cussler, E.L., Predicting the effect of membrane spacers on mass transfer. *J. Membr. Sci.*, 323, 2, 247–256, Oct. 2008, doi: 10.1016/j.memsci.2008.05.060.
34. Liu, J., Iranshahi, A., Lou, Y., Lipscomb, G., Static mixing spacers for spiral wound modules. *J. Membr. Sci.*, 442, 140–148, Sep. 2013, doi: 10.1016/j.memsci.2013.03.063.
35. Fritzmann, C., Hausmann, M., Wiese, M., Wessling, M., Melin, T., Microstructured spacers for submerged membrane filtration systems. *J. Membr. Sci.*, 446, 189–200, Nov. 2013, doi: 10.1016/j.memsci.2013.06.033.

36. Ali, S.M. *et al.*, Energy efficient 3D printed column type feed spacer for membrane filtration. *Water Res.*, 164, 114961, Nov. 2019, doi: 10.1016/j.watres.2019.114961.
37. Seo, J., Kushner, D.I., Hickner, M.A., 3D Printing of Micropatterned Anion Exchange Membranes. *ACS Appl. Mater. Interfaces*, 8, 26, 16656–16663, Jul. 2016, doi: 10.1021/acsami.6b03455.
38. Chowdhury, M.R., Steffes, J., Huey, B.D., McCutcheon, J.R., 3D printed polyamide membranes for desalination. *Science*, 361, 6403, 682–686, Aug. 2018, doi: 10.1126/science.aar2122.
39. Femmer, T., Kuehne, A.J.C., Torres-Rendon, J., Walther, A., Wessling, M., Print your membrane: Rapid prototyping of complex 3D-PDMS membranes via a sacrificial resist. *J. Membr. Sci.*, 478, 12–18, Mar. 2015, doi: 10.1016/j.memsci.2014.12.040.
40. Elliott, O. *et al.*, Design and Manufacturing of High Surface Area 3D-Printed Media for Moving Bed Bioreactors for Wastewater Treatment. *J. Contemp. Water Res. Educ.*, 160, 1, 144–156, Apr. 2017, doi: 10.1111/j.1936-704X.2017.03246.x.
41. Dong, Y. *et al.*, A Novel Bio-carrier Fabricated Using 3D Printing Technique for Wastewater Treatment. *Sci. Rep.*, 5, 1, 12400, Dec. 2015, doi: 10.1038/srep12400.
42. Czölderová, M. *et al.*, 3D printed polyvinyl alcohol ferrate(VI) capsules: Effective means for the removal of pharmaceuticals and illicit drugs from wastewater. *Chem. Eng. J.*, 349, 269–275, Oct. 2018, doi: 10.1016/j.cej.2018.05.089.
43. Figuerola, A. *et al.*, Metal–organic framework mixed-matrix coatings on 3D printed devices. *Appl. Mater. Today*, 16, 21–27, Sep. 2019, doi: 10.1016/j.apmt.2019.04.011.
44. Wang, Z., Wang, J., Li, M., Sun, K., Liu, C., Three-dimensional Printed Acrylonitrile Butadiene Styrene Framework Coated with Cu-BTC Metal-organic Frameworks for the Removal of Methylene Blue. *Sci. Rep.*, 4, 1, 5939, May 2015, doi: 10.1038/srep05939.
45. Li, Y. *et al.*, Graphene oxide-based evaporator with one-dimensional water transport enabling high-efficiency solar desalination. *Nano Energy*, 41, 201–209, Nov. 2017, doi: 10.1016/j.nanoen.2017.09.034.
46. He, P. *et al.*, Patterned Carbon Nitride-Based Hybrid Aerogel Membranes via 3D Printing for Broadband Solar Wastewater Remediation. *Adv. Funct. Mater.*, 28, 29, 1801121, Jul. 2018, doi: 10.1002/adfm.201801121.
47. Li, Y. *et al.*, 3D-Printed, All-in-One Evaporator for High-Efficiency Solar Steam Generation under 1 Sun Illumination. *Adv. Mater.*, 29, 26, 1700981, Jul. 2017, doi: 10.1002/adma.201700981.
48. Hu, Y., Liu, X., Bai, J., Shih, K., Zeng, E.Y., Cheng, H., Assessing heavy metal pollution in the surface soils of a region that had undergone three decades of intense industrialization and urbanization. *Environ. Sci. Pollut. Res.*, 20, 9, 6150–6159, Sep. 2013, doi: 10.1007/s11356-013-1668-z.

49. Appuhamillage, G.A. *et al.*, A biopolymer-based 3D printable hydrogel for toxic metal adsorption from water. *Polym. Int.*, 68, 5, 964–971, May 2019, doi: 10.1002/pi.5787.
50. Zhang, D., Xiao, J., Guo, Q., Yang, J., 3D-printed highly porous and reusable chitosan monoliths for Cu(II) removal. *J. Mater. Sci.*, 54, 8, 6728–6741, Apr. 2019, doi: 10.1007/s10853-019-03332-y.
51. Xu, Y., Wu, Q., Sun, Y., Bai, H., Shi, G., Three-Dimensional Self-Assembly of Graphene Oxide and DNA into Multifunctional Hydrogels. *ACS Nano*, 4, 12, 7358–7362, Dec. 2010, doi: 10.1021/nn1027104.
52. Bi, H. *et al.*, Spongy Graphene as a Highly Efficient and Recyclable Sorbent for Oils and Organic Solvents. *Adv. Funct. Mater.*, 22, 21, 4421–4425, Nov. 2012, doi: 10.1002/adfm.201200888.
53. Guo, L., Ye, P., Wang, J., Fu, F., Wu, Z., Three-dimensional Fe3O4-graphene macroscopic composites for arsenic and arsenate removal. *J. Hazard. Mater.*, 298, 28–35, Nov. 2015, doi: 10.1016/j.jhazmat.2015.05.011.
54. Valino, A.D., Dizon, J.R.C., Espera, A.H., Chen, Q., Messman, J., Advincula, R.C., Advances in 3D printing of thermoplastic polymer composites and nanocomposites. *Prog. Polym. Sci.*, 98, 101162, Nov. 2019, doi: 10.1016/j.progpolymsci.2019.101162.
55. Akhtar, M.M., Mohammad, A.D., Ehsan, M., Akhtar, R., J. ur Rehman, Z., Manzoor, Water resources of Balochistan, Pakistan—a review. *Arab. J. Geosci.*, 14, 4, 289, Feb. 2021, doi: 10.1007/s12517-021-06502-y.
56. Crini, G., Non-conventional low-cost adsorbents for dye removal: A review. *Bioresour. Technol.*, 97, 9, 1061–1085, Jun. 2006, doi: 10.1016/j.biortech.2005.05.001.
57. Ali, R.M., Hamad, H.A., Hussein, M.M., Malash, G.F., Potential of using green adsorbent of heavy metal removal from aqueous solutions: Adsorption kinetics, isotherm, thermodynamic, mechanism and economic analysis. *Ecol. Eng.*, 91, 317–332, Jun. 2016, doi: 10.1016/j.ecoleng.2016.03.015.
58. Liu, Z., Zhou, X., Liu, C., N-doped porous carbon material prepared via direct ink writing for the removal of methylene blue. *Diam. Relat. Mater.*, 95, 121–126, May 2019, doi: 10.1016/j.diamond.2019.04.010.
59. Wang, Y., Ling, X., Wang, Y., Zhang, J., Probing the effect of doped F and N on the structures and properties of fullerene-like hydrogenated carbon films. *Diam. Relat. Mater.*, 79, 32–37, Oct. 2017, doi: 10.1016/j.diamond.2017.08.015.
60. Mei, J.-Y., Qi, P., Wei, X.-N., Zheng, X.-C., Wang, Q., Guan, X.-X., Assembly and enhanced elimination performance of 3D graphene aerogel-zinc oxide hybrids for methylene blue dye in water. *Mater. Res. Bull.*, 109, 141–148, Jan. 2019, doi: 10.1016/j.materresbull.2018.09.034.
61. Ma, Z. *et al.*, Graphene oxide/chitin nanofibril composite foams as column adsorbents for aqueous pollutants. *Carbohydr. Polym.*, 144, 230–237, Jun. 2016, doi: 10.1016/j.carbpol.2016.02.057.

62. Wang, P., Emerging investigator series: the rise of nano-enabled photothermal materials for water evaporation and clean water production by sunlight. *Environ. Sci. Nano*, 5, 5, 1078–1089, 2018, doi: 10.1039/C8EN00156A.
63. Jun, Y.-S., Wu, X., Ghim, D., Jiang, Q., Cao, S., Singamaneni, S., Photothermal Membrane Water Treatment for Two Worlds. *Acc. Chem. Res.*, 52, 5, 1215–1225, May 2019, doi: 10.1021/acs.accounts.9b00012.
64. Kamyshny, A. and Magdassi, S., Conductive Nanomaterials for Printed Electronics. *Small*, 10, 17, 3515–3535, Sep. 2014, doi: 10.1002/smll.201303000.
65. Torrisi, F. *et al.*, Inkjet-Printed Graphene Electronics. *ACS Nano*, 6, 4, 2992–3006, Apr. 2012, doi: 10.1021/nn2044609.
66. Zhang, P. *et al.*, Three-dimensional water evaporation on a macroporous vertically aligned graphene pillar array under one sun. *J. Mater. Chem. A*, 6, 31, 15303–15309, 2018, doi: 10.1039/C8TA05412F.
67. Zhang, L., Li, H., Lai, X., Su, X., Liang, T., Zeng, X., Thiolated graphene-based superhydrophobic sponges for oil-water separation. *Chem. Eng. J.*, 316, 736–743, May 2017, doi: 10.1016/j.cej.2017.02.030.
68. Shin, J.H., Heo, J.-H., Jeon, S., Park, J.H., Kim, S., Kang, H.-W., Bio-inspired hollow PDMS sponge for enhanced oil–water separation. *J. Hazard. Mater.*, 365, 494–501, Mar. 2019, doi: 10.1016/j.jhazmat.2018.10.078.
69. Hu, W. *et al.*, An amphiphobic graphene-based hydrogel as oil-water separator and oil fence material. *Chem. Eng. J.*, 353, 708–716, Dec. 2018, doi: 10.1016/j.cej.2018.07.147.
70. Yuan, S., Strobbe, D., Kruth, J.-P., Van Puyvelde, P., Van der Bruggen, B., Super-hydrophobic 3D printed polysulfone membranes with a switchable wettability by self-assembled candle soot for efficient gravity-driven oil/water separation. *J. Mater. Chem. A*, 5, 48, 25401–25409, 2017, doi: 10.1039/C7TA08836A.
71. Rai, P.K. *et al.*, A critical review of ferrate(VI)-based remediation of soil and groundwater. *Environ. Res.*, 160, 420–448, Jan. 2018, doi: 10.1016/j.envres.2017.10.016.
72. Yuan, B., Chen, Y., Fu, M.-L., Degradation efficiencies and mechanisms of trichloroethylene (TCE) by controlled-release permanganate (CRP) oxidation. *Chem. Eng. J.*, 192, 276–283, Jun. 2012, doi: 10.1016/j.cej.2012.03.049.
73. Biswas, K., Taylor, M.W., Turner, S.J., Successional development of biofilms in moving bed biofilm reactor (MBBR) systems treating municipal wastewater. *Appl. Microbiol. Biotechnol.*, 98, 3, 1429–1440, Feb. 2014, doi: 10.1007/s00253-013-5082-8.
74. Tang, B., Zhao, Y., Bin, L., Huang, S., Fu, F., Variation of the characteristics of biofilm on the semi-suspended bio-carrier produced by a 3D printing technique: Investigation of a whole growing cycle. *Bioresour. Technol.*, 244, 40–47, Nov. 2017, doi: 10.1016/j.biortech.2017.07.132.
75. Hernández-Afonso, L. *et al.*, Ceramic-Based 3D Printed Supports for Photocatalytic Treatment of Wastewater. *J. Chem.*, 2017, 1–9, 2017, doi: 10.1155/2017/7602985.

76. Kudo, A. and Miseki, Y., Heterogeneous photocatalyst materials for water splitting. *Chem. Soc. Rev.*, 38, 1, 253–278, 2009, doi: 10.1039/B800489G.
77. Vyatskikh, A., Kudo, A., Delalande, S., Greer, J.R., Additive manufacturing of polymer-derived titania for one-step solar water purification. *Mater. Today Commun.*, 15, 288–293, Jun. 2018, doi: 10.1016/j.mtcomm.2018.02.010.
78. Martín de Vidales, M.J. *et al.*, 3D printed floating photocatalysts for wastewater treatment. *Catal. Today*, 328, 157–163, May 2019, doi: 10.1016/j.cattod.2019.01.074.
79. Zhou, X. and Liu, C., Three-dimensional Printing for Catalytic Applications: Current Status and Perspectives. *Adv. Funct. Mater.*, 27, 30, 1701134, Aug. 2017, doi: 10.1002/adfm.201701134.
80. Werber, J.R., Osuji, C.O., Elimelech, M., Materials for next-generation desalination and water purification membranes. *Nat. Rev. Mater.*, 1, 5, 16018, May 2016, doi: 10.1038/natrevmats.2016.18.
81. Ying, Y. *et al.*, Recent advances of nanomaterial-based membrane for water purification. *Appl. Mater. Today*, 7, 144–158, Jun. 2017, doi: 10.1016/j.apmt.2017.02.010.
82. Woo, Y.C. *et al.*, Hierarchical Composite Membranes with Robust Omniphobic Surface Using Layer-By-Layer Assembly Technique. *Environ. Sci. Technol.*, 52, 4, 2186–2196, Feb. 2018, doi: 10.1021/acs.est.7b05450.
83. Yao, M. *et al.*, Improving membrane distillation performance: Morphology optimization of hollow fiber membranes with selected non-solvent in dope solution. *Chemosphere*, 230, 117–126, Sep. 2019, doi: 10.1016/j.chemosphere.2019.05.049.
84. Yao, M. *et al.*, Effect of heat-press conditions on electrospun membranes for desalination by direct contact membrane distillation. *Desalination*, 378, 80–91, Jan. 2016, doi: 10.1016/j.desal.2015.09.025.
85. Yuan, S. *et al.*, Structure architecture of micro/nanoscale ZIF-L on a 3D printed membrane for a superhydrophobic and underwater superoleophobic surface. *J. Mater. Chem. A*, 7, 6, 2723–2729, 2019, doi: 10.1039/C8TA10249J.
86. Lin, W. *et al.*, Enhanced continuous liquid interface production with track-etched membrane. *Rapid Prototyping J.*, 25, 1, 117–125, Jan. 2019, doi: 10.1108/RPJ-12-2017-0251.
87. Da Costa, A.R., Fane, A.G., Wiley, D.E., Spacer characterization and pressure drop modelling in spacer-filled channels for ultrafiltration. *J. Membr. Sci.*, 87, 1–2, 79–98, Feb. 1994, doi: 10.1016/0376-7388(93)E0076-P.
88. Tan, W.S., Chua, C.K., Chong, T.H., Fane, A.G., Jia, A., 3D printing by selective laser sintering of polypropylene feed channel spacers for spiral wound membrane modules for the water industry. *Virtual Phys. Prototyping*, 11, 3, 151–158, Jul. 2016, doi: 10.1080/17452759.2016.1211925.

10

Future Perspectives and Challenges in Employing 3D-Assisted Biopolymers

Damanpreet Kaur and Anupreet Kaur*

Department of Biotechnology, University Institute of Engineering and Technology, Panjab University, Chandigarh, India

Abstract

Over the last decade, 3D-assisted bioprinting has enticed the researchers in the domain of biological scaffolds as well as tissue engineering, owing to their superior performance characteristics, ease of fabrication, and designability with precision. With these features, these 3D-printed constructs have been able to bridge the gap among the engineered and the natural tissues. In the article, insight of various aspects of the process is provided at first with an overview of biopolymers. Next, several future applications based on the 3DBP has been focused by giving an idea through important applications. This is followed by a discussion on inherent limitations or challenges while working with 3D-assisted biopolymers.

Keywords: 3D-assisted bioprinting, biopolymers, biological scaffolds, electrospinning, fabrication, nanofibers, nanomaterials, additive manufacturing

10.1 Introduction

3D printing (3DP) and nanotechnology has provided a pathway to create novel designs and application techniques. While 3D printing technologies can be used to obtain nanosized structures that cannot be easily materialized by other available methods, nanomaterials are utilized to enhance the features of printed materials such as electrical conductivity and sensing properties. Most of the general manufacturing techniques in materials engineering had relied on "top-down" methodologies. The requested

**Corresponding author*: anupreet410@gmail.com

Bhasha Sharma, Balaram Pani, Shashank Shekhar and Jude A. Okolie (eds.) Sustainable 3D Printing for Innovative Biopolymer Production and Applications, (215–232) © 2025 Scrivener Publishing LLC

shape is obtained via size reduction of the initial bulk material, thereby generating waste material as most of the bulk material is rendered in useless shapes. On the contrary, the "bottom-up" technique can easily address this problem [1, 2]. In this approach, the desired material is accomplished through the addition of the components in a sequential manner with substantial waste reduction. 3D printing, also referred to as additive manufacturing (AM), is basically the construction of physical objects in 3D from digital models. This is achieved through successive layer-by-layer build-up of materials such as metals, ceramics, plastic, or even living cells. Besides saving on waste material generation, 3D printing aids in the production of fine structures with absolute precision. Another compelling feature of 3D printing is design flexibility. Not only does this offer the designers with the freedom to create detailed structures owing to ease of operation but also this exhibits minimum physical challenges as in conventional techniques. 3D printing is broadly based on the utilization of polymers, plastics, and ceramic materials [3, 4].

At present, there exists several different 3DP or AM technologies classified according to ISO/ASTM standard, including the material extrusion, jetting, binder jetting, direct energy deposition, vat photo-polymerization, powder bed fusion, and sheet lamination. Depending on the printing technology and the manner of inks formulation, objects could be fabricated in a consecutive layer-by-layer fashion by dispensing the material through an extruder using chemical agents such as binder or a laser sintering/melting.

A three-dimensional object of almost any shape can be created using the 3D printing process from a digital model. The AM technique used in 3D printing involves laying down consecutive material layers in various configurations. With applications in the fields of industrial design, automated production, aerospace, architecture, the medical industry, food, and tissue engineering, 3D printing technology is utilized for both prototype as well as dispersed manufacturing [5–7].

Contemporary breakthroughs in 3D-printed ceramic membranes have gathered much recognition from researchers owing to their ambidexterity. Genesis of ceramic membranes using the AM approach has proven to be one of the recent techniques in ceramic membrane targeting water purification applications. With AM technique, ceramic membranes of various shapes, forms, and geometrically complex designs can be developed, exhibiting improved mass transfer and shear flow across the membrane. The same is anticipated to be ameliorated via the emerging 4D printing [8].

The article reviews recent AM-based technology and the utilities. The crux is that the cost of the downstream processes cannot be brought down due to the expensive basic setup compared to other available standard techniques. Apparently, the 3DP technique is a cost-intensive protocol for the production of ceramic membranes.

10.2 Principle of 3D Printing Technology

In 3DP, the virtual models from CAD or modeling software are "sliced" into desired digital cross-sections to be used by the machine as a guided procedure to print objects one at a time. Unless the layer deposition is finished, the material or a binding substance is deposited on the build bed or platform until the completion of the 3D model. Subsequently, the machine reads a stl file containing the design sequence and applies consecutive layers of solution, powder, or other materials to manufacture the 3D model from a series of cross-sections. Once the printing is finished, the supports could be removed so as to obtain the finished product. Furthermore, some selective 3D-printed objects might demand post-processing treatments or additions [9].

A myriad additive processes are available for 3D printing, such as stereolithography (SLA), selective laser sintering (SLS), direct metal laser sintering (DMLS), and fused deposition modeling (FDM). These can be differentiated on the basis of the materials used and in the method of layer deposition to create components. The commonly used processes for biopolymer processes are SLA, SLS, and FDM [9, 10].

A carbon dioxide (CO_2) laser beam is used in SLS to write the slices. According to the digital slices, selective sintering is applied to ensure that the slice geometries after the laser beam are scanned onto the layers of powder. The laser beam's interaction with the powder enables it to heat up to its melting point with the subsequent fusion of the particles together to create a solid mass. The article is taken out of the powder material and cleaned of any powder adhered to its walls during the last stage of the SLS process.

As per the sliced CAD model, thin successive layers in SLA are photocrosslinked by ultraviolet (UV) or visible light, leading to photopolymerization. Reactive or nonreactive diluents are commonly utilized as the method demands a liquid photo cross-linkable resin. Nevertheless, SLA has the ability to produce a large number of 3D designs that are widely different with a possibility to create nonlinear scaffold geometries offering a fine-tuning over the product's microstructure and geometry.

Intriguingly, FDM prints one layer at a time, with the material being directly deposited on the required surface. Each layer has a cylindrical, coiled shape owing to the extrusion through a nozzle. A small temperature-controlled nozzle called extruder is employed to push the thermoplastic filament and subsequent layer-by-layer deposition of semi-molten polymer onto a substrate. The subsequent layer is deposited after each completed layer as the foundation platform is lowered [11, 12].

A promising scaffold should possess an interconnected porous structure to provide sufficient space for the cells' uniform distribution and the delivery of oxygen and nutrients as well as partly mimic the topology and biological functions of the extracellular matrix (ECM). Electrospun nanofibers have an incredibly high surface-to-volume ratio, variable malleability, and porosity to adapt to a broad range of sizes and forms. Eelectrospinning combines elements of electro-spraying with traditional solution dry spinning of fibers. To manufacture solid threads from solution, no coagulation chemistry or high temperatures are required. Electrospinning is the method of spinning fibers with the assistance of electric forces. Recently, it was demonstrated that the electrospinning method is capable of creating fibers in the submicron range (Figure 10.1). Electrospinning has gained popularity in the previous decade not only because of its versatility in spinning a wide range of polymeric fibers but also because of its consistency in creating fibers in the submicron range. Fibers with dimensions less than 100 nm are commonly referred to as nanofibers in fiber science literature [13].

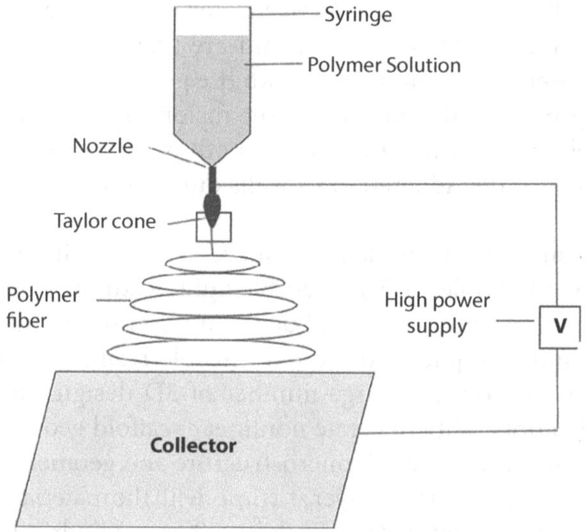

Figure 10.1 Components of the electrospinning procedure.

10.3 Biopolymers

Biopolymers are complex molecular moieties having defined 3D shapes and structures in contrast to synthetic polymers that have a profound random structure. This feature renders biopolymer molecules active *in vivo*. Biopolymers are covalently bonded monomeric moieties assembled into bigger molecules to generate polymeric biomolecules. These are biodegradable substances created by living organisms. The term "biopolymer" can be used to cover a broad range of substances that are often obtained from biological sources.

Biopolymers can be divided based on the nature of the constitutional unit they are made of—for instance, polysaccharides—sugars (e.g., cellulose found in plants), proteins—amino acids (e.g., myoglobin found in muscle tissues), and nucleic acids—nucleotides (DNA, genetic material). Based on their applications, these can be engaged as bioplastics, bioadhesive, biosurfactant, biodetergent, biofloculant, and so on [14]. Some of the major applications of biopolymers are summarized in Table 10.1.

Table 10.1 Applications and merits of some common biopolymers.

Biopolymers	Area of applications	Pros
Alginate	Cartilage tissue engineering as alginate hydrogels	Easy gelation of divalent cations
Chitosan	Cartilage tissue engineering as chitosan sponges, cell delivery	Produce enzymatically degradable soluble derivatives through chemical modification
Cellulose	Cell encapsulation, lysozyme and oral insulin delivery, 3D cell culturing	Ease of modification of hydrophilic groups to fabricate hydrogels
Hyaluronic acid	Scaffold for tissue engineering for treating osteochondral defects	Soluble in water, enzymatically degradable
Starch	Tissue engineering	Ease of functionalization, can be composited with synthetic polymers to incorporate mechanical strength

The most common renewable as well as biodegradable biopolymer universally known is cellulose. It is a homogenous polysaccharide made up of linear -(14)-glucan with intra- and intermolecular hydrogel networks. Altogether a new avenue for cellulose-based entities as a new class of raw materials for 3D printing procedures is prevalent. The cellulose nanofibrils (CNF) hydrogel is the natural cellulose-based 3D printing substance that is currently being sought the most. Since the CNF hydrogel-based inks are biocompatible and can replicate the extracellular matrix (ECM) milieu, they lately have been taken into consideration for use in a plethora of biomedical applications [15].

10.4 Applications

10.4.1 Medical and Pharmaceutical Products

3D printing has multi-array benefits for medical and pharmaceutical domains owing to its capacity to construct patient-focused and tailor-made medical devices, equipment, and medications. Contemporary research interests in the biomedical and pharmaceutical products through 3DP mainly include organ and tissue fabrication, customization of prostheses, implants, and anatomical models, personalized drug dosage forms, drug screening, and delivery in pharmaceutical research [16–20]. This is presented in Figure 10.2. Primarily, these could be developed from inkjet-based printing, extrusion-based printing particle-fused 3DP methodologies.

Wound Healing

Innovation in the field of wound therapy for the development of adequate wound dressing and skin regeneration materials is the need of the hour. Chitosan is effective in healing wounds. 3DP has made it simpler to precisely manage the spatial distribution and geometry of 3D chitosan hydrogel constructs [17].

Treatment of Cardiovascular Disease

For the aged population, cardiovascular disorders typically lead to a complicated etiology linked with the deterioration of vascular function. *In vitro* vascular tissue models must be developed immediately to address the vascularization problem. Vascular tissues made of collagen and sodium alginate have been printed using 3DP techniques via extrusion or inkjet printing. Such tissues can be used in bone fracture, bone pleats, ligaments,

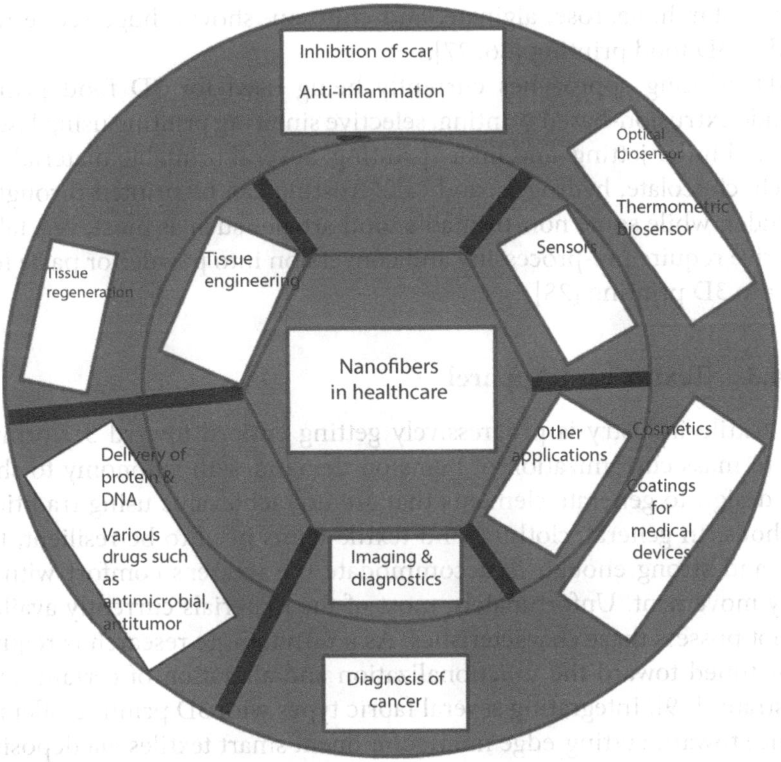

Figure 10.2 Applications of 3D biomaterials in the healthcare sector.

tendons, muscles, biosensors such as respiration sensors, cardiac sensors, sweat sensors, and blood pressure sensors [21–25].

10.4.2 Food Fabrication

The materials used for 3D food printing have to be edible, easily printable, and robust enough to hold their own weight as well as the weight of subsequently deposited layers without deforming or changing their shape. Besides that, these should be resistant to post-processing methods like baking, cooking, or frying. Guiding the physical, chemical, structural, rheological, and mechanical properties of the materials is the key to ensure a printing process with precision.

Along with the above-mentioned requisites, high abundance and bioactivity, a variety of naturally derived biopolymers, viz-a-viz, cellulose, pectin, starch, agarose, alginate, and chitosan, show a huge scope to be used in 3D food printing [26, 27].

3D printing approaches currently being used for 3D food printing include extrusion-based printing, selective sintering printing using laser or hot air, binder jetting, and inkjet printing. Several printable materials like starch, chocolate, hydrogels, and cake frosting can be printed through an extruder, while some non-printable food articles such as meat, vegetables, and rice require pre-processing and conversion into powder or paste form prior to 3D printing [28].

10.4.3 Textile and Apparel

The textile industry is progressively getting enticed toward 3D printing due to mass customization of items on demand with autonomy to shape and design to generate elements that are not achievable using traditional methods. In general, clothing and textile items need to be resilient, flexible, and strong enough to accommodate the wearer's comfort with the body movement. Unfortunately, most of the materials currently available do not possess these characteristics. As a result, more research is required to be tuned toward the functionalization and alteration of certain textile substrates [29]. Integrating several fabric types with 3D printing offers the chance toward cutting-edge multicomponent smart textiles via depositing intelligent materials or functional polymers onto textiles without surrendering the fabric's quality and flexibility.

In a more recent research, smart textiles with wearable electronics integrated with heart rate monitoring, high-end communications for soldier, firefighter, and astronaut protection, innovative sportswear with body temperature regulation, smart bandages, and virtual reality gloves are just a few exemplars of the many applications of 3D-printed textiles [30].

10.4.4 Effective Wastewater Treatment

Asian countries like China and India are growing in population and need their water framework to keep up. Large-scale conventional wastewater treatment could be expensive; therefore, cost-cutting 3D printing technology holds the potential to be revolutionary. Discharge of untreated water comprised of suspended solids, pathogenic bacteria, biodegradable organics, and nutrients can be harmful to the environment, making technological advancements in wastewater curing much more critical as water scarcity problems surge [31–34].

Components like pumps or items such as feed spacers are the areas where 3D printing can create quality and low-cost alternatives to traditional technologies.

Prolonged exposure to inorganic arsenic such as arsenite poses a serious threat to humans than arsenate and may even cause chronic arsenic poisoning, resulting in human diseases, including cancer, in the lung, bladder, liver, skin, kidney, and cardiovascular system. Arsenic confiscation via adsorption is potentially significant for small-scale water treatment due to its simplicity, cost-effectiveness, and ease of use [31]. Adsorption filtration is one of the adsorption processes for removing inorganic arsenic substances from water and is often combined with other technologies such as coagulation–flocculation, oxidation, and microbial sanitation. Adsorption filtration for arsenic removal is a convenient and simple after-treatment technique, making it suitable for routine application. Moreover, the adsorptive filtration system is reusable and offers many advantages, especially in terms of time and cost of operation. Therefore, iron oxide can be a very efficient alternative for domestic water treatment when used as an adsorbent in a filtration system. Iron oxide-coated sand, which consists of sand grains as a core and iron oxide as an outer layer for arsenic immobilization and removal, is a widely known adsorption medium for arsenic removal.

3D printing provides an energy-efficient technology that holds the potential to sequester arsenic from water at a minimal cost and that can be operated without electricity.

The versatility of 3DP allows filter properties such as size and shape to be easily changed according to the needs of any particular contaminated site. The filtration system comprised of a 3D-printed polylactic acid (PLA) filter coated with iron oxide to remove arsenite and width-controlled square channels to have the control over the water flow. This new approach has enabled the production of small, inexpensive, tailored water filtration systems that can perform without the need for electricity [35–38].

10.5 Nanofibers for Chiral Resolution

Currently, chiral medications make up about 50% of all pharmaceuticals in the market, with racemates making up the majority of these drugs. Despite the fact that enantiomers have a similar chemical structure, differences in biological activities such as toxicology, metabolism, pharmacology, and pharmacokinetics have been noted due to the isomers of the chiral drugs. Therefore, chiral separation is critical in order to eliminate the unnecessary isomer from a preparation and ensure the safety of drugs for its consumers [39]. Several drugs are still marketed in racemic form, such as β-blockers (carvedilol, atenolol, metoprolol, etc.), calcium channel antagonists (nimodipine, verapamil, felodipine, etc.), ACE inhibitors (benazepril, captopril, imidapril, etc.), hypnotics (thiopental, hexobarbital, pentobarbital, etc.), and antibiotic (ofloxacin).

Though there are different methods for chiral resolution, such as chromatographic resolution, crystallization separation, membrane-based separation, and kinetic resolution. Membrane-based separation, among all these chiral separation techniques, is economically and environmentally competitive to the other methods. Membrane-based separation can also be scaled up very easily, and under mild conditions it can be continuously operated [40].

As a result of rapid growth in nanotechnology from recent years, electrospun nanofibers are currently in trend for various applications. Electrospun nanofibers produced from natural and synthetic polymers are found suitable for the membrane-based separation of racemic compounds [41]. Cellulose tris(3,5-dimethylphenylcarbamate) (CDMPC) is known for its chiral recognition mechanism. The electrospun CDMPC membrane system has shown great potential for the efficient chiral separation of a racemic mixture, (R, S)-1-(naphthyl)-ethanol. During vacuum filtration, the enantioselective membrane allowed the enantioselective sorption of one enantiomer, while the other was enriched in filtrate [42].

Chitin is one of the most abundant green polymers. It is a polymer of N-acetyl glucosamine and has a chiral environment which can distinguish the absolute configuration of a specific enantiomer combination. Nanofibers prepared from chitin have been considered as strong candidates for membranes employed in chiral resolution due to their excellent properties such as low thermal expansion, antiparallel extended crystalline structure, and high fracture strength. The chitin nanofiber membrane system has proven to be effective for the selective transport of D-isomer of phenylalanine (Phe), glutamic acid (Glu), and lysine (Lys) as compared to the L-isomer from the respective racemic amino acid mixtures [43].

Nanoporous composite membranes prepared from micro- and nanofibrous material having a variable amount of chiral selector, (S, S)-1,2-diaminocyclohexane (DACH), have demonstrated the enantiomer separation of racemic D,L-tryptophan. The blank membranes, i.e., without the chiral selector exhibited no sorption activity, while the membrane having a chiral selector revealed a preferential sorption for one of the enantiomers [43].

Carbon nanofibers have better conductivity and high specific surface area, so they can be used in the fabrication of better-performing sensing devices. Cyclodextrins are popular as natural chiral ligands. SBE-β-CD (sulfobutyl ether-b-cyclodextrin), a derivative of β-CD, has great enantioselective properties. A highly selective electrochemical chiral sensor based on CNFs-SBE-β-CD has been introduced, which has the ability to rapidly and effectively identify CBL/AML/MET enantiomers (CBL, clenbuterol; AML, amlodipine; MET, metoprolol). This sensor is a cost-effective electrochemical tool which can be helpful for an effective analysis of different enantiomers in a specific racemic mixture [44].

Thus, these nanofibrous membranes can serve as a potential tool for the chiral resolutions of various racemic compounds.

10.6 Challenges

Challenges with respect to the material functionality, biocompatibility, printability, tissue biomimicry, degradation rates, and safe degradation products still need attention, both including therapeutic and non-therapeutic applications [45]. Another alternative to create composite inks with advantageous processability, mechanics, bioactivities, and printability is combining various biopolymer types with the purpose of achieving a complementary benefit. However, the variation in their rates of biodegradation might make it difficult for cells to replace them structurally with ECM.

Tissue biomimicry is another major deterrent for printing biopolymers for biomedical applications, especially in tissue engineering. This includes creating thick, vascularized, metabolically active tissues that could be replicated while maintaining the intricacy of the natural tissue, such as kidney, heart, lungs, and liver tissues [46].

Being cost-intensive and in the stage of infancy, 3D printing of food materials lag behind printable biopolymers compatible with regular food ingredients and conventional production procedures. The main disputes with 3D food printing using biopolymers that need to be established are compliance with stringent food safety regulations, suitable rheological properties for processing, printing, and self-supporting, printing precision and accuracy in the acceptable range, and easy manipulation of biopolymers customized nutrition or flavors. A spectrum of printable materials is limited only to specific 3D printing technologies. In other words, both the printing materials as well as the 3D printing schemes need to be broadened since the two are equally vital for membrane production performing separation and purification [47].

About 15% energy is used globally for the separation and purification of chemicals, reusable water, and gases that are of industrial significance, even though membrane processes offer minimal operating costs and energy consumption, being viable molecular separations which can be used over a wide range of applications including gas and liquid separations. Therefore, it is important to reduce the carbon footprint and cost factor in the industry via this technique in the future. Applications of AM techniques in membrane sector should anticipate further research and development in the near future [48, 49].

Like every other disruptive technology, 3D printing continues to face adaptability issues, especially on a large scale. This is largely due to the lack of infrastructure and facility.

10.7 Opportunities and Future Perspectives

Future research goals should aid in widening the selection of naturally derived biopolymer products as the novel 3D printing feedstock must include the following:

(1) maturing of innovative solvents/solutions competent of dissolving or dispersing formulations of natural biopolymers/cells as inks for various forms of 3D printing
(2) Inclusion of various biomolecules as composite inks with tunable and customized characteristics
(3) Ease of modifying or functionalizing naturally derived biopolymers to enhance the printability of inks
(4) Employment of components in biopolymers to induce chemical/physical stimuli, making the 3D-printed items programmable to external impetus such as temperature, light, pH, electric, and magnetic field, thereby advancing present 3D printing to future 4D printing

At the outset, water desalination seems to be revolutionized in the coming decades via advances in 3D printing technology simultaneously for both thermal as well as the membrane-based water desalination. 3D printing not only offers a unique combination of materials along with designs with unmatched process efficiency but also environmental sustainability. 3D printing provides the simple application of new nanostructured materials like graphene and CNTs to enhance solar absorption and increase the energy density of water evaporation for thermal desalination. From the design perspective, 3D printing exhibits limitless possibilities for systematic, "nature-encouraged" creation of solar collectors, heat exchangers, and other system components for the complete optimization of solar thermal desalination. Furthermore, 3D printing enables a more efficient use of other renewable resources, thereby promising a sustainable future for the water desalination industry. Furthermore, 3D printing has major two advantages: the potential for systematic membrane patterning: improved membrane execution and facile module design. These advantages of 3D printing for water desalination are evidenced by the increasing adoption in industries where the production of 3D-printed spacers and membranes is already commercialized. Several technical reports predict that the market size of RO systems and components will increase dramatically over the next years. 3D printing's role in water desalination will continue to grow rapidly as the cost of 3D printing infrastructure declines, accelerating academic research and increasing industrial adoption. However, widespread deployment of 3D printing in the desalination industry will require a strategic and collaborative role of researchers, industry, and policymakers. While many ideas can be applied to lab-scale, researchers need to focus on cheap, readily available materials that can be 3D-printed using cheaper, well-established technologies such as FDM [48, 49]. Additionally, there is a strong need to extend the material library that is 3D-printable with a variety of printing technologies. Apart from material extrusion that has exhibited limited application in membrane-assisted desalination, other 3D printing techniques are either material-specific, exorbitant, or not fully compatible with water-relevant applications. Polyjet printing, for example, shows great potential for manufacturing complex-shaped spacers. However, commercial Polyjet systems make use of their own standard resin for 3D printing. Therefore, researchers found new ways to integrate novel materials, such as carbon-based nanomaterials, into existing resins to maximize the functionality and cost-effectiveness of spacers while unlocking the potential for 3D printing. It is understood that such an approach is impractical when directly 3D printing the desired shape onto the surface of the membrane because of the incompatibility of the membrane material with the solvent. Production speed is another factor limiting the overall use of 3D printing. Conventional processes for manufacturing membrane and spacer are well developed and are capable of achieving production rates exceeding the demand rate. On the contrary, the 3D printing process is not yet improved in terms of speed, so matching the production speed of 3D printing with the actual demand requires serious attention from academic researchers and industry stakeholders is required. Nevertheless, the industry must join hands with science to create constructive, cohesive systems that can address real-time problems and practically solve them. Additionally, the desalination industry must demonstrate greater complacency and flexibility to integrate 3D printing into existing infrastructure. Ultimately, more funding and friendly policies from government agencies should render as the vital catalysts for the long-term positive impact of 3D printing on the desalination industry [2, 3].

10.8 Conclusion

Due to their exceptional physio-chemical properties, nanomaterials are critical in various fields. Loading microscopic nanomaterials onto macrostructures is necessary to move them from laboratory studies to commercial applications. However, the geometries of traditional support structures are generally limited and nanomaterials are susceptible to inhomogeneous spreading, aggregation, and loss. As a result, controlling the configuration of nanomaterials into complex 3D macroscopic structures without compromising their fundamental qualities remains a challenge. In addition, the researchers established that the 3D-printed adsorption structures were easily reusable up to 10 times and effective for contaminated groundwater samples. The findings suggest that 3D printing is a potential method for designing and fabricating custom macrostructures endowed with unique nanomaterial properties. Nevertheless, the surge in demand of additive manufacturing and shape-memory smart materials is anticipated to evolve a powerful 4D-based technology having time dimensions as well as stimuli-responsive materials.

References

1. Zennifer, A., Manivannan, S., Sethuraman, S., Kumbar, S.G., Sundaramurthi, D., 3D bioprinting and photocrosslinking: emerging strategies & future perspectives. *Biomater. Adv.*, 1, 134, 112576, 2022.
2. Saleh Alghamdi, S., John, S., Roy Choudhury, N., Dutta, N.K., Additive Manufacturing of Polymer Materials: Progress, Promise and Challenges. *Polym. (Basel)*, 13, 5, 753, 2021 Feb 28, doi: 10.3390/polym13050753. PMID: 33670934; PMCID: PMC7957542.
3. Alammar, A., Kois, J.C., Revilla-León, M., Att, W., Additive Manufacturing Technologies: Current Status and Future Perspectives. *J. Prosthodont.*, 31, S1, 4–12, 2022 Mar, doi: 10.1111/jopr.13477. PMID: 35313022.
4. Mohammed, A., Elshaer, A., Sareh, P., Elsayed, M., Hassanin, H., Additive Manufacturing Technologies for Drug Delivery Applications. *Int. J. Pharm.*, 580, 119245, 2020 Apr 30, doi: 10.1016/j.ijpharm.2020.119245. Epub 2020 Mar 20. PMID: 32201252.

5. Wang, J., Zhang, Y., Aghda, N.H., Pillai, A.R., Thakkar, R., Nokhodchi, A., Maniruzzaman, M., Emerging 3D printing technologies for drug delivery devices: Current status and future perspective. *Adv. Drug Deliv. Rev.*, 174, 294–316, 2021 Jul, doi: 10.1016/j.addr.2021.04.019. Epub 2021 Apr 23. PMID: 33895212.
6. Advincula, R.C., Dizon, J.R.C., Caldona, E.B., Viers, R.A., Siacor, F.D.C., Maalihan, R.D., Espera, A.H., Jr, On the progress of 3D-printed hydrogels for tissue engineering. *MRS Commun.*, 11, 5, 539–553, 2021, doi: 10.1557/s43579-021-00069-1. Epub 2021 Aug 3. PMID: 34367725; PMCID: PMC8330198.
7. Salmi, M., Additive Manufacturing Processes in Medical Applications. *Mater. (Basel)*, 14, 1, 191, 2021 Jan 3, doi: 10.3390/ma14010191. PMID: 33401601; PMCID: PMC7796413.
8. Pingale P, Dawre S, Dhapte-Pawar V, Dhas N, Rajput A. Advances in 4D printing: From stimulation to simulation. *Drug Deliv. Trans. Res.*, 13, 1, 164–88, 2023 Jan, doi: 10.1007/s13346-022-01200-y. Epub ahead of print. PMID: 35751000.
9. Villarreal-Gómez, L.J., Cornejo-Bravo, J.M., Vera-Graziano, R., Grande, D., Electrospinning as a powerful technique for biomedical applications: a critically selected survey. *J. Biomater. Sci., Polym. Ed.*, 27, 2, 157–76, 2016, doi: 10.1080/09205063.2015.1116885. Epub 2015 Dec 17. PMID: 26540235.
10. Che, H. and Yuan, J., Recent advances in electrospinning supramolecular systems. *J. Mater. Chem. B*, 10, 1, 8–19, 2021 Dec 22, doi: 10.1039/d1tb02304g. PMID: 34878489.
11. Yu, D.G., Wang, M., Li, X., Liu, X., Zhu, L.M., Annie Bligh, S.W., Multifluid electrospinning for the generation of complex nanostructures. *Wiley Interdiscip. Rev. Nanomed. Nanobiotechnol.*, 12, 3, e1601, 2020 May, doi: 10.1002/wnan.1601. Epub 2019 Nov 6. PMID: 31692241.
12. Hutmacher, D.W. and Dalton, P.D., Melt electrospinning. *Chem. Asian J.*, 6, 1, 44–56, 2011 Jan 3, doi: 10.1002/asia.201000436. PMID: 21080400.
13. Yilmaz, E.N. and Zeugolis, D.I., Electrospun Polymers in Cartilage Engineering-State of Play. *Front. Bioeng. Biotechnol.*, 8, 77, 2020 Feb 18, doi: 10.3389/fbioe.2020.00077. PMID: 32133352; PMCID: PMC7039817.
14. Diyana, Z.N., Jumaidin, R., Selamat, M.Z., Ghazali, I., Julmohammad, N., Huda, N., Ilyas, R.A., Physical Properties of Thermoplastic Starch Derived from Natural Resources and Its Blends: A Review. *Polym. (Basel)*, 13, 9, 1396, 2021 Apr 26, doi: 10.3390/polym13091396. PMID: 33925897; PMCID: PMC8123420.
15. Kanmani, P., Aravind, J., Kamaraj, M., Sureshbabu, P., Karthikeyan, S., Environmental applications of chitosan and cellulosic biopolymers: A comprehensive outlook. *Bioresour. Technol.*, 242, 295–303, 2017 Oct, doi: 10.1016/j.biortech.2017.03.119. Epub 2017 Mar 22. PMID: 28366689.

16. Hong, H., Seo, Y.B., Lee, J.S., Lee, Y.J., Lee, H., Ajiteru, O., Park, C.H., Digital light processing 3D printed silk fibroin hydrogel for cartilage tissue engineering. *Biomaterials, 232,* 119679, 2020.
17. Rajabi, M., McConnell, M., Cabral, J., Ali, M.A., Chitosan hydrogels in 3D printing for biomedical applications. *Carbohydr. Polym., 260,* 117768, 2021.
18. Gaurkhede, S.G., Osipitan, O.O., Dromgoole, G., Spencer, S.A., Di Pasqua, A.J., Deng, J., 3D printing and dissolution testing of novel capsule shells for use in delivering acetaminophen. *J. Pharm. Sci., 110,* 12, 3829–3837, 2021.
19. Rezvani Ghomi, E., Nourbakhsh, N., Akbari Kenari, M., Zare, M., Ramakrishna, S., Collagen-based biomaterials for biomedical applications. *J. Biomed. Mater. Res. Part B: Appl. Biomater., 109,* 12, 1986–1999, 2021.
20. Xu, K., Han, Y., Huang, Y., Wei, P., Yin, J., Jiang, J., The application of 3D bioprinting in urological diseases. *Mater. Today Bio,* 1, 16, 100388, 2022.
21. Chartrain, N.A., Gilchrist, K.H., Ho, V.B., Klarmann, G.J., 3D bioprinting for the repair of articular cartilage and osteochondral tissue. *Bioprinting,* 1, 28, e00239, 2022.
22. Yang, M., Sun, W., Wang, L., Tang, H., Xu, X., Yang, L., Ni, J., Zheng, K., Jiang, X., Xu, W., Zhao G., Curcumin loaded polycaprolactone scaffold capable of antiinflammation to enhance tracheal cartilage regeneration. *Mater. Des.*, 1, 224, 111299, 2022.
23. Yang, M., Zhang, Z.C., Yuan, F.Z., Deng, R.H., Yan, X., Mao, F.B., Yu, J.K., An immunomodulatory polypeptide hydrogel for osteochondral defect repair. *Bioact. Mater., 19,* 678–689, 2023.
24. Leberfinger, A.N., Dinda, S., Wu, Y., Koduru, S.V., Ozbolat, V., Ravnic, D.J., Ozbolat, I.T., Bioprinting functional tissues. *Acta Biomater., 95,* 32–49, 2019.
25. Bhattacharyya, A., Janarthanan, G., Kim, T., Taheri, S., Shin, J., Kim, J., Noh, I., Modulation of bioactive calcium phosphate micro/nanoparticle size and shape during *in situ* synthesis of photo-crosslinkable gelatin methacryloyl based nanocomposite hydrogels for 3D bioprinting and tissue engineering. *Biomater. Res., 26,* 1, 1–17, 2022.
26. Porta, R., Sabbah, M., Di Pierro, P., Biopolymers as food packaging materials. *Int. J. Mol. Sci., 21,* 14, 4942, 2020.
27. Mallakpour, S., Sirous, F., Hussain, C.M., A journey to the world of fascinating ZnO nanocomposites made of chitosan, starch, cellulose, and other biopolymers: Progress in recent achievements in eco-friendly food packaging, biomedical, and water remediation technologies. *Int. J. Biol. Macromol., 170,* 701–716, 2021.
28. Ng, S. and Kurisawa, M., Integrating biomaterials and food biopolymers for cultured meat production. *Acta Biomater., 124,* 108–129, 2021.
29. Butola, B.S., Recent advances in chitosan polysaccharide and its derivatives in antimicrobial modification of textile materials. *Int. J. Biol. Macromol., 121,* 905–912, 2019.
30. Wurm, F., Rietzler, B., Pham, T., Bechtold, T., Multivalent ions as reactive crosslinkers for biopolymers—a review. *Molecules, 25,* 8, 1840, 2020.

31. Fijoł, N., Abdelhamid, H.N., Pillai, B., Hall, S.A., Thomas, N., Mathew, A.P., 3D-printed monolithic biofilters based on a polylactic acid (PLA)–hydroxyapatite (HAp) composite for heavy metal removal from an aqueous medium. *RSC Adv.*, *11*, 51, 32408–32418, 2021.
32. Zhou, R., Han, R., Bingham, M., O'Rourke, C., Mills, A., 3D printed, plastic photocatalytic flow reactors for water purification. *Photochem. Photobiol. Sci.*, 21, 9, 1585–1600, 2022.
33. Nasr Esfahani, K., Zandi, M.D., Travieso-Rodriguez, J.A., Graells, M., Pérez-Moya, M., Manufacturing and application of 3D printed photo Fenton reactors for wastewater treatment. *Int. J. Environ. Res. Public Health*, *18*, 9, 4885, 2021.
34. Issac, M.N. and Kandasubramanian, B., Review of manufacturing three-dimensional-printed membranes for water treatment. *Environ. Sci. Pollut. Res.*, *27*, 29, 36091–36108, 2020.
35. Wu, L., Beirne, S., Cabot, J.M., Paull, B., Wallace, G.G., Innis, P.C., Fused filament fabrication 3D printed polylactic acid electroosmotic pumps. *Lab Chip*, *21*, 17, 3338–3351, 2021.
36. Gkantzou, E., Skonta, A., Vasios, A.G., Stamatis, H., 3D printed polylactic acid well-plate for multi-enzyme immobilization, in: *Multienzymatic Assemblies*, pp. 163–175, Humana, New York, NY, 2022.
37. Martinez, S.R., Palacios, Y.B., Heredia, D.A., Aiassa, V., Bartolilla, A., Durantini, A.M., Self-sterilizing 3D-printed polylactic acid surfaces coated with a BODIPY photosensitizer. *ACS Appl. Mater. Interfaces*, *13*, 10, 11597–11608, 2021.
38. Condi Mainardi, J., Rezwan, K., Maas, M., Genipin-crosslinked chitosan/alginate/alumina nanocomposite gels for 3D bioprinting. *Bioprocess. Biosyst. Eng.*, *45*, 1, 171–185, 2022.
39. Nguyen, L.A., He, H., Pham-Huy, C., Chiral drugs: an overview. *Int. J. Biomed. Sci.: IJBS*, *2*, 2, 85, 2006.
40. Sueyoshi, Y., Hashimoto, T., Yoshikawa, M., Ifuku, S., Chitin nanofiber membranes for chiral separation. *Sustain. Agric. Res.*, *1*, 1, 526-2016-37757, 2012, **CR2**.
41. Ifuku, S. and Saimoto, H., Chitin nanofibers: preparations, modifications, and applications. *Nanoscale*, *4*, 11, 3308–3318, 2012.
42. Otsuka, I., Pandey, K., Ahmadi-Nohadani, H., Nono-Tagne, S., Electrospun Cellulosic Membranes toward Efficient Chiral Resolutions via Enantioselective Permeation. *ACS Macro Lett.*, *10*, 7, 921–925, 2021.
43. Gaálová, J., Yalcinkaya, F., Cuřínová, P., Kohout, M., Yalcinkaya, B., Koštejn, M., Izák, P., Separation of racemic compound by nanofibrous composite membranes with chiral selector. *J. Membr. Sci.*, *596*, 117728, 2020.
44. Kalambate, P.K., Upadhyay, S.S., Shen, Y., Laiwattanapaisal, W., Huang, Y., Chiral nanocomposite of sulfobutyl ether-β-cyclodextrin embedded in carbon nanofibers for enantioselective electrochemical discrimination of

amlodipine, metoprolol and clenbuterol enantiomers. *J. Materiomics*, *7*, 2, 226–235, 2021.

45. Chinga-Carrasco, G., Rosendahl, J., Catalán, J., Nanocelluloses–Nanotoxicology, Safety Aspects and 3D Bioprinting, in: *Nanotoxicology in Safety Assessment of Nanomaterials*, pp. 155–177, Springer, Cham, 2022.
46. Hossain, N., Chowdhury, M.A., Shuvho, M., Ahmed, B., Kashem, M.A., Kchaou, M., 3D-printed objects for multipurpose applications. *J. Mater. Eng. Perform.*, *30*, 7, 4756–4767, 2021.
47. Murphy, S.V., De Coppi, P., Atala, A., Opportunities and challenges of translational 3D bioprinting. *Nat. Biomed. Eng.*, *4*, 4, 370–380, 2020.
48. Kahl, M., Gertig, M., Hoyer, P., Friedrich, O., Gilbert, D.F., Ultra-low-cost 3D bioprinting: modification and application of an off-the-shelf desktop 3D-printer for biofabrication. *Front. Bioeng. Biotechnol.*, *7*, 184, 2019.
49. Sanz-Garcia, A., Sodupe-Ortega, E., Pernía-Espinoza, A., Shimizu, T., Escobedo-Lucea, C., A versatile open-source printhead for low-cost 3D microextrusion-based bioprinting. *Polymers*, *12*, 10, 2346, 2020.

11

Economic and Environmental Assessment of Sustainable Polymer-Based 3D Printing

Brooke E. Rogachuk and Jude A. Okolie*

Engineering Pathways, Gallogly College of Engineering, University of Oklahoma, Norman, OK, United States

Abstract

In recent years, 3D printing has surged in popularity, primarily due to its more environmentally friendly attributes compared to conventional manufacturing techniques. This necessitates a thorough investigation into materials optimally suited for 3D printing applications. The chapter comprehensively reviews the economic and environmental impacts of three widely used 3D printing materials: polylactic acid (PLA), polyethylene terephthalate glycol (PETG), and acrylonitrile butadiene styrene (ABS). The analysis reveals that ABS is the least environmentally friendly option, while PLA stands out as the most eco-friendly material. However, it is important to note that PLA, despite its biodegradable properties, has significant negative effects on water depletion and eco-toxicity. This evaluation highlights the importance of balancing environmental concerns with the functional demands of 3D printing materials, and it is vital for steering future material choices and innovations toward sustainable manufacturing practices. This understanding is essential for the advancement of 3D printing technology in a manner that is consistent with environmental sustainability objectives.

Keywords: Polymer, techno-economic analysis, lifecycle assessment, polylactic acid, 3D printing

11.1 Introduction

3D printing is the process of making 3D objects from a digital file through additive manufacturing [1]. The digital file originates from computer-aided

**Corresponding author*: Jude.okolie@usask.ca

Bhasha Sharma, Balaram Pani, Shashank Shekhar and Jude A. Okolie (eds.) Sustainable 3D Printing for Innovative Biopolymer Production and Applications, (233–244) © 2025 Scrivener Publishing LLC

design (CAD). The early design stages and prototyping as well as documentation can all be done in CAD [2]. In this program, the user has access to a variety of shapes and tools to make a scaled-down model of what object the user would like to print. Next, the file gets spliced to be compatible with the 3D machine that is available. Next, the file goes to the 3D printer to be printed through the additive process. The printer heats up a filament that is added layer by layer to create the desired three-dimensional object. The ease of making new products with a fast fabrication speed makes 3D printing a popular new technology. There are many applications of 3D printing. The most recent advances are in the medical field, agriculture, construction, and automotive transport.

In the medical field, 3D printing has been used to customize surgical tools and protheses due to the flexible and easy customization of each print [3]. Furthermore, it has been used to make personalized drugs [3]. The 3D printer can print out the drug layer in powdered form to the exact dosage for each patient and allows the pill to dissolve faster. In the agriculture field, 3D printing allows for custom tools and replacement parts [1]. In the construction industry, houses have been made from large-scale 3D printing with concrete as the filament [4]. Finally, in the automotive industry, 3D printing can be used to print several parts as one fabricated piece [5]. The different applications of 3D printing between the different industries have to do with the variety of different 3D printing machines as well as the polymers that were used in the 3D printing filament [5].

The typical polymers include polylactic acid (PLA), acrylonitrile butadiene styrene (ABS), nylon, wood-polymer composite (WPC), and polyethylene terephthalate (PETG) [6]. Most of the traditional polymers are made from petroleum products or plastics. Petroleum products have negative environmental effects that will be discussed in further detail in the environment assessment section. Furthermore, there is an overabundance of plastic pollution on land and in waterways [7]. Sustainable polymers come from waste residues, recycled material, or low carbon biomass. The polymer goes through green manufacturing and then used in application, and then it can be recycled back into the process. The ability to recycle is essential for expensive polymers that beginning often make mistakes with. Additive manufacturing does produce less waste than subtractive manufacturing [8]. The process of 3D printing still produces a lot of waste, especially for personal use. Despite the advancement and arrays of benefits 3D printing presents to society, the economic and environmental assessment of sustainable materials 3D printing is scarcely reported in literature.

This chapter explores the economic and environmental assessment of sustainable polymer-based 3D printing which plays important roles in understanding the cost-effectiveness, environmental impact, and the overall sustainability of the technologies employed in manufacturing the polymers. These will undoubtedly further serve as a guide in formulating policies, making decisions, and adoption of environmentally friendly practices in the 3D printing industry. The economic and environmental assessment of sustainable polymers used in 3D printing is crucial for multiple reasons. Firstly, understanding the cost-effectiveness of these polymers ensures that the technology remains viable and accessible for widespread adoption, balancing financial feasibility with innovation. Secondly, evaluating the environmental impact is key to ensuring that these technologies align with global sustainability goals, minimizing their ecological footprint. This holistic assessment is not only fundamental for technological advancement but also vital for informing policy formulation, decision-making, and the adoption of eco-friendly practices in the 3D printing industry. By prioritizing both economic viability and environmental responsibility, this approach fosters a sustainable development pathway for 3D printing technologies, ensuring that they contribute positively to both industry growth and environmental conservation.

11.2 Economic Assessment of Sustainable Polymers in 3D Printing

One of the primary considerations when evaluating the economic viability of sustainable polymers in 3D printing is the cost comparison with traditional polymers. While sustainable polymers may have a higher upfront cost, their long-term benefits can offset these initial expenses. Sustainable polymers derived from renewable resources, such as bio-based polymers or recycled materials, can offer cost savings over time due to their lower dependence on fossil fuels and the potential for reduced waste disposal costs.

Moreover, sustainable polymers can contribute to the circular economy by enabling closed-loop material systems. This allows for the recycling and reusability of materials, reducing the need for raw material extraction and production. Although establishing the infrastructure for recycling and reuse may require upfront investment, the long-term cost savings and reduced environmental impact make it an economically attractive option.

Currently on Amazon ABS 3D printer filament (a filament made from petroleum products) costs $22.99 USD for 1 kg, whereas algae-based PLA printer filament costs $34.99 USD for 1 kg [8]. Prusament PLA recycled costs $55 USD for 1 kg [8]. Therefore, it is currently more cost-effective to use unsustainable raw materials for 3D printing. However, as environmental concerns continue to grow, consumers and businesses are increasingly seeking sustainable alternatives. This trend is reflected in the growing demand for eco-friendly products across various industries. By offering sustainable polymers in 3D printing, companies can tap into this market and differentiate themselves from competitors.

The recent development of open-source 3D printers, specifically self-replicating rapid prototypers (RepRaps), has made the distributed additive-based manufacturing of high-value objects more technically feasible and could significantly enhance mechatronics education and participation. These RepRaps, capable of manufacturing about half of their own parts, have applications in various fields, including prototyping and sustainable development. The study of Wittbrodt *et al.* [9] focuses on the life-cycle economic analysis (LCEA) of RepRap technology for an average US household, detailing the construction costs of a new low-cost RepRap and comparing the economic costs of printing 20 different open-source designs with their market prices. The findings reveal substantial cost savings for households using RepRaps, with payback times ranging from 4 months to 2 years and return on investment (ROI) between >200% and >40%. This study concludes that RepRaps represent an economically attractive investment for average households, with potential for widespread adoption as they become more reliable, cost-effective, and as the number and utility of open-source designs grow exponentially.

The rise of 3D printing has spurred the development of filament extruders and recyclebots, which can create 3D printer filament from both virgin polymer pellets and post-consumer shredded plastic. Recognizing the potential for ethical manufacturing in this domain, there is a growing market opportunity for producing ethical 3D printer filament. Feely *et al.* [10] proposes an "ethical product standard" for 3D filament, blending existing fair-trade standards with technical and life cycle analysis of recycled filament production and 3D printing manufacturing. These standards, aimed at businesses, focus on enabling economic development for waste pickers and cover a range of criteria: minimum pricing, fair trade premium, labor standards, environmental and technical standards, health and safety standards, and social standards. Social standards encompass issues like discrimination, harassment, freedom of association, collective bargaining, and discipline, ensuring comprehensive ethical considerations in the production of 3D printer filament.

11.3 Environmental Assessment of Sustainable Polymers in 3D Printing

Undoubtedly, plastic polymers, owing to their appealing features that include but not limited to reduced weight, excellent tensile strength, durability, and corrosion resistance, have been widely employed commercially and industrially [11]. The production of plastics has been on increase as its production increased to 381 million metric tons annually, which sees about 8 billion of them produced in 2015 [12]. One of the issues associated with the use of plastics is their inability to degrade, which has made their release into water and land have a negative impact on the environment, including greenhouse gas emissions, soil deterioration, and eco-toxicity [11]. Recently, it has been reported that only 20% of the 80% disposed non-biodegradable plastics have been recycled. The linear economy, which begins with extraction and ends with disposal, has exhausted natural resources and caused detrimental environmental effects [13]. If the current linear economic trajectory is sustained, it is predicted that, by 2050, the amount of plastics in the ocean will surpass the number of fish [11].

Fossil resources are currently running out and are not replenishable by nature. Finding naturally occurring fibers and polymers that are bio-based, biocompatible, and, most importantly, degradable instead of conventional ones need to be a top priority. These bio-based or sustainable polymers are created using renewable natural resources like corn, wheat, or trees [14]. These sustainable polymers could be synthetic, natural, or modified, and they are circular economy-compliant throughout their whole life cycle [15]. The life cycle of sustainable polymers is schematically shown in Figure 11.1 from feedstock through regeneration. Sustainable polymers have a lower energy consumption, a smaller environmental impact, and a balanced economic impact [16]. Moving forward, it is important for material scientists, policy makers, and engineers to take into account the circular economy's framework for the overall development of materials. A circular economy is a holistic approach to economic growth that avoids waste and the deterioration of resources while being ecologically conscientious. This approach allows for the recovery, recycling, and reuse of all manufactured goods, extending the useful life of the raw materials. While extending the useful life of materials that would otherwise be wasted, resource input, material waste, and carbon emissions are all lessened [17].

In comparison to non-biodegradable plastics, recycled plastics have been found to have a carbon footprint that is 3,000 times lower. In spite of this, only 14% of plastic is still recycled globally [11]. However, 3D

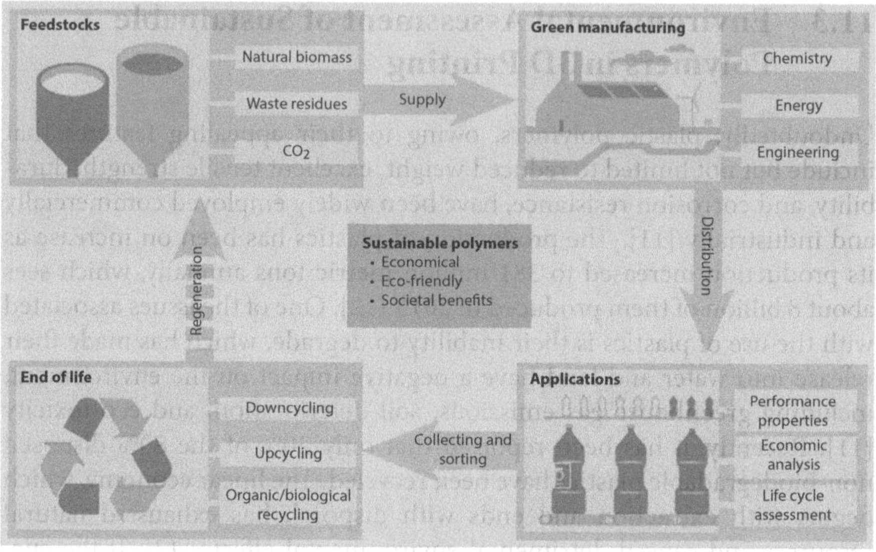

Figure 11.1 Life cycle of sustainable polymers. Adapted from [15].

printing has proven to be a viable option for these recycled plastics where they can be used as feedstock for filament. Polylactic acid (PLA), polyamide (PA/nylon), acrylonitrile butadiene styrene (ABS), polyethylene terephthalate glycol (PETG), and polycarbonate (PC) are the most frequently used plastics in 3D printing. PA, PC, PETG, and ABS, however, are non-biodegradable and petroleum-based [11].

The biopolymer PLA, which is from corn starch, is frequently used as a feedstock material for 3D printing filament. However, the increased utilization of sugarcane and corn starch for the synthesis of virgin PLA could compromise the food security of some developing nations [18]. The end-of-life disposal of commercial PLA parts has grown to be a substantial problem since a sizable portion of PLA does not degrade under typical circumstances, even with microbial action [11]. Therefore, extending the life of PLA through circular economy measures, such as reuse, recycling, and recovery, is crucial to lowering waste production and relieving demand on virgin materials [19]. The waste PLA materials could then be converted to 3D filaments and used to produce parts while minimizing the negative environmental effects.

It is very important to investigate the environmental impact of the recycled plastic wastes that are used in 3D printing. Through the use of life cycle assessment (LCA), the environmental effects of these wastes have been examined in several studies—for instance, Zhao *et al.* [20] did a study to compare the environmental effects of recycled PLA and those that are disposed to landfill and incinerator. Their report revealed that the recycled PLA had a significantly lower impact than its counterpart.

The LCA will be used in this chapter to examine the environmental impacts of PLA, PETG, and ABS in three phases (raw material extraction, manufacturing, and recycling) by using predetermined parameters. These parameters have been determined at each stage of the LCA either by experimental research or by obtaining the information from a database.

As shown in Figure 11.2, for PLA, the recycling phase has the greatest impact in six out of the eight parameters considered (climate change, particulate matter formation, ozone depletion, fossil depletion, human toxicity, and terrestrial acidification), while raw material extraction contributes most to water depletion and eco-toxicity [21].

For ABS, its recycling phase has the greatest impacts on the environments in three out of the eight parameters considered. This is because ABS's high specific heat causes it to spend more energy during recycling

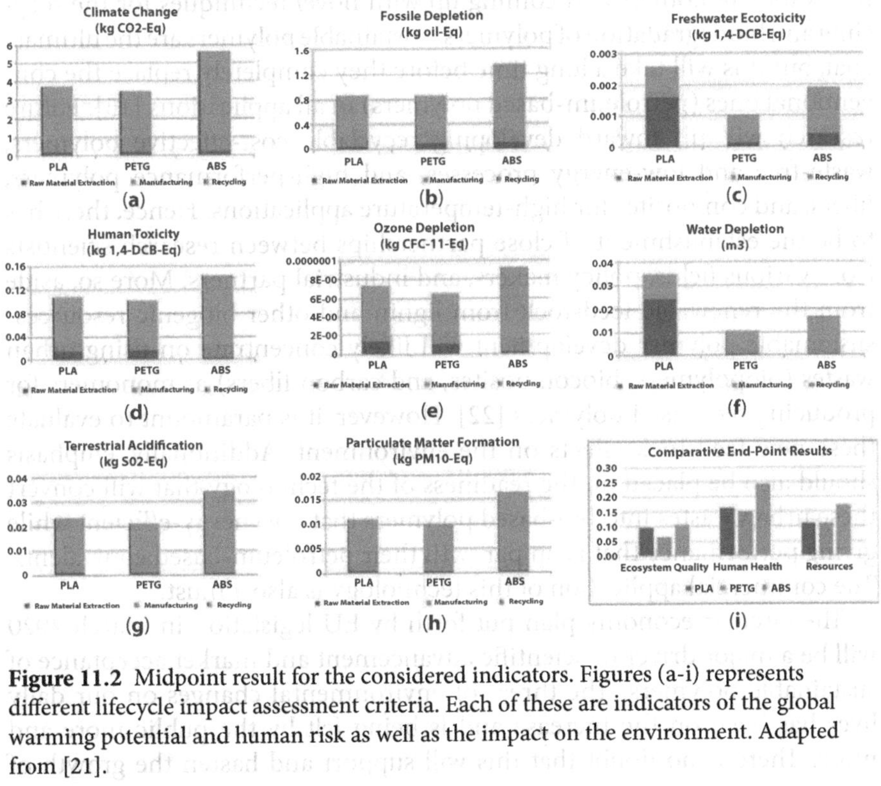

Figure 11.2 Midpoint result for the considered indicators. Figures (a-i) represents different lifecycle impact assessment criteria. Each of these are indicators of the global warming potential and human risk as well as the impact on the environment. Adapted from [21].

than the other. With the exception of freshwater toxicity, the raw material phase contributes the least to all parameters. Across all parameters, the manufacturing phase consistently has a 15% relative influence. Expectedly, ABS, a petroleum-based polymer, has the greatest environmental implication for all the considered parameters [21].

PETG has the least negative effects on the environment. The effects of PLA and PETG are relatively similar, with PETG having a little lower effect on human health and resource use. PLA material, on the other hand, has significant adverse effects on ecosystem quality, just somewhat less than ABS material [21].

11.4 Future Outlook

Sustainability in additive manufacturing is a multidimensional objective that includes but not limited to material sourcing, waste management, and energy use [14]. Tremendous efforts have been made in the last decades toward sustainability in polymer research through polymerization of renewable monomers and coming up with novel techniques for the recycling and biodegradation of polymers. Sustainable polymers are the ultimate goal, but this will take a long time before they completely replace the conventional ones (petroleum-based polymers) in all applications [15]. Future research will tilt toward developing recyclable cost-effective polymers, waste-free and low-energy processes, and high-performance polymers, fibers, and composites for high-temperature applications. Hence, there has to be the establishment of close partnerships between research scientists from various fields, policy makers, and industrial partners. More so, aside from the renewable feedstock from lignin and other biogenic resources-sustainable polymer development will likely concentrate on using urban wastes (biopolymers, biocomposites, and carbon fibers) as monomers for producing bio-based polymers [22]. However, it is paramount to evaluate these new feedstock effects on the environment. Additionally, emphasis should also be placed on the readiness of the technology that will convert these urban wastes into bio-based polymers that are energy-efficient while giving performance that is on par with their petroleum-based equivalents. The commercial application of this technology is also a must.

The circular economy plan put forth by EU legislation in March 2020 will be a major driver of scientific advancement and market acceptance of sustainable polymers. The threat of environmental changes on our daily lives has been on the increase and is being felt by the public more and more. There is no doubt that this will support and hasten the growth of a sustainable economy [15]. To overcome today's difficulties of developing sustainable polymers, research should start focusing on developing sustainable polymers while placing emphasis on catalytic technology, molecular design, promoting material recycling, and reducing waste disposal. The development of recycling technology (mechanical or chemical recycling) to transform polymer waste into valuable monomers, chemicals, and polymers using either mechanical or chemical recycling should be prioritized for the upcoming decade due to the massive environmental accumulation of plastic waste and the ongoing production of new plastics [15]. Technology-wise, we should all aspire to develop sustainable polymers with excellent performance features that will expand our industry. 3R (reduce, reuse, and recycle) should always come first if we want to maintain sustainability [23].

11.5 Conclusion

3D printing has become more and more popular in recent years owing to its environmentally friendly nature in comparison to the conventional one. Hence, it is paramount to search for materials that are most suited for 3D printing applications. This chapter examines the economic and environmental effects of the three most popular 3D printing materials (PLA, PETG, and ABS). ABS was found to be the least environmentally friendly polymer, while PLA is the most environmentally friendly one. PLA, despite being biodegradable, has the greatest adverse effects on water depletion and eco-toxicity.

References

1. Shahrubudin, N. *et al.*, An overview on 3D printing technology: Technological, materials, and applications. *Procedia Manuf.*, 35, 1286–1296, 2019, https://doi.org/10.1016/j.promfg.2019.06.089.
2. Oropallo, W. and Piegl, L.A., Ten challenges in 3D printing. *Eng. Comput.*, 32, 1, 135–148, 2015, https://doi.org/10.1007/s00366-015-0407-0.
3. Aimar, A. *et al.*, The role of 3D printing in medical applications: A state of the art. *J. Healthc. Eng.*, 2019, 1–10, 2019, https://doi.org/10.1155/2019/5340616.
4. Sakin, M. and Kiroglu, Y.C., 3D printing of buildings: Construction of the sustainable houses of the future by BIM. *Energy Procedia*, 134, 702–711, 2017, https://doi.org/10.1016/j.egypro.2017.09.562.

5. Lim, C.W.J. *et al.*, An overview of 3-D printing in manufacturing, aerospace, and automotive industries. *IEEE Potentials*, 35, 4, 18–22, 2016, https://doi.org/10.1109/mpot.2016.2540098.
6. Chadha, A. *et al.*, Effect of fused deposition modelling process parameters on mechanical properties of 3D printed parts. *World J. Eng.*, 16, 4, 550–559, 2019, https://doi.org/10.1108/wje-09-2018-0329.
7. Mohanty, A.K. *et al.*, Sustainable polymers. *Nat. Rev. Methods Primers*, 2, 1, 1–27, 2022, https://doi.org/10.1038/s43586-022-00124-8.
8. Frey, S., Best sustainable eco-friendly 3D printer filaments of 2022, All3DP, 2022, Retrieved from https://all3dp.com/1/best-sustainable-eco-friendly-3d-printer-filaments/.
9. Wittbrodt, B.T. *et al.*, Life-cycle economic analysis of distributed manufacturing with open-source 3-D printers. *Mechatronics*, 23, 6, 713–726, 2013.
10. Feeley, S.R., Wijnen, B., Pearce, J.M., Evaluation of potential fair trade standards for an ethical 3-D printing filament. *J. Sustain. Dev.*, 7, 5, 2014.
11. Jayawardane, H., Davies, I. J., Gamage, J. R., John, M., Biswas, W.K., Additive manufacturing of recycled plastics: a 'techno-eco-efficiency'assessment. *Int. J. Adv. Manuf. Technol.*, 126, 3, 1471–1496, 2023.
12. Singh, R. and Kumar, R., *Additive manufacturing for plastic recycling: Efforts in boosting a circular economy*, CRC Press, Boca Raton, Florida, 2022.
13. Sanchez, F.A.C., Boudaoud, H., Hoppe, S., Camargo, M., Polymer recycling in an open-source additive manufacturing context: Mechanical issues. *Addit. Manuf.*, 17, 87–105, 2017.
14. Wan, Y., Properties, preparation and application of nature fibers/sustainable polymers. *Highlights in Science, Engineering and Technology*, 52, 17–22, 2023.
15. Mohanty, A.K., Feng, W., Rosica, M., Hakkarainen, M., Raquez, J.-M., Mielewski, D.F., Narayan, R., Netravali, A.N., Misra, M., Sustainable polymers. *Nat. Rev. Methods Primers*, 2, 1, 46, 2022.
16. Ramesh, P. and Vinodh, S., State of art review on life cycle assessment of polymers. *Int. J. Sustain. Eng.*, 13, 6, 411–422, 2020, https://doi.org/10.1080/19397038.2020.1802623.
17. Chyr, G., Review of high-performance sustainable polymers in additive manufacturing. *Green Chem.*, 25, 2, 453–466, 2023, https://doi.org/10.1039/d2gc03474c.
18. Pellis, A., Malinconico, M., Guarneri, A., Gardossi, L., Renewable polymers and plastics: Performance beyond the green. *New Biotechnol.*, 60, 146–158, 2021.
19. Anderson, I., Mechanical properties of specimens 3D printed with virgin and recycled polylactic acid. *3D Print. Addit. Manuf.*, 4, 2, 110–115, 2017.
20. Zhao, P., Rao, C., Gu, F., Sharmin, N., Fu, J., Close-looped recycling of polylactic acid used in 3D printing: An experimental investigation and life cycle assessment. *J. Clean. Prod.*, 197, 1046–1055, 2018.

21. Kumar, R., Sharma, H., Saran, C., Shankar, T., Singh, K., A comparative study on the life cycle assessment of a 3D printed product with PLA, ABS & PETG materials. *Procedia CIRP*, 107, March, 15–20, 2023, https://doi.org/10.1016/j.procir.2022.04.003.
22. Fabbri, P., Viaggi, D., Cavani, F., Bertin, L., Michetti, M., Carnevale, E., ... & Fischer, P., Top emerging bio-based products, their properties and industrial applications. Berlin: Ecologic Institut., 2018, Available online: www.ecologic.eu (accessed on 22 March 2021).
23. Wei, R., Tiso, T., Bertling, J., O'Connor, K., Blank, L.M., Bornscheuer, U.T., *Nature Catal.*, 3, 867–871, 2020.

Index

Printed in the USA/Agawam, MA
September 16, 2026

910661.060

Endophytic Microbes: Isolation, Identification, and Bioactive Potentials

Edited by

A. Sankaranarayanan

Department of Life Sciences, Sri Sathya Sai University for Human Excellence, Navanihal, Kamalapura, Kalaburagi, Karnataka, India

Natarajan Amaresan and Mitesh Kumar Dwivedi

C.G. Bhakta Institute of Biotechnology, Uka Tarsadia University, Surat, Gujarat, India

Humana Press

Editors
A. Sankaranarayanan
Department of Life Sciences
Sri Sathya Sai University for Human Excellence
Navanihal, Kamalapura
Kalaburagi, Karnataka, India

Natarajan Amaresan
C.G. Bhakta Institute of Biotechnology
Uka Tarsadia University
Surat, Gujarat, India

Mitesh Kumar Dwivedi
C.G. Bhakta Institute of Biotechnology
Uka Tarsadia University
Surat, Gujarat, India

ISSN 1949-2448 ISSN 1949-2456 (electronic)
Springer Protocols Handbooks
ISBN 978-1-0716-2826-3 ISBN 978-1-0716-2827-0 (eBook)
https://doi.org/10.1007/978-1-0716-2827-0

This Humana imprint is published by the registered company Springer Science+Business Media, LLC, part of Springer Nature.
The registered company address is: 1 New York Plaza, New York, NY 10004, U.S.A.

Preface

The indiscriminate usage of fertilizers and pesticides not only destroys beneficial microbes in soil but also acts as a serious threat to the environment. The emerging threats by various abiotic and biotic factors, including drastic changes in climatic conditions, put agricultural industry throughout the globe at great risk. The increasing trends of green revolution have led to the development of promising candidates such as endophytic bacteria and fungi for sustainable development in agricultural field. Research reports and articles have listed the role of endophytic microbes in sustainable development in the agricultural field.

In this context, we have made an effort to compile a user-friendly laboratory manual on endophytic microbes. This manual provides the detailed procedures for the isolation and identification of endophytes, various biochemical methods and molecular methods for the identification of endophytes, various traditional methods, and modern methods to prove endophytism, bioactive potentials, and screening of their metabolites in 5 different sections with 29 protocols. The contents provided in the manual may be used to carry out research work from basic laboratories with elementary facilities to high-end laboratories.

We hope that the manual will be highly useful to the research and academic communities, especially young researchers/students. We are thankful to all contributors to this book, who spared their valuable time and cooperated with us for bringing this manual within the specified time period.

Healthy criticism and comments are highly welcome from the readers.

Kalaburagi, Karnataka, India — *A. Sankaranarayanan*
Surat, Gujarat, India — *Natarajan Amaresan*
Surat, Gujarat, India — *Mitesh Kumar Dwivedi*

Contents

Contributors

NATARAJAN AMARESAN • *C.G. Bhakta Institute of Biotechnology, UkaTarsadia University, Surat, India*

V. AMBIKAPATHY • *Department of Botany, A.V.V.M. Sri Pushpam College (Autonomous), (Affiliated to Bharathidasan University, Trichy), Thanjavur, Tamil Nadu, India*

ANBUKUMARAN ANANTHAN • *Department of Microbiology, Urumu Dhanalakshmi College, Trichy, Tamilnadu, India*

A. ANBUKUMARAN • *Department of Microbiology, Urumu Dhanalakshmi College, (Affiliated to Bharathidasan University), Trichy, Tamil Nadu, India*

PANNEERSELVAM ANNAMALAI • *PG and Research Department of Microbiology, A. V. V. M. Sri Pushpam College (Autonomous), Thanjavur, Tamilnadu, India; PG & Research Department of Microbiology, AVVM Sri Pushpam College (Auto), Thanjavur, Tamilnadu, India*

S. BABU • *Department of Botany, A.V.V.M. Sri Pushpam College (Autonomous), (Affiliated to Bharathidasan University, Trichy), Thanjavur, Tamil Nadu, India*

SOMA BARMAN • *Soil and Agro-Bioengineering Laboratory, Department of Environmental Science, Tezpur University, AssamTezpur, India*

SATYA SUNDAR BHATTACHARYA • *Soil and Agro-Bioengineering Laboratory, Department of Environmental Science, Tezpur University, Tezpur, Assam, India*

EMILIANA BOTTO • *Laboratorio de Biocatálisis y Biotransformaciones, DEPBIO-DQO, Universidad de la República (UdelaR), Montevideo, Uruguay*

C. BRÍGIDO • *MED-Mediterranean Institute for Agriculture, Environment and Development & CHANGE - Global Change and Sustainability Institute, Institute for Advanced Studies and Research, Universidade de Évora, Évora, Portugal*

SAPNA CHANDWANI • *C.G. Bhakta Institute of Biotechnology, UkaTarsadia University, Surat, India*

RAMACHANDRAN CHELLIAH • *Food Science and Biotechnology, School of Agriculture and Life Sciences, Kangwon National University, Chuncheon, Republic of Korea*

KI CHOON CHOI • *Grassland and Forage Division, National Institute of Animal Science, Seoul, South Korea*

RATAN CHOWDHURY • *Department of Botany, Rangapara College, Amaribari, Rangapara, Assam, India*

AMITA SUNEJA DANG • *Centre for Medical Biotechnology, M. D. University, Rohtak, Haryana, India*

EDILENE OLIVEIRA DA SILVA • *Instituto de CiênciasBiológicas, Universidade Federal doPará, Belém, PA, Brazil*

JULIANA FONSECA MOREIRA DA SILVA • *Universidade Federal do Tocantins, Palmas, TO, Brazil*

DEEPIKA • *Department of Microbiology, M. D. University, Rohtak, Haryana, India*

DHANASEKARAN DHARUMADURAI • *Department of Microbiology, School of Life Sciences, Bharathidasan University, Tiruchirappalli, Tamil Nadu, India*

MARIA DO CARMO PIMENTEL BATITUCCI • *Laboratório de Genética Vegetal e Toxicológica, Departamento de Ciências Biológicas, Universidade Federal do Espírito Santo, Vitória, Espírito Santo, Brazil*

José Luiz Martins do Nascimento • *Universidade Federal doAmapá, Macapá, Brazil; Instituto de CiênciasBiológicas, Universidade Federal doPará, Belém, PA, Brazil*

Drielly Dayanne Monteiro dos Santos Baliza • *Universidade Federal do Tocantins, Palmas, TO, Brazil*

Jean Carlos Vencioneck Dutra • *Secretaria de Estado da Educação do Espírito Santo, Vitória, Espírito Santo, Brazil; Laboratório de Genética Vegetal e Toxicológica, Departamento de Ciências Biológicas, Universidade Federal do Espírito Santo, Vitória, Espírito Santo, Brazil*

Eskálath Morganna Silva Ferreira • *Universidade Federal do Tocantins, Palmas, TO, Brazil*

Irlon M. Ferreira • *Universidade Federal doAmapá, Macapá, Brazil*

Harshida A. Gamit • *C.G. Bhakta Institute of Biotechnology, UkaTarsadia University, Surat, India*

Sonia Rodríguez Giordano • *Laboratorio de Microbiología Molecular, DEPBIO, Facultad de Química, Universidad de la República (Udelar), Montevideo, Uruguay; Laboratorio de Biocatálisis y Biotransformaciones, DEPBIO-DQO, Facultad de Química, Universidad de la República (Udelar), Montevideo, Uruguay*

David Gonzalez • *Laboratorio de Biocatálisis y Biotransformaciones, DEPBIO-DQO, Facultad de Química, Universidad de la República (Udelar), Montevideo, Uruguay*

Preethi Gopalsamy • *Askoscen Probionics (R & D), Trichy, Tamilnadu, India*

Senthilkumar Govindarajan • *PG & Research Department of Microbiology, AVVM Sri Pushpam College (Auto), Thanjavur, Tamilnadu, India*

K. R. Jeya • *Department of Biotechnology, Government Arts and Science College (Women), Perambalur, Tamil Nadu, India*

Gangadharaiah Jyothi • *Regional Plant Quarantine Station, Bangalore, Karnataka, India*

T. R. Kavitha • *University of Agricultural Sciences, Bangalore, Karnataka, India*

Manikyanahalli Chandrashekara Keerthi • *ICAR- Indian Grassland & Fodder Research Institute, Jhansi, UP, India*

Pradeep Kumar • *Department of Microbiology, Rohtak, Haryana, India; Department of Microbiology, M. D. University, Rohtak, Haryana, India*

Rishikesh Kumar • *ICAR-Indian Institute of Pulses Research, Kanpur, UP, India*

Leelavathi Lakshmanan • *Ministry of Environment Forest and Climate Change, Wildlife Institute of India, Dehradun, India; Forest, Ecology and Environment Department, Karnataka Biodiversity Board, Karnataka, India*

Thenmozhi Manivel • *Askoscen Probionics (R & D), Trichy, Tamilnadu, India*

Lakshmaiah Manjunatha • *Division of Crop Protection, ICAR-Indian Institute of Pulses Research, Kanpur, Uttar Pradesh, India*

Nanjundappa Manjunatha • *ICAR-National Research Centre on Pomegranate, Solapur, Maharashtra, India*

Raj Kumar Mishra • *ICAR-Indian Institute of Pulses Research, Kanpur, UP, India*

Karnan Muthusamy • *Grassland and Forage Division, National Institute of Animal Science, Seonghwan, South Korea; Askoscen Probionics (R & D), Trichy, Tamilnadu, India*

Lizzy A. Mwamburi • *University of Eldoret, Eldoret, Kenya*

G. Narasa Reddy • *College of Sericulture, Chintamani, University of Agricultural Sciences, Bangalore, Karnataka, India*

Gummeral Narasareddy • *College of Sericulture, Chintamani, University of Agricultural Sciences, Bangalore, Karnataka, India*
Joyce Prisca Bakuwa Njoloma • *Lilongwe University of Agriculture and Natural Resources (LUANAR), Bunda College Campus, Lilongwe, Malawi*
Deog-Hwan Oh • *Food Science and Biotechnology, School of Agriculture and Life Sciences, Kangwon National University, Chuncheon, Republic of Korea*
A. Paço • *MED-Mediterranean Institute for Agriculture, Environment and Development, CEBAL-Centro de Biotecnologia Agrícola e Agro-Alimentar do Alentejo, Beja, Portugal*
Tamilkani Pichai • *Department of Hospital Administration, Queen's College of Arts & Science, Thanjavur, Tamilnadu, India*
Raphael Sanzio Pimenta • *Universidade Federal do Tocantins, Palmas, TO, Brazil*
A. Prakash • *Indian Biotrack Research Institute, Thanjavur, Tamil Nadu, India*
Rinka S. Pramanik • *C.G. Bhakta Institute of Biotechnology, UkaTarsadia University, Surat, India*
Chikkabuddi Prashantha • *Regional Central Integrated Pest Management Centre, Faridabad, Haryana, India*
Reshma Devi Ramesh • *Forest, Ecology and Environment Department, Karnataka Biodiversity Board, Karnataka, India*
Simran Rani • *Department of Microbiology, Rohtak, Haryana, India; Department of Microbiology, M. D. University, Rohtak, Haryana, India*
R. Ranjini • *Kittur Rani Channamma College of Horticulture , University of Horticultural Sciences, Bagalkot, Karnataka, India*
Bankapura Mariyappa Ravikumara • *Plant Protection Division, Ministry of Agriculture & Farmers Welfare, Quarantine & Storage, KrishiBhawan, New Delhi, India*
Paula Rodríguez Bonnecarrere • *Laboratorio de Biocatálisis y Biotransformaciones, DEPBIO-DQO, Facultad de Química, Universidad de la República (UdelaR), Montevideo, Uruguay*
Shah Rushabh • *C.G. Bhakta Institute of Biotechnology, UkaTarsadia University, Surat, India*
A. Sankaranarayanan • *Department of Life Sciences, Sri Sathya Sai University for Human Excellence, Kamalapur, Kalaburagi, Karnataka, India*
Sarita • *Department of Microbiology, Rohtak, Haryana, India*
Vijayalakshmi Selvakumar • *Food Science and Biotechnology, School of Agriculture and Life Sciences, Kangwon National University, Chuncheon, Republic of Korea; Askoscen Probionics (R & D), Trichy, Tamilnadu, India*
N. Sengottaian • *Department of Microbiology, Urumu Dhanalakshmi College, (Affiliated to Bharathidasan University), Trichy, Tamil Nadu, India*
R. Shanmugapriya • *Department of Botany, A.V.V.M. Sri Pushpam College (Autonomous), (Affiliated to Bharathidasan University, Trichy), Thanjavur, Tamil Nadu, India*
A. S. Shijila Rani • *Department of Microbiology, Marudupandiyar College (Affiliated to Bharathidasan University, Trichy), Thanjavur, Tamil Nadu, India*
Simran • *Department of Microbiology, M. D. University, Rohtak, Haryana, India*
Bansa Singh • *ICAR-Indian Institute of Pulses Research, Kanpur, UP, India*
Kittan Subathra • *Plant Quarantine Station, Cochin, Kerala, India*
C. Suneetha • *University of Agricultural Sciences, Bangalore, Karnataka, India*
Pooja Suneja • *Department of Microbiology, M. D. University, Rohtak, Haryana, India*

T. R. Sunitha • *University of Agricultural Sciences, Bangalore, Karnataka, India*
Thirumagal Thirugnanam • *Department of Microbiology, School of Life Sciences, Bharathidasan University, Tiruchirappalli, Tamil Nadu, India*
Leela Saisree Uppala • *University of Massachusetts-Cranberry Station, East Wareham, MA, USA*
M. Veerapagu • *Department of Biotechnology & Microbiology, National College (Autonomous), Tiruchirapalli, Tamil Nadu, India*

Part I

Cultivation Methods of Bacterial Endophytes

Chapter 1

Isolation of Endophytic Bacteria from Leaves, Stems, or Roots

Joyce Prisca Bakuwa Njoloma

Abstract

Following the discovery of endophytic bacteria in potato, sugarcane, and rice, isolation and characterization of the same has advanced over the years. Studies of these endophytes in plant tissues require techniques that would enable researchers to either locate them in the plant tissues or isolate them from the tissues for subsequent visualization. Over the years, endophytic diazotrophic bacteria have been isolated from root, culm, stem, and leaf tissues and characterized for their activities in different crop plants, including plant growth promoters and biological nitrogen fixation. Various methods and culture media have been used depending on the crop plant under study and available resources for the laboratory procedures.

Key words Endophytic, Diazotrophic bacteria, Culture media, Biological fixation, Plant tissues

1 Introduction

Endophytic bacteria are a unique group of bacteria that colonize the interior tissues of a plant [1] and that do not visibly harm the plant [2]. The word endophyte is a combination of two terms (endon, "inner" or "within"; phyton = plant) to mean "in the plant" [3], so bacteria that reside in the plant tissues without inducing harmful reactions, disease symptoms, or organogenesis [4] are referred to as endophytic bacteria. The endophytic bacteria colonize niches that are more conducive to mutual interactions with healthy plant hosts [5] and were first reported in the internal tissues of a healthy potato plant [4]. They are capable of colonizing various plant organs such as roots, stems, leaves, flowers, fruits, and seeds [6, 7], indicating different capacities to colonize various internal plant tissues. They are reported to even colonize legume nodules [8] and tubercles of mycorrhizal fungi [9].

A. Sankaranarayanan et al. (eds.), *Endophytic Microbes: Isolation, Identification, and Bioactive Potentials*,
Springer Protocols Handbooks, https://doi.org/10.1007/978-1-0716-2827-0_1,

Some endophytic bacteria can provide nitrogen to the plants through biological nitrogen fixation and these are known as endophytic diazotrophic bacteria [10]. With the discovery of endophytic diazotrophic bacteria in sugarcane, other agronomically important crop species such as rice [11, 12], corn [13, 14], wheat [15], and Kallar grass (*Leptochloafusca* L.) [16] have been postulated to receive significant amounts of biologically fixed N from the atmosphere associated with the presence of bacteria. Thus, the endophytic diazotrophic bacteria have since been considered an important source of nitrogen input in agriculture and represent a promising alternative biofertilizer for chemical fertilizers [17, 18]. Besides fixing nitrogen [19, 20] endophytic bacteria can produce plant growth-promoting substances (hormones) like auxin and gibberellin in many field crops, help in nutrient uptake, and increase the plant's tolerance to biotic and abiotic stresses [21, 22].

Over the years, a number of scientific studies have led to the isolation of endophytic bacteria from a variety of plant species and evaluating their benefits using various methods and culture media depending on the crop plant under study and available resources for the laboratory procedures. This protocol highlights the isolation of endophytic bacteria from leaf, stem, and root tissues in general with reference to selected crop plants.

2 Materials

2.1 Isolation of Endophytic Bacteria from Tissues of Plant Leaves, Stems, and Roots

- Healthy fresh plant leaves, stems, and roots.
- Polythene bags (for sample collection).
- Ethanol.
- Sodium hypochlorite.
- Distilled water (washing purpose).
- Erlenmeyer flasks.
- Mortar and pestle.
- Beakers.
- Spatula.
- Test tubes.
- Petri dishes.
- Agar medium depending on availability in the various laboratories (N-free semi-solid and solidified media – Luria Bertani (LB), LGI-P, nutrient agar, potato dextrose agar, etc.).
- Yeast extract.
- Glycerol.

3 Methods

3.1 Isolation of Endophytic Bacteria from Tissues of Plant Leaves, Stems, and Roots

1. Collect fresh and healthy plant leaves, stems, and roots, thoroughly wash under running tap water, and rinse at least three times in distilled water.
2. Surface sterilize with ethanol followed by sodium hypochlorite (NaOCl) and then rinse at least three times in distilled water.
   - **Note**: Different concentrations and exposure duration have been used for different crop plants.
     (i) Rice stems and leaf sheaths surface-sterilize in 70% ethanol for 30 s followed by 2% NaOCl for 5 min.
     (ii) Leguminous plants (roots, nodules, leaves, and stems) surface-sterilize in 70% ethanol for 30 s followed by 3% NaOCl for 4 min.
     (iii) Sweet potato internode surface-sterilize in 95% ethanol for 1 min, followed by 5% NaOCl for 5 min.
     (iv) Sugarcane stems surface sterilized in 70% ethanol for 10 min, followed by 2% sodium hypochlorite (NaOCl) for 20 min.
3. After the surface sterilization, macerate 10 g of the plant materials in sterile distilled water.
   - For sugarcane stems, you peel using a sterile knife and then cut into small pieces about 3–5 cm from which sugarcane juice is extracted using a sterile mortar and pestle.
4. Dilute macerated tissues into 10^{-1} dilution by adding 9 volumes of sterile distilled water.
5. Make a serial dilution of up to 10^{-6} dilution by taking 1 mL of well-shaken suspension and adding into 9 mL water blank tubes.
6. Plate a 100 μL aliquot on semi-solid medium.
7. Incubate the Petri dishes at 30 °C for 4–6 days.
8. Purify single colonies of the bacteria isolates on new agar plates of the same medium and incubate for 3–5 days.
9. Distinct colonies grown on agar plates can be randomly picked up and transferred to a fresh semi-solid N-free medium for final purification.
10. Purified colonies can then be stored in 50% glycerol −80 °C for later characterization and identification.

References

1. Elbeltagy A, Nishioka K, Sato T, Suzuki H, Ye B, Hamada T, Isawa T, Mitsui H (2001) Endophytic colonization and in planta nitrogen fixation by a *Herbaspirillum* sp. isolated from wild rice species. Appl Environ Microbiol 67:5285–5293
2. Hallmann J, Quadt-Hallmann A, Mahafee WF, Kloepper JW (1997) Bacterial endophytes in agriculture crops. Can J Michrobiol 43:895–914
3. Chanway CP (1996) Endophytes: they're not just fungi. Can J Bot 74(3):321–322
4. Trevet IW, Hollis JP (1948) Bacteria in storage organs of healthy plants. Phytopathology 38: 960–967
5. Richardson A, Barea JM, McNeill A, Prigent-Combaret C (2009) Acquisition of phosphorus and nitrogen in the rhizosphere and plant growth promotion by microorganisms. Plant Soil 321(1–2):305–339
6. James EK, Gyaneshwar P, Mathan N, Barraquio WL, Reddy PM, Iannetta PPM, Olivares FL, Ladha JK (2002) Infection and colonization of rice seedlings by the plant growth- promoting bacterium Herbaspirillumseropedicae Z67. Mol Plant-Microbe Interact 15(9): 894–906
7. Trognitz F, Piller K, Nagel M, Borner A, Bacher C-F, Rechlik M, Mayrhofer H, Sessitsch A (2014) Isolation and characterization of endophytes isolated from seeds of different plants and the application to increase juvenile development. In Future seed – production, marketing, use and conservation [TagungZukünftigesSaatgut—Produktion, Vermarktung, Nutzung und Konzervierung]; 24–26 November 2014; Austria, pp 25–28. Irdning: HöhereBundeslehr- und ForschungsanstaltfürLandwirtschaftRaumberg-Gumpenstein
8. Benhizia Y, Benhizia H, Benguedouar A, Muresu R, Giacomini A, Squartini A (2004) Gamma proteobacteria can nodulate legumes of the genus Hedysarum. Syst Appl Microbiol 27(4):462–468
9. Paul LR, Chapman WK, Chanway CP (2013) Diazotrophic bacteria reside inside *Suillustomentosus/Pinus contorta* tuberculate ectomycorrhizae. Botany 91(1):48–52
10. Cavalcante VA, Döbereiner J (1988) A new acid tolerant nitrogen fixing bacterium associated with sugarcane. Plant Soil 108(1):23–31
11. Baldani VLD, Baldani JI, Döbereiner J (2000) (2000). Inoculation of rice plants with the endophytic diazotrophs *Herbaspirillumseropedicae* and Burkholderia spp. Biol Fertil Soils 30(5–6):485–491
12. Hurek T, Handley LL, Reinhold-Hurek B, Piche Y (2002) Azoarcus grass endophytes contribute fixed nitrogen to the plant in an unculturable state. Mol Plant-Microbe Interact 15(3):233–242
13. Olivares FL, Baldani VLD, Reis VM, Baldani JI, Döbereiner J (1996) Occurrence of the endophytic diazotrophs Herbaspirillum spp. in roots, stems, and leaves, predominantly of Gramineae. Biol Fertil Soils 21(3):197–200
14. Montañez A, Abreu C, Gill PR, Hardarson G, Sicardi M (2009) Biological nitrogen fixation in maize (Zea mays L.) by 15N isotope-dilution and identification of associated culturable diazotrophs. Biol Fertil Soils 45(3): 253–263
15. Sabry RS, Saleh SA, Batchelor CA, Jones J, Jotham J, Webster G, Kothari SL, Davey MR, Cocking EC (1997) Endophytic establishment of Azorhizobiumcaulinodans in wheat. Proc R Soc Lond B Biol Sci 264(1380):341–346
16. Barraquio WL, Revilla L, Ladha JK (1997) Isolation of endophytic diazotrophic bacteria from wetland rice. Plant Soil 194:15–24
17. Rao HC, Savalgi VP (2017) Isolation and screening of nitrogen fixing endophytic bacterium *Gluconacetobacterdiazotrophicus* GdS25. Int J Curr Microbiol App Sci 6(3):1364–1373
18. Puri A, Padda KP, Chanway CP (2018) Nitrogen-fixation by endophytic bacteria in agricultural crops: recent advances. IntechOpen, London
19. Kirchhof G, Reis VM, Baldani JI, Eckert B, Döbereiner J, Hartmann A (1997) Occurrence, physiological and molecular analysis of endophytic diazotrophic bacteria in gramineous energy plants. Plant Soil 194:45–51
20. Reinhold-Hurek B, Hurek T (1998) Interactions of gramineous plants with Azoarcus spp. and other diazotrophs: identification, localization, and perspectives to study their function. Crit Rev Plant Sci 17(1):29–54
21. Sturz AV, Matheson BG (1996) Populations of endophytic bacteria which influence host resistance to Erwinia induced bacterial soft rot in potato tubers. Plant Soil 184:265–271
22. Sturz AV, Christie BR, Matheson BG, Nowak J (1997) Biodiversity of endophytic bacteria which colonize red clover nodules, roots, stems and foliage and their influence on host growth. Biol Fertil Soils 25:13–19

Chapter 2

Heat-Treatment Method for Isolation of *Bacillus* spp. from Plant Tissues

T. R. Kavitha, C. Suneetha, and T. R. Sunitha

Abstract

Production of spore is an important feature of certain types of bacteria belonging to the members of genera such as *Bacillus*. Spores are metabolically inactive and in heat-resistant forms. This state occurs due to unfavorable environmental conditions and hinders the vegetative activity of the cell. Spore-forming *Bacillus* can resist excessive heat due to their chemical composition, therefore, to isolate spore-forming *Bacillus* species, heat shock and ethanol treatment could be the efficient ways.

Key words *Bacillus*, Isolation, Heat-resistant spore, Eco-friendly, Temperature

1 Introduction

Gram-positive aerobic Bacilli are represented by an increasing number of different genera and families. The most recent edition of Bergey's manual listed 19 different genera belonging to the family Bacillaceae [1], such as *Bacillus*, *Alkali bacillus*, *Halobacillus*, *Lentibacillus*, *Oceanobacillus*, and *Virgibacillus*. Many species, originally described as being members of the genus *Bacillus*, are now reclassified as being representatives of different genera and families.

Formulations based on *Bacillus* spores are especially suitable for industrial products since they remain alive for nearly unlimited time and can be stored like chemical products. *Bacillus* spp. are highly adapted to abiotic and biotic stress. They enhance plant tolerance against abiotic stress such as salinity, drought, and heavy metal toxicity. Due to their plant growth-promoting activity, many species of *Bacillus* are commercially used as phyto stimulators and biofertilizers. Furthermore, plant growth-promoting bacilli enable plants to suppress plant pathogens (biocontrol). Biocontrol products based on *Bacillus* and *Paenibacillus* spp. are increasingly used in agriculture as an environmentally friendly alternative to chemical pesticides.

A. Sankaranarayanan et al. (eds.), *Endophytic Microbes: Isolation, Identification, and Bioactive Potentials*,
Springer Protocols Handbooks, https://doi.org/10.1007/978-1-0716-2827-0_2,

Due to the presence of heat-resistant endospores, heat treatment (usually 10 min at 80 °C) is widely used to enrich the endospore-forming Bacilli [2]. While soil is a very common source for the isolation of Bacilli and one has also to consider that Bacilli could contaminate various foodstuffs and could poison different kinds of foods due to their production of heat-stable toxins (e.g., *Bacillus cereus*). In addition, Bacilli are also able to colonize plants and can be isolated from plant rhizosphere and from the surface of different plant organs, e.g., roots, stems, and leaves. In addition, *Bacillus* endophytes also occur in inner plant tissues.

Genus *Bacillus* is regarded as one of the classical genera for the biological control of plant diseases. The heat and desiccation-resistant spores make them more potent biological control agents. To ease of stable formulation, GRAS (generally regarded as safe) status by the US Food and Drug Administration are additional properties that make *Bacillus* the most used plant growth-promoting microbe. A number of *Bacillus* species such as *B. pumilus*, *B. amyloliquefaciens*, *B. cereus*, *B. subtilis*, *B. licheniformis*, *B. megaterium*, *B. mycoides*, and *B. velezensis* have been commercialized as bio-formulation.

2 Materials

2.1 Isolation of Bacillus from Roots

- Roots.
- Polythene bag (for sample collection).
- Beaker.
- NaCl.
- Mixer grinder.
- Petri plates.
- Micropipettes.
- Incubator.
- Microscope.

2.2 Isolation from the Cosmosphere, Microflora Associated with Corm Sheath

- Corm sheath.
- Normal saline.
- Test tubes.
- Petri plates.
- Luria-Bertani – 10 g of tryptone, 5 g of yeast extract, 10 g of NaCl, and 1 L of distilled water; adjust the pH to 7.0 with 1 N NaOH; and autoclave the mixture for 25 min at 120 °C.
- Nutrient agar – 0.5% peptone, 0.3% beef extract/yeast extract, 1.5% agar, 0.5% sodium chloride, distilled water, pH adjusted to neutral (6.8) at 25 °C (77 °F).

- Minimal media agar – dextrose 1 g, ammonium sulfate 1 g, dipotassium phosphate 7 g, monopotassium phosphate 2 g, sodium citrate 0.5 g, magnesium sulfate 0.1 g, agar 15 g, distilled water 1000 mL.
- Incubator (37 °C).
- Glycerol.

2.3 Isolation from the Solanaceous Phylloplane

- Healthy leaf samples.
- Rotary shaker.
- NaCl.
- Centrifuge.
- Nutrient agar.
- Petri plates.
- Glycerol nutrient broth.

2.4 Isolation from Cacao Leaf Tissue

- Leaf samples.
- Bleach (Clorox).
- Sterile distilled water.
- Pestle and mortar.
- 0.1 M Potassium phosphate buffer (pH 7.0).
- Test tubes.
- Incubator.
- Tryptic Soy Agar (TSA).
- Petri plates.

2.5 Isolation from Cacao Pods

- Cacao pods.
- Clorox.
- Test tubes.
- 0.1 M Potassium phosphate buffer (pH 7.0).
- Incubator (75 °C for 15 min).
- TSA (15 g tryptone, 5 g soytone-enzymatic digest of soybean meal, 5 g sodium chloride, 15 g agar).
- 20% Glycerol.
- Liquid nitrogen.

2.6 Isolation from Cacao Pods, Branch, and Flower

- Cacao pod, branch, and flower.
- 0.1 M Potassium phosphate buffer.
- Incubator.

- TSA (15 g tryptone, 5 g soytone-enzymatic digest of soybean meal, 5 g sodium chloride, 15 g agar).
- 20% glycerol.
- Liquid nitrogen.

2.7 Isolation from White Asparagus

- White asparagus.
- Tryptone glucose extract agar (TGE) (5 g – Casein peptone, 3 g – beef extract, 1 g – dextrose, 15 g – agar, 1000 mL – demineralized water, pH 7 ± 2 @ 25 °C).
- Incubator.
- Casoy broth supplemented with glycerol.
- −80 °C.

2.8 Isolation from the Spermosphere of Peas and Dwarf French Beans

- Seeds.
- Nutrient broth.
- Sonicator.
- Incubator.
- Water bath.
- Thin-walled test tubes.
- −80 °C.
- NA.

3 Methods

3.1 Isolation of Bacillus from Roots

- Immerse 10 g of root sample in a beaker containing 100 mL of saline solution (0.8% NaCl), grind the samples by using mixer.
- Heat the root extracts at 80 °C for 12 min to kill non-sporulating bacteria, as well as vegetative forms of sporulating bacteria, which allow for the recovery of only heat-resistant spores.
- Plate the aliquots (100 μL) onto three Petri dish plates containing NA.
- Incubate the plates in an incubator chamber at 29 °C for 5 days.
- Estimate the number of colonies by considering the typical morphological features of *Bacillus*.

3.2 Isolation from the Cormosphere, Microflora Associated with Corm Sheath

- Remove the soil adhered to corms by vigorous shaking and then separate the corm sheath (cormosphere) from corm by peeling it off under sterile conditions.
- Add 1 g of corm sheath in 10 mL of normal saline and heat for 20 min at 80 °C to select *Bacillus*.

- Spread the 10^{-6} dilution of the suspension on three media, i.e., Luria–Bertani, Nutrient Agar, and Minimal Media agar and incubate at 37 °C for 24 h.
- Purify the isolated *Bacillus* species by streak plate method and store in 60% glycerol at −80 °C till further screening [3].

3.3 Isolation from the Solanaceous Phylloplane

- Collect pieces of leaf (5 g) from healthy solanaceous plants (tomato, eggplant, and pepper), wash for 1 h by stirring them on a rotary shaker in 100 mL of physiological saline solution (0.85% NaCl).
- Centrifuge 50 mL of the suspension at 10,000 g for 10 min and suspend the precipitate in 1 mL of water.
- Heat the aliquots of the final suspension to 90 °C for 10 min, and then inoculate on nutrient agar at 28 °C.
- Select some representative colonies from the countable plates and streak twice on new NA plates to obtain pure colonies.
- Store purified bacterial strains at −80 °C in 20% glycerol nutrient broth.

3.4 Isolation from Cacao Tissue

- Collect the cacao tissue and sample and perform the analysis within 4 h.
- Remove four leaf disks (3.0 cm) from leaves, disinfect for 5 min in 20% bleach (Clorox), rinse three times in sterile distilled water, triturate with a mortar and pestle in 3 mL 0.1 M potassium phosphate buffer (pH 7.0).
- Then transfer to a test tube, and incubate at 75 °C for 15 min.
- Plate 50 μL of the suspension onto tryptic soy agar (TSA) and incubate for 48 h at 24 °C.
- Streak individual colonies on TSA to purify bacteria.

3.5 Isolation from Cocoa Pods, Branches, and Flowers

- Surface disinfect the Pods with 20% Clorox.
- Aseptically remove the exocarp and remove the sections (0.42 cm^3) of mesoderm. Remove the periderm aseptically from flower cushions and branches and sample the internal sections (0.5 cm^3).
- Place the tissue into test tubes with 2.0 mL of 0.1 M potassium phosphate buffer (pH 7.0) and incubate at 75 °C for 15 min.
- Directly place the pod, branch, and flower cushion sections on TSA, and incubate for 24–48 h at 24 °C.
- Select individual bacterial colonies and maintain on TSA.
- Store bacteria on TSA slants for shipping and in 20% glycerol under liquid nitrogen for long-term storage.

3.6 Isolation from Asparagus

- Collect white asparagus samples (perform sampling on three occasions during a period of 6 months, with the crop in different growth stages – vegetative, hoeing, and harvesting stage).
- Eliminate vegetative cells by heat treatment of the samples.
- Make serial decimal dilutions and inoculate onto tryptone glucose extract agar (TGE) plates.
- Incubate the plates at 55 °C under anaerobic conditions for isolation of thermophilic microorganisms.
- Chose the colonies of strains that represent all of the colony types that could be distinguished visually, and purify these by reculturing on the same medium.
- Identify the isolates as *Bacillus* sp. by morphological analysis.
- Store pure cultures in Casoy broth supplemented with glycerol (2:1, v/v) at −80 °C.

3.7 Isolation from the Spermosphere of Peas and Dwarf French Beans

- Wash the seeds by agitating seeds individually in nutrient broth for a minute using a whirl mixer.
- Sonicate the single seed individually in nutrient broth for 10s at the lowest setting using an MSE 10/75 Mk2 bench sonicator.
- Incubate the seeds for 5 days at either 10, 25, or 37 °C in 5 mL broth to encourage any *Bacillus* species present to produce spores.
- Place 1 mL samples of these treatments in a water bath in thin-walled test tubes at 80 °C for 10 min to select the spore-producing isolates by heat-killing the vegetative cells.
- Prepare a dilution series and plate the selected dilutions out onto NA plates [4].

3.8 Isolation from Sugarcane Roots

- Wash the root samples using a sponge and soap to remove large debris and then surface disinfest to eliminate the epiphytotic microbial community.
- Immerse 100 g. of root sample in a beaker containing 100 mL of saline solution (0.8% NaCl).
- Isolation of *Bacillus* by two methods.

World Health Organization Protocol

- Heat the root extract at 80 °C for 12 min.
- Plate the aliquots (100 μL) onto three Petri dish plates containing NA supplemented with penicillin G (100 mg/L).
- Incubate at 29 °C for 5 days.
- Estimate the number of colonies based on typical morphological features of *Bacillus*.

Polanczyk Protocol

- Heat the root extracts at 80 °C for 12 min.
- Plate the aliquots (100 μL) onto three Petri dishes containing NA growth medium.
- Incubate at 29 °C for 5 days.
- Inoculate the characterized and identified colonies in casein-casein-yeast (CCY) broth supplemented with penicillin G (100 mg/L) and place under agitation (200 rpm) at 28 °C for 72 h [5].

References

1. Logan NA, de Vos P (2015) Bacillus. In: Bergey's Manual Trust (ed) Bergey's manual of systematics of archaea and bacteria. Wiley in association with Bergey's Manual Trust, Hoboken, pp 1–163. https://doi.org/10.1002/9781118960608.gbm00530
2. Fritze D (2002) *Bacillus* identification – traditional approaches. In: Fédération européenne des sociétés de microbiologie (ed) Applications and systematics of bacillus and relatives. Blackwell Publishing, Oxford, pp 8–22
3. Magotra S, Bhagat N, Ambardar S, Ali T, Hurek BR, Hurek T, Verma PK, Vakhlu J (2021) Field evaluation of PGP Bacillus sp. strain D5 native to Crocus sativus, in traditional and non-traditional areas, and mining of PGP genes from its genome. Sci Rep 11:5454. https://doi.org/10.1038/s41598-021-84585-z
4. Walker R, Powell AA, Seddon B (1998) Bacillus isolates from the spermosphere of peas and dwarf French beans with antifungal activity against Botrytis cinerea and Pythium species. J App Microbiol 84:791–801
5. Ferreira MEP, Leite LG, Polanczyk RA, Suzuki MT, de Paula Freitas R, da Silva Bueno RN, da Silva RSA (2018) Two methods for isolation of endophytic and edaphic *Bacillus* spp. from sugarcane fields. Arq Inst Biol 84:1–7

Chapter 3

Isolation and Detection of Endophytic *Streptomyces* from Actinorhizal *Casuarina* Root Nodules

Thirumagal Thirugnanam and Dhanasekaran Dharumadurai

Abstract

Streptomyces have received much attention due to their significant role in the decomposition of organic matter and the production of biologically active primary and secondary metabolites. *Streptomyces* are saprophytic, soil-dwelling bacteria, and most *Streptomyces* are rhizosphere colonizers and reported for their plant growth-promoting property by phytohormone (IAA) production, increasing nutrient assimilation, and antibiosis. After the collection and pretreatment of root nodules, serial dilution is generally employed for the isolation of endophytic *Streptomyces* bacterium from actinorhizal root nodules. Several physiological, biochemical, and molecular identification methods are used for the detection of *Streptomyces* species.

Key words Actinorhizal plant, Endophytes, Actinobacteria, Amphotericin-B

1 Introduction

Plants show an intrinsic relationship with a broad range of microbial populations that colonize the rhizosphere (rhizobacteria), the phyllosphere (epiphytic bacteria), and plant tissue (endophytes) [1]. An endophytic microorganism resides within its host plant, for at least a part of its life without causing any apparent disease symptom or infection. Actinorhizal plants have been seen in eight genera belonging to three orders Fagales, Rosales, and Cucurbitales. These plants have root nodules inhabited by bacteria identified as nitrogen-fixing actinobacteria [2]. Actinobacteria are Gram-positive, typically filamentous bacteria in the phylum actinobacteria [3]. The actinobacteria particularly *Streptomyces* species are well-known saprophytic soil-dwelling bacteria that decompose organic matter, notable polymers such as lignocellulose, starch, and chitin in soil [4]. The presence of L, L isomer of 2,6-diaminopimelic acid (L, LDAP) and the absence of any diagnostic sugar in the cell wall are the salient features of the genus *Streptomyces* [5]. They have a

A. Sankaranarayanan et al. (eds.), *Endophytic Microbes: Isolation, Identification, and Bioactive Potentials*,
Springer Protocols Handbooks, https://doi.org/10.1007/978-1-0716-2827-0_3,

competitive advantage above many other microorganisms in soil ecosystems, because of their filamentous and sporulating lifestyle, which provides them to persist during harsh environmental conditions, and the most striking property of *Streptomyces* is the ability to produce bioactive secondary metabolites such as antibacterial, antifungal, antiviral, antitumoral, anti-hypertensives, and mainly antibiotics and immunosuppressives [6]. Nearly two-thirds of natural antibiotics have been isolated from actinobacteria, and around 75% of them are from the *Streptomyces* genus. Amphotericin B is widely used to prevent fungal contamination in actinobacteria isolation. Most *Streptomyces* are effective rhizosphere and rhizoplane colonizers [7]. They can also be endophytes colonizing the inner tissues of host plants. Endophytic *Streptomyces* may improve the growth of their host plants by the production of important plant hormones, increasing nutrient assimilation and antibiosis mechanism [8]. Endophytic Streptomyces are reported from several plants in actinorhizal plants *Casuarina* spp. and others including *Curcuma phaeocaulis, Catharanthes roseus, Azadiracta indica, Ocimum sanctum,* and *Phyllanthus amarus* (Table 1). The endophytic *Streptomyces* are present in various plant parts including leaf, stem, flower, bark, root, and root nodules.

Table 1
List of endophytic *Streptomyces* spp. isolated from various plants [9–22]

S. no	Streptomyces spp.	Host plant	Specimen	Country	Reference
1	*Streptomyces cavouensis* *Streptomyces flaveus* *Streptomyces mutabilis* *Streptomyces yeochonensis* *Streptomyces racemochromogenes* *Streptomyces polychromogenes* *Streptomyces antibioticus* *Streptomyces lannensis* *Streptomyces collinus*	*Emblica officinalis* *Terminalia arjuna* *Terminalia chebula* *Murraya koenigii* *Azadirachta indica* *Rauwolfia serpentina*	Root, Stem, Leaf	India, Ireland	[9]
2	*Streptomyces longisporoflavus*	*Rauwolfia densiflora*	Stem, Leaf	India	[10]
3	*Streptomyces phytohabitans*	*Curcuma phaeocaulis*	Root	China	[11]
4	*Streptomyces* spp. NRRL 30566	*Grevillea pteridifolia*	Stem	USA	[12]

(continued)

Table 1 (continued)

S. no	Streptomyces spp.	Host plant	Specimen	Country	Reference
5	*Streptomyces cavourensis* AB184264.1	*Catharanthes roseus*	Leaf	India	[13]
6	*Streptomyces olivochromogenes*	*Tinospora crispa Phaleria macrocarpa, Curcuma aeruginosa, Andrographis paniculata, Caesalpinia sappan, Curcuma xanthoriza, Gynura procumbens, Physalis peruviana, Hibiscus sabdariffa*	Root, Leaf, Stem	Indonesia	[14]
7	*Streptomyces* spp.	*Azadiracta indica, Ocimum sanctum, Phyllanthus amarus*	Root, Leaf	India	[15]
8	*Streptomyces aureofaciens* CMUAc130	*Zingiber officinale, Alpinia galanga*	Root	Thailand	[16]
9	*Streptomyces* spp.	*Mondo grass (Ophiopogon japonicus ker-Gawler), Kinginso (Goody procera), Common Sorrel (Rumex acetosa), Chusan Palm (Trachycarpus fortunei), Mugwort (Artemisia indica var.maximowiczii), Dokudami (Houttuynia), Prickly Sow-thistle (Sonchus asper), Fuki (Petasites japonicus), Hellebore (Helleborus orientalis), Pteridophyta*	Root	Japan	[17]
11	*Streptomyces halophytocola* spp. *nov.*	*Tamarix chinensis Lour*	Stem	China, Japan	[18]
16	*Streptomyces sampsonii, Streptomyces olivaceus, Streptomyces tempisquensis, Streptomyces anulatus.*	*Schima wallichii*	Leaf, Root, Bark, Flower	India	[19]
17	*Streptomyces griseorubiginosus*	*Musa acuminata (*Banana)	Root, leaf	China	[20]
18	*Streptomyces* spp.	Pigeon pea	Root	India	[21]

(continued)

Table 1 (continued)

S. no	Streptomyces spp.	Host plant	Specimen	Country	Reference
19	*Streptomyces niveus* *Streptomyces camponoticapitis* *Streptomyces diastaticus* *Streptomyces yanii* *Streptomyces laceyi* *Streptomyces hyalinus* *Streptomyces sanglieri* *Streptomyces cyslabdanicus* *Streptomyces corchorusii* *Streptomyces kebangsaanensis* *Streptomyces pulveraceus* *Streptomyces hyalinus*	*C. sinensis*, *C. japonica*, *C. sasanqua*, *C. rosiflora*, *C. drupifera*, *Eurya japonica*	Root, leaf	India	[22]

2 Materials

2.1 Collection and Pre-treatment of Root Nodules

- Fresh root nodules.
- Ziplock bags (collection purpose).
- Absolute alcohol.
- 3% sodium hypochlorite.
- Distilled water (washing purpose).

2.2 Isolation of Endophytic Streptomyces

- Root nodule sample of *Casuarina* spp.
- Mortar & pestle.
- 250 mL Erlenmeyer flask.
- GLM agar media.
- Actinomycetes isolation agar.
- *Streptomyces* agar.
- International *Streptomyces* project (ISP) medium 3.
- ISP 7.
- Starch casein agar.
- Amphotericin B (75 mg/mL) and rifampicin (2.5 mg/mL).
- Glycerol.

2.3 Characterization of Streptomyces

Characterization will be performed by the International *Streptomyces* Project by the method of Shirling and Gottlieb [27]. The comprehensive characterization and detection and findings for *Streptomyces* identification are shown in Table 3.

2.4 Microscopic Analysis

- ISP 4 agar.
- Glutaraldehyde (2.5%v/v).
- Osmium tetroxide (1%w/v).
- Ethanol.
- Gold.
- Electron microscope.

2.5 Molecular Identification

- PCR reaction fine chemicals as per standard protocols.
- Primers – PA: 5′-AGA GTT TGA TCC TGG CTC AG-3′ and PH: 5′-AAG GAG GTG ATC CAG CCG CA-3′.
- Thermocycler.
- 1.2% agarose.
- TAE Buffer.
- Transilluminator.

3 Methods

3.1 Collection and Pre-treatment of Root Nodules [23–25]

1. Pick the fresh nodules from the actinorhizal plant and mix with moist sandy soil in ziplock cover bags and store at −4 °C.
2. Rinse nodules in absolute alcohol for 10 s.
3. Sterilize with 3% sodium hypochlorite for 2–5 min.
4. Rinse 5–8 times with distilled water.
5. To validate the surface sterilization procedure, spread the aliquots of the sterile distilled water used in the final wash on GLM agar media (Table 2).
6. Examine for bacterial growth after incubation at 28 °C for 7 days.

3.2 Isolation of Streptomyces *from Root Nodules [22, 23]*

1. Crush the nodules in normal saline solution (0.85%) with mortar & pestle.
2. Prepare the tissue extract for tenfold serial dilution.
3. Weigh 1 gm of sample and mix in 9 mL sterile distilled water (stock solution).
4. Perform serial dilution (use 1 mL from the stock solution to prepare the final volume of 10^{-1} to 10^{-8} dilutions).
5. Finally spread 0.1 mL of suspension onto five selective isolation media, namely actinomycetes Isolation agar, Streptomyces agar, International Streptomyces Project (ISP) medium 3, ISP7, and Starch casein agar (Table 2).

Table 2
List of medium used in this study [22]

Medium	Composition	Reference
GLM agar	Yeast extract, 3 g; malt extract, 3 g; peptone, 5 g; starch, 10 g; agar, 20 g; distilled water, 1000 mL; pH 7.4 [Gms/liter]	[22]
AIA	Sodium caseinate 2.000, L-asparagine 0.100, sodium propionate 4.000, dipotassium phosphate 0.500, magnesium sulfate 0.100, ferrous sulfate 0.001, agar 15.000 final pH (at 25 °C) 8.1 ± 0.2 [Gms/liter]	[22]
SA	Malt extract 10.000, yeast extract 4.000, dextrose 4.000, calcium carbonate 2.000, agar 12.000 [Gms/liter]	[22]
ISP 3	Oat meal 20.000, agar 18.000, ferric sulfate heptahydrate 0.001, trace salts, manganese chloride tetrahydrate 0.001, zinc sulfate heptahydrate 0.001. Final pH (at 25 °C) 7.3 ± 0.2 [Gms/liter]	[22]
ISP 7	L-asparagine 1.000, L-tyrosine 0.500, dipotassium hydrogen phosphate 0.500, magnesium sulfate heptahydrate 0.500, sodium chloride 0.500, trace salt solution (mL) 1.000, agar 20.000. Trace salt solution contains ferrous sulfate heptahydrate 1.360 mg, copper chloride, $2H_2O$ 0.027 mg, cobalt chloride, $6H_2O$ 0.040 mg, sodium molybdate, dihydrate 0.025 mg, zinc chloride 0.020 mg, boric acid 2.850 mg, manganese chloride, tetrahydrate 1.800 mg, sodium tartarate 1.770 mg. Final pH (at 25 °C) 7.3 ± 0.1[Gms/liter]	[22]
ISP 4	Starch, soluble 10.000, dipotassium hydrogen phosphate 1.000, magnesium sulfate heptahydrate 1.000, sodium chloride 1.000, ammonium sulfate 2.000, calcium carbonate 2.000, ferrous sulfate heptahydrate 0.001, manganous chloride, heptahydrate 0.001, zinc sulfate heptahydrate 0.001, agar 20.000 final pH (at 25 °C) 7.2 ± 0.2 [Gms/liter]	[22]
SCA	Casein powder 1.00, starch 10.00, sea water 37.00, agar 15.00, distilled water 1000 mL, pH (at 25 °C) 7.2 ± 0.2 [Gms/liter]	[22], HiMedia

6. Supplement the media with amphotericin B (75 mg/mL) & rifampicin (2.5 mg/mL) to inhibit the growth of other endophytic fungi and bacteria.
7. Incubate the culture plates for 7–21 days at 28 °C.
8. Sub-culture the isolates on GLM agar medium.
9. Store in 20% glycerol at −80 °C for future use.

3.3 Characterization of Streptomyces spp.

3.3.1 Morphological and Physiological Characteristic Analysis [5, 26, 27, 31]

1. Examine the morphological characteristics of the isolates according to the methods recommended by Shirling and Gottlieb for the International *Streptomyces* Project by the method of Shirling and Gottlieb [27].
2. Tests for various physiological and biochemical characteristics as per standard procedures.
3. Interpret the results with previous findings of *Streptomyces* spp.

Table 3
Different morphological and biochemical characteristics of selected *Streptomyces* spp. [5]

Characteristics	1	2	3	4	5	6	7	8	9	10	11	12	13	14
Spore chains rectiflexiles	–	+	–	–	–	+	+	+	–	+	–	–	+	+
Spore chains spirales	+	–	+	+	+	–	–	–	+	–	+	+	–	–
Spore mass red	–	–	–	–	+	–	–	–	–	–	–	–	–	–
Spore mass grey	–	+	–	+	–	+	–	+	–	–	+	+	–	–
Spore mass yellow	+	–	+	–	–	–	–	–	–	–	–	–	–	–
Mycelia pigment red/orange	–	–	–	–	+	–	–	–		–	–	–	–	–
Diffusible pigment produced	+	–	+	–	–	–	–	–	–	–	+	–	–	–
Diffusible pigment yellow/brown	+	–	+	–	–	–	–	–	–	–	+	–	–	–
Melanine on tyrosine agar	–	–	–	–	+	–	–	–	+	+	–	–	–	–
Diaminopimelic acid	+	+	+	+	–	+	–	+	–	–	+	+	+	+
Coagulation of milk	+	+	–	+	+	+	–	+	+	+	+	+	+	–
Catalase test	+	+	+	+	+	+	+	+	+	+	+	+	+	+
Oxidase	+	+	+	+	+	+	+	–	+	+	+	+	+	+
H_2S production	–	+	+	+	+	+	+	+	+	+	+	–	+	+
Nitrate reduction	+	+	+	+	–	+	–	+	+	+	–	–	+	–
Citrate utilization	–	+	+	+	+	+	+	+	+	+	+	+	+	+
Hydrolysis of														
Starch	+	+	+	+	+	+	+	+	+	+	+	+	+	+
Lipid	+	+	+	–	+	+	+	–	–	+	+	–	+	+
Gelatin	+	+	+	+	+	+	+	+	+	+	+	+	+	+
Lecithin	+	–	+	–	+	–	+	–	–	+	–	+	–	–
Urease	+	+	+	+	–	+	–	+	+	+	+	+	+	+
Degradation of														
Guanine	+	+	–	–	+	+	+	+	+	+	–	–	+	+
Xanthine	+	+	–	+	+	+	+	–	+	+	–	+	+	–
Pectin	–	–	–	+	–	+	+	–	–	–	–	–	+	–
Arbutin	–	–	–	–	–	–	–	+	–	–	–	+	–	–
Adenine	+	+	+	+	+	+	+	+	+	+	–	–	–	–
Esculin	–	+	–	+	–	+	+	+	+	–	+	+	–	+
Utilization of														
Adonitol	+	–	–	–	+	–	–	–	+	–	–	+	+	–

(continued)

Table 3
(continued)

Characteristics	1	2	3	4	5	6	7	8	9	10	11	12	13	14
D-xylitol	+	_	_	_	_	_	_	_	_	_	_	+	+	_
Sucrose	_	+	_	+	+	_	_	+	_	_	+	+	+	_
D-glucose	+	+	+	+	+	+	+	+	_	+	+	+	+	+
D-galactose	+	_	+	+	+	+	_	+	_	_	_	_	+	_
D-fructose	+	+	+	+	+	+	_	+	_	+	_	+	+	_
D-mannitol	+	+	+	+	+	+	_	+	+	_	+	+	+	+
D-sorbitol	+	+	+	+	+	+	+	+	+	+	+	_	+	+
D-lactose	+	+	+	+	+	+	+	+	+	+	+	_	+	+
D-maltose	+	+	+	+	+	+	_	+	+	+	+	+	+	+
Raffinose	+	+	+	+	+	_	+	+	+	+	+	+	+	_
L-Rhamnose	_	+	_	+	_	+	_	+	+	_	+	_	+	+
Cellubiose	+	+	+	+	+	+	+	+	+	_	_	_	_	+
L-arabinose	_	+	+	+	_	+	+	+	_	+	_	+	+	_
Meso-inositol	+	+	+	+	+	_	_	+	+	+	+	+	+	+
Dextran	+	+	_	+	_	_	_	+	+	_	_	_	+	+
D-Melezitose	+	+	_	+	+	_	_	+	+	+	_	+	+	_
D-Milibiose	+	+	+	+	+	_	_	_	+	+	_	+	+	_
L-cysteine	_	+	+	+	+	+	+	+	+	_	+	_	_	+
L-histidine	+	+	_	+	+	+	+	+	+	_	+	+	+	+
L-valine	+	_	_	_	+	_	_	+	+	+	+	_	+	_
L-Hydroxyproline	_	_	_	_	+	_	+	_	_	+	_	+	+	_
L-arginine	+	+	+	+	+	+	+	+	+	+	+	+	+	+
L-asparagine	+	+	+	+	+	+	+	+	+	+	+	+	+	+
L-tyrosine	_	_	_	_	+	_	+	_	+	+	+	_	+	+
L-serine	+	+	+	+	+	+	+	+	+	+	+	+	+	+
L-phenylalanine	+	_	+	_	+	+	+	_	+	+	+	+	+	_
DL-a-amino-n-butyric acid	_	_	+	_	_	+	+	_	_	+	+	_	+	+

+ Positive; _ Negative

1. *Streptomyces romosus*; 2. *Streptomyces antibioticus*; 3. *Streptomyces chromogenus*; 4. *Streptomyces rouchi*; 5. *Streptomyces filipinensis*; 6. *Streptomyces atroolivaceus*; 7. *Streptomyces exfloliatus*; 8. *Streptomyces halstedii*; 9. *Streptomyces cyaneus*; 10. *Streptomyces violaceus*; 11. *Streptomyces olivaceoviridis*; 12. *Streptomyces lydicus*; 13. *Streptomyces phaeochromogenus*; 14. *Streptomyces chromofuscus*

3.3.2 Microscopic Analysis [28]

1. For electron microscopic analysis, inoculate spores of *Streptomyces* spp. on ISP 4 agar.
2. Incubate for 7 days at 28 °C.
3. Remove a plug of the culture and fix in glutaraldehyde (2.5%v/v).
4. Wash with water and post-fixed in osmium tetroxide (1%w/v) for 1 h.
5. Wash the sample twice with water and dehydrate in ascending ethanol before drying in a critical point drying apparatus (PolaronE3000).
6. Finally coat in gold and examine under a scanning electron microscope at 15 kV.

3.3.3 Molecular Identification [28–30, 32]

1. Extract the genomic DNA from *Streptomyces* as per standard protocol using phenol-chloroform extraction method.
2. Amplify the fragments of 16S rRNA gene by PCR using universal primers – PA: 5′-AGA GTT TGA TCC TGG CTC AG-3′ and PH: 5′-AAG GAG GTG ATC CAG CCG CA-3′.
3. Perform PCR amplifications in a total volume of 25 μL, containing 50 ng of genomic DNA, 2.5 μL (10×) Taq Buffer, 1.5 μL (15 mM) MgCl2, 2.0 (2.5 mM) dNTPs, 0.5 μL (10 pmol) each primer and 1 μL, *Taq* DNA polymerase.
4. The PCR conditions consist an initial denaturation at 94 °C for 5 min, followed by 30 amplification cycles of denaturation at 94 °C for 1 min, annealing at 57 °C for 40 s, extension at 72 °C for 1 min, and a final extension of 10 min at 72 °C.
5. Perform gel electrophoresis to check the amplified product on 1.2% agarose gel using TAE buffer along with desired DNA markers.
6. Separate the amplified PCR product and examine under gel documentation system.

References

1. Passari AK, Mishra VK, Gupta VK, Yadav MK, Saikia R, Singh BP (2015) *In vitro* and *In vivo* plant growth promoting activities and DNA fingerprinting of antagonistic endophytic actinomycetes associates with medicinal plants. PLoS One 10(9):e0139468. https://doi.org/10.1371/journal
2. Carro L, Pujic P, Trujillo ME, Normand P (2013) *Micromonospora* is a normal occupant of actinorhizal nodules. J Biosci 38:685–693. https://doi.org/10.1007/s12038-013-9359-y
3. Panitch B, Takuji N, Yoko T, Sompop P (2017) Diversity of endophytic actinomycetes isolated from roots and root nodules of *Pueraria candollei* Grah. ex Benth. and the analyses of their secondary metabolites. Chiang Mai J Sci 44(1): 1–14
4. Cao L, Qiu Z, You J, Tan H, Zhou S (2004) Isolation and characterization of endophytic Streptomyces strains from surface-sterilized tomato (*Lycopersicon esculentum*) root. Lett Appl Microbiol 39:425–430. https://doi.org/10.1111/j.1472-765X.2004.01606.x

5. Singh N, Rai V (2011) Isolation and characterization of *Streptomyces* spp. from durg district of Chhattisgarh for antimicrobial activity. Curr Trend Biotechnol Pharm 5(1):1043–1053
6. de Rudi Emerson LP, da Ingrid Reis S, Mayra KM, de João Lúcio A, de Janete Magali A (2012) Antibiotics produced by *Streptomyces*. Braz J Infect Dis 16(5):466–471
7. Oluwaseyi SO, Olubukola O (2019) Babalola1 *Streptomyces*: implications and interactions in plant growth promotion. Appl Microbiol Biotechnol 103:1179–1188. https://doi.org/10.1007/s00253-018-09577-y
8. Ryan FS, Martin K, Matthew H (2011) *Streptomyces* as symbionts: an emerging and widespread theme? FEMS Microbiol Rev:1–15. https://doi.org/10.1111/j.1574-6976.2011.00313.x
9. Anwesha G, Animesh G, Rajal D, Archana Y, Bhim PS, Vijai KG, Rajeev S, Ratul S (2015) Antimicrobial biosynthetic potential and genetic diversity of endophytic actinomycetes associated with medicinal plants. FEMS Microbiol Lett 362:fnv158. https://doi.org/10.1093/femsle/fnv158
10. Akshatha VJ, Nalini MS, D'Souza C, Prakash HS (2014) Streptomycete endophytes from anti-diabetic medicinal plants of the Western Ghats inhibit alpha-amylase and promote glucose uptake. Lett Appl Microbiol 58:433–439
11. Bian G-K, Qin S, Yuan B, Zhang Y-J, Xing K, Ju X-J, Li W-J, Jiang J-H (2012) *Streptomyces phytohabitans* spp. nov., a novel endophytic actinomycete isolated from medicinal plant *Curcuma phaeocaulis*. Ant van Leeuw 102: 289–296
12. Castillo U, Harper J, Strobel G, Sears J, Alesi K, Ford E et al (2003) Kakadumycins, novel antibiotics from *Streptomyces* spp. NRRL 30566, an endophyte of *Grevillea pteridifolia*. FEMS Microbiol Lett 224(2): 183–190
13. Kafur A, Khan AB (2011) Isolation of endophytic actinomycetes from *Catharanthes roseus* (L.) G. Don leaves and their antimicrobial activity. Iran J Biotechnol 9(4):301–306
14. Pujiyanto S, Lestari Y, Suwanto A, Budiarti S, Darusman LK (2012) Alpha-glucosidase inhibitor activity and characterization of endophytic actinomycetes isolated from some Indonesian diabetic medicinal plants. Int J Pharm Pharm Sci 4(1):327–333
15. Shenpagam NH, Kanchana D, Sinduja G, Sandhya R (2012) Isolation of endophytic actinomycetes from medicinal plants and its mutational effect in biocontrol activity. Int J Pharm Sci Res 3(11):4338–4344
16. Taechowisan T, Lumyong S (2003) Activity of endophytic actinomycetes from roots of *Zingiber officinale* and *Alpinia galanga* against phytopathogenic fungi. Ann Microbiol 53(3): 291–298
17. Matsumoto A, Takahashi Y (2017) Endophytic actinomycetes: a promising source of novel bioactive compounds. J Antibiot 70:514–519. https://doi.org/10.1038/ja.2017.20
18. Qin S, Bian GK, Tamura T, Zhang YJ, Zhang WD, Cao CL, Jiang JH (2013) *Streptomyces halophytocola* spp. nov., an endophytic actinomycete isolated from the surface-sterilized stems of a coastal halophyte *Tamarix chinensis* Lour. Int J Syst Evol Microbiol 63:2770–2775
19. Ajit Kumar P, Vineet Kumar M, Vijai Kumar G, Ratul S, Bhim Pratap S (2016) Distribution and identification of endophytic *Streptomyces* species from *Schima wallichii* as potential biocontrol agents against fungal plant pathogens. Pol J Microbiol 65(3):319–329
20. Lixiang C, Zhiqi Q, Xin D, Hongming T, Yongcheng L, Shining Z (2004) Isolation of endophytic actinomycetes from roots and leaves of banana (*Musa acuminata*) plants and their activities against *Fusarium oxysporum* f. Spp. Cubense. World J Microb Biotechnol 20:501–504
21. Bind M, Nema S (2019) Isolation and molecular characterization of endophytic bacteria from pigeon pea along with antimicrobial evaluation against *Fusarium udum*. Appl Microbiol Open Access 5:163. https://doi.org/10.35248/2471-9315.1000163
22. Borah A, Thakur D (2020) Phylogenetic and functional characterization of culturable endophytic actinobacteria associated with *Camellia spp.* for growth promotion in commercial tea cultivars. Front Microbiol 11:318. https://doi.org/10.3389/fmicb.2020.00318
23. Solomon L, Fassil A (2014) Symbiotic and phenotypic characteristics of Rhizobia nodulating faba bean *(Vicia faba) from* Tahtay Koraro, Northwestern zone of Tigray Regional State, Ethiopia. Int J Technol Enhance Emerg Eng Res 2(11):2347–4289
24. Narayanasamy M, Dhanasekaran D, Thajuddin N, Akbarsha MA (2020) Morphological, molecular characterization and biofilm inhibition effect of endophytic *Frankia* sp. from root nodules of Actinorhizal plant Casuarina sp. S Afr J Bot 134:72–83
25. Kuster E, Williams ST (1964) Selection of media for isolation of *Streptomyces*. Nat Publ Group 202
26. Prashith Kekuda TR (2016) Isolation, characterization and antimicrobial potential of

endophytic Actinomycetes. Int J Curr Microbiol App Sci 5(7):100–116. https://doi.org/10.20546/ijcmas.2016.507.008

27. Shirling EB, Gottlieb D (1966) Methods for characterization of *Streptomyces* species. Int J Syst Bacteriol 16(3):313–340
28. Higginbotham SJ, Murphy CD (2008) Identification and characterization of a *Streptomyces* spp. isolate exhibiting activity against methicillin-resistant *Staphylococcus aureus*. Microbiol Res 165(1):82–86. https://doi.org/10.1016/j.micres.2008.12.004
29. Wilson K (1997) Preparation of genomic DNA from bacteria. Curr Protoc Mol Biol 1997: 2.4.1–2.4.5
30. Ghodhbane-Gtari F, Nouioui I, Salem K, Ktari A, del Carmen Montero-Calasanz M, Tisa LS, Klenk H-P, Gtari M (2014) *Nocardia casuarinae* spp. nov., an actinobacterial endophyte isolated from root nodules of *Casuarina glauca*. Ant van Leeuw. https://doi.org/10.1007/s10482-014-0168-6
31. Narayanasamy M, Lavania R, Dhanasekaran D, Thajuddin N (2020) Plant growth promoting active metabolites from *Frankia* spp. of Actinorhizal *casuarina* spp. Appl Biochem Biotechnol 191:74–91. https://doi.org/10.1007/s12010-020-03243-8
32. Nithya K, Muthukumar C, Biswas Br, Alharbi N. S, Shine K., Khaled Jamal M and D. Dhanasekaran. 2018. Desert Actinobacteria as a source of bioactive compounds production with a special emphases on Pyridine-2,5-diacetamide a new pyridine alkaloid produced by *Streptomyces* sp.DA3-7. Microbiol Res, 207, 116–133. https://doi.org/10.1016/j.micres.2017.11.012

Chapter 4

Isolation of Endophytic Fungi from Stems, Leaves, or Roots

V. Ambikapathy, N. Sengottaian, A. S. Shijila Rani, S. Babu, and A. Anbukumaran

Abstract

Endophytic microorganisms are buried companions of plants living mutually beneficial life inside the host plant. Beneficial effects of endophytes are attaining importance with the possibility of obtaining novel medicinally important compounds as well as their role in increasing crop productivity, because they produce a variety of compounds and interact with other microorganisms, pathogenic and non-pathogenic. With the development of modern tools and techniques of molecular biology, the identity of these microorganisms and their interactions with the host and other organisms may be potentially used for human welfare.

Key words Endophytes, Stems, Leaves or roots, Potato dextrose agar, Malt extract

1 Introduction

Endophytes are microorganisms that inhabit the plant tissue without causing any harmful effects at least for one period of their life cycle as opposed to the epiphytic microorganisms that live on a plant's surface [1, 2]. Endophytes exist at intervals in plant tissues and will grow within roots, stems, and/or leaves, rising to sporulate at a plant or aged host tissue [3]. The characterization, population dynamics, and diversity of endophytes used to boost plant growth and plant health may be studied in detail for their novel biologically active secondary metabolites [4–6]. Endophytes usually comprise either bacteria or fungi. Fungal-mediated endophytes provide numerous benefits to their host plants in crop improvement [7].

Endophytic fungi have the ability to develop biotic and abiotic stress tolerance of their plant hosts. Endophytic fungi mitigating drought, salinity, and nutrient shortfall are reported in meta-analysis [8]. The diversity of endophytic colonizers depends on various specific factors, their host specificity, and mechanisms of

A. Sankaranarayanan et al. (eds.), *Endophytic Microbes: Isolation, Identification, and Bioactive Potentials*, Springer Protocols Handbooks, https://doi.org/10.1007/978-1-0716-2827-0_4,

plant growth promotion. Some endophytes can have a broad host range and can be used as bioinoculants in developing a safe and sustainable agriculture system [9].

2 Materials

- Healthy fresh leaves/stem/roots.
- Sterile polythene.
- Methanol.
- Distilled water.
- Potato dextrose agar (PDA) (peeled potato – 200 gm, dextrose – 20 gm, agar – 16 gm, distilled water – 1000 ml).
- Malt extract agar (MEA) (malt extract – 30 gm, mycological peptone – 5 gm, agar – 16 gm, distilled water – 1000 ml).
- Petri plate.
- Conical flask.
- Spirit lamp.
- Hot air oven.

3 Methods

3.1 Surface Sterilization Technique [10]

The surface sterilization of leaves/stem/roots consists of the following steps:

- Wash the healthy leaves/stem/roots under tap water to remove adhering soil particles and epiphytes.
- After that the healthy leaves/stems/roots sample is cut into small pieces of approximately 0.5 cm × 0.5 cm.
- Immerse the leaves/stem/roots samples in 0.1% mercuric chloride solution for 10 min. Followed by 2 min in 70% ethanol.
- Wash the samples with distilled water for several times and keep them in a sterile blotting paper for dryness.

3.2 Isolation of Endophytic Fungi [11]

- Prepare the potato dextrose agar medium or malt extract medium by sterilizing at 121 °C for 15 min in an autoclave.
- Pour the medium in a sterile Petri plate supplemented with streptomycin sulfate (20 mg/L) to suppress the growth of bacteria.
- After solidification, place a small piece of leaves/stem/roots on the surface of potato dextrose agar or malt extract agar by using sterilized forceps.

- Incubate the plates at room temperature for at least 10–15 days.
- After the incubation period, observe the plates for fungal growth, purify them by using the standard method and maintain them as individual colonies in an agar slant.

4 Note

The surface sterilization procedure used should be effective in killing any epiphytic microbe on the plant surface without affecting the host tissue.

References

1. Anbukumaran A, Ambikapathy V, Panneerselvam A (2016) Isolation and characterization of endophytic fungi from *Crinum asiaticum* Lin. Kodiyakarai, Tamilnadu. India Der Pharm Lett 8(9):269–275
2. Shijila Rani AS, Ambikapathy V, Panneerselvam A (2018) Diversity of endophytic fungi from leaves of *Centella asiactica* L. and *Melia azedarch* L. Int J Recent Sci Res 9(1): 23083–23086
3. Velmurugan P, Kamala-kannan S, Balachandar V, Lakshmanaperumalsamy P, Chae J, Oh B (2010) Natural pigment extraction from five filamentous fungi for industrial applications and dyeing of leather. Carbohydr Polym 79(2):262–268
4. Schulz B, Boyle C, Draeger S, Rommert AK, Krohn K (2002) Endophytic fungi: a source of biologically active secondary metabolites. Mycol Res 106:996–1004
5. Strobel GA (2003) Endophytes as sources of bioactive products. Microbes Infect 5:535–544
6. Schulz B, Boyle C (2005) The endophytic continuum. Mycol Res 109:661–687
7. Chitnis VR, Suryanarayanan TS, Nataraja KN, Rajendra Prasad S, Oelmuller R, Shaanker RU (2020) Fungal-mediated crop improvement: the way ahead. Front Plant Sci 11:561007. https://doi.org/10.3389/fpls.2020.561007
8. Rho H, Hsieh M, Kandel SL, Cantillo J, Doty SL, Kim SH (2018) Do endophytes promote growth of host plants under stress? A meta-analysis on plant stress mitigation by endophytes. Microb Ecol 75:407–418. https://doi.org/10.1007/s00248-017-1054-3
9. Afzala I, Zabta Khan S, Shomaila S, Shaheen S (2019) Plant beneficial endophytic bacteria: mechanisms, diversity, host range and genetic determinants. Microbiol Res 221:36–49
10. Gamboa MA, Laureano S, Bayman P (2002) Measuring diversity of endophytic fungi in leaf fragments: does size matter? Mycopathologia 156:41–45
11. Guo LD, Hyde KD, Liew ECY (1998) A method to promote sporulation in palm endophytic fungi. Fungal Divers 1:109–113

Part II

Biochemical Methods for Identification of Endophytes

Chapter 5

Detection of Endophytes by Cultivation Method

Paula Rodríguez, David Gonzalez, and Sonia Rodríguez Giordano

Abstract

Detection of endophytes can be performed through several techniques, among them, the so-called "Fragment Culture technique" is one of the oldest and more used ones. Until the advent of molecular biology techniques, fragment culture plating was the major technique for evaluating endophyte diversity and detecting endophytes. This technique is still universally used for the detection of fungi and actinomycetes. This technique is simple, inexpensive, and can be performed in a routine microbiology laboratory with no special equipment. This chapter provides the essentials to be undertaken for the endophyte detection through this methodology. This chapter also emphasizes the importance of plant material disinfection procedure and the use of adequate controls to ensure that detected microorganisms are true endophytes.

Key words Endophytes, Fungi, Actinomycetes, Fragment plating, Fragment culturing, Sampling, Disinfection, Vegetal culture media

1 Introduction

Endophytes are microorganisms that live inside the plant in a symbiotic relationship with its vegetable host [1]. They provide the host with important benefits, while the plant offers them nutrients and a protected ecological niche [2]. The symbiotic relationship between the plant and its microbiome has been well documented in agricultural studies. It includes natural phytopathogen, inhibition, nitrogen fixation, phytohormones production, mineral acquisition, and production of growth promoter compounds [3]. Nonetheless, in some cases, what is beneficial for the plant is harmful to animal or human health. An example of this is the presence of endophytic fungi of the genera *Neotyphodium* and *Epichloe* in agricultural important fescues which are detrimental to livestock due to the production of alkaloids compounds. The production of alkaloids leads to poor animal performance, poor weight gain, reduced conception rate, and in some cases, causes severe health problems such as ryegrass staggers or fescue toxicosis

A. Sankaranarayanan et al. (eds.), *Endophytic Microbes: Isolation, Identification, and Bioactive Potentials*,
Springer Protocols Handbooks, https://doi.org/10.1007/978-1-0716-2827-0_5,

[4, 5]. For this reason, reliable methods that allow endophyte detection during the breeding process, seed testing, and marketing have been pursued.

Different techniques are available, that fulfill the requirement for rapid detection and identification of diverse strains, including microscopic observation, differential staining techniques, immunological analysis, and molecular methodologies. In the abovementioned case, the immunological detection has been the most useful technique [4, 5], although molecular analysis targeting specific biosynthetic genes has also proved useful [6]. Culturing of surface disinfected tissue fragments or seeds, named "Fragment plating method", has also been used for detection of these fungi [7]. Although it is a time-consuming method to be used for routine analysis, it is inexpensive, and it allows morphological studies for fungi identification.

With the upcome of molecular biology-based methodologies, culturing techniques have been questioned. Molecular methodologies such as 16S PCR amplification or high-throughput sequencing (HTS) provide a direct assessment of endophytic microbiome with unprecedented details of community composition, and it has been preferred over culturing methodologies for its ability to detect obligate symbionts, as well as slow growing or rare microorganisms that do not grow readily on standard media [6, 8–11]. Nonetheless, some commonly cultured clades were never found by PCR-based approaches, indicating that both methods can yield complementary information [9].

Culture-based methods are frequently considered in studying the endophytic community, since they provide cultures for further experiments. This is particularly true for fungi and actinomycetes [12, 13]. Indeed, most knowledge on endophyte diversity, overall fungi diversity, is still based on traditional culturing approaches [8, 11]. The fragment plating method relies on the recovery of endophytes after surface disinfection by direct plating of a vegetable tissue fragment on different solid media. This procedure does not imply isolation, as it does the dilution to extinction method following sample maceration; and thus, the developed colonies could be associated with more than one microorganism. However, fragmentation is less harmful than maceration, and a larger endophyte recovery rate has been associated with this technique [14]. Sampling, sample conservation, proper manipulation, as well as media and growing conditions have a strong impact on endophyte recovery and detection.

A rationale for the sampling procedure was proposed by Strobel [1], who suggested to guide the selection of plant species and habitat based on the particular trait that is pursued. Endophytic community varies with plant species, but also with geographical location as well as abiotic or biotic stress factors affecting the plant. Additionally, areas with greater biodiversity on the plant side are more likely to also contain a larger variety of endophytes [13–15].

Samples should be carefully conserved when translated to the laboratory, if possible, by use of thermally isolated containers. Preferably, they should be processed within the next 24 h after collection. It has proven better to use fresh samples than frozen samples to capture endophytes that do not survive freezing conditions [16].

Surface treatment largely influences the diversity of endophytic fungi, and it is extremely important to adjust exposure time to disinfectant in order to exterminate epiphytic flora without compromising endophytes. Most studies on endophytes have been performed in leaves, and leaves are the hardest organs for proper disinfection techniques. Thick leaves require more exposure time to surface disinfection agents compared to thin leaves [11]. In general, a more stringent procedure can be used with older and lignified plant tissue [17]. Roots, stems, and barks are easily disinfected, and, in many cases, the outer layer can be removed to further guarantee that isolates are indeed endophytic microorganisms. Different disinfectants have been used for surface cleansing, comprising among others formaldehyde, hydrogen peroxide, mercuric chloride, ethanol, and sodium hypochlorite, the two later being the most universal ones [17]. Exposure time to the different chemical agents varies from a few seconds to several minutes depending on the characteristics of the particular tissue to be treated and concentration of the chemical agent. Leaves should be briefly exposed to disinfectants since penetration of the chemical agent can kill part of the endophytes to be recovered. On the other hand, if the vegetal sample is resistant to the penetration of the chemical agent, longer exposure times can be adequate. The importance of adequate control experiments should be emphasized since complete surface disinfection is a must to guarantee the endophytic nature of the isolates. Three methodologies for disinfection control have been proposed: imprinting the surface treated plant tissue onto nutrient media, inoculating a nutrient broth with a representative sample of disinfected plant material, or culturing an aliquot of the final rinsing water [17]. Le Cocq has stressed that the final wash should be thoroughly evaluated for presence of viable cells [18]. This could be achieved by filtering the total volume of the final wash and culturing the filter on appropriate solid or liquid media.

The preparation of sample fragments after disinfection can be done by the use of a sterilized blade or knife to cut small fragments of the vegetal tissue. Small samples are preferred for endophyte recovery, and several samples (9–12) can be cultured on the same Petri dish.

Finally, particular attention should be addressed to culture media and incubation conditions. Zhang et al reported the use of ten-fold diluted media to slow the growth of fast-growing microorganisms. They also indicated the use of low-nutrient media, such as Reasoner's 2A to detect bacteria that require oligotrophic environments [16].On the other hand, Lacerda reported a two-fold recovery of fungal endophytes when using dicloran rose bengal media compared to the poor nutrient media synthetic nutrient agar, both complemented with chloramphenicol [19]. If a particular phylogenetic group is targeted, the use of an adequate media for that particular taxon is highly recommended. Our group has previously demonstrated that the use of diverse media allows for diverse detection and isolation of endophytes: a rich media favoring bacteria recovery, an adequate media for recovering yeast and fungi, and a third minimal media based on triturated sterile plant material. The latter has been designed to mimic the natural environment on the host plant [2]. Incubation temperature also has a great influence on endophyte recovery. Hagh et al reported improvement on the recovery of endophyte diversity by low-temperature incubation [12]. The use of different culture media and incubation conditions will warrant the recovery of a larger diversity of endophytes.

2 Materials

2.1 Sampling

- Sterile bags.
- Sterile scissors or knife (oven sterilized is preferable).
- Geo-localization device.
- Photo camera (cell phone can provide both geo-localization and camera).
- Cooler or thermally isolated recipient for sample transport to the lab.

2.2 Pre-treatment (Optional)

- Tap running water.
- Soft brush.
- Beaker.
- Sonicator bath.

2.3 Disinfection

- Sodium hypochlorite solution of adequate concentration (see Table 1) (*see* **Note 1**).
- 70% ethanol.
- Deionized sterile water in a washing bottle.
- Laminar flow chamber (if not available, disinfected lab bench with two Bunsen burners).
- Empty sterile recipient (Falcon tube or similar) to collect final wash.
- Sterile paper towels.

Table 1
Different disinfection procedures depending on plant/organ characteristics

Organ type	Plant	Disinfection procedure	Ref.
Leaves	*F. sylvatica*	(a) 70% EtOH, 2 min (b) 1% NaClO, 5 min (c) 70% EtOH, 1 min	[20]
Leaves	*Pinus taeda*	(a) 96% ethanol,[a] 5 s (b) 0.5% NaClO, 2 min (c) 70% ethanol, 2 min	[9]
Leaves	*Oxalis acetosella* L.	(a) 96% ethanol,[a] 5 s (b) 0.5% NaClO, 1 min (c) 70% ethanol, 1 min	[21]
Leaves	*Paulliniacupana*	(a) 70%, 30 s (b) 2.5% NaClO, 5 min	[14]
Leaves	*Quercus brantii* L.	(a) 70% EtOH, 1 min (b) 1% NaClO, 30 s (c) 70% EtOH, 1 min	[12]
Stem	*Quercus brantii* L.	(a) 70% EtOH, 1 min (b) 1% NaClO, 1 min (c) 70% EtOH, 1 min	[12]
Stem	*Glycine max*	(a) 70% EtOH, 5 s (b) 0,525% NaClO, 2 min	[22]
Seeds	*Paulliniacupana*	(a) 70%, 30 s (b) 2.5% NaClO, 20 min	[14]
Roots	*Paulliniacupana*	(a) 70%, 30 s (b) 2.5% NaClO, 8 min	[14]
Roots	*Daucus carota*	(a) 10% NaClO, 3 min (b) 70% EtOH, 3 min	[23]
Roots	*Triticum aestivum*	(a) 99% EtOH, 1 min (b) 3.1% NaClO, 6 min (c) 99% EtOH, 30 s	[17]
Roots	*Aphelandra tetragona* *Pteridium aquilinum*	(a) 75% EtOH, 3 min (b) 3–5% NaClO, 1–3 min (c) 75% EtOH	[17]
Roots	*Oryza sativa*	(a) 75% EtOH, 1 min (b) 20% NaClO, 3 min (c) 75% EtOH, 30 s	[17]
Fruits	*Citrus japonica*	(a) 10% NaClO, 3 min (b) 70% EtOH, 3 min	[23]

[a]Note: 95% ethanol is not recommended since contraction of plant tissue can trap epiphytic flora

2.4 Disinfection Control

- Sterile 200 μL tips.
- Micropipette 200 μL.
- Sterile stainless-steel tweezer.
- Sterile Drigalski spatula.
- Petri dishes with PDA and TSA.
- Erlenmeyer with 100 mL sterilized nutrient broth (TSB or PDB).
- 0.45 μm sterile filter membrane (optional).
- Sterile filtration device (optional).
- Vacuum pump (optional).

2.5 Culture Media (see Table 2)

- Commercial culture media or its components.
- Antibiotics (optional).
- Plant material (optional for plant-based media).
- Laboratory balance.
- Autoclave.
- Sterile empty Petri dishes.
- Laminar flow chamber (if not available, disinfected lab bench with two Bunsen burners).

2.6 Fragment Culture

- Sterile knife or blade.
- Sterile empty Petri dishes for sample processing.
- Laminar flow chamber (if not available, disinfected lab bench with two Bunsen burners).
- Culture media Petri dishes (adequate for the desired detection purpose).

3 Methods

3.1 Sampling

1. Register geographical coordinates.
2. Confirm plant identity (botanical assistance is highly recommended).
3. Register characteristics of the surrounding area (nature conservation area, pristine or cultivated area, altitude, temperature, humidity, surrounding vegetation, other biotic and abiotic factors).
4. Take pictures of the sampled plants and surrounding areas.
5. Take samples from several specimens of the same plant if available (avoid plants with symptoms of disease, damage, or senescence).

Table 2
Recommended media for different endophytic microorganisms

Target microorganism	Culture media and supplements[a]	Ref
Fungi	Potato dextrose agar (PDA) with 100 U/mL penicillin	[17, 24]
	Malt extract agar (MEA) with 0.1 g/L tetracycline and 10 μL/L cyclosporin A	[10]
	Malt extract agar (MEA) with 100 mg/L tetracycline hydrochloride	[20]
	Dicloran rose bengal (DRB) with 150 μg/mL chloramphenicol	[19]
Bacteria	Trypticase soy agar (TSA)	[14]
Actinomycetes	Starch casein agar (SCA) with cycloheximide 50 μg/mL and nystatin 50 μg/mL (nalidixic acid and K2Cr2O7, 60 μg/mL can be added to restrict the grown of non-actinomycetes)	[25, 26]
	Yeast extract malt extract agar (YEMA) with cycloheximide 50 μg/mL and nystatin 50 μg/mL (nalidixic acid and K2Cr2O7, 60 μg/mL can be added to restrict the grown of non-actinomycetes)	[25, 26]
	Actinomycetes isolation agar (AIA) with cycloheximide 50 μg/mL and nystatin 50 μg/mL (nalidixic acid and K2Cr2O7, 60 μg/mL can be added to restrict the grown of non- actinomycetes)	[25, 26]
Fastidious or biotrophic strains	Vegetal based media	[21]
	Tape water agar (TWA) supplemented with plant extract	[21]
	Leaf-based culture media (disrupted leaf pad with agar overlayer)	[27]
	Leaf tea bag infusion-based media	[27]
	Triturated plant material-based media	[23]
Slow growing microorganisms	Synthetic nutrient agar supplemented with 150 μg/mL of chloramphenicol and bengal rose	[19]
	10-fold diluted nutrient media	[16]
	Reasoner's 2A agar (R2A)	[16, 17]
Endophytic diazotrophs	Nitrogen-free synthetic media	[17]
Detection of particular enzymatic activity	Minimal media with no carbon source supplemented with chitin, cellulose, or pectin for chitinolytic, cellulolytic, and pectinolytic activity respectively	[17]

[a]Aqueous antibiotic solutions should be filter-sterilized and added to the media after cooling autoclaved media to 45 °C. Ethanol-based solutions do not need sterilization

6. Use sterile gloves and scissors or knives (70% ethanol surface disinfection could be enough if oven sterilization is not possible).
7. Place samples in sterile plastic bags.
8. Label samples adequately.
9. Place sample bags in an isolated container.
10. Bring samples to the lab, keep refrigerated (freezer not recommended) until lab processing.
11. Process samples within the next 24 h preferably, at most within 48 h from the collection.

3.2 Pre-treatment (Optional for Roots)

1. Thoroughly wash under a tap running water.
2. Softly brush surface to eliminate soil particles.
3. Sonicate in a water bath for a few minutes if soil particles are still adhered to root surface.

3.3 Disinfection (Standard Procedure)

1. Thoroughly wash the vegetal surface with sterile distilled water.
2. Soak in 70% ethanol for 1 min (*see* **Note 2**).
3. Rinse with deionized sterile water.
4. Soak in (0.5–20%) NaClO freshly made solution for 0,5–20 min (depending on organ and plant type). See Table 1.
5. Rinse with deionized sterile water.
6. Soak in 70% ethanol for 1 min.
7. Rinse with deionized sterile water.
8. Save the later washing water for microbiological analysis.
9. Dry the disinfected organ in a sterile paper towel.

3.4 Control of Disinfection Procedure

Control by Surface Imprinting:

1. Take a leave or other disinfected sample fragment with sterile tweezers.
2. Place on top of the corresponding nutrient agar (PDA, TSA) for 30 s.
3. Retire the sample fragment with sterile tweezers.
4. Incubate the Petri dishes at adequate conditions for 48 h to 4 weeks.

Control by Broth Enrichment:

1. Take a representative sample of the disinfected plant organ surface.
2. Aseptically cut it into smaller pieces.
3. Inoculate into a nutrient broth (TSB, PDB).
4. Incubate at adequate conditions for 48 h to 5 days.

Control by Sampling and Plating Final Washing Water:

1. Take 200 μL of the final wash using an automatic pipette and a sterile tip.
2. Aseptically transfer it to a Petri dish with the adequate culture media (TSA, PDA).
3. Spread on Petri dish surface using a sterile Drigalski spatula.
4. Incubate at adequate conditions for 48 h to 4 weeks.

Control by Filtering Final Washing Water:

1. Aseptically transfer the total volume of the final wash to a microbiology filtration funnel equipped with a sterile filter.
2. Apply vacuum for filtration.

3. Take out the membrane filter with sterile tweezers.
4. Place membrane filter on adequate culture media (TSA, PDA).
5. Incubate at adequate conditions for 48 h to 4 weeks.

3.5 Culture Media (see Table 2)

1. Prepare commercial media following supplier instructions.
2. Properly sterilize and cool it down to 45 °C.
3. Add antibiotics when necessary.
4. Distribute in sterile Petri dishes (in laminar flow chamber or lab bench between two Bunsen burners).
5. Let them dry in laminar flow chamber for 30 min (slightly open) or in lab bench overnight (closed).
6. Use it for endophyte recovery.

3.6 Fragment Culture

1. Properly disinfect the laminar flow chamber or lab bench.
2. Turn on laminar flow and place all the needed material inside.
3. Place the disinfected plant organ in an empty sterile Petri dish (glass preferred) or another similar container.
4. Cut into small pieces using a knife or blade (2 mm^2 for leaves, 3 mm long fragments for roots, cut in half for seeds).
5. Place in appropriate culture media (9–12 fragments per plate, or 2–4 seed halves per plate for adequate nutrient supply).
6. Incubate under desired conditions.
7. Observe colony development over time.
8. Save plate for further work (characterization, isolation, DNA sampling).

4 Notes

1. 0.3–20% active chlorine has been reported. Note: sodium hypochlorite solutions decompose over time. Be careful to prepare new diluted solutions from fresh commercial hypochlorite.
2. Notice that more concentrated ethanol solutions should be avoided since it can lead to contraction of the plant tissue trapping some epiphytic flora.

References

1. Strobel G, Daisy B (2003) Bioprospecting for microbial endophytes and their natural products. Micriobiol Mol Biology Rev 67(4): 491–502
2. Rodriguez P, Gonzalez D, Rodríguez Giordano S (2016) Endophytic microorganisms: a source of potentially useful biocatalysts. J Mol Catal B Enzym 133:S569–S581
3. Firdous J, Lathif NA, Mona R, Muhamad N (2019) Endophytic bacteria and their potential application in agriculture: a review. Indian J Agric Res 53(1):1–7

4. Hahn H, Huth W, Schöberlein W, Diepenbrock W (2003) Detection of endophytic fungi in *Festuca* spp. by means of tissue print immunoassay. Plant Breed 122(3):217–222
5. Trento S, Elias S, Garay A, Zavala J (2007) Comparison of endophyte detection in fescue and ryegrass seeds using an immunoblot assay and a microscopic method. Seed Sci Technol 35(1):65–74
6. Mishra VK, Passari AK, Leo VV, Singh BP (2017) Molecular diversity and detection of endophytic fungi based on their antimicrobial biosynthetic gene. In: Singh BP, Gupta VK (eds) Molecular markers in mycology: diagnostics and marker developments. Springer, Cham, pp 1–35
7. Mirlohi A, Sabzalian MR, Sharifnabi B, Nekoui MK (2006) Widespread occurrence of *Neotyphodium* -like endophyte in populations of *Bromus tomentellus* Boiss in Iran. FEMS Microbiol Lett 256(1):126–131
8. Siddique AB, Khokon AM, Unterseher M (2017) What do we learn from cultures in the omics age? High-throughput sequencing and cultivation of leaf-inhabiting endophytes from beech (*Fagus sylvatica* L.) revealed complementary community composition but similar correlations with local habitat conditions. Myco Keys 20:1–16
9. Arnold AE, Henk DA, Eells RL, Lutzoni F, Vilgalys R (2007) Diversity and phylogenetic affinities of foliar fungal endophytes in loblolly pine inferred by culturing and environmental PCR. Mycologia 99(2):185–206
10. Unterseher M, Gazis R, Chaverri P, Guarniz CFG, Tenorio DHZ (2013) Endophytic fungi from Peruvian highland and lowland habitats form distinctive and host plant-specific assemblages. Biodivers Conserv 22(4):999–1016
11. Rashmi M, Kushveer JS, Sarma VV (2019) A worldwide list of endophytic fungi with notes on ecology and diversity. Mycosphere 10(1): 798–1079
12. Hagh Doust N, Safaie N, Schmitt I, Otte J, Bálint M (2019) Culture-based methods using low-temperature incubation revealed cold-adapted fungal endophytes from semiarid forests. For Pathol 49(3):1–11
13. Nalini MS, Prakash HS (2017) Diversity and bioprospecting of actinomycete endophytes from the medicinal plants. Lett Appl Microbiol 64(4):261–270
14. Liotti RG, da Silva Figueiredo MI, da Silva GF, de Mendonça EAF, Soares MA (2018) Diversity of cultivable bacterial endophytes in *Paulliniacupana* and their potential for plant growth promotion and phytopathogen control. Microbiol Res 207:8–18
15. Unterseher M, Siddique AB, Brachmann A, Peršoh D (2016) Diversity and composition of the leaf mycobiome of beech (*Fagus sylvatica*) are affected by local habitat conditions and leaf biochemistry. PLoS One 11(4):1–16
16. Zhang J et al (2021) High-throughput cultivation and identification of bacteria from the plant root microbiota. Nat Protoc 16(2): 988–1012
17. Hallmann J, Berg G, Schulz B (2007) Isolation procedures for endophytic microorganisms in microbial root endophytes. Soil Biol 9:299–319
18. Le Cocq K, Gurr SJ, Hirsch PR, Mauchline TH (2017) Exploitation of endophytes for sustainable agricultural intensification. Mol Plant Pathol 18(3):469–473
19. Lacerda LT, LFP G, Rodrigues A (2018) Diversity of endophytic fungi in *Eucalyptus microcorys* assessed by complementary isolation methods. Mycol Prog 17(6):719–727
20. Unterseher M, Schnittler M (2009) Dilution-to-extinction cultivation of leaf-inhabiting endophytic fungi in beech (*Fagus sylvatica* L.). Different cultivation techniques influence fungal biodiversity assessment. Mycol Res 113(5):645–654
21. Prior R, Görges K, Yurkov A, Begerow D (2014) New isolation method for endophytes based on enzyme digestion. Mycol Prog 13(3): 849–856
22. Impullitti AE, Malvick DK (2013) Fungal endophyte diversity in soybean. J Appl Microbiol 114(5):1500–1506
23. Rodríguez P et al (2011) Stereoselective biotransformation of α-alkyl-β-keto esters by endophytic bacteria and yeast. J Mol Catal B Enzym 71(3–4):90–94
24. Du W et al (2020) Diversity and antimicrobial activity of endophytic fungi isolated from *Securinega suffruticosa* in the Yellow River Delta. PlosONE 3:1–18
25. Waheeda K, Shyam KV (2017) Formulation of novel surface sterilization method and culture media for the isolation of endophytic actinomycetes from medicinal plants and its antibacterial activity. J Plant Pathol Microbiol 8(2): 1–9
26. Passari AK, Mishra VK, Saikia R, Gupta VK, Singh BP (2015) Isolation, abundance and phylogenetic affiliation of endophytic actinomycetes associated with medicinal plants and screening for their in vitro antimicrobial biosynthetic potential. Front Microbiol 6:1–13
27. Nemr RA et al (2020) In situ similis' culturing of plant microbiota: a novel simulated environmental method based on plant leaf blades as nutritional pads. Front Microbiol 11(4):1–15

Chapter 6

Detection of Endophytes by Visualization Methods

Vijayalakshmi Selvakumar, Karnan Muthusamy, Thenmozhi Manivel, Preethi Gopalsamy, Panneerselvam Annamalai, Ramachandran Chelliah, and Deog-Hwan Oh

Abstract

The plant endosphere is full of an assorted group of the microbial population. There is general concurrence that this microbial population makes considerable effects to plant health. Detection of these endophytic microbes can be performed by both classical (culturing methods and visualization methods) and recently adopted genomic approaches. Among them, visualization-based method is one of the most frequently used methods. Colonization by microorganisms in plant tissues has been analyzed by such a microscope, which can be separated into three basic types: light microscope, electron microscope, and fluorescence microscope. Individual or combinations of these methods are the most frequently used visualization-based approaches. This section affords the indispensables for carrying out endophyte detection through this line of attack, heightening the importance of sample preparation and also the importance of fluorescent labeling used to ensure the presence of endophytes.

Key words Endosphere, Visualization, Electron microscope, Fluorescence microscope, Light microscope

1 Introduction

Endophytes are an endosymbiotic group of microorganisms, frequently bacteria of fungi, which are able to inhabit, inter and/or intracellular locations of the plants [1] [2]. Endophytic microbes are known to enhance plant growth and nutrition, may improve the plant's ability to tolerate various types of stresses (both biotic and abiotic) and enhance the plant resistance to insects and pests [3]. They are ubiquitous in nature and exhibit complex interactions with their host plants, which include mutualism, antagonism, and rarely parasitism [4]. Many researchers have stated the presence of one or more types of endophytes in every single plant studied up till now. Endophytes can systematically colonize different parts of host plants and this colonization may be organ and

A. Sankaranarayanan et al. (eds.), *Endophytic Microbes: Isolation, Identification, and Bioactive Potentials*, Springer Protocols Handbooks, https://doi.org/10.1007/978-1-0716-2827-0_6,

tissue-specific due to selective pressure, hence it gives tissue-specific protection from diseases [5]. Endophytes are ahead biotechnological and industrial consequence as a result of their capability to secrete secondary metabolites, serve as biocontrol agents, antimicrobials, anti-tumor agents, and development of natural antioxidants, antidiabetic agents, and insecticidal products [6]. However, commercialization of these natural products obtained from endophytic species (bacteria, fungi, and actinomycetes) is not yet prevalent.

Studying about endophytic microorganisms is consistently hard to experts for the reason that usually microbes go concealed in plants, many cannot be isolated, and it is difficult to coordinate assessments, where endophytes are unable to be cultured or plants lacking endophytes cannot be found [7]. Generally, four sets of methods such as microscopy (visualization methods), isolation of pure culture, biochemical approaches, and molecular methods were used for their detection. All these techniques have their merits and demerits and the results were constrained based on the methods chosen [8]. Although the study of endophytes is basically a method-dependent process, their detection and identification are difficult, today, a wide scope of microscopy-based strategies are accessible and have been utilized to recognize microorganisms in plant tissues and to assess the colonization patterns of bacterial endophytes through molecular collaborations and elements inside living cells in explicit vegetative tissues [9]. Hence, the objective of this work was to establish the different protocols for individual microscopic technique and their relevant sample preparation and in a combination of one or two of these techniques for the visualization, localization, and quantification of the endophytic microbial population (bacteria, fungi, and actinomycetes) in plant tissues.

2 Materials

2.1 Sampling

- Sterile bags.
- Sterile scissors or knife (oven sterilized is preferable).
- Geo-localization device.
- Photo camera (cell phone can provide both geo-localization and camera).
- Cooler or thermally isolated recipient for sample transport to the lab.

2.2 Pre-treatment (Optional)

- Sonicator bath.

2.3 Visualization of Endophytes in Plant Cell Lines by FISH Technique

Probe Preparation

- 50-FAM labeled probe Eub 338 (target for most bacteria).
- LGC35ab (for all Firmicutes).
- 50-Cy3 labeled HCG69a (target Actinobacteria).
- ALP969 (for α-proteobacteria).
- BET42a (for β-proteobacteria).
- GAM42a (for γ-proteobacteria).
- HCG69/ns (for non-sense negative controls) prepare as per the protocol stated in [10].

Disinfection

- Sodium hypochlorite solution of adequate concentration.
- 70% ethanol.
- Deionized sterile water in a washing bottle.
- Laminar flow chamber (if not available, disinfected lab bench with two Bunsen burners).
- Sterile paper towels.

3 Methods

3.1 Sampling

(i) Record geographical coordinates.

(ii) Authenticate plant identity (botanical assistance is highly recommended).

(iii) List characteristics of the surrounding area (nature conservation area, cultivated area, altitude, temperature, humidity, surrounding vegetation, and other biotic and abiotic factors).

(iv) Capture the pictures of the sampled plants and surrounding areas.

(v) Take samples from several specimens of the same plant if available (avoid plants with symptoms of disease, damage, or senescence).

(vi) Use sterile gloves and sterile scissors or knives (70% ethanol surface disinfection could be enough if oven sterilization is not possible).

(vii) Put the samples in sterile plastic bags.

(viii) Properly label the samples.

(ix) Put sample bags in isolated container.

(x) Transport samples to lab, keep refrigerated (freezer not recommended) until lab processing.

(xi) Process samples within the next 24 h preferably, at most within the 48 h from collection.

3.2 Pre-treatment (Optional for Roots)

(i) Entirely wash under tap running water.

(ii) Gently brush surface to eliminate soil particles.

(iii) Sonicate in water bath for a few minutes in order to remove if soil particles are still adhered to root surface.

3.3 Disinfection

Samples were disinfected with standard procedure. As narrated in Subheading 1.

3.4 Light Microscope

Hood and Shew Protocol: KOH – Aniline Blue Staining Method [2]

(i) Autoclave the samples for 15 min at 121 °C in 50 ml of 1 M KOH solution.

(ii) Wash the samples with deionized water (3 times).

(iii) Mount the specimen on a clean glass slide in several drops of stain solution (KOH – aniline blue) (0.05% aniline blue in 0.067 MK_2HPO_4 at pH 9).

(iv) Analyze the specimen under research microscope.

ROS Staining [11]

(i) Wash the roots properly with sterilized H_2O.

(ii) Put the root of seedlings into DAB (3, 3′-diaminobenzidine tetrachloride) solution (2.5 mm) overnight.

(iii) Wash with deionized water.

(iv) Counterstain with aniline blue and mount on slide.

(v) Observe under a light microscope.

3.5 Electron Microscope

Scanning Electron Microscope (SEM) [14]

(i) Cut the leaf disks (6 mm in diameter).

(ii) Store the leaf disks in 1% glutaraldehyde and transfer to half-strength Karnovsky's fixative for 4 h.

(iii) Rinse with 0.1 M phosphate buffer three times (for each in 20 min).

(iv) Dehydrate in ethanol series (in sequence: 30%, 50%, 70%, 80%, 90%, 95%, and 100%) (each step in 25 min).

(v) Then immerse in 100% ethanol overnight.

(vi) After mount the leaf disks on SEM stubs using carbon conductive tabs and then coat with platinum using a Hammer 6 sputtering coater.

(vii) Examine the samples under SEM.

3.6 Transmission Electron Microscope (TEM) [15]

(i) Cut and trim the roots and leaves of plant samples into 1.0–1.5 mm sizes.

(ii) Fix the samples by immersing in 2–3% (X/V) glutaraldehyde in 0.1 M sodium cacodylate buffer (pH 7.2) at 4 °C overnight.

(iii) Post fix the samples in 2% osmium tetroxide buffer solution and embed in epoxy resin.

(iv) Subsequently, section the sample (0.1 μm) with ultramicrotome.

(v) Stain the sample with a saturated solution of uranyl acetate and lead acetate.

(vi) Subject the samples to TEM micrographs.

3.7 Fluorescence Microscope [13]

(i) Cut the leaf into pieces [approx... (1 mm × 2 mm)].

(ii) Fix the leaf pieces with 2% paraformaldehyde and 2.5% glutaraldehyde on 0.1 M phosphate buffer (pH 7.2) under vacuum for 1 h.

(iii) Wash them thoroughly in buffer (3 times).

(iv) After washing dehydrate in an ethanol series and embedded in LR white resin (London Resin).

(v) Dry the 200 nm thick sections onto poly-l-lysine-coated slides.

(vi) Perform immuno labeling.

- Rinse the slide sections with phosphate buffered saline/ Tween (PBS-T).
- Place in blocking buffer (0.1% BSA-C) for 15 min.
- Incubate with β-D-glucan antibody (1:100 dilutions) overnight at 4 °C.
- Wash thoroughly with PBS-T and incubate with alexa goat antimouse 488 (1:600 dilutions) for 1 h.
- Wash thoroughly.
- Detect the endophytes by fluorescence.

3.8 Combined Confocal Microscopy and FISH (Fluorescence In Situ Hybridization) [12]

Sample Preparation

(i) Soak the leaf disks in 80% ethanol for 2 weeks.

(ii) Then boil the disk in KOH solution (10%w/v) for 20 min.

(iii) Transfer this to 30% (v/v) H_2O_2 solution at room temperature until complete clarification (20–40 min).

(iv) Then rapid immersion in 10% (v/v) HCl and rinse with sterile distilled water (3 times).

(v) Keep it in sterile distilled water for 15 days until carrying out the FISH experiment.

FISH Experiment

(i) Use EUB338 (universal oligonucleotide probe) synthesized by Invitrogen (Milan, Italy) and label with fluorescein at 5′ end.

(ii) Clear leaf disks are placed onto poly-L-lysine-coated slides and fix by using 4% (v/v) paraformaldehyde for 150 min.

(iii) Fix the foliar disks in PBS (Oxoid) and dry in an oven at 46 °C for 10 min and dehydrate successively in 50, 80, and 100% ethanol solutions for 3 min each.

(iv) Perform pretreatment of the specimens with proteinase –K (10 mg /ml) at 37 °C for 5 min and wash the specimens in ice-cold 1X PBS.

(v) Carry out hybridization process (hybridization buffer 25% formamide, 0.9 M NaCl, 0.01% SDS, 20 mM Tris HCl, pH 7.2) 30 μL containing 10 ng of EUB338.

(vi) The hybridization buffer (25% formamide, 0.9 M NaCl, 0.01% SDS, 20 mM Tris-HCl pH 7.2) (30 μL), containing 10 ng of EUB338, is spotted onto dry specimens, and the slides are incubated in a dark humid chamber at 46 °C overnight.

(vii) After hybridization, a washing step in a prewarmed washing buffer (20 mM Tris-HCl pH 7.2, 0.01% SDS, 40 mM NaCl, 5 mM EDTA) is performed at 46 °C for 30 min.

(viii) The slides are then rinsed with distilled water and stored in the dark.

(ix) Visualize under confocal microscope.

Note

- The choice, concentration, and detergent exposure durations (SDS, Triton X-100) during pre-treatments, hybridization washing step appear critical in improving the permeability of the plant cells as well as the microbial cells in fresh plant tissue without impairing significant effects on the former.
- Cells are also tried in FISH with EUB338, as above, after fixation with 2% formalin/paraformaldehyde, or formalin-glacial acetic acid-ethanol fixative.

3.9 Visualization of Endophytes in Plant Cell Lines by FISH Technique [16]

Pre-treatment of Plant Cells

(i) Use cellulose Y-C (0.5%) for pre-treating the plant cells in MS medium with 3% sucrose (pH 6.0) and 0.01% TritonX-100 for 30 min.

(ii) Then wash with hybridization buffer (900 mM NaCl, 20 mM Tris HCl (pH 7.4), 5 mM EDTA pH 8.0; no use of detergents) with slow spin at 500 rpm in a tabletop centrifuge.

Hybridization

(i) Carry out hybridization in 500 μL centrifuge tubes with 100 μL permeabilized cell aliquots and a single probe (200 ng) at 46 °C for 3–4 h in a water bath under dark conditions.

(ii) Wash with a buffer with 0.01% Triton X-100 at 48 °C for 15 min (3 times).

(iii) Disperse the cells in a hybridization buffer.

(iv) Mount in Carbonate buffered Glycerol (1:2).

(v) Observe under Fluorescence Microscope (treated samples are used to check the auto-fluorescence of plant cells).

(vi) Eliminate the non-specific binding by incubation and wash the samples overnight in a wash solution containing 0.001% Triton X-100 at 4 °C.

References

1. Gouda S, Das G, Sen SK, Shin HS, Patra JK (2016) Endophytes: a treasure house of bioactive compounds of medicinal importance. Front Microbiol 7:1538. https://doi.org/10.3389/fmicb.2016.01538
2. Verma SK, Kharwar RN, Gond SK, Kingsley KL, White JF (2019) Exploring endophytic communities of plants: methods for assessing diversity, effects on host development and potential biotechnological applications. Seed endophytes. Springer, Cham, pp 55–82
3. Parthasarathi S, Sathya S, Bupesh G, Samy RD, Mohan MR, Kumar GS et al (2012) Isolation and characterization of antimicrobial compound from marine Streptomyces hygroscopicus BDUS 49. World J Fish Mar Sci 4(3): 268–277. https://doi.org/10.5829/idosi.wjfms.2012.04.03.5658
4. Nair DN, Padmavathy S (2014) Impact of endophytic microorganisms on plants, environment and humans. Sci World J 2014:1–11. https://doi.org/10.1155/2014/250693
5. Mengistu AA (2020) Endophytes: colonization, behaviour, and their role in defense mechanism. Int J Microbiol. https://doi.org/10.1155/2020/6927219
6. Yadav AN, Kumar V, Dhaliwal HS, Prasad R, Saxena AK (2018) Microbiome in crops: diversity, distribution, and potential role in crop improvement. In: Prasad R, Gill SS, Tuteja N (eds) Crop improvement through microbial biotechnology. Elsevier, New York, pp 305–332
7. Fadiji AE, Babalola OO (2020) Elucidating mechanisms of endophytes used in plant protection and other bioactivities with multifunctional prospects. Front Bioeng Biotechnol 8: 467. https://doi.org/10.3389/fbioe.2020.00467
8. Raja S, Subhashini P, Thangaradjou T (2016) Differential methods of localisation of fungal endophytes in the seagrasses. Mycology 7(3): 112–123. https://doi.org/10.1080/21501203.2016.1218966
9. Kandel SL, Joubert PM, Doty SL (2017) Bacterial endophyte colonization and distribution within plants. Microorganisms 5(4):77. https://doi.org/10.3390/microorganisms5040077
10. Amann RI, Ludwig W, Schleifer KH (1995) Phylogenetic identification and in situ detection of individual microbial cells without cultivation. Microbiol Rev 59(1):143–169. https://journals.asm.org/journal/mr
11. White JF Jr, Torres MS, Somu MP, Johnson H, Irizarry I, Chen Q et al (2014) Hydrogen peroxide staining to visualize intracellular bacterial infections of seedling root cells. Microsc Res Tech 77(8):566–573. https://doi.org/10.1002/jemt.22375
12. Piccolo SL, Ferraro V, Alfonzo A, Settanni L, Ercolini D, Burruano S, Moschetti G (2010) Presence of endophytic bacteria in Vitis vinifera leaves as detected by fluorescence in situ hybridization. Ann Microbiol 60(1):161–167. https://doi.org/10.1007/s13213-010-0023-6
13. Johnston PR, Sutherland PW, Joshee S (2006) Visualising endophytic fungi within leaves by detection of (1→ 3)-ß-d-glucans in fungal cell walls. Mycologist 20(4):159–162. https://doi.org/10.1016/j.mycol.2006.10.003
14. Huang YL, Zimmerman NB, Arnold AE (2018) Observations on the early establishment of foliar endophytic fungi in leaf discs and living leaves of a model woody angiosperm, Populus trichocarpa (Salicaceae). J Fungi 4(2):58. https://doi.org/10.3390/jof4020058
15. Nongkhlaw FMW, Joshi SR (2017) Microscopic study on colonization and antimicrobial property of endophytic bacteria associated with ethnomedicinal plants of Meghalaya. J Microscop Ultrastruct 5(3):132–139. https://doi.org/10.1016/j.jmau.2016.09.002
16. Thomas P, Franco CM (2021) Intracellular bacteria in plants: elucidation of abundant and diverse cytoplasmic bacteria in healthy plant cells using in vitro cell and callus cultures. Microorganisms 9(2):269. https://doi.org/10.3390/microorganisms9020269

Chapter 7

Detection of Endophytes by Immunological Methods

Bankapura Mariyappa Ravikumara, Lakshmaiah Manjunatha, Kittan Subathra, Gummeral Narasareddy, Gangadharaiah Jyothi, and Chikkabuddi Prashantha

Abstract

Endophytes are hidden companion of plant species and reside within healthy plant tissues without causing any apparent symptoms and noticeable injury to the host. Among the different detection techniques of endophytes, immunological methods of detection are the more promising as compared to other modern techniques. The principle behind immunological techniques is to apply antibodies that bind specifically with endophytic organism. If the endophyte occurs in a tested sample, we can notice the different color, compared with control samples. Although molecular methods are replacing many conventional techniques, still an immunological technique is playing a crucial role by its accuracy and rapidity in detection. There are some commercially available immunological diagnostic kits that facilitate the use of these methods in the majority of endophytes detection in laboratories and at the field level across the world.

Key words Endophytes, Immunological, Antibodies, ELISA, DBIA, TPIA and Immunogold labeling

1 Introduction

The association of plants with microorganisms may be classified as mycorrhiza, pathogenic, epiphytic, saprotrophic, and endophytic based on the type of colonization and their roles [1]. Endophytic microorganisms are referred as microbes that inhabit the internal parts of a plant. Endophytic microorganisms such as bacteria, fungi, eukaryote, and archaea reside in plant tissues [2], they are known not to cause any harm to the host plant. They exhibit a symbiotic association with tissues of most plants and sometimes can be slightly pathogenic. They gain entrance into the seed, leaf, stem, and root of a plant [3]. The associated endophytes generally improve plant growth by secreting phytohormones and consequently help in nutrition improvement using bidirectional nutrient

A. Sankaranarayanan et al. (eds.), *Endophytic Microbes: Isolation, Identification, and Bioactive Potentials*, Springer Protocols Handbooks, https://doi.org/10.1007/978-1-0716-2827-0_7,

transfer and enhancement of the health of plants by protecting them against phytopathogens [4, 5]. The presence of endophytes in the plant system provides protection, persistence and improved yield against herbivores, insects, pathogens, drought, and several other biotic and abiotic stresses [6]. Endophytes are generally isolated from the internal tissues of plants after surface sterilization and they have been isolated from many plants and tree species [7]. Endophytes are a rich and reliable source of genetic diversity with biological novelty in agricultural applications [8]. Endophytes serve as biocontrol agents, antimicrobial agents, antitumor agents, and immunosuppressants [9, 10].

Endophytes are associated with plants in various forms, including bacteria (actinomycetes or mycoplasma) or fungi that have been colonized inside the plant tissues. More than 200 genera from 16 phyla of bacterial species have been reported to be associated with endophytes and among them, most of the species belong to the phyla actinobacteria, proteobacteria, and firmicutes [11]. The diversity of endophytic bacteria ranges from gram-positive to gram-negative bacteria, such as *Achromobacter, Acinetobacter, Agrobacterium, Azorhizobium, Bacillus, Bradyrhizobium Brevibacterium, Burkholderia, Microbacterium, Pseudomonas, Xanthomonas,* etc. [12–15]. Bacterial endophytes are diverse in nature and are known to produce different bioactive metabolites that act as antimicrobial and anticancer compounds [16].

Actinomycetes possess mycelium-like fungus and form spores [17, 18] and are known to produce various chemical substances of medicinal significance [19, 20]. Among them, *Streptomyces* is one of the dominant genera [11, 21] with 76% of antimicrobial compounds of biological interest have been reported from the single genus, *Streptomyces* [11, 21, 22]. Endophytic mycoplasma species were also reported as plant endophytes based on the symbiotic relationship with some red algae, such as *Bryopsis pennata, B. hypnoides,* and also in arcobacter [23]. Fungi are multicellular eukaryotic heterotrophic group of organisms with various life cycles that include symbiotic relationships with a wide variety of autotrophic organisms [24]. Based on the phylogeny and life-history traits endophytic fungi have been classified into two broad groups. These include the clavicipitaceous, which infect some grasses confined to cool regions and the non-clavicipitaceous endophytes, which are from asymptomatic tissues of non-vascular plants, ferns and allies, conifers and angiosperms and are limited to the Ascomycota or Basidiomycota group [25, 26]. The fungi such as *Torreya mairei, Fusarium* sp., and *Chloridium* sp. are known to possess strong antibacterial and antifungal properties against numerous food-borne infectious agents [25].

2 Materials Required

2.1 ELISA Technique

2.1.1 Standard Solutions

10,000 pg/mL: Add 1 mL of sample diluent buffer into one tube of standard (10 ng per tube) and mix thoroughly. Note: Store this solution at 4 °C for up to 12 h (or −20 °C for 48 h) and avoid freeze-thaw cycles. 5000 pg/mL: Mix 0.3 mL of 10,000 pg/mL with 0.3 mL of sample diluent buffer and mix thoroughly. 2500 pg/mL: Mix 0.3 mL of 5000 pg/mL with 0.3 mL of sample diluent buffer and mix thoroughly. Perform similar dilutions until the standard solutions with these concentrations (pg/mL) are made: 1250, 625, 312, 156, and 78. Add 100 μL of each of the diluted standard solutions to the appropriate empty wells. Repeat in duplicate or triplicate for accuracy (Note: The standard solutions are best used within 2 h).

2.1.2 Biotinylated Antibody

Calculate the total volume needed for the assay by multiplying 0.1 mL/well and the number of wells required. Add 2–3 extra wells to the calculated number of wells to account for possible pipetting errors. Generate the required volume of diluted antibody by performing a 1:100 dilutions (for each 1 μL concentrated antibody, add 99 μL antibody dilution buffer) and mixing thoroughly.

2.1.3 Avidin-Biotin-Peroxidase (ABC)

Calculate the total volume needed for the assay by multiplying 0.1 mL/well and the number of wells required. Add 2–3 extra wells to the calculated number of wells to account for possible pipetting errors. Generate the required volume of diluted ABC solution by performing 1:100 dilutions (for each 1 μL concentrated ABC solution, add 99 μL ABC dilution buffer) and mixing thoroughly (Note: The diluted ABC solution should not be prepared more than 1 h prior to the experiment). Production of the monoclonal and polyclonal antibodies.

2.2 Indirect ELISA

Bicarbonate/carbonate antigen-coating buffer (100 mM NaHCO3 in deionized water; pH adjusted to 9.6), microtiter plate, BOD incubator, PBS (Phosphate Buffered Saline) buffer (10 mM Na2HPO4 and 1.8 mM NaH2PO4 in deionized water with 0.2% Tween 20; pH Adjusted to 7.4), blocking buffer (5% w/v non-fat dry milk in PBS buffer), Bovine serum albumin (BSA). Substrate preparation, signal detection, and data analysis are similar to Sandwich ELISA and Indirect ELISA.

2.3 Dot Blot Immunoassay (DBIA)

Compared with ELISA, DBIA has essential advantages: it is less laborious, significantly reduces costs and shortens the time for analysis of the endophytic microbes [27]. Reagents: TBS: 20 mM Tris-HCl; 150 mM NaCl; pH 7.5, TBS-T: 0.05% Tween20 in TBS, BSA/TBS-T: 0.1% BSA in TBS-T, Nitrocellulose membrane (BIO-RAD, Trans-Blot, etc.).

2.4 Tissue Print Immunoassay(TPIA)

Nitrocellulose membranes, goat antibodies to biotin labeled with alkaline phosphatase [28], 0.05 M PBS, pH 7.4, containing 25 μL/L of Tween 20 and 1 mg/l of Polyvinyl alcohol 72,000 (PVA), 0.2 M Tris buffered saline, pH 7.4 (TBS; containing 0.5 M NaCl), 4 mM MgCl2. Staining solution (a mixture of 15 mL of Aqua dest laboratory water, containing 6 mg of Napthol AS-MX phosphate, and 15 mL of 0.2 M Tris-HCl, pH 8.0, containing 90 mg of Fast Red TR Salt).

2.5 Immunogold Labeling Technique

Immunogold labeling or Immunogold staining (IGS) is a technique used in electron microscopy [29]. Gold is used for its high electron density which increases electron scatter to give high contrast dark spots [30]. The solutions and reagents required are: TBS-Tween (0.05 M TBS, 0.05% Tween 20, pH 7.6): Trizma base (6.1 g), NaCl (9 g), distilled water (1000 mL). Mix to dissolve and adjust pH to 7.6 using 1 N HCl and then add 0.5 mL Tween 20. 0.1 M Glycine in 0.1 M PB: To prepare 100 mL, add 0.75 g of glycine to 100 mL of 0.1 M PB. 0.2 M Sucrose Solution: To prepare 100 mL, add 8 g of sucrose to 100 mL of 0.1 M PB. Post-fixative (2% glutaraldehyde in 0.1 M PB): To prepare 100 mL, add 4 mL of 50% glutaraldehyde to 95 mL of 0.1 M PB. L.R. White: Kit from EMS, Blocking Buffer: (1% BSA, 3% Nomal Serum, 0.1% Fish Gelatin, 0.05% Sodium Azide in 0.05 M TBS, pH 7.6). 5% Uranyl Acetate Solution: To prepare 50 mL, add 2.5 g of uranyl acetate to 50 mL of distilled water. Cover with foil and stir overnight. Add 10 drops of glacial acetic acid. Store solution in 4 °C. Reynold's Lead Citrate Solution: To prepare 50 mL, add lead nitrate (1.33 g), Sodium citrate, dihydrate (1.76 g), 1 N NaOH (5 mL), and distilled water (30 mL). All the chemicals are stirred for 10 minutes to dissolve and add additional 15 mL of distilled water. Store solution for 3–6 months at 4 °C. Note: the amount of NaOH is very important. Too much will make the solution cloudy.

3 Methods

3.1 Isolation of Fungal Endophytes from Different Sources

1. Wash the plant specimens thoroughly in running tap water for 5 min to remove dust and debris and then air dried.
2. The cleaned leaves, petioles, roots, and seeds are surface sterilized with 75% ethanol for 1 min and sodium hypochlorite (3% available chlorine) for 3–5 min and then remove outer epidermal tissues and cuticle before cut under sterile conditions into small pieces.
3. Clean the tissue bits in sterilized water and 75% ethanol for 30 s and then clean in sterilized water again and place the air-dried tissue bits on water agar (WA) medium, incubate at room temperature for growth of endophytic fungi.

4. Transfer the endophytic fungi into potato dextrose agar (PDA) plates and incubate for two to 6 days. Subculture the plates until getting pure culture [31].

3.2 Isolation of Bacterial Endophytes from Different Sources

1. Cut the leaves and roots samples into small pieces and macerate separately in phosphate buffer (pH 7.2) in a sterile pestle and mortar.
2. Prepare the tissue extract for tenfold dilution in sterile saline solution [and, 10-7] and use 0.1 mL of the aliquot for inoculations on Nutrient Agar medium.
3. The inoculations should be in triplicates separately for both roots and leaf tissue extract and incubate the plates at 37 °C.
4. Differentiate the bacterial colonies on the basis of morphological colony characters and purify the bacterial colonies by streaking techniques. Repeat the isolation process till monocultures were obtained for further experimentations [32]. For the bacterial endophytes isolation, Yeast Manitol Agar, Ashbeys and Picovasky's medium are to be used.

3.3 Enzyme-Linked Immunosorbent Assay (ELISA)

ELISA (enzyme-linked immunosorbent assay) is a plate-based assay technique designed for detecting and quantifying peptides, proteins, antibodies, and hormones [33]. The most common types are **direct, indirect, sandwich,** or **competitive** ELISAs (Table 1).

1. **Direct ELISA**

   For direct detection, an endophytes antigen coated to a multiwell plate is detected by an antibody that has been directly conjugated to an enzyme.

Table 1
Summary of key steps in different ELISA types

	Indirect	Direct	Sandwich	Competitive
Capture ab coating	X	X	√	X
Antigen coating	√	√	X	√
Blocking	√	√	√	√
Sample (antigen) incubation	X	X	√	√
Primary ab incubation	√	√	√	√
Secondary ab incubation	√	X	√	√
Substrate prep	√	√	√	√
Signal detection	√	√	√	√
Data analysis	√	√	√	√

2. **Indirect ELISA**

   Coat the antigen to a multi-well plate which is detected in two stages or layers. First, an unlabeled primary antibody, which is specific for the antigen, is applied. Next, an enzyme-labeled secondary antibody is bound to the first antibody. The secondary antibody is usually an anti-species antibody and is often polyclonal.

3. **Sandwich ELISA**

   Sandwich ELISAs typically require the use of matched antibody pairs, where each antibody is specific for a different, non-overlapping part (epitope) of the antigen molecule.

4. **Competitive ELISA**

   In this process, first the primary antibody is incubated with the sample antigen and the resulting antibody–antigen complexes are added to wells that have been coated with the same antigen. After an incubation period, any unbound antibody is washed off.

Interpretation of ELISA Data Output

The ELISA assay yields three different types of data output:

1) **Quantitative**: ELISA data can be interpreted in comparison to a standard curve in order to precisely calculate the concentrations of antigen in various samples.
2) **Qualitative**: ELISAs can also be used to achieve a yes or no answer indicating whether a particular antigen is present in a sample, as compared to a blank well containing no antigen or an unrelated control antigen.
3) **Semi-quantitative**: ELISAs can be used to compare the relative levels of antigen in assay samples, since the intensity of signal will vary directly with antigen concentration.

(a) **Production of Polyclonal Antibodies**

1. Homogenize the endophytes containing tissue (20 g) in cold suspension at a ratio of 1 g tissue to 5 mL of suspension.
2. Filter the extract and centrifuge at 2000 g for 15 min followed by high-speed centrifugation at 39000 g for 1 h and resuspension of the pellet in 1.5 mL suitable buffer.
3. Incubate the resuspended pellet for 1 h at 4 °C with 4 mL of undiluted antiserum prepared against endophytes-free plant extract.
4. Repeat the centrifugation at low and high speed as described above, the final pellet resuspended in 1 mL of PBS buffer (0.01 M phosphate buffer saline) and use the resuspended pellet for immunization and production of polyclonal antiserum.

5. Inject the white rabbit intramuscularly in legs and subcutaneously in the neck with an emulsion of equal volumes of partially purified endophytes and Freund's incomplete adjuvant (FIA).
6. Inject four times at an interval of 7 days. The rabbit will bleed four times at daily intervals beginning 5 weeks after primary immunization. Use this antiserum after absorption with endophyte-free plant sap for serological detection and identification [34, 35].

(b) **Production of Monoclonal Antibodies**

1. Isolate the endophytes from the host plant and maintain on corn meal malt agar (CMM) or bacterial-specific medium [36].
2. Remove the endophytic growth from the culture plate and ground in a sterile tissue homogenizer. Transfer the extract to a 1 L Erlenmeyer flask containing 500 mL of Schenk and Hildebrandt basal salts and vitamin (Sigma Chemical Company, St. Louis, MO) liquid media. After mixing, dispense the 50 mL of the resulting suspension into 125-mL tissue culture flasks.
3. Place the flasks on a G10 Gyratory orbital shaker at 125 rpm under room temperature until endophytic cultures develop the clumps. Further, suspensions should be vacuum filtered through a Buchner funnel and wash with 1–2 L of distilled H_2O. Remove the mycelial debris from the funnel by washing with phosphate buffered saline (PBS; pH 7.3; 2.76 g NaH_2PO_4, and 8.77 g NaCI -1 distilled H_2O) and place in a tissue homogenizer.
4. The homogenizer placed on ice and mycelia/growth will ground until homogenous and then centrifuged for 10 min at 3000 g to remove cellular debris.
5. Add the polyethylene glycol (PEG) to the supernatant until PEG concentration equaled 100 g/L and precipitation of the proteins will be achieved through chilling the mixture overnight at 4 °C.
6. Then centrifuge the solution for 20 min at 12000 g and discard the supernatant. Resuspend the protein pellet in 5 mL PBS containing 1 g/L sodium dodecyl sulfate (SDS) and incubated for 5 min at 60 °C. Centrifuge the SDS solution for 20 min at 12000 g to remove any undissolved material, and dilute the supernatant with 1:4 in cold acetone (−20 °C).
7. After mixing, the protein will be precipitated by chilling the solution overnight at −20 °C. The solution will centrifuge

for 20 min at 12000 g, the supernatant discarded, and the protein pellet washed with 5 mL of cold acetone. Following a second spin, the pellet dried under a stream of N_2 gas.

8. Resuspend the protein pellet in PBS and record the absorbance values (OD_{260} and OD_{280}), calculate the protein concentration [37]. Dialyze the protein for 24 h against 3× 3 L of distilled water at 4 °C. Lyophilize the protein and stored at −70 °C.
9. Monoclonal antibodies are generated to the lyophilized protein extract by suspending 150 μg of the protein extract in sterile distilled water, emulsifying with an equal volume of Fruend's complete adjuvant (Sigma Chemical Company), and injecting three Balb/c mice with 50 μg of adjuvant emulsified protein extract per mouse. Mice are boosted with 50 μg of protein extract emulsified in Freund's incomplete adjuvant (Sigma Chemical Company) 28 days later.
10. Test the sera of each mouse for antibody production 7 days' post-booster immunization by means of an indirect ELISA assay to the protein extract. The mouse demonstrating the highest titer is re-immunized with 25 μg of protein extract emulsified in Freund's incomplete adjuvant 14 days after the booster immunization.
11. Three days after the final injection euthanatized the mouse by cervical dislocation, its spleen will be removed, macerated, and the suspended lymphocytes hybridized with SP 2/0 myeloma cells.
12. The hybridoma cells are resuspended in 90 mL of Iscove's modified Dulbecco's medium (Sigma Chemical Company) (IMDM) containing 200 μL/mL fetal calf serum (Sigma Chemical Company) (FCS). One hundred microliters of cell suspensions are transferred to 96-well microtiter ELISA plates. Cell lines are nourished on Day 7 by adding 75 μL of IMDM-FCS medium. Beginning on Day 10, hybridoma cell lines are tested for antibody production.

1. **Sandwich ELISA**

   1. Dilute the capture antibody to a final concentration of 1–10 μg/mL in bicarbonate/carbonate antigen-coating buffer (100 mM $NaHCO_3$ in deionized water; pH adjusted to 9.6).
   2. Pipette 100 μL of diluted antibody to each well of a microtiter plate.

3. Cover the plate with adhesive plastic and incubate at 4 °C overnight (or 37 °C for 30 min).
4. Remove the coating solution and wash the plate 3× with 200 μL PBS (Phosphate Buffered Saline) buffer (10 mM Na_2HPO_4 and 1.8 mM NaH_2PO_4 in deionized water with 0.2% Tween 20; pH adjusted to 7.4) with for 5 min each time. The coating/washing solutions can be removed by flicking the plate over a sink. The remaining drops can be removed by patting the plate on a paper towel or by aspiration. Do not allow the wells to dry out at any time.
5. Pipette 200 μL blocking buffer (5% w/v non-fat dry milk in PBS buffer) per well to block residual protein-binding sites. Alternatively, BSA or BlockACE can be used to replace non-fat dry milk.
6. Cover the plate with adhesive plastic and incubate for 1–2 h at 37 °C (or at 4 °C overnight).
7. Remove the blocking solution and wash the plate 2× with 200 μL PBS for 5 minutes each time. Flick the plate and pat the plate as described in the coating step.
8. Prepare for the diluted standard solutions, biotinylated antibody, and ABC solutions.
9. Serially dilute the sample with blocking buffer immediately before use. The optimal dilution should be determined by a titration assay according to the antibody manufacturer.
10. Pipette 100 μL of each of the diluted sample solutions and control to each empty well. Repeat in duplicate or triplicate for accuracy. The negative control should be species- and isotype-matched as well as non-specific immunoglobulin diluted in PBS buffer.
11. Cover the plate with adhesive plastic and incubate for 2 h at room temperature.
12. Remove the content in the wells and wash them 3× with 200 μL PBS buffer for 5 min each time. Flick the plate and pat the plate as described in the coating step.
13. Pipette 100 μL of diluted antibody to the wells with control, standard solutions, and diluted samples.
14. Cover the plate with adhesive plastic and incubate for 1 h at 37 °C (or 2 h at room temperature). These incubation times should be sufficient to receive a strong signal. However, if a weak signal is observed, perform incubation overnight at 4 °C for a stronger signal.
15. Remove the content in the wells and wash them 3× with 200 μL PBS for 5 min each time. Flick the plate and pat the plate as described in the coating step.

16. Pipette 100 μL of diluted ABC solution to the wells with control, standard solutions, and diluted samples.
17. Cover the plate with adhesive plastic and incubate for 0.5 h at 37 °C.
18. Remove the content in the wells and wash them 3× with 200 μL PBS buffer for 5 min each time. Flick the plate and pat the plate as described in the coating step.

Substrate Preparation
Prepare the substrate solution immediately before use or bring the pre-made substrate to room temperature. The two widely used enzymes for signal detection are horse radish peroxidase (HRP) and alkaline phosphatase (AP), and their corresponding substrates, stopping solutions, detection absorbance wavelengths and color developed are as follows:

Enzyme	Substrate[a]	Stop solution	Absorbance (nm)	Color developed
HRP	TMB	2 M H_2SO_4	450	Yellow
AP	pNPP	0.75 M NaOH	405	Yellow

[a]TMB: 3,3′,5,5′-tetramethylbenzidine; pNPP: p-nitrophenyl-phosphate

Note

(a) The TMB substrate must be kept at 37 °C for 30 min before use.
(b) Hydrogen peroxide can also act as a substrate for HRP.
(c) Sodium azide is an inhibitor of HRP. Do not include the azide in buffers or wash solutions if HRP-labelled conjugate is used for detection.

Signal Detection

I. Pipette 90 μL of substrate solution to the wells with the control, standard solutions and diluted samples.
II. Incubate the plate at 37 °C in the dark. If TMB is used, shades of blue will be observed in the wells with the most concentrated solutions. Other wells may show no obvious color.
III. Color should be developed in positive wells after 15 min. After sufficient color development, pipette 100 μL of stop solution to the appropriate wells (if necessary).
IV. Read the absorbance (OD: Optical Density) of each well with a plate reader.

Data Analysis

(a) Prepare a standard curve using the data produced from the diluted standard solutions. Use absorbance on the Y-axis (linear) and concentration on the X-axis (log scale).

(b) Interpret the sample concentration from the standard curve.

2. **Indirect ELISA**

1. Dilute purified antigens to a final concentration of 1–10 μg/mL in bicarbonate/carbonate antigen-coating buffer (100 mM NaHCO3 in deionized water; pH adjusted to 9.6).
2. Pipette 100 μL of diluted antigen to each well of a microtiter plate.
3. Cover the plate with adhesive plastic and incubate at 4 °C overnight (or 37 °C for 30 min).
4. Remove the coating solution and wash the plate 3× with 200 μL PBS (Phosphate Buffered Saline) buffer (10 mM Na_2HPO_4 and 1.8 mM NaH_2PO_4 in deionized water with 0.2% Tween 20; pH Adjusted to 7.4) with for 5 min each time. The coating/washing solutions can be removed by flicking the plate over a sink. The remaining drops can be removed by patting the plate on a paper towel or by aspiration. Do not allow the wells to dry out at any time.
5. Pipette 200 μL blocking buffer (5% w/v non-fat dry milk in PBS buffer) per well to block residual protein-binding sites. Alternatively, BSA or BlockACE can be used to replace non-fat dry milk.
6. Cover the plate with adhesive plastic and incubate for 1–2 h at 37 °C (or at 4 °C overnight).
7. Remove the blocking solution and wash the plate 2× with 200 μL PBS for 5 min each time. Flick the plate and pat the plate as described in the coating step.
8. Prepare for the diluted standard solutions as shown on above reagent preparation row.
9. Serially dilute the primary antibody of choice with blocking buffer. The optimal dilution should be determined by a titration assay according to the antibody manufacturer.
10. Pipette 100 μL of each diluted antibody per well. Repeat in duplicate or triplicate for accuracy. The negative control should be species- and isotype-matched as well as non-specific immunoglobulin diluted in PBS buffer.
11. Cover the plate with adhesive plastic and incubate for 1 h at 37 °C (or 2 h at room temperature). These incubation times should be sufficient to receive a strong signal. However, if a

weak signal is observed, perform incubation overnight at 4 °C for a stronger signal.

12. Remove the diluted antibody solution and wash the wells 3× with 200 μL PBS for 5 min each time. Flick the plate and pat the plate as described in the coating step.
13. Serially dilute the conjugated secondary antibody with blocking buffer immediately before use. The optimal dilution should be determined by a titration assay according to the antibody manufacturer.
14. Pipette 100 μL of diluted secondary antibody solution to each well.
15. Cover the plate with adhesive plastic and incubate for 2 h at room temperature.
16. Remove the content in the wells and wash them 3× with 200 μL PBS buffer for 5 min each time. Flick the plate and pat the plate as described in the coating step.

    Substrate Preparation, signal detection, and data analysis are similar to Sandwich ELISA.

3. **Direct ELISA**

    1. Dilute purified antigens to a final concentration of 1–10 μg/mL in bicarbonate/carbonate antigen-coating buffer (100 mM $NaHCO_3$ in deionized water; pH adjusted to 9.6).
    2. Pipette 100 μL of diluted antigen to each well of a microtiter plate.
    3. Cover the plate with adhesive plastic and incubate at 4 °C overnight (or 37 °C for 30 min).
    4. Remove the coating solution and wash the plate 3× with 200 μL PBS (Phosphate Buffered Saline) buffer (10 mM Na_2HPO_4 and 1.8 mM NaH_2PO_4 in deionized water with 0.2% Tween 20; pH adjusted to 7.4) with for 5 min each time. The coating/washing solutions can be removed by flicking the plate over a sink. The remaining drops can be removed by patting the plate on a paper towel or by aspiration. Do not allow the wells to dry out at any time.
    5. Pipette 200 μL blocking buffer (5% w/v non-fat dry milk in PBS buffer) per well to block residual protein-binding sites. Alternatively, BSA or BlockACE can be used to replace non-fat dry milk.
    6. Cover the plate with adhesive plastic and incubate for 1–2 h at 37 °C (or at 4 °C overnight).

7. Remove the blocking solution and wash the plate 2× with 200 μL PBS for 5 min each time. Flick the plate and pat the plate as described in the coating step.
8. Prepare for the diluted standard solutions as shown on above reagent preparation row.
9. Serially dilute the conjugated primary antibody with blocking buffer immediately before use. The optimal dilution should be determined by a titration assay according to the antibody manufacturer.
10. Pipette 100 μL of diluted secondary antibody solution to each well.
11. Cover the plate with adhesive plastic and incubate for 2 h at room temperature.
12. Remove the content in the wells and wash them 3× with 200 μL PBS buffer for 5 min each time. Flick the plate and pat the plate as described in the coating step. Substrate preparation, signal detection, and data analysis are similar to Sandwich ELISA and Indirect ELISA.

4. **Competitive ELISA**

1. Dilute purified antigens to a final concentration of 20 μg/mL in bicarbonate/carbonate antigen-coating buffer (100 mM $NaHCO_3$ in deionized water; pH adjusted to 9.6).
2. Pipette 100 μL of diluted antigen to each well of a microtiter plate.
3. Cover the plate with adhesive plastic and incubate at 4 °C overnight (or 37 °C for 30 min).
4. Remove the coating solution and wash the plate 3× with 200 μL PBS (Phosphate Buffered Saline) buffer (10 mM Na_2HPO_4 and 1.8 mM NaH_2PO_4 in deionized water with 0.2% Tween 20; pH adjusted to 7.4) with for 5 min each time. The coating/washing solutions can be removed by flicking the plate over a sink. The remaining drops can be removed by patting the plate on a paper towel or by aspiration. Do not allow the wells to dry out at any time.
5. Pipette 200 μL blocking buffer (5% w/v non-fat dry milk in PBS buffer) per well to block residual protein-binding sites. Alternatively, BSA or BlockACE can be used to replace non-fat dry milk.
6. Cover the plate with adhesive plastic and incubate for 1–2 h at 37 °C (or at 4 °C overnight).

7. Remove the blocking solution and wash the plate 2× with 200 μL PBS for 5 min each time. Flick the plate and pat the plate as described in the coating step.
8. Prepare for the diluted standard solutions as shown on above reagent preparation row.
9. Serially dilute the sample with blocking buffer immediately before use. The optimal dilution should be determined by a titration assay according to the antibody manufacturer.
10. Pipette 100 μL of diluted sample to each well.
11. Cover the plate with adhesive plastic and incubate for 2 h at room temperature.
12. Remove the content in the wells and wash them 3× with 200 μL PBS buffer for 5 min each time. Flick the plate and pat the plate as described in the coating step.
13. Serially dilute the primary antibody of choice with blocking buffer. The optimal dilution should be determined by a titration assay according to the antibody manufacturer.
14. Pipette 100 μL of each diluted antibody per well. Repeat in duplicate or triplicate for accuracy. The negative control should be species- and isotype-matched as well as non-specific immunoglobulin diluted in PBS buffer.
15. Cover the plate with adhesive plastic and incubate for 1 h at 37 °C (or 2 h at room temperature). These incubation times should be sufficient to receive a strong signal. However, if a weak signal is observed, perform incubation overnight at 4 °C for a stronger signal.
16. Remove the diluted antibody solution and wash the wells 3× with 200 μL PBS for 5 min each time. Flick the plate and pat the plate as described in the coating step.
17. Serially dilute the conjugated secondary antibody with blocking buffer immediately before use. The optimal dilution should be determined by a titration assay according to the antibody manufacturer.
18. Pipette 100 μL of diluted secondary antibody solution to each well.
19. Cover the plate with adhesive plastic and incubate for 2 h at room temperature.
20. Remove the content in the wells and wash them 3× with 200 μL PBS buffer for 5 min each time. Flick the plate and pat the plate as described in the coating step. Substrate preparation, signal detection, and data analysis are similar to Sandwich ELISA and Indirect ELISA.

5. **Tissue Print Immunoassay (TPIA)**

1. The plant tissue like tillers, as well as individual branches and parts of inflorescences, are printed cut side-down onto a nitrocellulose membrane. As a control a mycelial sample or any endophytes suspension is squashed onto each membrane.
2. After printing, the membrane is incubated for 1 min in 0.05 M PBS, pH 7.4, containing 25 μL/l of Tween 20 and 1 mg/l of polyvinyl alcohol 72,000 (PVA), which works as a blocking agent.
3. The membrane will be washed twice with the same buffer without PVA (washing buffer) and incubated for 2 h in the enzyme-conjugated immunoglobulin solution.
4. The conjugate, prepared as described by Clark and Adams (1977), is diluted 1: 150 to 1: 300 (dilution depending on the quality of the immunoglobulin) in PBS buffer containing 10 g of PVP 15 (33,422, SERVA; Heidelberg, Germany) and 1 g of ovalbumin (Grade III, SIGMA A-5378) per liter.
5. After incubation, the membrane is washed twice with washing buffer for 5 min.
6. The washing buffer is replaced by 0.2 M Tris buffered saline, pH 7.4 (TBS; containing 0.5 M NaCl) for 5 min.
7. Then the membrane is equilibrated for 5 min with 0.2 M Tris-HCl, pH 8.0, containing 4 mM $MgCl_2$.
8. Finally, the membrane is stained in a staining solution (a mixture of 15 mL of Aqua dest laboratory water., containing 6 mg of Napthol AS-MX phosphate, and 15 mL of 0.2 M Tris-HCl, pH 8.0, containing 90 mg of Fast Red TR Salt) for approximately 20 min, until a red color appeared in prints of E+.
9. After air-drying the membrane at room temperature, it is storable and always available for evaluation.
10. For detection of endophyte infection by means of TPIA, the prints are evaluated by use of a Zeiss Stemi 2000 C stereomicroscope at 30–50 magnification.

6. **Dot Blot Immunoassay (DBIA)**

1. Have nitrocellulose membrane ready, draw grid by pencil to indicate the region you are going to blot.
2. Using narrow-mouth pipette tip, spot 2 μL of samples onto the nitrocellulose membrane at the center of the grid. Minimize the area that the solution penetrates (usually 3–4 mm diam.) by applying it slowly.
3. Let the nitrocellulose membrane dry.

4. Block non-specific sites by soaking in 5% BSA in TBS-T (0.5–1 h). Use 10 cm Petri Dish for reaction chamber.
5. Incubate with primary antibody (0.1–10 μg/mL for purified antibody, 1:1000 to 1:100000 dilutions for antisera, 1: 100 to1:10000 for hybridoma supernatant) dissolved in BSA/TBS-T for 30 min at RT.
6. Wash three times with TBS-T (3 × 5 min).
7. Incubate with secondary antibody conjugated with HRP (for optimum dilution, follow the manufacturer's recommendation) for 30 min at RT.
8. Wash three times with TBS-T (15 min × 1, 5 min × 2), then once with TBS (5 min).
9. Incubate with ECL reagent for 1 min, cover with Saran-wrap (remove excessive solution from the surface), and expose X-ray film in the dark room. Try several different lengths of exposure.
10. Compare the signal from your unknown sample to that of standard and estimate the concentration.

7. **Immunogold Labeling Technique**
   1. Fix the specimen with 4% formaldehyde in 0.1 M PB, pH 7.4 for 12–24 h. For free cells or cell suspension, fix cell pellet for 2–4 h.
   2. Rinse in 0.1 M PB for 3 × 10 min.
   3. Incubate specimen in 0.1 M Glycine Solution for 20 min to quench the free aldehyde groups.
   4. Incubate in 0.2 M Sucrose Solution for 2 × 15 min or overnight at 4 °C.
   5. Rinse in 0.1 M PB briefly.
   6. Samples are dehydrated with rising ethanol concentrations (70%, 90%, and 100%, 10 min each) and thereafter embedded in a series of Unicryl dilutions (ethanol: Unicryl 2: 1, 1: 2, and pure Unicryl, 1 h each). Or incubate specimen in two changes of 70% ethanol 30 min each. Then incubate in L.R. White and 70% ethanol (2:1) mixture: slowly add one part of 70% ethanol drop by drop to two parts of L.R. White, and shake gently to avoid mixture becomes milky. Incubate specimen in the mixture for 1 h.
   7. Transfer specimen in three changes of pure L.R. White for 1 h each. Keep specimen in the last change of L.R. White overnight at room temperature (specimen can be stored in L.R. White at 4 °C for weeks if necessary).
   8. Place specimen in bottom of Gelatin Capsule, fill up with L.R. White to the brim, and slide the other half of the capsule on. Polymerize in 50 °C oven for 48 h or longer.

9. Trim EM blocks and cut 0.5 um thick sections and stain with toluidine blue if necessary, and then select the region of interest and trim blocks further to the appropriate size.
10. Cut ultrathin sections at 60–90 nm and collect sections on formvar-carbon coated nickel grids (5 grids per specimen), and allow the grids to dry overnight prior to staining.
11. Steam grids in Epitope Unmasking Solution (based on LM test or use citrate buffer pH 6.0) for 20 min, and allow grids to cool for 20 min. This step may not be necessary.
12. Rinse sections for 2 × 2 min by place grids on large droplets of TBS-Tween.
13. Incubate sections in IHC Ultra Serum Block – Species match secondary antibody for 30 min.
14. Incubate sections on droplets of primary antibody diluted in Universal Antibody Diluent for 2 h at room temperature.-Note: primary antibody dilution is usually 10 times more concentrated than immune staining at LM level.
15. Rinse sections on large droplets of TBS-Tween for 6 × 2 min.
16. Incubate sections on droplets of biotinylated secondary antibody (1:50, Vector Lab) in Universal Antibody Diluent for 1 h at room temperature.
17. Rinse on large droplets of TBS-Tween for 6 × 2 min.
18. Incubate sections with gold conjugated streptavidin (1:20, EMS) in TBS, pH 7.6 for 1 h. Note: do not use BSA or serum containing solution/reagent to dilute streptavidin-gold since BSA or serum may contain biotin, therefore reduce streptavidin-gold affinity to biotinylated secondary antibody.
19. Rinse on large droplets of TBS-Tween for 6 × 2 min.
20. Post-fix sections on droplets of 2% glutaraldehyde in 0.1 M PB for 10 min and then rinse in TBS-Tween for 6 × 2 min.
21. Rinse sections in 3 × 20 dips of distilled water, and then stain with uranyl acetate for 15 min and lead citrate for 1–2 min.
22. Observe sections under electron microscopy.

4 Conclusion

Immunological techniques are the rapid and accurate methods to study the detection and colonization of the endophytes in various parts of plant species specifically internal tissues by using the polyclonal and monoclonal antibodies. Practically immunological techniques and traditional methods provide comparable accuracy in endophytes detection and requires less training. The cost of the immunological assay is slightly higher than the cost of the conventional methods but cheaper than the molecular techniques.

References

1. Brader G, Corretto E, Sessitsch A (2017) Metagenomics of plant microbiomes. In: Charles T, Liles M, Sessitsch A (eds) Functional metagenomics: tools and applications. Springer, Cham, pp 179–200
2. de Tender C (2017) Microbial community analysis in soil (rhizosphere) and the marine (plastisphere) environment in function of plant health and biofilm formation, pp 1–274. Doctor Ph.D. in Biotechnology Thesis. Ghent University, Ghent
3. Yadav AN (2018) Biodiversity and biotechnological applications of host specific endophytic fungi for sustainable agriculture and allied sectors. Acta Sci Microbiol 1:44
4. Andreozzi A, Prieto P, Mercado-Blanco J, Monaco S, Zampieri E, Romano S (2019) Efficient colonization of the endophytes *Herbaspirillum huttiense* RCA24 and *Enterobacter cloacae* RCA25 influences the physiological parameters of *Oryza sativa* L. cv. Baldo rice. Environ Microbiol 21:3489–3504
5. Shen FT, Yen JH, Liao CS, Chen WC, Chao YT (2019) Screening of rice endophytic biofertilizers with fungicide tolerance and plant growth promoting characteristics. Sustainability 11:1133. https://doi.org/10.3390/su11041133
6. Rana ZA, Fuad A, Khalid R, Alghuthaymi MA, Alsalmi R, Li C (2019) A brief history of endophyte detection techniques in grasses. Sustain Agri Res 8(3)
7. Rosenblueth M, Martinez-Romero E (2006) Bacterial endophytes and their interactions with hosts. Acta Pharmacol Sin 19(8):827–837
8. Strobel G, Daisy B (2003) Bioprospecting for microbial endophytes and their natural products. Microbiol Mol Biol Rev 67(4):491–502
9. Gouda S, Das G, Sen SK, Shin HS, Patra JK (2016) Endophytes: a treasure house of bioactive compounds of medicinal importance. Front Microbiol 7:1538. https://doi.org/10.3389/fmicb.2016.01538
10. Yadav AN, Kumar V, Dhaliwal HS, Prasad R, Saxena AK (2018) Microbiome in crops: diversity, distribution, and potential role in crop improvement. In: Prasad R, Gill SS, Tuteja N (eds) Crop improvement through microbial biotechnology. Elsevier, New Delhi, pp 305–332
11. Golinska P, Wypij M, Agarkar G, Rathod D, Dahm H, Rai M (2015) Endophytic actinobacteria of medicinal plants: diversity and bioactivity. Antonie Van Leeuwenhoek 108:267–289. https://doi.org/10.1007/s10482-015-0502-7
12. Sun H, He Y, Xiao Q, Ye R, Tian Y (2013) Isolation, characterization, and antimicrobial activity of endophytic bacteria from *Polygonumcuspidatum*. Afr J Microbiol Res 7:1496–1504. https://doi.org/10.5897/AJMR12.899
13. Engelhard M, Hurek T, Reinhold-Hurek B (2000) Preferential occurrence of diazotrophic endophytes, *Azoarcus* spp., in wild rice species and land races of *Oryza sativa* in comparison with modern races. Environ Microbiol 2:131–141
14. Chantreuil C, Giraud E, Prin Y, Lorquin J, Ba A, Gillis M, de Lajudie P, Dreyfus B (2000) Photosynthetic bradyrhizobia are natural endophytes of the African wild rice *Oryza breviligulata*. Appl Environ Microbiol 66: 5437–5447
15. McInroy JA, Kloepper JW (1995) Survey of indigenous bacterial endophytes from cotton and sweet corn. Plant Soil 173:337–342
16. Berdy J (2012) Thoughts and facts about antibiotics: where we are now and where we are heading. J Antibiot 65:385–395. https://doi.org/10.1038/ja.2012.27
17. Chaudhary HS, Soni B, Shrivastava AR, Shrivastava S (2013) Diversity and versatility of actinomycetes and its role in antibiotic production. Int J Pharm Sci 3:83–94
18. Barka EA, Vatsa P, Sanchez L, Gaveau-Vaillant-N, Jacquard C, Klenk HP (2016) Taxonomy, physiology, and natural products of Actinobacteria. Microbiol Mol Biol Rev 80:1–43. https://doi.org/10.1128/MMBR.00019-15
19. Gayathri P, Muralikrishnan V (2013) Isolation and characterization of endophytic actinomycetes from mangrove plants for antimicrobial activity. Int J Curr Microbiol Appl Sci 2:78–79
20. Singh R, Dubey AK (2015) Endophytic actinomycetes as emerging source for the rapeutic

compounds. Indo Global J Pharm Sci 5:106–116

21. Zhao K, Penttinen P, Guan T, Xiao J, Chen Q, Xu J (2011) The diversity and anti-microbial activity of endophytic actinomycetes isolated from medicinal plants in Panxi Plateau China. Curr Microbiol 62:182–190. https://doi.org/10.1007/s00284-010-9685-3
22. Gangwar M, Dogra S, Gupta UP, Kumar RN (2014) Diversity and biopotential of endophytic actinomycetes from three medicinal plants in India. Afr J Microbiol Res 8:184–191. https://doi.org/10.5897/AJMR2012.2452
23. Hollants J, Leroux O, Leliaert F, Decleyre H, De Clerck O, Willems A (2011) Who is in there? Exploration of endophytic bacteria within the siphonous green seaweed Bryopsis (Bryopsidales, Chlorophyta). PLoS One 6: e26458. https://doi.org/10.1371/journal.pone.0026458
24. Dayle ES, Polans NO, Paul DS, Melvin RD (2001) Angiosperm DNA contamination by endophytic fungi: detection and methods of avoidance. Plant Mol Biol Rep 19:249–260. https://doi.org/10.1007/BF02772897
25. Jalgaonwala RE, Mohite BV, Mahajan RT (2011) Natural products from plant associated endophytic fungi. J Microbiol Biotechnol Res 1:21–32
26. Bhardwaj A, Agrawal P (2014) A review fungal endophytes: as a store house of bioactive compound. World J Pharm Pharm Sci 3:228–237
27. Huth W (1997) Tissue print immunoassay – a rapid and reliable method for routinely evidence of graminae viruses. Newl DeutPhytosanitary 49(8):185–191
28. Hahn H, Huth W, Scho Berlein W, Diepenbrock W (2003) Detection of endophytic fungi in *Festuca* spp. by means of tissue print immunoassay. Plant Breed 122:217–222
29. Iborra FJ, Kimura H, Cook PR (2004) The functional organization of mitochondrial genomes in human cells. BMC Biol 2:9. https://doi.org/10.1186/1741-7007-2-9
30. Alberts B, Johnson A, Lewis J, Raff M, Roberts K, Walter P (2008) Molecular biology of the cell, 5th edn. Garland Science, New York. ISBN 978-0-8153-4106-2
31. Song JJ, Pongnak W, Soytong K (2016) Isolation and identification of endophytic fungi from 10 species palm trees. Int J Agri Technol 12(2):349–363
32. Gupta RM, Prathamesh SK, Madhuri LR, Nikhil NJ (2015) Isolation, characterization and identification of endophytic bacteria by 16S rRNA partial sequencing technique from roots and leaves of Prosopis cineraria plant. Asian J Plant Sci Res 5(6):36–43
33. ELISA Handbook: Principle, Trouble shooting, sample preparation and Assay protocols by Boster Biological technology, pp 1–31
34. Clark MF, Barbara DJ, Davies DL (1983) Production and characteristics of antisera to *Spiroplasma citri* and clover phyllody-associated antigens derived from plants. Ann Appl Biol 103:251–259
35. Salehi M, Izadpanah K, Siampour M, Esmailzadeh-hosseini SA (2011) Polyclonal antibodies for the detection and identification of Fars alfalfa witches' broom phytoplasma. Bull Insectol 64:59–60
36. Hiatt EE, Hill NS, Bouton JH, Mims CW (1997) Monoclonal antibodies for detection of *Neotyphodium coenophialum*. Crop Sci 37: 1265–1269
37. Segel IH (1976) Biochemical calculations, 2nd edn. Wiley, New York

Chapter 8

Detection of Endophytes by Electron Microscope

Pradeep Kumar, Simran Rani, Sarita, Amita Suneja Dang, and Pooja Suneja

Abstract

Investigation of endophytes has come a long way from typical isolation and cultivation to much sophisticated molecular biology techniques. However, microscopy is the only direct method to visualize/detect the endophytes and reveal their precise location. Electron microscopy has made ultrastructural analysis possible by making further addition to the already available techniques. TEM (transmission electron microscopy) is used to view the specimens viz. tissue sections, molecules, etc., through which electrons pass generating a projection image as a result it offers valuable information about the inner structure of the sample. On the other hand, more conventional SEM (scanning electron microscopy) works on the principle of secondary electrons emission from the surface of a specimen, thereby providing primary information about the sample's surface. This chapter describes the detailed protocol of TEM & SEM for detecting the endophytes and studying the pattern of their colonization within intracellular as well as intercellular spaces of plant tissues.

Key words Endophytes, Detection, Electron microscopy, TEM, SEM

1 Introduction

The endophytes living inside plants are naturally selected and more bioactive than epiphytes because they interact more closely with the host plant in its internal microenvironment. The significant contribution of these endophytic microbes toward plant growth promotion, stress tolerance, improving nutrient acquisition, drug development, etc., has boosted the research interest in this field. It is very challenging for researchers to study endophytes as methods used for detection, isolation, and identification of endophytes are still insufficient to provide a complete picture of this complex host-parasite interaction [1–3]. However, these methods have been innovated and improvised over time. Four basic sets of techniques used for their detection and identification purposes include microscopy, isolation of pure cultures, biochemical, and DNA-based analysis [4]. Among these techniques, only microscopy supports in

A. Sankaranarayanan et al. (eds.), *Endophytic Microbes: Isolation, Identification, and Bioactive Potentials*, Springer Protocols Handbooks, https://doi.org/10.1007/978-1-0716-2827-0_8,

providing direct evidence for the presence of endophytes [5]. Although the presence of endophytes can be detected indirectly via the culture method but direct observation helps in the understanding mode of infection, the extent of colonization within the tissue and the response of plants toward microbe [6]. Moreover, it is often impossible to reproduce the optimal conditions required for the growth of a number of endophytes in the case of simple isolation and cultivation [7]. Light and electron microscope can reveal the accurate location of endophytes within the plant tissue [8]. The traditional methods of tissue fixation, microtomy, and staining have been limited by their inability to differentiate the microbial cells (endophytes) from the constituents of plant tissues/cellular inclusions [9]. Transmission electron microscopy (TEM), fluorescent in situ hybridization (FISH), triphenyl tetrazolium chloride vital staining, etc., are some additional methods, enriching our understanding of endophytes and plant-endophyte interactions [10]. Electron microscopy is employed mainly for the ultrastructural analysis of the bacterial and fungal endophytes [11]. Scanning electron microscope and transmission electron microscope both working on slightly different principles have been a lot of assistance in detection of endophytes, their extent of colonization, interaction with the host plant, and establishment within plant environment [12].

2 Materials

- Sterile distilled water.
- 1% Sodium hypochlorite solution.
- Ethanol.
- Acetone.
- Mortar and pestle.
- Nutrient Agar medium or Tryptic soy agar.
- Glycerol.
- Potato dextrose agar (PDA)/malt extract agar (MEA) media.
- Antibiotics.
- Agar gel plates.
- 0.1 M Sodium cacodylate buffer (pH 7.4).
- Phosphate buffer (pH 7.2).
- 2.5% Glutaraldehyde.
- 1% (w/v) Osmium tetroxide (OsO_4).
- Epoxy resin.
- 1% uranyl acetate.
- Reynolds Lead citrate.

- Plastic bags.
- Blades.
- Incubator.
- Slides.
- Oven.
- Glass knife.
- Ultra microtome.
- Transmission electroscope.
- Degasser.
- Double-sided tape.
- Gold-palladium.
- Scanning electron microscope.

3 Methods

3.1 Sample Collection

Collect healthy plants and separate them into leaves, stems, and roots. Use sterile zip-lock plastic bags for packing and bring the collected samples to the laboratory.

3.2 Sample Sterilization

1. Wash the plant sample under tap water to eliminate dust.
2. Treat with 1% sodium hypochlorite solution (NaOCl) for 5–10 min with continuous shaking and then wash with sterile distilled water.
3. Continue the surface sterilization process with 95% ethanol for 30 s, followed by rinsing with sterile distilled water for 5 times.
4. For endophytic colonization study, plant parts such as leaves, roots, and shoots can be used directly whereas for the study of morphologically unique characteristics, isolation of endophytes is carried out.

3.3 Isolation of Bacterial Endophytes

1. Crush the sterilized plant tissues in sterile mortar and pestle.
2. Prepare serial dilutions (10^{-5}, 10^{-6}, and 10^{-7}) from this extract.
3. Take 0.1 mL solution as inoculum and pour it on Nutrient Agar medium or Tryptic soy agar. Inoculate in triplicates for better results.
4. Incubate the plates at 30 °C for 24–48 h and observe daily for the appearance of colonies.
5. Subculture the colonies on agar media many times to achieve pure cultures.
6. Keep these cultures in 50% glycerol at 80 °C for future usage.

3.4 Isolation of Endophytic Fungi [13, 14]

1. Place the sterilized plant tissue onto potato dextrose agar (PDA)/malt extract agar (MEA) media and incubate at 25–30 °C for 7–14 days.
2. Monitor to make sure that there is no saprophytic contamination.
3. Isolate hyphal tips of fungi from plant tissues and culture in potato dextrose agar for pure isolates.

3.5 Electron Microscopic Analysis

3.5.1 Fixation of Sample [15, 16]

1. For bacteria, excise agar gel plates containing bacterial colonies and trim to 5 mm × 5 mm as thin as possible; for fungi, cut fungal disk (5 mm × 5 mm) from a fully grown mycelia culture plates and cut plant sample into 1.0–1.5 mm size.
2. Place the sample (plant tissue/bacterial isolate/fungal isolate) in 0.1 M sodium cacodylate buffer (pH 7.4)/phosphate buffer (pH 7.2) and 2.5% glutaraldehyde at room temperature for 2 h. Repeat the above step 2–3 times to ensure proper fixation.
3. Wash the sample thrice in the phosphate buffer and refix it in 1% (w/v) Osmium tetroxide (OsO_4) for 2 h (Optional: Cacodylate buffer for 4 h at 4 °C).
4. Then wash the sample with distilled water twice for 30 min each.
5. Dehydrate the sample with alcohol (acetone/ethanol) series (30–100%) for 10 min with each concentration except 20 min with 100% alcohol (make uniform gradient of alcohol concentration to prevent sudden dehydration of sample).

3.5.2 Scanning Electron Microscopy [17, 18]

1. During dehydration, remove carbon dioxide by degasser so that no gas-liquid interference occurs during drying.
2. Transfer the sample to 100% acetone for 20 min and then freeze dry.
3. Mount the specimen stub with conductive double-sided tape and coat it with gold–palladium.
4. Examine with a scanning electron microscope (5X, 10X, 45X, and 100X) at different magnification at an accelerating voltage.

3.5.3 Transmission Electron Microscopy [12, 16]

1. After dehydration in a series of alcohol, embed the sample in epoxy resin and place the block overnight at 60 °C in an oven.
2. Prepare ultra-thin sections of about 60 nm from resin blocks using a glass knife with an Ultramicrotome.
3. Mount sections on copper grids and perform primary staining with 1% uranyl acetate, followed by counterstaining with Reynold's lead citrate (LC).
4. Take the micrograph in a transmission electroscope at an operating voltage of 60–200 kV (Fig. 1).

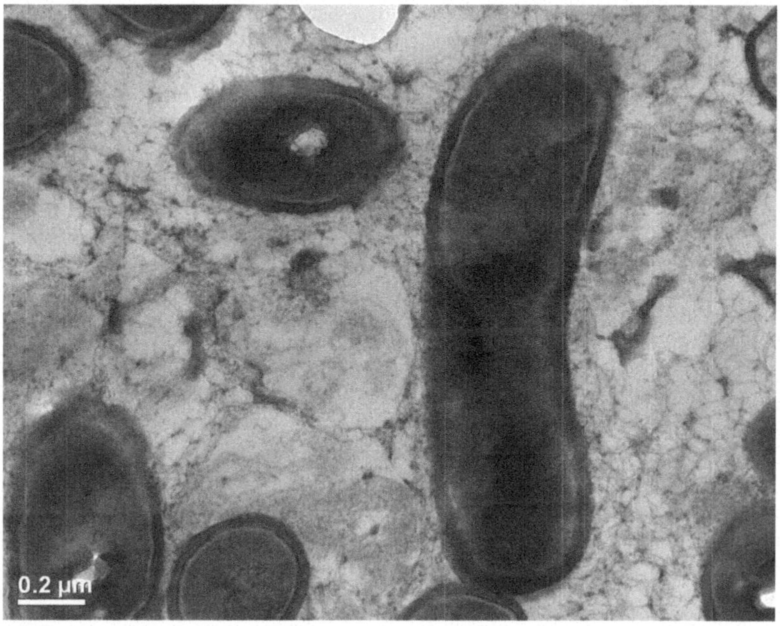

Fig. 1 Transmission electron microscope image of endophytic *B. licheniformis* MHN12

Acknowledgments

The authors acknowledge the Department of Microbiology and Centre for Medical Biotechnology, Maharshi Dayanand University, Rohtak, India for infrastructural support. PK & SR acknowledge Council of Scientific & Industrial Research, New Delhi, India for CSIR fellowship.

References

1. Rodriguez RJ, White JF Jr, Arnold AE, Redman AR (2009) Fungal endophytes: diversity and functional roles. New Phytol 182:314–330
2. Shahzad R, Khan AL, Bilal S, Asaf S, Lee IJ (2018) What is there in seeds? Vertically transmitted endophytic resources for sustainable improvement in plant growth. Front Plant Sci 9:24
3. Bhutani N, Maheshwari R, Negi M, Suneja P (2018) Optimization of IAA production by endophytic *Bacillus* spp. from *Vigna radiata* for their potential use as plant growth promoters. Isr J Plant Sci 65:83–96
4. Bayman P (2007) The mycota IV: environmental and microbial relationships. Springer, Berlin/Heidelberg
5. James EK (2000) Nitrogen fixation in endophytic and associative symbiosis. Field Crops Res 65:197–209
6. Johnston PR, Sutherland PW, Joshee S (2006) Visualising endophytic fungi within leaves by detection of (1→3)-ß-d-glucans in fungal cell walls. Mycologist 20:159–162
7. Santoyo G, Moreno-Hagelsieb G, del Carmen Orozco-Mosqueda M, Glick BR (2016) Plant growth-promoting bacterial endophytes. Microbiol Res 183:92–99
8. Kuldau GA, Yates IE (2000) Evidence for Fusarium endophytes in cultivated and wild plants. In: Bacon CW, White JF Jr (eds) Microbial endophytes. Dekker, New York, pp 85–117

9. Thomas P, Soly TA (2009) Endophytic bacteria associated with growing shoot tips of banana (*Musa* sp.) cv. Grand Naine and the affinity of endophytes to the host. Microb Ecol 58:952–964
10. Thomas P, Reddy KM (2013) Microscopic elucidation of abundant endophytic bacteria colonizing the cell wall–plasma membrane perispace in the shoot-tip tissue of banana. AoB Plants 5:plt011
11. Esposito-Polesi NP, de Abreu-Tarazi MF, de Almeida CV, Tsai SM, de Almeida M (2017) Investigation of endophytic bacterial community in supposedly axenic cultures of pineapple and orchids with evidence on abundant intracellular bacteria. Curr Microbiol 74:103–113
12. Nongkhlaw FM, Joshi SR (2017) Microscopic study on colonization and antimicrobial property of endophytic bacteria associated with ethnomedicinal plants of Meghalaya. J Microsc Ultrastruct 5:132–139
13. Raja S, Subhashini P, Thangaradjou T (2016) Differential methods of localisation of fungal endophytes in the seagrasses. Mycology 7:112–123
14. Tazik Z, Rahnama K, White JF, Soltanloo H, Hasanpour M, Iranshahi M (2020) LC-MS based identification of stylosin and tschimgine from fungal endophytes associated with *Ferulaovina.* Iran J Basic Med Sci 23:1565–1570
15. Kaláb M, Yang AF, Chabot D (2008) Conventional scanning electron microscopy of bacteria. Infocus Mag 10:42–61
16. Ramanujam B, Renuka S, Poornesha B, Shylesha AN (2016) Electron microscopic studies for confirmation of endophytic colonization of *Beauveria bassiana* in Maize. J Pure Appl Microbiol 10:3017–3022
17. Lou B, Peng B, Rong N, Li Y, Chen H, Sree KS, Gao Q, Varma A (2014) Root and root endophytes from the eyes of an electron microscopist. In: Morte A, Varma A (eds) Root engineering. Springer, Berlin/Heidelberg
18. Castillo U, Myers S, Browne L, Strobel G, Hess WM, Hanks J, Reay D (2005) Scanning electron microscopy of some endophytic *Streptomycetes* in snakevine- *Kennedia nigricans.* Scanning 27:305–311

Chapter 9

Detection of Endophytes by Reactive Oxygen Staining

Simran Rani, Pradeep Kumar, Deepika, Amita Suneja Dang, and Pooja Suneja

Abstract

Endophytes (bacteria or fungi) are colonizers of the internal plant tissues without causing any disease or hostile symptoms. Endophytes have attracted a lot of attention due to the enormous benefits provided by them to the respective host plants. However, various factors like culturing hitches and lack of pioneering identification techniques have restricted the understanding of endophytes. Their study is challenging as many of these remain unisolated, unnoticed, and have difficulty in obtaining gnotobiotic plants to perform experiments. Moreover, due to the lack of proper staining methods, it is tough to visualize the microorganisms in living plant cells and tissues. This chapter summarizes the detection of endophytes by the reactive oxygen staining method in which ROS (reactive oxygen species) like hydrogen peroxide (H_2O_2) and superoxide (O_2^-) associated with the microbial invasion of eukaryotic cells are stained by DAB (3,3′-diaminobenzidine tetrachloride) and NBT (nitroblue tetrazolium) microscopic methods respectively.

Key words Endophytes, Detection, ROS, Reactive Oxygen Staining, DAB, NBT

1 Introduction

Endophytes, either fungi or bacteria are a bonafide residents of plant tissues without exhibiting any negative symptoms of their existence, and are isolated after surface sterilization [1, 2]. A tissue blotting technique was introduced to ensure surface sterilization and establish their endophytic nature [3]. Direct visualization of endophytism helps to study their pattern of growth, density, and tissue-specific concentration. It is an uphill task to visualize the endophytes in plant tissues and categorize them taxonomically due to the absence of any identifying feature in-vivo in most of the cases. Also, these methods need to be optimized depending upon the plant part (root, stem, seeds, and leaves) and endophytes (bacteria or fungi). Evaluation of endophytism can be performed by histological investigations using tissue or endophyte-specific

A. Sankaranarayanan et al. (eds.), *Endophytic Microbes: Isolation, Identification, and Bioactive Potentials*, Springer Protocols Handbooks, https://doi.org/10.1007/978-1-0716-2827-0_9,

stains and visualizing them under the microscope (light microscope, confocal microscope, electron microscope, and fluorescence microscope) [4, 5].

Reactive oxygen species are the oxygen derivatives such as singlet oxygen, superoxide, peroxide ion, hydroxyl radical, and hydrogen peroxide, more reactive than the oxygen molecule (O_2). They have been recognized for their role under stress in plants for a long time, but recent researches have thrown light on their significance in the regulation of plant growth and development [6]. Reactive oxygen species are scavenged by a full-fledged antioxidant machinery (enzymatic as well as non-enzymatic) so that there is a balance between the production and breakdown of reactive oxygen species. All eukaryotes use reactive oxygen species secretion to kill the microbes, being the part of their innate defense system.

Reactive oxygen staining is a method to detect the endophytes by staining these reactive oxygen species formed in the plant tissue as an outcome of plant-microbe interaction (Fig. 1). To degrade the intracellular endophytes, plant cells secrete superoxides that are transformed into water and hydrogen peroxide by superoxide dismutase (SOD). These methods of endophyte detection are not only easy to perform but also cost-effective than the other visualization methods like TEM and fluorescent microscopy. Besides this, they also provide cues into the host's defense response and process of endophyte degradation. In DAB method, 3, 3′-diaminobenzidine tetrachloride dye is used to stain H_2O_2 forming brown/red coloration. However, to selectively stain endophytes inhabiting deeper into the root cells, DAB is used in conjunction with counterstain,

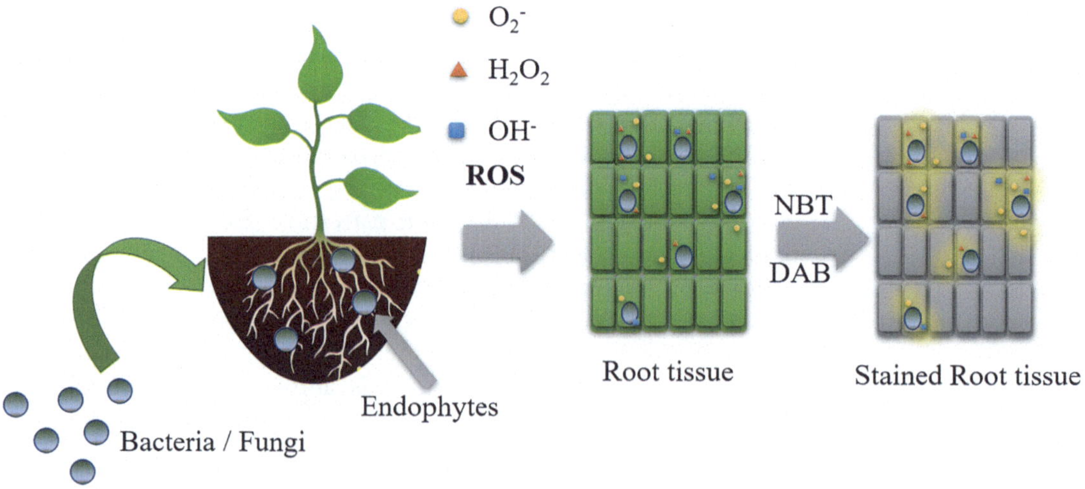

Fig. 1 Detection of endophytes by reactive oxygen staining

aniline blue/lactophenol, leading to more effective penetration into the root cells. The bacterial lysis process is visualized by this combination of stains. DAB results into brown coloration, indicating the presence of endophytes in tissues, whereas cells without endophytes do not undergo significant H_2O_2 activity, therefore, remain unstained. NBT method is used to stain superoxide radicals which reduce NBT and results in dark blue and water-insoluble formazan [7]. It can be used to detect endophytes in different plant tissues unlike DAB method, which is not able to detect endophytes in shoot tissues because of the inability of DAB to penetrate the same. NBT staining is associated with the differential distribution of superoxide radicals in different zones of plant tissues. Highly intense regions correspond to maximum number of endophytes whereas lighter color of stain is associated with comparatively lesser colonization of endophytes [8, 9].

2 Materials

- Sterilized water.
- 2.5 mM DAB (3,3′-diaminobenzidine tetrachloride).
- HCl (0.1 N).
- 100 mM potassium phosphate buffer (pH 6.9).
- 5 purpurogallin units/mL of horseradish peroxidase.
- Bleaching solution:- Ethanol: Acetic acid: Glycerol (3:1:1).
- Aniline blue.
- Phenol crystals.
- Glycerol.
- Lactic acid.
- 0.05 M Sodium phosphate buffer (pH 7.5).
- 0.05% NBT (Nitroblue tetrazolium).
- Ethanol.
- Amber-colored bottle.
- Aluminum foil.
- Blades.
- Plastic bags.
- Magnetic stirrer.
- pH meter.
- Slides.
- Light microscope.

3 Methods

3.1 DAB Microscopic Method to Visualize Endophytes in Plant Tissues

3.1.1 Preparation of Staining Solution [10]

1. Preparel mg/mL DAB staining solution by dissolving 50 mg DAB in 45 mL distilled water in an amber-colored bottle.
2. Adjust the pH to 3.8 by adding 0.1 N HCl slowly while mixing well on a magnetic stirrer.
3. Make the volume to 50 mL.

(Note: It must be prepared fresh before use)

3.1.2 Staining with DAB Solution [8]

1. Collect plant sample (except shoot).
2. Wash the sample properly with sterilized water to remove any exogenous material associated with them.
3. Stain by flooding the water agar plates with 5 mL of a solution of 100 mM potassium phosphate buffer, pH 6.9, 2.5 mM DAB and 5 purpurogallin units/mL of horseradish peroxidase. Cover with aluminum foil and incubate for 4–5 h followed by washing with deionized water.

   (Note: In case of leaf tissue, treat sample with bleaching solution after incubating with DAB solution to remove chlorophyll content.)
4. For counterstaining, excise the sample with blade and put it on a slide containing aniline blue/lactophenol stain (0.05 g aniline blue dye, 20 g phenol crystals, 40 mL glycerol, 20 mL lactic acid, 20 mL water).
5. Examine under the light microscope.

3.2 NBT Microscopic Method to Visualize Endophytes in Plant Tissues [11–13]

3.2.1 Preparation of Staining Solution [9]

1. To get a 0.2% staining solution dissolve 0.1 g NBT in 50 mM sodium phosphate buffer (pH 7.5) in an amber-colored bottle.
2. Make the volume up to 50 mL.
3. Mix the solution well with magnetic stirrer.

(Note: It must be prepared fresh before use)

3.2.2 Staining with NBT Solution

1. Collect the plant samples.
2. Wash the samples with sterilized water to remove any exogenous material associated with them.
3. Trim the plant sample into small pieces (5 X 5 mm) with a razor blade.
4. Incubate the sample in 0.05 M sodium phosphate buffer (pH 7.5) containing 0.05% NBT for 1 h (Note: In case of leaf tissue, put the tissue into ethanol overnight to remove chlorophyll content after NBT staining).

5. Fix the incubated specimens in ethanol in order to stop the NBT reaction.
6. Photograph/examine the stained samples under a light microscope.

Acknowledgments

The authors acknowledge the Department of Microbiology and Centre for Medical Biotechnology, Maharshi Dayanand University, Rohtak, India for infrastructural support. PK & SR acknowledge Council of Scientific & Industrial Research, New Delhi, India for CSIR fellowship.

References

1. Maheshwari R, Bhutani N, Bhardwaj A, Suneja P (2019) Functional diversity of cultivable endophytes from *Cicer arietinum* and *Pisum sativum*: bioprospecting their plant growth potential. Biocatal Agric Biotechnol 20: 101229
2. Dudeja SS, Giri R, Saini R, Suneja-Madan P, Kothe E (2012) Interaction of endophytic microbes with legumes. J Basic Microbiol 52: 248–260
3. Schulz P, Kirschbaum C, Prüßner J, Hellhammer D (1998) Increased free cortisol secretion after awakening in chronically stressed individuals due to work overload. Stress Med 14:91–97
4. Stone JK, Polishook JD, White JF (2004) Endophytic fungi. In: Foster M, Bills G (eds) Biodiversity of fungi, 1st edn. Elsevier Academic Press, Burlington, pp 241–270
5. Sun X, Guo LD (2012) Endophytic fungal diversity: review of traditional and molecular techniques. Mycology 3:65–76
6. Mhamdi A, Van Breusegem F (2018) Reactive oxygen species in plant development. Development 145:dev164376
7. Munkres KD (1990) Histochemical detection of the secretion of superoxide radicals and hydrogen peroxide by age-1 mutants of Neurospora. Fungal. Genet Rep 37:12
8. White JF Jr, Torres MS, Somu MP, Johnson H, Irizarry I, Chen Q, Zhang N, Walsh E, Tadych M, Bergen M (2014) Hydrogen peroxide staining to visualize intracellular bacterial infections of seedling root cells. Microsc Res Tech 77:566–573
9. Kumar D, Yusuf MA, Singh P, Sardar M, Sarin NB (2014) Histochemical detection of superoxide and H2O2 accumulation in *Brassica juncea* seedlings. Bio Protocol 4:e1108
10. Daudi A, O'Brien JA (2012) Detection of hydrogen peroxide by DAB staining in *Arabidopsis* leaves. Bio Protocol 2:e263
11. Tanaka A, Christensen MJ, Takemoto D, Park P, Scott B (2006) Reactive oxygen species play a role in regulating a fungus–perennial ryegrass mutualistic interaction. Plant Cell 18: 1052–1066
12. Shinogi T, Suzuki T, Kurihara T, Narusaka Y, Park P (2003) Microscopic detection of reactive oxygen species generation in the compatible and incompatible interactions of *Alternaria alternata* Japanese pear pathotype and host plants. J Gen Plant Pathol 69:7–16
13. Doke N (1983) Involvement of superoxide anion generation in the hypersensitive response of potato tuber tissues to infection with an incompatible race of *Phytophthora infestans* and to the hyphal wall components. Physiol Plant Pathol 23:345–357

Part III

Molecular Methods for Identification of Endophytes

Chapter 10

Identification of Bacterial Endophytes by 16S rRNA

V. Ambikapathy, S. Babu, R. Shanmugapriya, A. Prakash, and A. S. Shijila Rani

Abstract

Phenotypic profiles for microbial identification are unusual for rare, slow-growing, and fastidious microorganisms. The genotypic identification of microorganisms was widespread through the use of PCR and DNA sequencing. 16S rRNA sequencing has played a pivotal role in the accurate identification of microorganisms. The 16S rRNA region is universally distributed among microorganisms and is species-specific.

Key words Endophytic bacteria, PCR, 16S rDNA

1 Introduction

The 16S rRNA gene has been used to evaluate the diversity of endophytic bacteria with outstanding success. Carl Woese pioneered the use of 16S rRNA more than 30 years ago [1]. Among the advantages of using 16S rRNA is its universality, because it is present in almost all bacteria, allowing its use in taxonomic and phylogenetic identification [2]. The high degree of conservation of the 16S rRNA sequence within species is an important factor for differentiating organisms, in addition to its size, because it is large enough to be amplified using PCR, and thus, can be used in phylogenetic analyses [3]. A number of bioinformatic tools can be used to analyze the 16S rRNA gene when identifying organisms. One of the most prominent ones is BLAST (Basic Local Alignment Search Tool), which compares all the sequences deposited in public domain databases, such as the National Center for Biotechnology Information (NCBI). Another tool used in bioinformatics is the MEGA (Molecular Evolution Genetics Analysis) software, which, in addition to sequence alignment and phylogenetic inference, has many parameters available to the user for analysis [4].

A. Sankaranarayanan et al. (eds.), *Endophytic Microbes: Isolation, Identification, and Bioactive Potentials*,
Springer Protocols Handbooks, https://doi.org/10.1007/978-1-0716-2827-0_10,

2 Materials

- Luria-Bertani (LB) broth (tryptone – 10 g, yeast extract – 5 g, NaCl – 10 g, distilled water – 1000 mL; pH to 7.0).
- Endophytic bacteria.
- Conical flask.
- TE buffer (mM Tris-HCl, 1 mM EDTA, pH 8.0).
- PCR instrument.
- Centrifuge.
- Sodium chloride.
- 2 mL Centrifuge Tubes.

3 Methods

3.1 Isolation of DNA from Endophytic Bacteria [5]

- Two 2 mL Centrifuge Tubes with your name and transfer 0.2 mL Cell Suspension, a suspension of bacteria, to one of your tubes.
- Add 0.2 mL DNA Release Buffer to the tube containing the Bacterial Suspension. Invert the tube several times to slowly mix. The DNA Release Buffer breaks open the bacterial cells releasing the DNA.
- Add 0.02 mL Protease to the tube to digest and remove the cellular material and protein and release the genomic DNA.
- Gently mix by inverting the tube 5–6 times and then place in a 50–55 °C waterbath or heating block for 1 h.
- After 1 h, add 0.1 mL DNA Salt Solution to the tube and mix by inverting the tube several times. The salt solution aids in the precipitation of the DNA.
- Centrifuge the tube for 5 min at 5000xg to pellet the cell debris. Transfer the supernatant to your other labeled tube.
- Add 0.8 mL Precipitation Solution, close the tube and, while watching, slowly invert the tube several times to mix. White DNA strands may appear. (OPTIONAL: The genomic DNA can be visualized on an agarose gel).

3.2 16S Gene Amplification [6]

- The 25 μL of PCR mixture (Bioline) contained 12.5 μL MyTaq Red Mix 2x, 1 μL primer forward, 1 μL primer reverse, and 1 μL of genomic DNA with concentration 100 ng/μL as a template, and PCR grade water to make up the volume.
- A negative control (PCR mixture without DNA template) was included for each PCR reaction. Amplifications were carried out in an Agilent Sure cycle using the following conditions: 95 °C

for 3 min, 35 denaturation cycles at 95 °C for 15 s, annealing temperatures 52 °C 30 s, primer extension at 72 °C for 45 s followed by a final extension at 72 °C for 3 min.

3.3 Identification of Endophytic Bacteria [7]

- The PCR product was run on 1% agarose gel containing floroafe, isolated, and purified using Zymo Spin Column PCR purification kit (Zymo Research).
- Sequencing process was made at 1st Base Asia, Malaysia, furthermore Sequence of DNA was edited using Chromas Pro software and alignment sequence by DNA MAN software and aligned with the sequences in the GeneBank by basic local alignment search tool (BLAST) program (http://blast.ncbi.nlm.nih.gov/blast/Blast.cgi) to reveal the sequence homology with the closely related organism.

References

1. Woo PC, Lau SK, Teng JL, Tse H (2008) Then and now: use of 16S rDNA gene sequencing for bacterial identifications and discovery of novel bacteria in clinical microbiology laboratories. Eur J Clin Microbiol Infect Dis 14:908–934
2. Kembel SW, Wu M, Eisen JA, Green JL (2012) Incorporating 16S gene copy number information improves estimates of microbial diversity and abundance. PLoS Comput Biol 8:e1002743
3. Arief I, Jakaria ST, Wulandari Z, Andreas E (2013) Isolation and characterization of plantaricin produced by lactobacillus plantarum strains (IIA-1A5, IIA-1B1, IIA-2B2). Media Peternakan 36(2):91
4. Tamura K, Peterson D, Peterson N, Stecher G (2011) MEGA5: molecular evolutionary genetics analysis using maximum likelihood, evolutionary distance, and maximum parsimony methods. Mol Biol Evol 28:2731–2739
5. Chomczynski P, Mackey K, Drews R, Wilfinger W (1997) DNAzol: a reagent for the rapid isolation of genomic DNA. BioTechniques 22:550–553
6. Bej AK, Mahbubani MH, Atlas RM (1991) Amplification of nucleic acids by polymerase chain reaction (PCR) and other methods and their applications. Crit Rev Biochem Mol Biol 26:301–334
7. Ghyselinck J, Pfeiffer S, Heylen K, Sessitsch A, De Vos P (2013) The effect of primer choice and short read sequences on the outcome of 16S rRNA gene based diversity studies. PLoS One 8:1–14

Chapter 11

Identification of Fungal Endophytes by ITS rDNA Technique

Pooja Suneja, Pradeep Kumar, Simran Rani, Simran, and Amita Suneja Dang

Abstract

Fungi are diverse in terms of their morphology, ecology, metabolism, and phylogeny. They exhibit the production of various bioactive metabolites, which renders them useful for their application in numerous fields. The traditional methods of fungal identification based on phenotypic characteristics involve the expertise of a skilled mycologist. The fungal classification based on morphology can be misleading due to the presence of too many species in a single genus and may not be able to accurately deduce the evolutionary relationships. Molecular methods based on rDNA amplification have emerged as new tools for more reliable and rapid identification of fungal species. The molecular approaches for fungal classification are either based on three nuclear ribosomal genes (LSU, SSU, ITS) or one or multiple protein-coding genes. ITS rDNA, a universal barcode marker for fungi, is widely used as the molecular technique for identifying the fungal endophytes and comprises DNA extraction, PCR amplification, DNA sequencing, and phylogenetic analysis. Its power to elucidate intraspecific and interspecific variation among congeneric species proves it extremely valuable regarding the determination of phylogenetic relationships at lower levels.

Key words Fungal endophytes, Molecular techniques, Barcode marker, ITS rDNA, ITS Primers, Fungal identification

1 Introduction

Endophytes are the microorganisms, either bacteria or fungi, that inhabit the plant tissues without causing any apparent symptoms of the disease. There are no plant species reported on the earth without one or more endophytes in them. Their association can be obligate or facultative displaying complex interactions with their hosts, most of which have not been fully understood. Fungal endophytes symbiotically reside inside their plant host and produce numerous bioactive metabolites that play a significant role in promoting their growth and combating stresses. Mycologists used conventional methods based on morphological characteristics and their ability to produce asexual or sexual spores for their identification. Morphological identification of these fungal endophytes is

A. Sankaranarayanan et al. (eds.), *Endophytic Microbes: Isolation, Identification, and Bioactive Potentials*, Springer Protocols Handbooks, https://doi.org/10.1007/978-1-0716-2827-0_11,

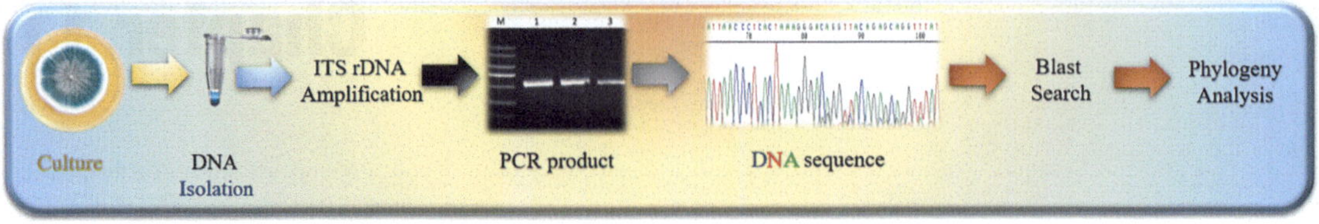

Fig. 1 Identification of fungal endophytes by ITS rDNA technique

very important to understand their evolution. However, identification by these phonetic characteristics has their own limitations like limited number of morphological characteristics, fossil records, and inability to perform well for lower level classification [1, 2]. Non-sporulating isolates cannot be identified in any taxonomic position based on morphology studies alone [3]. Molecular techniques based on DNA sequencing along with morphological studies have enhanced the ability to recognize and understand fungal diversity in natural ecosystems. Their ease to use, high specificity, more sensitivity, and reproducibility have made them popular and widely used options for the identification purposes [4].

The use of molecular approaches for fungal identification came into existence about two decades ago [5]. For molecular identification (Fig. 1), primers are designed for large (28S) and small subunits (18S) of ribosomal RNA as well as internal transcribed spacer region (ITS). The differences in the rate of evolution in these regions have been observed and found to be slower in the case of SSU (small subunit rRNA gene) as compared to the ITS region. The SSU-based amplification is therefore used for classification at higher taxonomic levels, while ITS region being the sequence of choice for analysis at lower taxonomic levels.

Internal transcribed spacers (ITS) region is the non-coding sequence in the nuclear ribosomal DNA (rDNA) of the genome recommended as a universal barcode for fungal sequencing by the International Fungal Barcoding Consortium [6]. ITS region as depicted in Fig. 2 consists of a 600 bp long sequence having two variable spacers, namely, ITS-1 and ITS-2, separated by a highly conserved 5.8S rRNA gene. The whole ITS region is flanked by 18S rRNA at 5′ end of ITS-1 spacer and by 28S rRNA gene at 3′ end of ITS-2 spacer [5]. It has been possible to design universal primers to amplify the complete ITS region or ITS-1/ITS-2 in the vast majority of fungi due to the presence of highly conserved 18S, 5.8S, and 28S rRNA genes [7]. ITS region has a higher degree of variation, enabling it to distinguish among closely related species as well as within the species [8]. A high copy number of rRNA genes makes it easier to be amplified even from a lesser DNA sample. Quicker and precise results are another advantage of the ITS rDNA technique making it to be more preferred over traditional techniques.

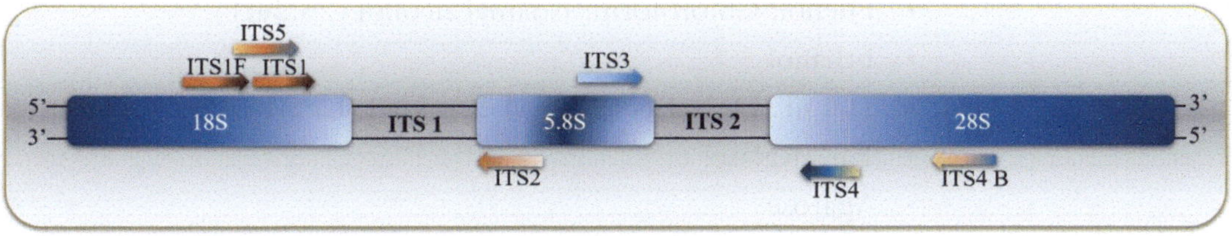

ITS1F-5'CTTGGTCATTTAGAGGAAGTAA3'

ITS1 - 5'TCCGTAGGTGAACCTGCGG3'

ITS2 - 5'GCTGCGTTCTTCATCGATGC3'

ITS3 - 5'GCATCGATGAAGAACGCAGC3'

ITS4 - 5'TCCTCCGCTTATTGATATGC3'

ITS5 - 5'GGAAGTAAAAGTCGTAACAAGG3'

Fig. 2 ITS region primers

2 Materials

2.1 Isolation of Endophytic Fungi

- Healthy and fresh plant sample.
- Sterile surgical blade.
- Ethanol.
- Sodium Hypochlorite (NaOCl).
- Distilled water.
- Sterile tissue paper.
- Potato Dextrose Agar (PDA).
- Petri plates.
- Incubator.
- Laminar air flow.
- Ampicillin and Streptomycin.
- Glycerol.

2.2 Molecular Characterization

- Fungal mycelia on PDA plate.
- Mortar and pestle.
- Liquid nitrogen.
- Eppendorf tubes.
- 2X CTAB buffer (700–1000 μL preheated) (2% w/v CTAB, 100 mM Tris-HCl, 1.4 M NaCl, 20 mM EDTA, pH 8.0).
- Water bath (60 °C).
- Centrifuge.

- Phenol: Chloroform: Isoamyl alcohol (25:24:1).
- Ethanol.
- 100 μL TE buffer (10 mM Tris-HCl, 1 mM EDTA, pH 8).
- 1 μL RNase.
- Agarose.
- Ethidium bromide.
- 2.5 μL 10X PCR buffer (100 mM Tris-HCl, pH 8.8, 50 mM KCl, 0.1% Triton-X-100).
- Thermal cycler.
- dNTPs.
- $MgCl_2$.
- Template DNA.
- Taq DNA Polymerase.
- Universal Primers – ITS1, ITS2, ITS3, ITS4, and ITS5 [1, 5].

3 Methods

3.1 Isolation of Endophytic Fungi [9]

1. Wash the plant sample gently under running tap water for 5 min to remove soil and sediments from the surface.
2. Separate the whole plant into root, stem, petiole, leaf, and flower stalk.
3. Cut stem, root, flower stalk, and petiole into 10 × 10 mm segments and leaves into 5 × 2 mm segments with the help of a sterile surgical blade.
4. Sterilize the tissue surface by immersing in 70–95% ethanol solution and sodium hypochlorite (NaOCl) for 30–120 s and 1–10 min, respectively.
5. Finally, wash with 75% ethanol for 30 s.
6. After surface sterilization, rinse the samples in distilled water thrice for 1 min each time and blot dry on sterile tissue paper to remove excess moisture.
7. Once dry, place 6–7 tissue segments on Potato Dextrose Agar (PDA) plate containing 200 mg/L ampicillin and streptomycin to suppress bacterial growth.
8. Incubate inoculated plates for 2–3 weeks at 28 ± 2 °C.
9. Observe the plates everyday for the fungal growth appearance.
10. Collect the fungal mycelia growing from tissue segments and maintain them on fresh PDA plates (without antibiotics).
11. Use 5–8 days old fungal mycelia for DNA extraction or store them in 10–15% glycerol (v/v) at −80 °C for further use.

3.2 Molecular Characterization of Endophytic Fungi

3.2.1 DNA Extraction [10]

1. Scrap the 50 mg of 5–8 days old fungal mycelia from the surface of agar plate and homogenize in liquid nitrogen in sterile mortar and pestle for 30–60 min.
2. Transfer the fine powder into a 1.5 mL eppendorf tube containing 2 X CTAB buffer (700–1000 μL preheated at 60 °C).
3. Incubate the mixture in a water bath at 60 °C for 30–40 min with occasional gentle swirling.
4. Centrifuge at 12,000 rpm for 5–10 min.
5. Transfer the supernatant into a fresh eppendorf tube and to this add an equal volume of Phenol:Chloroform:Isoamyl alcohol (25:24:1), mix gently.
6. Centrifuge again at 12,000 rpm for 5 min.
7. Take the supernatant in another fresh eppendorf tube and to this add an equal volume of Chloroform:Isoamyl alcohol (24: 1).
8. Centrifuge the mixture at 10,000 rpm for 5 min and take the supernatant.
9. Add 2 volumes of chilled ethanol, gently invert the tube to mix, and store overnight at −20 °C for the complete precipitation of DNA.
10. Separate the precipitate from the mixture by centrifugation at 10,000–12,000 rpm for 5 min.
11. Discard the supernatant slowly and wash the DNA pellet in 70% ethanol.
12. Dry using speed vacuum for 10 min, or air dry for 30 min.
13. Resuspend the pellet in 100 μL TE buffer.
14. Add 1–2 μL of RNase and incubate at 37 °C for 1 h.
15. Perform electrophoresis to check DNA in 0.8–1% (w/v) agarose gel.
16. Stain the gel with 0.5 μg/mL ethidium bromide to visualize DNA under a UV transilluminator.

3.2.2 PCR Amplification [1, 11–13]

1. Amplify rDNA ITS1–5.8S-ITS2 region with universal primers, ITS1 & ITS2 for ITS-1, ITS3 & ITS4 for ITS-2 and ITS1 & ITS4/ITS4 & ITS5 for entire ITS region in a PTC-100™ thermal cycler.
2. Perform amplification in 50 μL reaction volume containing PCR buffer, 2.0 mM $MgCl_2$, 0.2 mM dNTP, 0.2 μM of each primer, 50 ng template DNA and 2.5 U Taq DNA polymerase.
3. The thermal cycling program involves initial denaturation at 94 °C for 5 min, followed by 35 cycles of 40 s each at 94 °C, primer annealing at 45–55 °C for 50 s, 1 min extension at 72 ° C and final 10 min extension at 72 °C.

4. Set water as a negative control.
5. Examine the amplified products by electrophoresis in 2% (w/v) agarose gel with 0.5 µg/mL ethidium bromide in TAE buffer (0.4 M Tris, 50 mM NaOAc, 10 mM EDTA, pH 7.8) and visualize under UV light.

3.3 DNA Sequencing

Purify the PCR products and sequence them in an automated sequencer directly.

3.4 Phylogenetic Analysis

1. Use EMBOSS merger tool to merge the sequenced fungal ITS sequences and get the entire length of the ITS sequence.
2. Compare the sequence with previously deposited sequences in National Centre for Biotechnology Information (NCBI) database using BLAST.
3. Download the ITS sequences to analyze phylogenetic relationships.
4. Align ITS sequence of sample and retrieved sequences from NCBI, using CLUSTAL W program in MEGA X software with adjustment of required gap score.
5. Perform phylogenetic analysis by both maximum parsimony and neighbor-joining methods with 1000 bootstraps for better results.

Acknowledgments

The authors acknowledge the Department of Microbiology and Centre for Medical Biotechnology, Maharshi Dayanand University, Rohtak, India for infrastructural support. PK & SR acknowledge the Council of Scientific & Industrial Research, New Delhi, India for CSIR fellowship.

References

1. Raja HA, Miller AN, Pearce CJ, Oberlies NH (2017) Fungal identification using molecular tools: a primer for the natural products research community. J Nat Prod 80:756–770
2. Hoyos-Carvajal L, Orduz S, Bissett J (2009) Growth stimulation in bean (*Phaseolus vulgaris* L.) by *Trichoderma*. Biol Control 51:409–416
3. Guo LD, Hyde KD, Liew EC (2001) Detection and taxonomic placement of endophytic fungi within frond tissues of *Livistonachinensis* based on rDNA sequences. Mol Phylogenet Evol 20:1–3
4. Maheshwari R, Bhutani N, Suneja P (2020) Isolation and characterization of ACC deaminase producing endophytic *Bacillus mojavensis* PRN2 from *Pisumsativum*. Iran J Biotechnol 18:11–20
5. White TJ, Bruns T, Lee SJ, Taylor J (1990) Amplification and direct sequencing of fungal ribosomal RNA genes for phylogenetics. In: Innis MA et al (eds) PCR protocols: a guide to methods and applications. Elsevier, St. Louis, pp 315–322
6. Schoch CL, Seifert KA, Huhndorf S, Robert V, Spouge JL, Levesque CA, Chen W, Fungal Barcoding Consortium (2012) Nuclear ribosomal internal transcribed spacer (ITS) region as a universal DNA barcode marker for Fungi. Proc Natl Acad Sci 109:6241–6246

7. Xu J (2016) Fungal DNA barcoding. Genome 59:913–932
8. Ihrmark K, Bödeker I, Cruz-Martinez K, Friberg H, Kubartova A, Schenck J, Strid Y, Stenlid J, Brandström-Durling M, Clemmensen KE, Lindahl BD (2012) New primers to amplify the fungal ITS2 region–evaluation by 454-sequencing of artificial and natural communities. FEMS Microbiol Ecol 82:666–677
9. Park YH, Kim Y, Mishra RC, Bae H (2017) Fungal endophytes inhabiting mountain-cultivated ginseng (Panax ginseng Meyer): diversity and biocontrol activity against ginseng pathogens. Sci Rep 7:1–10
10. Lee SB, Milgroom MG, Taylor JW (1988) A rapid, high yield mini-prep method for isolation of total genomic DNA from fungi. Fungal Genet Rep 35:23–24
11. Sun X, Guo LD, Hyde KD (2011) Community composition of endophytic fungi in *Acer truncatum* and their role in decomposition. Fungal Divers 47:85–95
12. Al-Naemi FA, Nishad R, Ahmed TA, Radwan O (2014) First report of *Thielaviopsispunctulata* causing black scorch disease on date palm in Qatar. Plant Dis 98:1437–1437
13. Martin KJ, Rygiewicz PT (2005) Fungal-specific PCR primers developed for analysis of the ITS region of environmental DNA extracts. BMC Microbiol 5:1–11

Chapter 12

Identification of Fungal Endophytes by 28SrDNA Technique

Karnan Muthusamy, Tamilkani Pichai, Vijayalakshmi Selvakumar, Anbukumaran Ananthan, Reshma Devi Ramesh, Leelavathi Lakshmanan, Senthilkumar Govindarajan, Panneerselvam Annamalai, and Ki Choon Choi

Abstract

Fungi that live in plants without causing apparent symptoms of infection are referred to as endophytic fungi. Over 1 million species of endophytic fungi are estimated to stay alive based on a proportion of vascular plants to fungal species of 1:4 or 1:5. The improvement in the field of molecular biology brings an innovative perception to endophyte diversity researches. Application of molecular techniques, such as DNA fingerprinting and methods of sequencing, has the potential to overcome the obstacles in traditional cultivation-dependent methods. This study shows the identification of fungal endophytes by the 28SrDNA technique. The molecular study protocol normally includes the following steps (1) total genomic DNA isolation (including fungi and plants) from surface-sterilized plant tissues, (2) amplification of DNA fragments (28Sgenes) from total DNA with fungal-specific primers, (3) separation of PCR products (bands) and (4) cloning of PCR products directly into plasmids (e.g., pMD20-T vector) and transferring into *E. coli* DH5α, and (5) positive clones for different taxa are screened by DNA fingerprinting.

Key words Endophytic fungi, Molecular techniques, Sequencing methods, 28sDNA, PCR amplification

1 Introduction

Endophytes may offer significant benefits to their host plants by producing a surfeit of materials that afford defense and endurance value, for instance, tolerance on stress, insect, and disease resistance, yield enhancement, and herbicide resistance [1]. Fungal endophytes are microfungi that colonize and live within healthy plant tissues without inducing symptoms of disease [2].

Exclude pathogenic fungi and mycorrhizal fungi, all endophytes as mutualists that colonize aboveground divisions of living plant tissues and do not cause disease symptoms [3]. Therefore, endophytic fungi, as an important component of natural

A. Sankaranarayanan et al. (eds.), *Endophytic Microbes: Isolation, Identification, and Bioactive Potentials*, Springer Protocols Handbooks, https://doi.org/10.1007/978-1-0716-2827-0_12,

ecosystems, can play a key role in material and energy recycle [4]. The association is essentially invigorated by the endophyte fungi, yielding an excess of natural substances immediately cultured by the endophytic fungi in the external environment of their regular hosts or ecological positions under in vitro conditions [5].

Endophytic fungi are pathogenic, restrictive pathogenic, or non-pathogenic for the host plant is essential. DMEs can be passing on horizontal or vertically. When the seeds are contacted through the fungal endophytes, a vertical transmission occurs. Horizontal transmission involves the formation of exterior spores and their airborne dispersal infects many other hosts [6]. Dependent on the hostgeno type, a few endophytic fungi may perhaps be pathogenic in stressed hosts, while they can be helpful in other circumstances due to restrictive pathogenic characteristics [7].

Fungi encompass erratic antifungal susceptibility patterns. Even though a commercial DNA sequencing method, the D2 region of the 28S ribosomal DNA (rDNA) has shown a number of effectiveness for recognizing filamentous fungi in a clinical laboratory [8]. The endophytic fungi united with *Artemisia nilagirica* were scrutinized and one strain of *Pythium intermedium* (Oomycota) and one strain of *Rhizopus oryzae* (Mucoromycota) were isolated along with the common clade of Ascomycota [9]. Since their host and take part in crucial roles *in vivo* for example signaling, defense, and symbiosis regulation. Mainly investigations are based on their use as biochemical tools and the end products are to be used in pharmaceutics, industry, and agriculture [10].

A comparative study related to infection frequency between cultivated plants and wild plants of *Artemisia annua* was performed. The results had shown a somewhat high infection rate of occurrence of the endophytic fungi in cultivated roots (20.9%) than in native roots (16.7%). Further, the authors described that the naturally regenerated roots harbored richer fungal genotypes, which maintains the theory that natural plant species are influenced to host rich and new fungal species [11].

Artemisia is a huge plant genus among species producing a range of fascinating and active substances and its endophytic communities are under-examined. The recognition of the fungal endophytes in *Artemisia* spp. is made generally based on morphological traits and molecular investigation by means of rDNA sequences, together with the internal transcribed spacers (ITS) and the 5.8S region. To the best of our understanding, exploration and phylogenetic studies of the *Artemisia* spp. fungal endophytes there have been only four studies [12].

The reproductive and morphological traits are chiefly passionate and come up short with respect to provided that unambiguous species delimitations. Consequently, the molecular traits are embraced in fungal organization and phylogenetics [13].

Our nation is wealthy in plant assets, and endophytic bacteria sources are also furthermore improved in like way. Therefore, many biological functions of endophytic bacteria have become another treasury for human to exploit microbial resources. For example, industrialized production of fermentative active substances from endophytic fungi in medicine could increase the yield of paclitaxel and other precious anticancer medications [14].

2 Materials

2.1 Sampling

- Paper bags inside zip-locked bags.
- Sterile scissors or knife (oven sterilized is preferable).
- Sterile surgical blade.
- Photo camera (cell phone can provide both geo-localization and camera).

2.2 Pre-treatment (Optional)

- Tap running water.
- Soft brush.
- Beaker.
- Sonicator bath.

2.3 Disinfection

- Sterile water.
- 70% ethanol.
- 15% sodium hypochlorite.
- Laminar Air flow chamber.
- Sterile paper towels.

2.4 Materials for DNA Purification by Paper Slurry Method

- 50 cm 2 piece of Whatman 3 mm filter paper.
- TE Buffer (10×) prepare by dissolving 186 mg EDTA and 605 mg Tris in 50 mL dH2O (adjust the pH to 8.0 with HCl). Maintain at 4 °C until use (before using it is diluted to ten fold).
- Paper slurry (it can be prepared by cutting filter paper into tiny (1–2 mm^2) pieces and add into a 50 mL tube containing 40 mL of TE buffer). Mix well by hand for at least 5 min (maintain at 4 °C).
- Agarose gel with isolated DNA.
- Clean razor blade.
- Filter column: Punch a small hole in the bottom of a 500 μL tube with a 23gauge needle. Piecewise remove the paper slurry with tweezers and put on the opening in the 500 μL tube. Pack with a 200 μL pipette tip and remove excess liquid. Do again awaiting paper column is around 3 mm high. Put it in a 1.5 mL tube together eluent.

2.5 Solutions Used for DNA Purification from Acrylamide

- Crush and Soak Solution.
- 500 mm NH4OAc 3.3 g.
- 0.1% SDS (0.1 g SDS/100 mL).
- 0.1 mm EDTA.
- 20 mL of 500 mM EDTA up to 100 mL with at room temperature.
- 3 M NaOAc pH 5.2.
- 24.6 g sodium acetate (anhydrous).
- pH to 5.2 with acetic acid and bring up to 100 mL with at room temperature.

2.6 Other Reagents

- DMCS treated glass wool (50 g).
- 0.22 mm disposable micro tip filters (syringe type) blue tips with melted tips to serve as pestle for crushing acrylamide.

3 Methods

3.1 Sampling and Isolation of Endophytes

(Kindly refer protocols in Subheading 1 for sampling and isolation of endophytes and see Fig. 1 in this section).

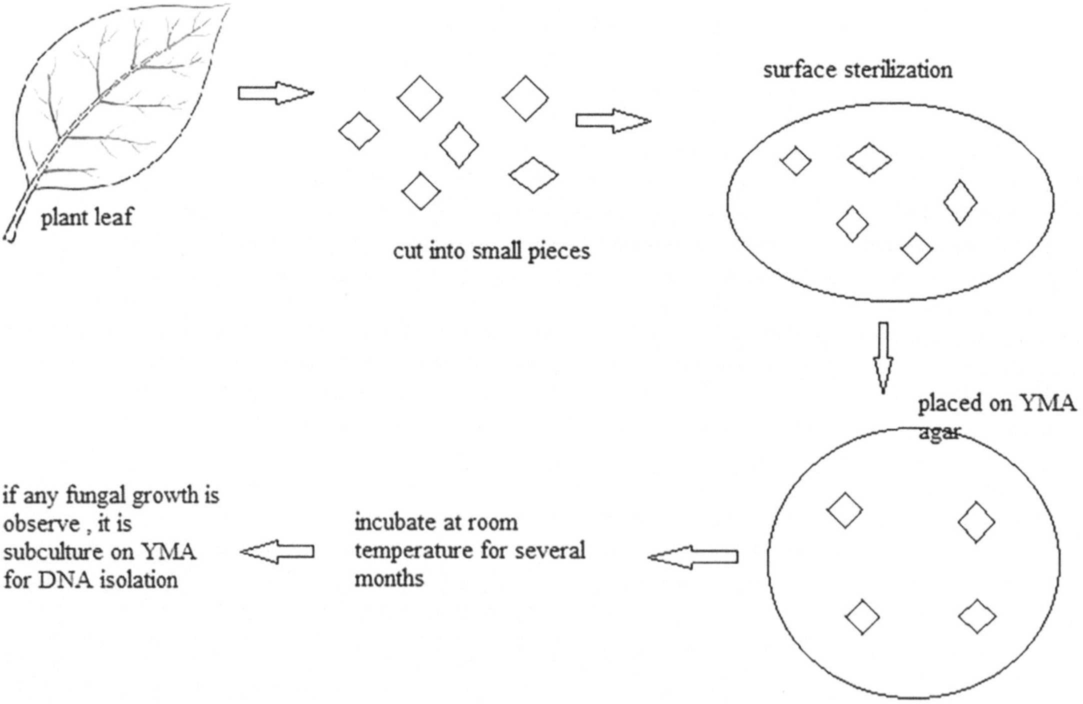

Fig. 1 Overview of isolation of endophytes

3.2 DNA Extraction

(i) Inoculate the endophytic fungi into the potato dextrose broth and incubate at 125 rpm on an Incubator Shaker at 25 °C for 1 week.

(ii) Harvest the fungal tissue, lyophilize, and stored at −80 °C.

(iii) Place 50 mg of lyophilized mycelium into a 2 mL microcentrifuge tube with ca. 0.2 mL 1.0 mM zirconia-glass beads.

(iv) Crush the mycelium with a plastic pestle and further ground in a Fast-Prep-120 sample grinder for 3 s at the maximum speed setting of 6.

(v) Suspend the powdered mycelium in 700 μL detergent solution (2 M NaCl, 0.4% w/v deoxycholic acid–sodium salt, 1% w/v polyoxyethylene 20 acetyl ether and then agitate for 14 s in the Fast-Prep at maximum speed.

(vi) Incubate the vials for 5 min at 55 °C in a heat block and then centrifuge at 7000 g for 5 min follow by emulsion with 700 μL chloroform/isopropyl and centrifuge at 7000 g for 5 min.

(vii) Carefully transfer the aqueous phase to a clean tube to which add an equal volume of 6 M guanidinium thiocyanate.

(viii) Gently mix the 15 μL of silica powder with the solution and incubate at room temperature for 5 min, follow by 3 s centrifugation, after discarding the supernatant.

(ix) Rinse the glass powder twice by suspending in 750 μL ethanol buffer (10 mM Tris-HCl, pH 8.0, 0.1 mM EDTA, 50% ethanol) with a disposable transfer pipette, and then collect by centrifugation.

(x) Discard the supernatant and dry the glass powder pellet on a heat block at 55 °C for 5–10 min.

(xi) Re-hydrate the glass powder with 105 μL ultra-pure water and elute the genomic DNA by incubating on a heat block at 55 °C for 5–10 min.

(xii) Follow vortex and centrifugation; collect the 100 μL of the aqueous supernatant into a clean tube (see Fig. 2).

3.3 Gel Purification Methods

A simple, rapid, and inexpensive method is as follows:

(i) Take 2 Eppendorf tubes, little glass wool, a knife or blade, pair of Forceps and a needle or pin. Sterilize by autoclaving.

(ii) A little amount of glass wool is filled aseptically in one of the tubes. Flame the needle using Bunsen burner (which also sterilizes it) until it is red hot and a small puncture is made at the bottom of one tube asset the glass wool. The puncture is made from the exterior at the bottom.

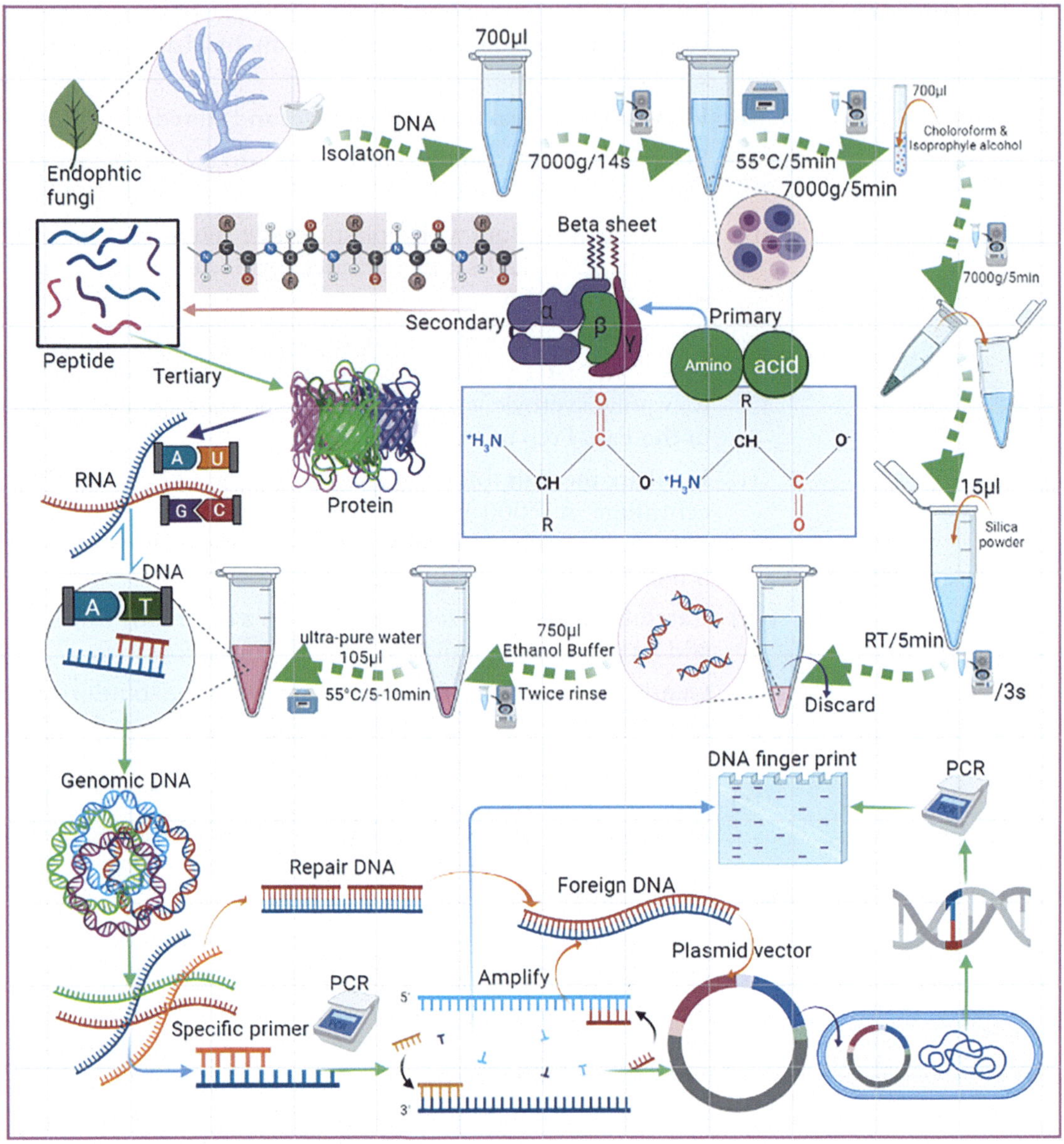

Fig. 2 Extraction of genomic DNA from endophytic fungi

(iii) Place this tube into the other tube (thus the lid remains open). To fix the joint by means of sellotape.

(iv) Now cut out the small portion of the gel with the required DNA fragment, slice it into more and more tiny pieces by sterile knife and obtain the fractions and fill into the upper tube.

(v) Spin centrifuge at 5000 or 10,000 rpm for 5 min and collect the elute as concentrated and pure DNA ready for any use.

3.3.1 Purification of DNA from Agarose Gels (Paper Slurry Method)

This procedure isolates DNA from agarose gels by filtration through a filter-paper column. The column is made in a 500 μL tube from slurry of filter paper in TE buffer.

Steps

(i) Excise the DNA band from the surrounding gel with a clean razor blade; be sure to remove as little gel as possible.

(ii) Excise gel fragment dice into tiny sections with the same razor and shift onto filter column.

(iii) Centrifuge it for 10 min at maximum speed (approximately 20,000 g) using microcentrifuge.

(iv) Transfer eluent to a fresh tube and re-centrifuge agarose.

(v) Mix all the eluent with that from previous centrifugation.

(vi) Arrange a second spin column and shift the remaining agarose to the new column. Spin once more.

(vii) All eluent fractions are mixed and concentrated by means of cold ethanol precipitation.

3.3.2 Gene-Kleen Protocol

The following method describes a variation of the method of Vogelstein and Gillespie, 1979 [17]. Excise the band of interest from the TAE gel using a clean scalpel blade and place it in a pre-weighed eppendorf tube.

(i) Melt the agarose by adding 3 volumes of 6 M NaI, 0.1% sodium thiosulfate solution (approx. 5 min with vortexing). For TBE gels, 0.1 volumes of 1 M mannitol should also be added to aid gel solubilization.

(ii) Vortex glass suspension (flimsily mashed glass scintillation vial suspended 1:1 in sterile H_2O or Fluka analytical filter aid resuspended 1:4 in sterile nano-pure H_2O) and add 7.5 μL to agarose solution (7.5 μL should be used for up to 5 μg DNA and thereafter an extra 1ul should be used for each extra μg DNA).

(iii) Permit the DNA to combine for 15–20 min at room temperature.

(iv) Whirl down glass-DNA for 30 s. Using microfuge at highest speed and suspiciously take away supernatant and discard it.

(v) Perform washing of pellet by 500ul 4.5 M NaI, 0.1% sodium thiosulfate, re-pellet, and discard supernatant.

(vi) Cleanse pellet 2 × 500ul in 1 × TE, pH 7.5, 200 mm NaCl, and 60% EtOH. After last centrifugation, remove all traces of the supernatant and allow air-drying for 5 min.

(vii) To elute DNA at 37–55 °C in 25ul TE, pH 8.0 for 5 min.

(viii) N.B: It is vital to elute in a buffer with a pH of 8 as elution in water or lesser pH buffers reduces the yield noticeably.

(ix) Spin along glass for 5 min at highest speed and suspiciously take out supernatant to a fresh eppendorf tube.

(x) Repeat it and pool supernatants to get rid of pellet. Re-spin collective sample to eliminate traces of glass for 5 mins and shift fresh supernatant to a new, sterile eppendorf tube.

Note All reagents are made from the highest grade chemicals available (esp. vital for the NaI). NaI solutions were prepared with sterile, nano-pure H_2O and final ethanol was made into smaller amount of ethanol, sterilized, and made up to a final concentration of 70% (v/v) after sterilization. Glass "beads" were prepared from a flimsily crushed scintillation vial (i.e., high-quality glass) by overwhelming by means of a pestle and mortar.

3.3.3 Elution of DNA Fragments from Low-Melting Agarose Gel by Dr. Roe Method

Protocol

(i) Separate DNA fragments through an LMP agarose gel containing ethidium bromide (0.5 μg/mL).

(ii) Incise out the area of the gel having desired DNA.

(iii) Put the excised DNA-containing agarose gel slice in a 1.5 mL tube and freeze at −70 °C for 15 min or in anticipation of frozen. It is feasible to recess at this phase in the elution process and put down the gel slice frozen at −70 °C.

(iv) Incubate the tube at 65 °C in order to melt the slice.

(v) One volume of TE-saturated phenol is adding, spin for 30 s, and freeze the sample at −70 °C for 15 mins.

(vi) To separate the phases, thaw the sample, and centrifuge in a microcentrifuge at 12,000 g for 5 min at room temperature.

(vii) Remove the aqueous phase to a clean tube, extract twice with equal volume of ether, ethanol precipitation, and rinse and dry the DNA pellet.

3.3.4 Freezing Method

(i) Remove gel slice containing DNA fragment and place in 10 volumes of 300 mm NaOAc, pH 7.0 (300 mL 1 M NaOAc, pH 7.0) 1 mM EDTA (2 mL of 500 mM EDTA, pH 8.0) (698 mL ddH2O).

(ii) Incubate at 22 °C for 30 min. Shift gel slice to a clean tube.

(iii) Put the tube in a Dry Ice/Ethanol bath for 5 min.

(iv) Pierce the bottom of a 0.5 mL tube by means of a needle. Put the gel slice into this tube and put it inside a 1.5 mL microcentrifuge tube.

(v) Spin for 15 min.

(vi) Gather the Eluent from the 1.5 mL eppendorf tube. Take out and precipitate the DNA.

3.3.5 DNA Fragment Purification from Acrylamide

For smaller DNA fragments (20–400 bp) the acrylamide purification is preferred.

Steps

(i) Run a 4–6% acrylamide gel in 1× TBE, stain in EthBr (1–10 mg/mL), and cut out the desired band.

(ii) Crush the acrylamide with a p1000 tip with a melted end to resemble a pestle for the eppendorf "mortar."

(iii) Add 1 mL crush and soak solution and incubate overnight at 37 °C.

(iv) Spin in the microfuge for 10 min at 14,000 g. Remove as much liquid as possible and add another 500 mL of crush and soak solution.

(v) Repeat the spin and pool the recovered supernatant.

(vi) Add 0.1volume of 3 M NaOAc, 2.5 volumes of EtOH and carrier (glycogen or tRNA).

(vii) Spin as usual, wash, and dry. Resuspend in 20 μL TE.

3.3.6 Rapid Elution of DNA from Agarose Gels

A method of quickly purifying agarose gel DNA fragments for use in subsequent reactions such as further restriction enzyme modification, T4 ligase, *Taq* polymerase, Klenow, or polynucleotide kinase. Although higher yields of purified DNA can be obtained from commercially available purification kits, greater ligation efficiencies per given amount of DNA have been seen with the use of these described spin columns.

Glass Wool Spin Column Preparation

Steps

(i) Using a clean hypodermic needle, drive a hole in the end of a 650 μL Eppendorf tube from the inside then cut off the cap.

(ii) Wearing gloves, pull off a small piece of glass wool and roll it into a ball between hands. Place the glass wool plug in the bottom of the 650 mL tube. You may need to use a 1 mL pipet tip to pack the glass wool in the bottom of the tube. Do not cut the glass wool, as this may produce small glass fibers that can pass through the hole in the small tube.

(iii) Place the tube with the glass wool in a 1.7 mL Eppendorf tube. This completed spin column is now ready for use.

Eluting DNA from Agarose Gel Fragments

(i) Visualize ethidium bromide-stained agarose gel with a transilluminator on low setting. Working quickly to minimize UV exposure to the DNA, excise the fragment of interest with a clean razor blade. After removing excess liquid with a Kimwipe, place the agarose fragment in the spin column.

(ii) Centrifuge the tube at 2874× *g* for no more than 45 s to elute the DNA. Spinning longer co-elutes substances that are inhibitory to further enzymatic reactions. 2874× *g* corresponds to 5500 *g* for Eppendorf centrifuge model 5415; 5000 *g* for model 5417.

(iii) Using a transilluminator, check eluent in the bottom of the outer tube for the presence of ethidium bromide-stained DNA.

(iv) The eluted DNA can now be used directly in enzymatic reactions. For ligations, no more than 40% of the total reaction volume should be eluted DNA, as this reduces ligation efficiencies.

(v) The DNA fragments can be further purified by ethanol precipitation. The addition of 2–4 mg of carrier tRNA prior to precipitation can increase yield, and its presence in ligation reactions is non-inhibitory.

3.4 PCR Amplification of the Fungal 28S Regions

(i) The 28S rRNA genes were amplified with the primers (see Table 1).

(ii) The PCR reaction mixture (25 μL) contained 1× PCR buffer (Takara), 200 μM dNTP, 0.2 μM of each primer, 3 mM $MgCl_2$, and 2.5 U *Taq* DNA polymerase (Takara).

(iii) After initial denaturation at 94 °C for 5 min, each thermal cycling was as follows: denaturation at 94 °C for 30 s, annealing at 56 °C for 40 s, and elongation at 72 °C for 1 min.

(iv) At the end of 30 cycles, the last extension step was performed at 72 °C for 10 min.

(v) The PCR products of five reactions from the same sample were pooled together to minimize PCR bias [15].

Table 1
PCR primers for 28S rDNA region amplification

S. no	Forward primers (3′ to 5′)	Reverse primers (5′ to 3′)	Reference
1	AGGGGCGAAAGACYAATCGAA	TTCGATTRGTCTTTCGCCCCT	[16]
2	TTTTGGTAAGCAGAACTGGYG	CRCCAGTTCTGCTTACCAAAA	[16]
3	GATCTYRGTGGYAGTACRAVT	ABTYGCTACTRCCACYRAGATC	[16]
4	GGGAATCYRACTGTHTAATTAAA	TTTAATTADACAGTYRGATTCCC	[16]
5	TGATTTCTGCCCAGTGCTYWGAA WGT	ACWTTCWRAGCACTGGGCAGAAA TCA	[16]
6	AACGGCGGRRGTAACTATGACTYT	ARAGTCATAGTTACYYCCGCCGTT	[16]
7	GGGAAAGAAGACCCTGTTGAG	CTCAACAGGGTCTTCTTTCCC	[16]
8	AAGACCCTGTTGAGYTTGACTCT	AGAGTCAARCTCAACAGGGTCTT	[16]
9	GGGAGTTTGRCTGGGGCGG	CCGCCCCAGYCAAACTCCC	[16]

(vi) The pooled PCR products were electrophoretically separated, and the band of approximate size in the electrophoresis pattern was excised and purified (as already mentioned in the DNA purification steps).

3.5 Construction of 28S rDNA Region Clone Library [15]

(i) Ligate the purified PCR products into the pMD20-T Vector and then transform them into competent *E. coli* DH5α cells. Screen the colonies on LB agar supplemented with ampicillin (100 μg/mL^{-1}), X-gal (100 μg/mL^{-1}), and IPTG (0.5 mm).

(ii) Randomly pick the white colonies for further studies.

(iii) The 28S rDNA region genes are re-amplified using the above primers, and verify the correct size using agarose gel electrophoresis.

(iv) PCR aliquots (10 μL) were characterized by single digestion with the restriction endonucleases *Taq*I and *Msp*I.

(v) Restriction fragment length polymorphism (RFLP) patterns are classified based on the agarose gel electrophoresis patterns.

(vi) Clones with similar restriction patterns are grouped together, and choose the clones randomly from each RFLP group for sequencing.

(vii) Examine the sequences for possible chimeras by programs UCHIME (http://drive5.com/uchime/uchime_download.html) to remove chimeric sequences, then deposited in the GenBank database.

3.6 Phylogenetic Analysis and Diversity Analysis [15]

(i) Environmental gene sequences were compared initially with those in GenBank using BLAST analysis to determine their approximate phylogenetic affiliation.

(ii) Phylogenetic and diversity analysis was inferred by the neighbor-joining method.

(iii) Evolutionary distances were calculated with the Kimura two-parameter model (transition/transversion ratio = 2.0).

(iv) The diversity calculation analyses and rarefaction curves to the number of observed operational taxonomic units (OTUs) were carried out using the program Estimates S v.8.0 (http://viceroy.eeb.uconn.edu/estimates/).

(v) The library coverage values were calculated using the formula [1-(n/N)], in which n is the number of OTUs represented only by a single clone and N is the total number of clones in the library.

(vi) Rarefaction curves to numbers of observed OTUs were calculated within the fungal assemblage at each dataset.

References

1. Strobel G, Daisy B, Castillo U, Harper J (2004) Natural products from endophytic microorganisms. J Nat Prod 67(2):257–268
2. Petrini O (1991) Fungal endophytes of tree leaves. In: Andrews JH, Hirano SS (eds) Microbial ecology of leaves. Springer, New York, pp 179–197
3. Carroll GC (1986) The biology of endophytism in plants with particular reference to woody perennials. Microbiol Phyllos:203–222
4. Sun X, Guo LD, Hyde KD (2011) Community composition of endophytic fungi in Acer truncatum and their role in decomposition. Fungal Divers 47(1):85–95
5. Kusari S, Singh S, Jayabaskaran C (2014) Biotechnological potential of plant-associated endophytic fungi: hope versus hype. Trends Biotechnol 32(6):297–303
6. Sarsaiya S, Shi J, Chen J (2019) A comprehensive review on fungal endophytes and its dynamics on Orchidaceae plants: current research, challenges, and future possibilities. Bioengineered 10(1):316–334
7. Jain A, Sarsaiya S, Wu Q, Lu Y, Shi J (2019) A review of plant leaf fungal diseases and its environment speciation. Bioengineered 10(1): 409–424
8. Hall L, Wohlfiel S, Roberts GD (2004) Experience with the MicroSeq D2 large-subunit ribosomal DNA sequencing kit for identification of filamentous fungi encountered in the clinical laboratory. J Clin Microbiol 42(2):622–626
9. Myrchiang P, Dkhar MS, Devi HR (2014) Studies on endophytic fungi associated with medicinally important aromatic plant Artemisia nilagirica (CB Clarke) Pamp. and their antagonistic activity against Phytophthorainfestans. J Adv Lab Res Biol 5(4):112–119
10. Schulz B, Boyle C (2005) The endophytic continuum. Mycol Res 109(6):661–686
11. Qian Y, Kang J, Geng K, Wang L, Lei B (2014) Endophytic fungi from Artemisia argyiLevl. etVant. and their bioactivity. Chiang Mai J Sci 41(4):910–921
12. Gashgari R, Gherbawy Y, Ameen F, Alsharari S (2016) Molecular characterization and analysis of antimicrobial activity of endophytic fungi from medicinal plants in Saudi Arabia. Jundishapur J Microbiol 9(1)
13. Nilsson RH, Kristiansson E, Ryberg M, Hallenberg N, Larsson KH (2008) Intraspecific ITS variability in the kingdom Fungi as expressed in the international sequence databases and its implications for molecular species identification. Evol Bioinforma 4:EBO-S653
14. Ijaz A, Imran A, Haq MA u, Khan QM, Afzal M (2016) Phytoremediation: recent advances in plant-endophytic synergistic interactions. Plant Soil 405:179–195
15. Tan H, Cao L (2014) Fungal diversity in sheep (Ovisaries) and cattle (Bostaurus) feces assessed by comparison of 18S, 28S and ITS ribosomal regions. Ann Microbiol 64(3):1423–1427
16. Machida RJ, Knowlton N (2012) PCR primers for metazoan nuclear 18S and 28S ribosomal DNA sequences. PLoS One 7(9):e46180
17. Vogelstein B, Gillespie D (1979) Preparative and analytical purification of DNA from agarose. Proc Natl Acad Sci 76(2):615–619

Chapter 13

Detection of Endophytes by Molecular Fingerprinting Techniques

Lakshmaiah Manjunatha, Gummeral Narasareddy, Bankapura Mariyappa Ravikumar, Leela Saisree Uppala, Rishikesh Kumar, Nanjundappa Manjunatha, Manikyanahalli Chandrashekara Keerthi, Raj Kumar Mishra, and Bansa Singh

Abstract

Endophytic fungi and bacteria are ubiquitous and occur within all known plants, including a broad range of hosts in various ecosystems, and therefore play an important role in the natural environment. Million species of endophytic fungi and bacteria are present in plant tissues. Nevertheless, our recognition of endophyte diversity is limited at present. In surveys of endophyte diversity, traditional techniques, such as culture-dependent methods, have been routinely used in since long. The discovery of endophytes in natural environments, however, has been limited by traditional methodology due to some non-culturable endophytes. Molecular techniques, such as DNA fingerprinting and sequencing methods, have been successfully employed in the detection and identification of endophytes fungi, and different endophyte diversity and community composition have been documented by cultivation-dependent and molecular techniques. This chapter summarizes the use of molecular fingerprinting protocols in the study of endophytic fungi and bacteria.

Key words Endophyte, DNA fingerprinting, Identification, Marker, PCR

1 Introduction

Endophytes are the most important organisms that can develop beneficial associations with plants [1]. Endophytic microorganisms such as bacteria and fungi with enormous production of secondary metabolites have biological activity features, which occur within plants' system without any threat for their living. Molecular techniques coupled with PCR amplification and sequencing have just been intensely applied in the characterization of endophytic fungi and bacteria based on phylogenetic and sequence similarity analogy.

A. Sankaranarayanan et al. (eds.), *Endophytic Microbes: Isolation, Identification, and Bioactive Potentials*, Springer Protocols Handbooks, https://doi.org/10.1007/978-1-0716-2827-0_13,

Molecular fingerprinting techniques are potential tools in revealing the population genetic alignment and diversity of the endophytes. Internal transcribed spacer (ITS) and rDNA gene sequencing methods and phylogenetic evolutionary relationships have been effectively employed for microbial identification at various taxonomic levels [2]. Theoretically identifying the sequences into various taxonomic levels based on phylogenetic analysis and sequence similarity comparison [3]. The influence of the introduction of selected endophytes and shifts in endophytic populations on bacterial populations have recently been studied using molecular fingerprinting techniques. Diverse methods have been employed, such as PCR-DGGE [4] and PCR followed by terminal restriction fragment length polymorphism (PCRT-RFLP) [5]. Other molecular fingerprinting techniques, such as PCR followed by temperature gradient gel electrophoresis (PCRTGGE) [6] and PCR-single strand conformational polymorphism (PCR-SSCP) [7], have not yet been applied to the microbial community in plants. To discriminate genuine endophytes from epiphytes, contamination of samples from the outside of plants is avoided by the tissue imprint method [8, 9]. In practice, bands representing more than one taxon have been reported [7]. Also, single isolates represented by more than one band in fingerprints have been observed, as shown for *Bacillus* sp. strain Sal1 [9]. The use of a single restriction enzyme may be sufficient for species differentiation by PCR-RFLP [10]. However, often the ability to differentiate between closely related species requires the application of more than one restriction enzyme [11]. Terminal restriction fragment length polymorphism (T-RFLP) is a molecular genetic profiling technique for the assessment of species diversity in a sample [12]. The GC clamp is required for immobilizing denatured DNA fragments within the gel matrix. T-RFLP is recommended for studying habitats with low species richness [13], which is often the case for internal plant environments [14]. T-RFLP has appeared to be more sensitive for the detection of weak bands than PCR-DGGE [15]. To obtain extra details about the techniques, we refer readers to the *Molecular microbial ecology manual* [16].

1.1 Molecular Fingerprinting

A classic fingerprinting technique produces molecular markers that should be polymorphic sufficient to differentiate between microorganisms. Fingerprinting techniques vary depending on how the DNA fragments/banding patterns that are generated and examined. DNA fingerprinting techniques, such as random amplified polymorphic DNA (RAPD), simple sequence repeat (SSR) or inter-SSR (ISSR), amplified fragment length polymorphism (AFLP), restriction fragment length polymorphism (RFLP), terminal-RFLP (T-RFLP), denaturing gradient gel electrophoresis (DGGE), temperature gradient gel electrophoresis (TGGE), and single-stranded conformation polymorphism (SSCP), are well recognized and have been effectively useful to evaluate genetic diversity of microbes in the natural environment [17–24]. These methods have recently been employed to assess the population and community structure of endophytic fungi and bacteria as well. Other important fingerprinting techniques are arbitrary primed PCR (AP-PCR), allele-specific associated primers (ASAP), cleaved amplified polymorphic sequences (CAPS), DNA amplification fingerprinting (DAF), sequence characterized amplified region (SCAR), SSR-single primer amplification reaction (SSR-SPAR), and sequence tagged microsatellite loci (STMS). The data from these methods are essential to be converted to computer-readable format. Once changed, the data can then be analyzed using various kinds of computer software. The collection (scoring and entry of band information into the computer) and analysis of data using univariate and multivariate statistical approaches; and interpretation of the data are the three key stages that are involved in the molecular data analysis for DNA fingerprinting. Similarly, laboratory measures must be identical between diverse personnel.

1.2 Data Collection and Analysis

The scoring of the data can be done based on the presence/absence (1 or 0) in the case of dominant markers (such as RAPDs or AFLPs or ISSRs) or as allele frequencies for SSRs/RFLPs or as biallelic for SNPs. SSRs and RFLPs can also be scored as presence/absence, but some genetic information will be lost, so more markers should be used if markers will be scored this way. For presence/absence data, the data should be entered into a spreadsheet (such as EXCEL) in the required format (Table 1).

2 Materials Required

2.1 RAPD

Refrigerator (4 °C), Deep freezer (−20 °C), Microwave or hot plate, Laminar air flow Fume hood, pH meter, Centrifuge, Standard balance, Thermocycler, Gel electrophoresis units, Power supply units, UV transilluminator/gel documentation system, Vortexer, Pipetman (2–20 μL, 20–200 μL, 200–1000 μL).

Supplies and Solutions Needed 10× Taqpoymerase buffer, 25 mM magnesium chloride, 25 mM dNTP's (mixture of dATP, dCTP, dGTP, and dTTP), Taqpoymerase (5 Units/μL), 5 μM RAPD primers, template or genomic DNA (10 ng/μL), sterile double distilled water, TAE buffer, agarose, gel loading dye, DNA size standard, ethidium bromide, tips, 0.5 mL and 1.5 mL microcentrifuge tubes, 0.2 mL PCR tubes, staining trays.

2.2 Diversity Analysis Using Denaturing Gradient Gel Electrophoresis (DGGE)

Solutions Required 40% Polyacrylamide solution, Urea, Formamide (deionized) from Sigma Aldrich, 50× TAE Buffer, TEMED, Ammonium persulfate, GelStar, Sybr Green I, Spacer Grease, Distilled Water, Glycerol (optional).

Table 1
The 16S and 26S rRNA gene-targeted primers for detection of a specific group of actinomycetes, bacteria, and fungi

Proposed target	Primer name, primer sequence (5′–3′)	Reference	Size of amplified product (bp)	Initial denaturation	Amplification	Final extension
Archaea	Arch344f (344–363a) 5′- ACGGGGCGCAG CAGGCGCGA-3′, 927r (927–948) 5′-CCCGCCAA TTCCTTTAAGTTTC-3′	[25, 26]	605	94 °C, 5 min	10 cycles at 94 °C, 30 s; 61–56 °C 30 s; 72 °C, 1 min;	72 °C, 7 min
Alphaproteobacteria	F203 (174–203a), 5′- CCGCATACGCCCTACGGGG GAAAGATTTAT-3′ R1492 (1492–1513), TACGGY TACCTTGTTACGACTT-3′	[27, 28]	1340	94 °C, 5 min	25 cycles at 94 °C, 1 min; 56 °C, 1 min; 72 °C, 2 min	72 °C, 10 min
Betaproteobacteria	F948 (931–948), 5′- CGCACAAGCGGTGGATGA-3′, R1492 (1492–1513), 5′-TACGGYTACCTTGTTACGACTT-3′	[27]	583	94 °C, 5 min	25 cycles at 94 °C, 1 min; 61 °C, 1 min; 72 °C, 2 min	72 °C, 10 min
Gammaproteobacteria enteric bacteria	fD2 (8–27), 5′- AGAGTTTGATCATGGCTCAG-3′ rP1 (1492–1512), ACGGTTACCTTGTTACGACTT-3′	[32]	1505	4 °C, 2 min	30 cycles at 94 °C, 15 s; 52 °C, 30 s; 72 °C, 1 min	72 °C, 7 min
Actinomycetales	F243HGC (226–243), 5′-GGATGAGCCCGCGGCCTA-3′, R1492 (1492–1513),5′-TACGGYTACCTTGTTACGACTT-3′	[28]	1288	95 °C, 5 min	35 cycles at 94 °C, 1 min; 63 °C, 1 min; 72 °C, 2 min	72 °C, 10 min
Yeast	NL1 (63–86b), 5′-GCATATCAATAAGCGGAGGAAAAG-3′, NL4 (624–642), 5′- GGTCCGTGTTTCAAGACGG-3′	[33]	580	95 °C, 5 min	35 cycles at 94 °C, 1 min; 52 °C, 1 min; 72 °C, 2 min	72 °C, 10 min

Pseudomonas spp.	First step: *Pseudomonas* spp. specific PsR: 5′-GGTCTGAGAGGATGATCAGT-3′ and pSf: 5′-TTAGCTCCACCTCGCGGC-3′ Second step: *Pseudomonas* spp. specific	[34]	760	94 °C for 5 min	35 cycles at 94 °C for 1 min, 66 or 68 °C (for soil DNA and pure culture, respectively) for 1 min, and 72 °C for 2 min,	72 °C for 10 min
Burkholderia spp.	First step: *Burkholderia* spp. specific Burk3: 5′-CTGCGAAAGCCGGAT-3′ and BurkR: 5′-TGCCATACTCTAGCYYGC-3′ Second step: *Burkholderia* spp. specific Burk3 with GC clamp and r1378: 5′- CGGTGTGTACAAGGCCCGGGAACG-3′	[35]	500	95 °C for 4 min	Touchdown PCR as followed 94 °C for 1 min annealing 62 °C decreasing by 1 °C every fifth cycle up to 60 °C after that 25 cycles at 58 °C for 90 sec, and extension 72 °C for 2 min	72 °C for 10 min
Bacillus spp.	First step: *Bacillus* spp. specific Bacf: 5′-GGGAAACCGGGGCTAATACCGGAT-3′ and r1378 Second step: Eubacterial f968 with GC clamp and r1378	[36]	1300	94 °C for 5 min	35 cycles at 94 °C, 1 min; 65 °C, 90 sec; 72 °C, 2 min	72 °C for 10 min
Corynespora spp. (LSU region)	V9G- 5′- TTACGTCCCTGCCCTTTGTA-3′ LR5-5′-TCCTGAGGGAAACTTCG-3′	[37, 38]	–	–	30 cycles at 94 °C, 60 sec; 50 °C, 45 sec; 72 °C, 1 min -	72 °C for 7 min --
Colletotrichum spp. (GAPDH)	GDF1-5′- GCCGTCAACGACCCCTTCATTGA-3′, GDR2–5′- GGGTGGAGTCGTACTTGAGCATGT-3′	[39]		95 °C for 4 min	35 cycles at 95 °C, 30 sec; 60 °C, 30 sec; 72 °C, 45 sec	72 °C for 7 min

(continued)

Table 1
(continued)

Proposed target	Primer name, primer sequence (5′–3′)	Reference	Size of amplified product (bp)	Initial denaturation	Amplification	Final extension
Fusarium spp. (B-tubulin)	T1 5′- AAC ATG CGT GAG ATT GTA AGT-3′ T2 5′- TAG TGA CCC TTG GCC CAG TTG-3′	[40]	600	–	35–40 cycles at 94 ° C, 35 sec; 52 °C, 55 sec; 72 °C, 2 min	–
Penicillium spp. (B-tubulin)	Bt2a-5′- GGT AAC CAA ATC GGT GCT TTC -3′ Bt2b -5′- ACC CTC AGT GTA GTG ACC CTT GGC-3′	[41]	550	94 °C for 5 min	35 cycles at 94 °C, 45 sec; 55 °C, 45 sec; 72 °C, 1 min	72 °C for 10 min
Non-sporulating fungi (LSU region)	LROR-5′- ACC CGC TGA ACT TAA GC-3′ LR5–5′- TCC TGA GGG AAA CTT CG-3′	[38]	–		30 cycles at 94 °C, 60 sec; 50 °C, 45 sec; 72 °C, 1 min	72 °C for 7 min -
Trichoderma spp.	TEF1-728F 5′- CAT CGA GAA GTT CGA GAA GG-3′ TEF1-rev GCC ATC CTT GGA GAT ACC AGC-3′	[42]	350	–95 °C for 8 min -	35 cycles at 95 °C, 55 sec; 58 °C, 45 sec; 72 °C, 60 sec--	–72 °C for 5 min -
Xylaria sp. *Cladosporium* sp. (ACT region)	ACT-512F-5′- ATGTGCAAGGCCGGTTTC G-3′ ACT-783R-5′- TACGAGTCCTTCTGGCCCAT -3′	[42]	300	–95 °C for 8 min	35 cycles at 95 °C, 55 sec; 61 °C, 45 sec; 72 °C, 60 sec---	–72 °C for 5 min -
All fungal genera (ITS region)	ITS1–5′-TCCGTAGGTGAACCTGCGG-3′ ITS4- 5′-TCCGCTTATTGATATGC-3′	[31]	350–880	–95 °C for 2–3 min	25 cycles at 95 °C, 30 sec; 50–60 °C, 30 sec; 72 °C, 30 sec –2 min depends upon product size	–72 °C for 10 min --

Other Materials 10 mL Disposable serological pipets, Conical centrifuge tubes for polyacrylamide solutions (15 mL, 50 mL), 1.5 mL tubes for APS solution, Gel-loading tips from standard company, kimwipes, Teflon stir-bar, Powder-free nitrile gloves.

2.3 PCR-RFLP

Solutions Required 5× Taqpoymerase buffer (Promega), 25 mM magnesium chloride, 25 mM dNTP's (Promega) Taqpoymerase (5 Units/μL) (Promega), 5 μM required primers (IDT), template DNA (10 ng/μL), sterile double distilled water, Restriction Enzymes with buffer (NEB).

Other Materials Andy Gold™ Nucleic Acid Gel Stain (Applied BioProbes), Agarose (Ameresco) Gel-loading tips (Thermo), Powder-free nitrile gloves (Kimberly), Electrophoresis system (Biorad), Gel documentation system (Biorad), Thermocycler, Gel electrophoresis units, Pipetman (1–10 μL, 20–200 μL, 200–1000 μL).

2.4 Box-PCR Finger Printing

Thermal cycler (Applied Biosystem), 50 ng of genomic DNA, 10× Taq Buffer, 25 mM $MgCl_2$, 2.5 mMdNTPs, 10 pmol BOXAIR primer, DMSO (100%), 10 mg/mL BSA, 2 U Taq DNA Polymerase. Other materials: ethidium bromide (0.5 μg/mL) solution (Himedia), Agarose (Ameresco) Gel-loading tips (Thermo), TAE buffer (Himedia), Powder-free nitrile gloves (Kimberly), Electrophoresis system (Biorad), Gel documentation system (Biorad), NTsys version 2.2.

2.5 Its-RFLP

5× Taqpoymerase buffer (Promega), 25 mM magnesium chloride, 25 mM dNTP's (Promega) Taqpoymerase (5 Units/μL) (Promega), 5 μM ITS primers (IDT), template DNA (10 ng/μL), sterile double distilled water, Hae III, Hinf I and Rsa I (Promega) Other materials: ethidium bromide (0.5 μg/mL) solution (Himedia), Agarose (Ameresco) Gel-loading tips (Thermo), Tris-borate-EDTA buffer (Himedia), Powder-free nitrile gloves (Kimberly), Electrophoresis system (Bio-rad), Gel documentation system (Biorad), Thermocycler, Gel electrophoresis units, Pipetman (1–10 μL, 20–200 μL, 200–1000 μL).

2.6 AFLP

50 μg concentration DNA, 10× Assay Restriction Buffer (NEB), EcoR I (20 U/μL) (NEB), MseI (10 U/μL) (NEB), Sterile distilled water, T4 DNA ligase buffer with 10 mM ATP (NEB), *EcoR I* adapter (10 pmol/μL), *Mse I* adapter (50 pmol/μL), T4 DNA *ligase* (NEB), Template DNA, 1 mM each dNTP (NEB), 10 X Taq buffer (NEB), 1 U Taq polymerase (5 U/μL) (NEB), Polyacrylamide stock solution of 4.5%, TEMED (Tetramethylethylenediamine), Ammonium persulfate (Sigma), ethanol, bind silane, acetic acid (Himedia), Silver stain, sodium carbonate, NTSYS-pc version 2.0.

3 Protocols of Different Fingerprinting Techniques

3.1 Fungal DNA Isolation

DNA isolation protocol, a modified method of Gul et al. [29], is presented here.

1. Weigh 0.2–0.5 g of 5–6 days old fresh fungal sample (any endophytic fungal sample) and crush with the help of mortar and pestle in 500 μL of extraction buffer solution, containing 5% sodium dodecyl sulfate, 0.1 M NaCl and 0.5 M Tris-HCl, (pH 8) in 2.0 mL centrifuge tube containing glass beads of different diameters.
2. Vortex the sample for about 10–15 min at maximum speed. No incubation is required in this method and then heat at 65 °C for 30 min. During heating, invert the tubes carefully at every 5 min interval.
3. Vortex the tubes for 10–15 s after incubation. Then centrifuge the tubes after incubation for 10 min at 11,000 g.
4. Transfer the supernatant to 2.0 mL clean tubes, then add equal volume of chloroform, phenol, isoamyl alcohol mixture (24: 25:1).
5. Again, vortex the tubes for 10–15 s and centrifuge for 5 min at 10,000 g.
6. Take the aqueous layer in a new Eppendorf tube and add Chloroform:isoamyl alcohol (24:1) mixture in equal volume. Then centrifuge the mixture at 10,000 g for 5 min and obtain only supernatant.
7. Mix the resultant supernatant with equal volume of absolute ethanol in a new Eppendorf to get precipitated DNA.
8. To improve the efficiency of DNA precipitation, incubate the Eppendorf containing mixture of DNA and absolute ethanol at 4 °C for 1 h and separate the precipitated DNA from the mixture by centrifugation at 14,000 g for 10 min.
9. Wash the DNA pellets with cold/chilled ethanol (70%) and air dry the pellets.
10. The DNA pellet can be taken in 40 μL TE buffer.
11. The DNA obtained is checked for quality by running agarose gel electrophoresis. Load the 0.5–1.0 μL DNA solution onto agarose gel (0.8%) for electrophoresis and stain the gel with 0.6 gm ethidium bromide to visualize DNA in the UV.

3.2 Molecular Characterization of Endophytic Fungi

Morphologically characterized endophytes can be identified by sequencing of the translation elongation factor-1α (TEF-1α) gene and internal transcribed spacer (ITS) regions [30].

3.2.1 ITS Region Amplification

(a) Amplify the ITS regions using ITS1 and ITS4 primer pairs [31].

(b) Perform PCR amplifications in a total volume of 25 μL containing 0.5 μL of 0.5 μM template DNA, 5 μL 5× PCR buffer, 4.0 mM MgCl2, 0.8 mM dNTP mix (Promega, WI, USA) of both forward and reverse primers and 0.625 U of Taq polymerase (Promega, WI, USA), in a PTC-100 Peltier Thermal Cycler (MJ Research Inc., MA, USA).

(c) Set up a PCR program with an initial denaturation at 95 °C for 3 min, followed by 34 cycles of denaturation at 95 °C for 1 min, annealing at 52 °C for 30 s, extension at 72 °C for 1 min, and a final extension at 72 °C for 10 min.

3.2.2 TEF-1α Gene Amplification

1. Amplify the TEF-1α gene using EF1 (ATGGGTAAGGAGGACAAGAC) and EF2 (GGAAGTACCAGTGATCATGTT) primers modified from O'Donnell et al. [40].
2. Perform the PCR amplification in a 25 μL reaction mixture containing 5 μL 5× PCR buffer, 2.5 mM MgCl2, 0.2 μL template DNA, 0.64 mM of dNTPs mix (Promega), 0.625 unit Taq polymerase (Promega), and 0.8 μM of both forward and reverse primers, using a Thermal Cycler (Eppendarf). The PCR cycles are based on O'Donnell et al. (1998) with some modifications: the initial denaturation at 94 °C for 1 min followed by 36 cycles of denaturation at 95 °C of 35 s, annealing at 59 °C for 55 s, and extension at 72 °C for 90 s, followed by a final extension at 72 °C for 10 min.
3. Visualize the PCR products using gel electrophoresis.
4. Purify the amplified PCR products using a purification kit (Genei labs, Bangalore, India) according to the manufacturer's protocol, and send the PCR products to a service provider (Genei labs, Bangalore, India).
5. For ITS and TEF-1α sequences, perform a BLAST search NCBI database.

Sequence Analysis The sequence data generated during the study are subjected to BLAST searches in the nucleotide database of GenBank (www http://blast.ncbi.nlm.nih.gov/) to determine their most probable closely related taxa. Sequence data can be retrieved from GenBank based on recent publications. Raw forward and reverse sequences assembled using Geneious Pro.v4.8.5. Sequence alignments were carried out with MAFFT v.6.864b [43] and alignments are manually improved where necessary. The sequence datasets are combined using BioEdit v.7.2.5 [44]. Maximum Likelihood (ML) and Bayesian Inference (BI) analyses for the sequence dataset. The phylogenetic trees are configured in Figure Tree v. 1.4 (Rambaut and Drummond 2008) and edited using Microsoft Office PowerPoint 2007 and Adobe Illustrator CS3 (Adobe Systems Inc., USA).

(a) **Bacterial DNA Isolation**

(i) **Method-I**

- Collect 1.5 mL of a 48-h bacterial culture and centrifuge for 5 min at 14000 g.
- Resuspend in 1 mL of TE buffer (mM Tris-HCl, 1 mM EDTA, pH 8.0) and centrifuge for 5 min at 14000 g.
- Resuspend the obtained pellet in 500 μL of TE buffer and finally add 0.5 g of glass pearls (0.1 mm in diameter) and add 15 μL of 20% SDS.
- Homogenized the cells for 30 s in a vortex mixer and add 500 μL of buffered phenol.
- Mixed the solution properly and centrifuged for 5 min at 14000 g.
- Extract aqueous phase once with phenol-chloroform (1: 1) and once more with chloroform.
- Following the extraction of the aqueous phase, add 20 μL of 5 M NaCl and isopropanol for precipitation of DNA (5 min at room temperature).
- Centrifuge for 10 min at 14000 g to collect the precipitated DNA.
- Wash the DNA pellet with 70% ethanol, air dry and resuspend in 30 μL of autoclaved, ultrapure water.
- Use obtained DNA for further studies.

(ii) **Method-II (Cell Lysis Method)** [45]

- Inoculate purified culture of endophytic bacteria in nutrient broth medium (5 mL) at 28–30 °C.
- Harvest bacterial cells from medium by centrifugation (5 min at 15,000) rpm at the late-log phase and washed with 1 mL NaCl (1% w/v).
- Resuspend the pellet in 200 uL distilled water and store at −20 °C.
- Prepare cell lysates by mixing a 40 uL cell suspension with a 10 uL proteinase K solution (1 mg mL −1) and 50 uL buffer (pH 8; 40 mM tris(hydroxymethyl)-aminomethane), 1% Tween 20 and 1 mM EDT A.
- Incubate the mixture at 60 °C for 20 min, then at 95 °C for 5 min and centrifuge.
- Use obtained crude DNA in the supernatant as a template for PCR amplification.
- Sequence analysis as discussed earlier.

(b) **Diversity Analysis Using Denaturing Gradient Gel Electrophoresis (DGGE): Extraction of Total DNA**

1. The surface disinfection of samples is performed as described for the isolation of endophytic fungi [46]. The total DNA is extracted from leaves and roots with the commercial kit according to the manufacturer's recommendations.
2. The DNA is then quantified using a Thermo Scientific™ NanoDrop™ 2000/2000c Spectrophotometer and standardized to 10 ng/μL DNA.
3. Partial amplification of the SSU rDNA region for DGGE analysis: The specific fungal primers FF390 (5′- CGATAAC GAACGAGACCT-3′) and FR1 GC (5′-CCCCCGCGC G CGGGCGGGGCGGGGGCACGGGCCGAICCATT-CAATCGGTAIT-3′) (I = Inosine) described by Vainio and Hantula [47] are used to amplify part of the SSU (small subunit or 18S) rDNA region.
4. Conduct three reactions of each sample (leaf and root) separately with the proper modifications for each reaction. The carry out the amplifications directly with the primer containing the GC clamp (nested) using the Eppendorf Mastercycler® thermal cycler (Eppendorf, Germany) programmed to perform an initial denaturation at 94 °C for 4 min, followed by 35 cycles at 94 °C for 1 min, 50 °C for 1 min and 72 °C for 1 min, with a final extension at 72 °C for 10 min. The 50 μL amplification reaction contained: 1× Colorless Go Taq® Flexi Buffer, 0.25 mM $MgCl_2$, 0.1 mMdNTPs, 0.2 μM FF390 primer, 0.2 μM FR1 GC primer, 0.5 mg mL^{-1} of bovine serum albumin (BSA), 2.0 μL DMSO, 1.0 μL formamide, Go Taq® DNA polymerase 1.25 units (Promega, Madison, USA), 10 ng of total genomic DNA for each sample.
5. The negative controls consisted of PCR reaction without genomic DNA to evaluate the presence of contaminants. Use sterile and nuclease-free materials in the preparation of the reactions. After amplification, analyze the PCR products by gel electrophoresis on a 1.2% agarose gel.

   Comparative Analysis Between Leaves and Roots by DGGE

6. DGGE can be performed in a C.B.S. Scientific® model 240 DGGE system for access to the endophytic fungal communities from leaves and roots.
7. Perform the electrophoresis on a 1-mm thick 6% polyacrylamide gel, with an acrylamide–bisacrylamide ratio of 37.5: 1, and the vertical denaturing gradient of urea and formamide from 35% to 55%.

8. The running buffer should be 1× TAE (40 mM Tris, 40 mM acetic acid, 1 mM EDTA, pH 7.4). For the DGGE analysis, approximately, 50 ng of PCR product mixed with the same volume of staining buffer (2% bromophenol blue, 2% xylene cyanol, 100% glycerol).
9. Separated the samples at a constant temperature of 60 °C for 16 h at 100 V, stained with SYBR Gold (50 μg/mL) (Invitrogen) by stirring gently for 15 min, and visualize under UV light.
10. Analyzes the DGGE profiles using the software Bionumerics® 5.1 (Applied Maths, Kortrijk, Belgium).
11. Calculate similarities in the banding patterns among the three replicates of leaves and roots based on the presence and absence of bands and expressed as a coefficient of similarity. The Dice similarity coefficient is used to compare the DGGE profiles.
12. This similarity coefficient can be calculated using the following formula: $D_{sc} = [2j/(a + b)]$, where a = number of bands or OTUs (operational taxonomic units) in DGGE on lane 1, b = number of bands on lane 2, j = number of common bands in lanes 1 and 2, and $D_{sc} = 1$ indicates an identical profile.
13. Dendrograms showing a cluster can be constructed using the UPGMA. Diversity of the fungal communities are usually analyzed using the Richness index (S). Based on the binary matrix, variable richness can be estimated that is generated by software BioNumerics® version 5.1. BioNumerics® analyzes a scanned image of the gel where the presence of a band is coded as one (1) and absence as zero (0).

(c) **Random Amplified Polymorphic DNA (RAPD) Protocol**

1. Synthesize the randomly amplified polymorphic DNA (RAPD) primers (10 base pairs) from any molecular-based company.
2. For RAPD-Polymerase chain reaction, total volume of reaction mixture should be 25 μL containing 50 ng/μL DNA template, 2 μL of 5× Taq buffer, 1 μL of 2.5 mM dNTP mix, 2 μL of 25 mM $MgCl_2$, 2 μL of 0.1 μM primer and 0.3 μL of 5 U/μL Taq DNA polymerase (Genei Labs Pvt. Ltd., Bangalore, India).
3. Perform PCR amplification in Thermal cycler (Eppendorf, Germany) with the following conditions: one cycle of initial denaturation at 94 °C for 5 min followed by 40 cycles of denaturation at 94 °C for 30 sec, primer annealing at 36.7 ° C for 1 min, extension at 72 °C for 2 min and final extension at 72 °C for 5 min.

4. Separate the amplified PCR products on 1.5% agarose gel in 1× TAE buffer stained with ethidium bromide (10 mg/ mL) using capillary electrophoresis unit (Genetix Biotech Asia Pvt. Ltd., New Delhi) and visualize the gel under UV light in gel documentation system (Bio rad, USA).
5. For RAPD marker analysis, score the polymorphic RAPD markers as 1 and 0 for the presence and absence of DNA bands to construct binary matrices. Analyze the binary matrices of RAPD by NTSYS-pc v.2.0 software. Construct dendrogram with the help of Jaccard's coefficient by using SHAN clustered program by selecting the unweighted pair group method with arithmetic mean (UPGMA) in NTSYS-pc v.1.8 [48].

(d) **PCR-Restriction Fragment Length Polymorphism (PCR-RFLP)**

(i) **16S rRNA RFLP Fingerprinting Protocol**

1. Take crude DNA (2 μL) of endophytic bacteria for a PCR amplification template.
2. Use single primer 879F (5′ -GCCTGGGGAG-TACGGCCGCA-3′) [49] for grouping of isolates at species level.
3. Obtain RAPD patterns according to Rivas et al. [49]. PCR conditions are as follows: preheating at 95 ° C for 5 min; 35 cycles of denaturing at 95 °C for 1 min, annealing at 54 °C for 1 min, extension at 72 °C for 2 min, and a final extension at 72 °C for 7 min.
4. To study the diversity of the isolates, follow PCR-RFLP analysis on 16S rRNA genes.
5. Select two different restriction enzymes (HinfI and MspI) for RFLP analysis.
6. 10 μL of each 16S rRNA PCR products are subjected to restriction enzyme digestion by HinfI and MspI (Takara).
7. For DNA cleavages, incubate overnight at 37 °C to achieve the complete restriction of DNA fragments.
8. Analyze RAPD and RFLP results in 2.5% agarose gel by following the user instructions and staining with 10,000× Andy Gold™ Nucleic Acid Gel Stain (Applied BioProbes).

(ii) **ITS-RFLP analysis** [50]

1. Digest the PCR products using the restriction endonucleases Hae III, Hinf I, and Rsa I (Promega).
2. To perform each digestion, reaction volume of 20 μL should be containing 10.4 μL of sterile distilled H_2O,

0.5 μL of restriction endonuclease, 1 μL 10r reaction buffer (Promega), 0.1 μL BSA and 8 μL of amplified ITS product.

3. Carry out the digestions at 37 °C for 4 h and the resulting restriction fragments are separated by electrophoresis on a 3–4% agarose gel at a constant current in Tris-borate-EDTA buffer for approximately 1.5 h containing ethidium bromide (0.5 μg/mL) and visualized under UV light.
4. Isolates for which identical RFLP patterns obtained with all endonucleases are considered belong to the same RFLP group.

(e) **BOX-PCR Fingerprinting**

1. BOX PCR fingerprinting is carried out using primer sequences BOXA1R (5′-CTACGGCAAGGCGACGCTG ACG-3′) as previously described by Rademaker et al. [51].
2. Perform PCR reaction in a thermal cycler (Applied Biosystem, Singapore) in 25 μL total volume, containing 50 ng of genomic DNA, 10× Taq Buffer, 25 mM $MgCl_2$, 2.5 mM dNTPs, 10 pmol BOXAIR primer, 1 μL of DMSO (100%), 10 mg/mL BSA, 2 U *Taq* DNA Polymerase.
3. The PCR conditions for amplification are carried out in the following: initial denaturation at 95 °C for 7 min followed by 30 cycles at 94 °C for 30 s, at 57 °C for 1 min, and at 65 °C for 8 min and a final extension step at 65 °C for 16 min.
4. The PCR fragments are separated on 1.2% agarose gel using 1× TAE buffer and visualize under gel documentation system Gel Doc XR+ system (BioRad).
5. For BOX markers, record the different banding pattern in the binary form, i.e., "1" for the presence of a band, it should be clearly visualized in the gel and assign "0" when there is no band or not possible to determine, to generate a binary matrix [52].
6. The binary matrix can be calculated by Jaccard's similarity coefficient for each pair-wise comparison. Constructed dendrogram from the similarity matrix by UPGMA method given by Lopez and Alippi [53] supported by NTsys version 2.2.

(f) **Amplified Fragment Length Polymorphism (AFLP)**

(a) **General Protocol for AFLP Analysis (Fig. 1)**

The AFLP protocol involves five important steps viz.

1. Digestion of genomic DNA using restriction enzymes.

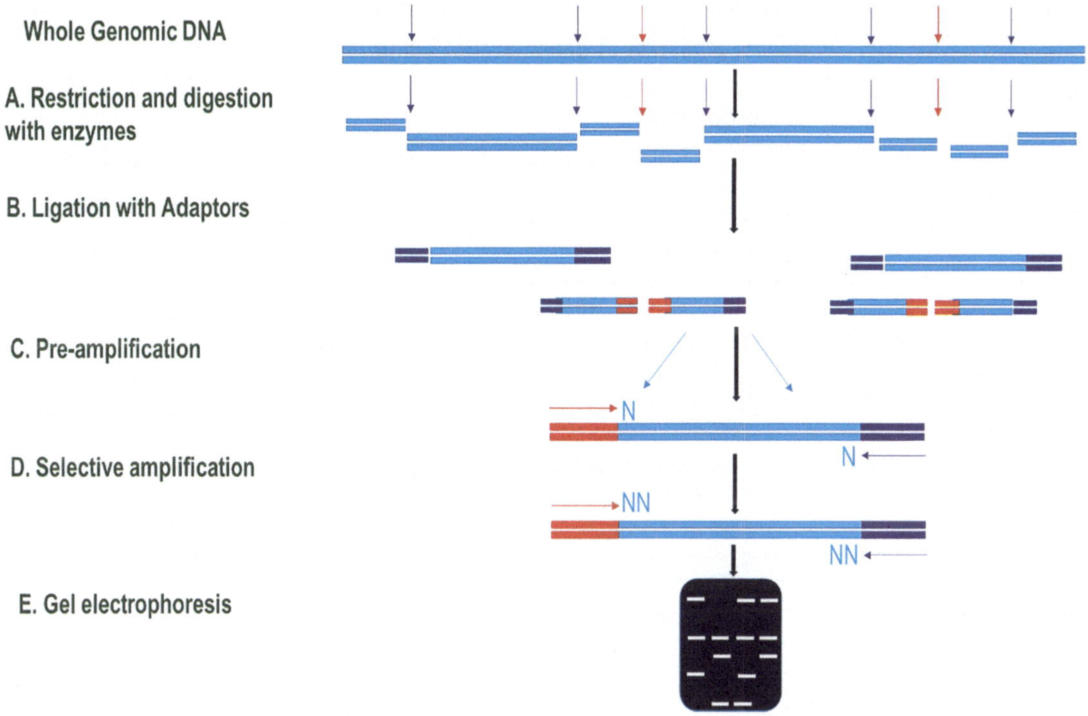

Fig. 1 General outline of the AFLP procedure (Chial 2008)

2. Ligation of double-stranded adapters for cut ends of restricted fragments.
3. Pre-amplification PCR involves primers with single nucleotide extension.
4. Re-amplification PCR involves primers with three-nucleotide extension.
5. Electrophoretic separation of amplicons on a gel matrix, followed by visualization of the band pattern.

(b) **Double Digestion of Genomic DNA Using Restriction Enzymes (Table 2)**

The double digestion reaction mix is prepared in microtubes and incubated at 37°C for 3 h in a PCR machine. The digestion components are as follows.

The digested product is loaded onto 1.2% agarose gel to confirm the complete digestion of the genomic DNA.

(c) **Preparation and Ligation of Adapters**

Single-stranded adaptor sequences are annealed to prepare double-stranded adapter sequence. Both *EcoR I* and *Mse I* adapters are finally used at the concentration of 5 pmol/μL and 50 pmol/μL, respectively.

Table 2
Double digestion of genomic DNA using restriction enzymes

Digestion mix	1× (μL)
0.5 μg concentration DNA	–
10× assay restriction buffer (NEB)	5
EcoR I (20 U/μL) (NEB)	0.5
MseI (10 U/μL) (NEB)	0.5
Sterile distilled water	34
Total	50

- *EcoRI* adapter 1 and 2, are diluted to 10 pmol/μL concentration by using sterile nuclease free/deionized water.
- Equal volumes of *EcoRI* adapter 1 and 2 (10 pmol/μ L) are added to a fresh tube to give a final concentration of 5 pmol/μL.
- The steps 1 and 2 are repeated for *Mse I* adapter to get 50 pmol/μL final concentration by adding equal volumes of each 100 pmol/μL stock.
- The mixed adapters are incubated at 65 °C for 15 minutes and again at 37 °C for 30 minutes and finally left at room temperature for 1 h.
- The annealed double-stranded adapters are stored at −20 °C.

(d) **Oligonucleotide Adapter Sequences**

EcoR I Adapter

5′ – CTCGTAGACTGCGTACC - 3′.

3′- CTGACGCATGGTTAA -5′

Mse I Adapter

5′ – GACGATGAGTCCTGAG-3′

3′ – TACTCAGGACTCAT -5′

(e) **Ligation of Double-stranded Adapters**

The cut ends of restriction fragments generated after digestion are ligated with adapter sequences. Twenty microliters of digested reaction mix is used to prepare the ligation reaction mix (Table 3).

Ligation reaction mix is incubated at 37 °C overnight. The digested/ligated product is diluted to 1:10 with $T_{10}E_1$ (pH 8) and stored at −20 °C.

(f) **Pre-amplification**

Table 3
Ligation mix

Ligation mix	1× (μL)
Digested DNA	20.0
T4 DNA ligase buffer with 10 mM ATP (NEB)	2.5
EcoR I adapter (10 pmol/μL)	0.5
Mse I adapter (50 pmol/μL)	0.5
T4 DNA *ligase* (NEB)	0.5
Sterile distilled water	1
Total	25

Table 4
Pre-amplification reaction mix

Pre-amplification reaction mix	1× (μL)
Template DNA	3
EcoR I + N * (10 pmol/μL)	2.5
Mse I + N * (10 pmol/μL)	2.5
1 mM each dNTP (NEB)	2.5
10 X *Taq* buffer (NEB)	1.5
1 U *Taq* polymerase (5 U/μL) (NEB)	0.4
Sterile distilled water	2.6
Total	15

Where N* is a primer with a single nucleotide extension

An initial round of PCR (pre-amplification) is carried out to enrich a subset of the AFLP template; the primers used in pre-amplification have a single base selection. The bases of selection chosen depend on the final primer combination for re-amplification.

Pre-amplification mix involves the following components (Table 4).

Pre-amplification PCR is carried out with the following PCR program.

$$\left.\begin{array}{l} 94\,^{\circ}\mathrm{C}/30\ \mathrm{s} \\ 56\,^{\circ}\mathrm{C}/1\ \mathrm{min} \\ 72\,^{\circ}\mathrm{C}/1\ \mathrm{min} \\ 10\,^{\circ}\mathrm{C}/30\ \mathrm{min} \end{array}\right\}20\ \text{cycles}$$

Table 5
Re-amplification reaction mix

Re-amplification reaction mix	1× (μL)
Template DNA	2.5
EcoR I + N * (10 pmol/μL)	2
Mse I + N * (10 pmol/μL)	2
1 mM each dNTP (NEB)	2
10 X *Taq* buffer (NEB)	1
1 U *Taq* polymerase (5 U/μL) (NEB)	0.3
Sterile distilled water	0.2
Total	10

Where N* is a three-nucleotide extension primer

The pre-amplified products are diluted to different dilutions with $T_{10}E_1$ pH 8. Template of 2 μL each from these dilutions is used for re-amplification PCR. The pre-amplified product is loaded onto 1.2% agarose gel to check the amplification. Depending on the amplification intensity the working dilution is used for further amplifications. All the diluted products are stored at −20 °C

(g) **Re-amplification**

In re-amplification PCR, selection of fragments is made using primers with three-nucleotide extension. The reaction mixture involves the following components (Table 5).

Touchdown PCR Program Used for Re-amplification

94 °C/30 s
65 °C/30 s reducing by 0.7 °C/cycle to 56 °C
72 °C/1 min
} 12 cycles

94 °C/30 s
56 °C/30 s
72 °C/1 min
10 °C/30 min
} 24 cycles

Denaturation of AFLP Products

Four microliters of stop loading dye is added to each sample. Denaturation is carried out at 94 °C for 5 min and then cooled to 10 °C for 5 min. The product is stored at −20 °C.

(h) **Polyacrylamide Gel Electrophoresis (PAGE)**

Polyacrylamide stock solution of 4.5% should be prepared in advance and stored at 4 °C. The glass plates should be thoroughly cleaned and salinized before pouring polyacrylamide gel.

Plate and Solutions Preparation

Bind silane stock solution: 40 mL of 100% ethanol +150 μL of bind silane.

Bind silane working solution: 10 mL stock bind solution +1 mL 10% acetic acid.

Repel coat

1. Clean both the plates using detergent. Then rinse the plate with millipore water and dry with paper towels.
2. Wipe the working side of the large plate with acetone and alcohol with 1000 μL of repel by spreading with paper towels in even sweeps across the whole plate. Give ethanol wipe after 5 min to spread evenly and final polish with second ethanol wipe.
3. Treat the small plate twice with bind silane by spreading 500 μL of working solution evenly over the whole surface.
4. Clean the spacers with ethanol and aligned on the edges of large plate. Place the smaller plate on top and taped on both sides to secure both plates together.

Pouring of Polyacrylamide Gel

1. Take sixty milliliters of chilled 4.5% polyacrylamide mix in a pouring bottle and 30 μL of chilled TEMED (Tetramethylethylenediamine) and 400 μL of freshly prepared 10% ammonium persulphate (APS) are added.
2. Pour polyacrylamide slowly between the plates. Place the comb in reverse direction between plates and the gel will be allowed to polymerize for 1.5 h.
3. Remove the comb by sliding it out horizontally.
4. Mount the plates onto the vertical electrophoresis unit, with the longer plate in the outer position; one-liter 1× TBE buffer is filled in the top and bottom tanks.
5. The gel sticking in between two plates near comb insertion area should be washed free of gel and comb and put in its position.
6. Load the loading dye randomly into few wells and allow the plate to run at 1000 V for 15 min to pre-heat the plates.
7. Load the three microliters of each denatured re-amplified product.

8. Carry out the electrophoresis at 1200–1400 V for 1 to 2 h until the tracking dye (bromophenol blue) moves into the lower tank buffer.
9. Unmount the gel and separate the two plates using plastic wedge. The gel remains in complete contact with the smaller plate (bind plate).
10. The bind plate should be silver stained for visualization of AFLP fragments.

(i) **Silver Staining of the Gel**

1. **Solutions prepared**
   a) **Developer**- Three percent sodium carbonate is prepared (60 g sodium carbonate in 2 L of distilled water) and stored at 4 °C
   b) **Fixer**- Ten percent acetic acid (200 mL acetic acid in 1800 mL of distilled water).
   c) **Silver stain**- 1.01 N Silver nitrate (12 mL 1.01 N Silver nitrate in 2 L of distilled water) or dissolve 3 g of silver nitrate powder in 2 L of distilled water. Add three milliliters of formaldehyde (40%) to this solution.
   d) **Developing solution** – Immediately prior to developing the gel, add 300 μL of sodium thiosulphate (0.1 N) and 3 mL formaldehyde to pre-chilled sodium carbonate solution.
2. **Silver staining procedure**
   - Place the bind plate in fixer (10% acetic acid) for half an hour on a shaker.
   - Wash the bind plate 3 times in fresh Millipore water and further place on a shaker for 10 min in millipore water to remove greasiness.
   - Then place the bind plate in silver stain solution for half an hour on a shaker.
   - Rinse the bind plate in 2 L distilled water for 10 s.
   - This is to be transferred immediately into the developing solution and immersed up and down holding the top end of the plate.
   - Stop the development reaction when the bands near the bottom of the gel start to appear by adding 2 L fixer.
   - Then soak the gel in a tray of distilled water for 10 min and left to dry overnight in vertical position.

(j) **Band Scoring and Data Analysis**

The separation of AFLP amplicons through gel electrophoresis results in banding pattern at different levels (loci). The amplicons generated in AFLP inherit as dominant markers. Therefore, the bands are scored in binary system; the score 1 is assigned for the presence of band and 0 for the absence of band at each locus. The variation in band intensity is not taken into consideration to avoid confusion in scoring. The data so obtained could be used for linkage map construction, tagging, genes controlling, economic importance, identifying duplicates in the germplasm, and assessing genetic variability. In this training course the binary data is used to estimate pairwise genetic distance based on Jaccard's coefficients using NTSYS-pc version 2.0 software. Construction of dendrogram using unweighted Paired Group Arithmetic Mean (UPGMA) algorithm based on distance matrix is demonstrated.

(a) **Technical AFLP Protocol for Endophytic Microbes [54]**

1. Take 200 ng of genomic DNA and digest with 5 units of *Eco*RI (Fermentas) for 2 h at 37 °C in 2× Tango™ buffer (66 mMTris-acetate, pH 7.9), 20 mM Mg-acetate, 132 mM K-acetate, 0.1 mg/mL BSA) in a 20 μL reaction volume.
2. Add 5 units of *Tru*I (isoschizomer of *Mse*I; Fermentas) after *Eco*RI digestion and adjust the buffer concentration again to 2× Tango™ buffer in a total volume of 30 μL.
3. Incubate the reaction mixture at 65 °C for 2 h. Add 10 μL ligation reaction mix consisting of 1 unit T4 DNA ligase (Fermentas), 1× T4 DNA ligase buffer, 50 μM *Mse*I adapter and 5 μM *Eco*RI adapter (Integrated DNA Technologies [IDT], directly to the digestion reaction that targets the frequent and rare cut fragments respectively.
4. Incubate the digestion-ligation reaction mixture overnight at 4 °C. Following ligation, dilute the digestion-ligation reaction mix @ 1 : 5 with sterile distilled water and use as template for the pre-selective PCR.
5. Each 15 μL pre-selective PCR reaction should contain 2.5 μL of the diluted digestion-ligation reaction mix, 1 μM *Mse*I + 0 primer, 1 μM *Eco*RI + 0 (IDT), 1× Kapa *Taq* Readymix (contains 0.2 μM of each dNTP, 1.5 μM $MgCl_2$, 0.3 unit *Taq* Polymerase, 1× Kapa Buffer A, (KapaBiotech).
6. Set up pre-selective PCR amplification program with an initial denaturing step of 94 °C for 5 min, followed by 23 cycles consisting of denaturation at 94 °C for 30 s, annealing at 56 °C for 30 s, elongation at 72 °C for 30 s, and a final elongation step at 60 °C for 30 min.

7. For confirmation of successful amplification run 5 μL of the PCR product on a 1% agarose gel and observe a smear between 100 and 500 bp.
8. Following successful amplification, dilute the pre-selective PCR products with sterile distilled PCR-grade water (1 : 19 dilution) of which 5 μL should use as template for selective PCR amplification.
9. For each 20 μL selective PCR reaction contains 0.25 μM of fluorescently labelled *EcoR*I + NNN and 1 μM unlabeled *Mse*I + CTT (IDT) and 1× Kapa *Taq* Readymix. PCR reactions can be done without a step-down PCR step, following pre-selective PCR conditions described above but with 30 repeat cycles.
10. After amplification, for each DNA sample mix 5 μL of each fluorescently labeled PCR product and purify using a commercially available purification System (NucleoFast Purification System, Germany). Perform the electrophoresis on the DNA Analyser (Applied Biosystems, USA) with the ROX500 size standard (Applied Biosystems).
11. Perform an automated fragment size calling and scoring with Genemarker Version 2.2.0 (SoftGenetics, LLC, CA, USA) with the manufacturer's default settings. Manually confirm the presence or absence of all fragments.
12. Locus-specific variability can be measured with the polymorphic information content (PIC) for dominant markers.
13. Generate a binary presence-absence data matrix for each species and calculate the total number of loci generated per primer pair as well as the percentage of polymorphic loci for each taxon in GenAlEx version 6.4 [55]. Expected heterozygosity (H_E), under the assumption of Hardy-Weinberg equilibrium, can also be calculated in GenAlEx version 6.4.

4 Conclusion

Molecular fingerprinting techniques are potent tools in revealing the population genetic configuration and diversity of endophytes. Development of these markers will allow the sensitive detection of endophyte occurrence in plant tissues. Molecular sequencing methods suggest an effective method for the identification of culturable and non-culturable endophytes by directly amplified rDNA fragments from plant tissues. The conventional PCR-based techniques can be able to identify endophytes specifically, but they cannot be used to quantify endophytic biomass within plant tissues. However, real-time PCR can detect small quantities of DNA in environmental samples and has been successfully used to determine the population density of endophytes. For accurate detection and quantification of endophytic diversity, development of molecular methods like DNA fingerprinting, DNA sequencing, real-time PCR, and DNA barcoding will become more predictable, precise, speedy, and understandable techniques in the future prospects.

Acknowledgments

The authors are grateful to Dr. N.P. Singh, The Director of ICAR-Indian Institute of Pulses Research, Kanpur, India for constant encouragement and support to draft this protocol.

References

1. Santoyo G, Moreno-Hagelsieb G, del Carmen Orozco-Mosqueda M, Glick BR (2016) Plant growth-promoting bacterial endophytes. Microbiol Res 183:92–99
2. Amann R, Ludwig W (2000) Ribosomal RNA-targeted nucleic acid probes for studies in microbial ecology. FEMS Microbiol Rev 24: 555–565
3. Sun X, Guo L-D (2012) Endophytic fungal diversity: review of traditional and molecular techniques. Mycology 3:165–176. https://doi.org/10.1080/21501203.2012.656724
4. Reiter B, Wermbter N, Gyamfi S, Schwab H, Sessitsch A (2003) Endophytic *Pseudomonas* spp. populations of pathogen-infected potato plants analysed by 16S rDNA- and 16S rRNA-based denaturing gradient gel electrophoresis. *Plant Soil* 257:397–405
5. Thies JE (2007) Soil microbial community analysis using terminal restriction fragment length polymorphisms. Soil Sci Soc Am J 71(2):579–591. https://doi.org/10.2136/sssaj2006.0318
6. Felske A, Wolterink A, Van Lis R, Akkermans AD (1998) Phylogeny of the main bacterial 16S rRNA sequences in Drentse a grassland soils (The Netherlands). Appl Environ Microbiol 64:871–879
7. Schmalenberger A, Tebbe CC (2003) Bacterial diversity in maize rhizospheres: conclusions on the use of genetic profiles based on PCR-amplified partial small subunit rRNA genes in ecological studies. Mol Ecol 12:251–261
8. Araujo WL, Marcon J, Maccheroni W Jr, Elsas JD, VanVuurde JW, Azevedo JL (2002) Diversity of endophytic bacterial populations and their interaction with Xylellafastidiosa in citrus plants. Appl Environ Microbiol 68:4906–4914
9. Garbeva P, van Overbeek LS, van Vuurde JWL, van Elsas JD (2001) Analysis of endophytic bacterial communities of potato by plating and denaturing gradient gel electrophoresis (DGGE) of 16S rDNA based PCR fragments. Microb Ecol 413:69–383
10. Fajardo V, González I, Dooley J et al (2009) Application of polymerase chain reactionrestriction fragment length polymorphism analysis and lab-on-a-chip capillary electrophoresis for the specific identification of game and domestic meats. J Sci Food Agric 89(5): 843–847
11. Rojas M, González I, Fajardo V et al (2009) Identification of raw and heat-processed meats from game bird species by polymerase chain reaction-restriction fragment length polymorphism of the mitochondrial D-loop region. Poult Sci 8(3):669–679. ISSN 0032-5791
12. Liu WT, Marsh TL, Cheng H et al (1997) Characterization of microbial diversity by determining terminal restriction fragment length polymorphisms of genes encoding 16S rRNA. Appl Environ Microbiol 63(11): 4516–4522. ISSN 0099-2240
13. Engebretson JJ, Moyer CL (2003) Fidelity of select restriction endonucleases in determining microbial diversity by terminal-restriction

fragment length polymorphism. Appl Environ Microbiol 69:4823–4829

14. Sessitsch A, Reiter B, Pfeifer U, Wilhelm E (2002) Cultivation-independent population analysis of bacterial endophytes in three potato varieties based on eubacterial and Actinomycetes-specific PCR of 16S rRNA genes. FEMS Microbiol Ecol 39:23–32
15. Conn VM, Franco CMM (2004) Analysis of the endophyticactinobacterial population in the roots of wheat (Triticumaestivum L.) by terminal restriction fragment length polymorphism and sequencing of 16S rRNA clones. Appl Environ Microbiol 70:1787–1794
16. Akkermans ADL, van Elsas JD, de Bruijn JJE (2001) Molecular microbial ecology manual. Kluwer, Dordrecht
17. Bock CH, Thrall PH, Brubaker CL, Burdon JJ (2002) Detection of genetic variation in*Alternariabrassicicola* using AFLP fingerprinting. Mycol Res 106:428–434. https://doi.org/10.1017/S0953756202005762
18. Klamer M, Roberts MS, Levine LH, Drake BG, Garland JL (2002) Influence of elevated CO2 on the fungal community in a coastal scrub oak forest soil investigated with terminal restriction fragment length polymorphism analysis. Appl Environ Microbiol 68:4370–4376. https://doi.org/10.1128/AEM.68.9.4370-4376.2002
19. Anderson IC, Campbell CD, Prosser JI (2003) Diversity of fungi in organic soils under a moorland – scots pine (*Pinussylvestris* L.) gradient. Environ Microbiol 5:1121–1132. https://doi.org/10.1046/j.1462-2920.2003.00522.x
20. Anderson IC, Cairney JWG (2004) Diversity and ecology of soil fungal communities: increased understanding through the application of molecular techniques. Environ Microbiol 6:769–779. https://doi.org/10.1111/j.1462-2920.2004.00675.x
21. Huai WX, Guo LD, He W (2003) Genetic diversity of an ectomycorrhizal fungus *Tricholomaterreumin Larixprincipis-rupprechtii* stand assessed using RAPDs. Mycorrhiza 13:265–270. https://doi.org/10.1007/s00572-003-0227-8
22. Jansa J, Mozafar A, Kuhn G, Anken T, Ruh R, Sanders IR, Frossard E (2003) Soil tillage affects the community structure of mycorrhizal fungi in maize roots. Ecol Appl 13:1164–1176
23. Liang Y, Guo LD, Ma KP (2004) Genetic structure of a population of the ectomycorrhizal fungus *Russulavinosa* in subtropical woodlands in Southwest China. Mycorrhiza 14:235–240. https://doi.org/10.1007/s00572-003-0260-7
24. Liang Y, Guo LD, Ma KP (2005) Population genetic structure of an ectomycorrhizal fungus *Amanita manginiana* in a subtropical forest over 2 years. Mycorrhiza 15:137–142. https://doi.org/10.1007/s00572-004-0311-8
25. Raskin L, Stromley J, Rittmann B, Stahl D (1994) Group-specific 16S ribosomal-RNA hybridization probes to describe natural communities of methanogens. Appl Environ Microbiol 60(4):1232–1240
26. Jurgens G, Lindstrom K, Saano A (1997) Novel group within the kingdom Crenarchaeota from boreal forest soil. Appl Environ Microbiol 63(2):803–805
27. Gomes N, Heuer H, Schonfeld J, Costa R, Mendonca-Hagler L, Smalla K (2001) Bacterial diversity of the rhizosphere of maize (*Zea mays*) grown in tropical soil studied by temperature gradient gel electrophoresis. Plant Soil 232(1–2):167–180. https://doi.org/10.1023/A:1010350406708
28. Heuer H, Krsek M, Baker P, Smalla K, Wellington EMH (1997) Analysis of actinomycete communities by specific amplification of genes encoding 16S rRNA and gel electrophoretic separation in denaturing gradients. Appl Environ Microbiol 63:3233–3241
29. Gul F, Hussain A, Jan J, Hamayun M (2017) Genomic DNA extraction for molecular identification of endophytic fungi: an easy and efficient protocol. Biosci Biotechnol Res Asia 14(2):2. https://doi.org/10.13005/bbra/2492
30. Schulz BJE, Boyle CJC, Sieber TN (2006) Microbial root endophytes, Soil biology, vol 9. Springer, Berlin/Heidelberg. https://doi.org/10.1007/3-540-33526-9
31. White T, Bruns T, Lee S, Taylor J (1990) Amplification and direct sequencing of fungal ribosomal RNA genes for phylogenetics. In: Innis MA, Gelfand DH, Sninsky JJ, White TJ (eds) PCR protocols: a guide to methods and applications. Academic Press, San Diego, pp 315–322
32. Weisburg W, Barns S, Pelletier D, Lane D (1991) 16S ribosomal DNA amplification for phylogenetic study. J Bacteriol 173(2):697–703
33. Cocolin L, Bisson L, Mills D (2000) Direct profiling of the yeast dynamics in wine fermentations. FEMS Microbiol Let 189(1):81–87. https://doi.org/10.1016/S0378-1097(00)00257-3

34. Garbeva P, van Veen JA, van Elsas JD (2004) Assessment of the diversity, and antagonism towards *Rhizoctonia solani*AG3, of *Pseudomonas* species in soil from different agricultural regimes. FEMS Microbiol Ecol 47:51–64
35. Salles JF, De Souza FA, van Elsas JD (2002) Molecular method to assess the diversity of *Burkholderia*species in environmental samples. Appl Environ Microbiol 68:1595–1603
36. Garbeva P, van Veen JA, van Elsas JD (2003) Predominant *Bacillus* spp. in agricultural soilunder different management regimes detected via PCR-DGGE. Microb Ecol 45:302–316
37. De Hoog GS, Gerrits Van Den Ende AHG (1998) Molecular diagnostics of clinical strains of filamentous Basidiomycetes. Mycoses 189: 183–189
38. Vilgalys R, Hester M (1990) Rapid genetic identification and mapping of enzymatically amplified ribosomal DNA from several Cryptococcus species. J Bacteriol 172:4238–4246
39. Templeton MD, Rikkerink EHA, Solon SL, Crowhurst RN (1992) Cloning and molecular characterization of the glyceraldehyde3-phosphate dehydrogenase encoding gene and cDNA from the plant pathogenic fungus *Glomerellacingulata*. Gene 122:225–230
40. O'Donnell K, Cigelnik E (1997) Two divergent intragenomic rDNA ITS2 types within a monophyletic lineage of the fungus *Fusarium* are nonorthologous. Mol Phylogenet Evol 7: 103–116
41. Glass NL, Donaldson GC (1995) Development of primer sets designed for use with the PCR to amplify conserved genes from filamentous ascomycetes. Appl Environ Microbiol 61: 1323–1330
42. Carbone I, Kohn LM (1999) A method for designing primer sets for speciation studies in filamentous ascomycetes. Mycologia 91:553–556
43. Katoh K, Standley DM (2016) A simple method to control over-alignment in the MAFFT multiple sequence alignment program. Bioinformatics:btw108. https://doi.org/10.1093/bioinformatics/btw108
44. Hall T (2004) Bioedit version 6.0.7
45. Hiraishi HT, Razandi A, Sugimoto M, Harada T, Ivey T, K.J. (1992) Role of cellular superoxide dismutase against reactive oxygen metabolite injury in cultured bovine aortic endothelial cells. J Biol Chem 267(21): 14812–14817. https://doi.org/10.1016/S0021-9258(18)42112-6
46. Fernandes EG, Pereira OL, Silva CC, Bento CBP, Queiroz MV (2015) Diversity of endophytic fungi in *Glycine max*. Microbiol Res 181:84–92. https://doi.org/10.1016/j.micres.2015.05.010
47. Vainio EJ, Hantula J (2000) Direct analysis of wood-inhabiting fungi using denaturing gradient gel electrophoresis of amplified ribosomal DNA. Mycol Res 104(8):927–936. https://doi.org/10.1017/S0953756200002471
48. Rohlf FJ, NTSYS-pc (1993) Numerical taxonomy and multivariate analysis system version 1.80. Exter software. Setauket, New York
49. Rivas R, Peix A, Mateos PF, Trujillo ME, Martinez-Molina E, Velazquez E (2006) Biodiversity of populations of phosphate solubilizing rhizobia that nodulates chickpea in different Spanish soils. Plant Soil 287(1–2): 23–33. https://doi.org/10.1007/s11104-006-9062-y
50. Bougoure JJ, Bougoure DS, Cairney JWG, Dearnaley JDW (2005) ITS-RFLP and sequence analysis of endophytes from *Acianthus, Caladenia* and *Pterostylis* (Orchidaceae) in South-Eastern Queensland. Mycol Res 109(4):452–460. https://doi.org/10.1017/S095375620500225X
51. Rademaker JLW, Louws FJ, Bruijn FJ (1998) Characterisation of diversity of ecologically important microbes by rep-PCR fingerprinting. In: Akkermans ADL, Van Elsas JD, Bruijn FJ (eds) Molecular microbial ecology manual, supplement, 3rd edn. Kluwer Academic Publishers, Dordrecht, pp 1–26
52. Sneath PHA, Sokal RR (1973) Numerical taxonomy: the principles and practice of numerical classification. W. H. Freeman and Company, San Francisco
53. Lopez AC, Alippi AM (2009) Diversity of Bacillus megaterium isolates cultured from honeys. LWT Food Sci Technol 42:212–219. https://doi.org/10.1016/j.lwt.2008.05.001
54. Blignaut M, Ellis AG, Le Roux JJ (2013) Towards a transferable and cost-effective plant AFLP protocol. PLoS One 8(4):e61704. https://doi.org/10.1371/journal.pone.0061704
55. Peakall R, Smouse PE (2006) GENALEX 6: genetic analysis in Excel. Population genetic software for teaching and research. Mol Ecol Notes 6:288–295. https://doi.org/10.1111/j.1471-8286.2005.01155.x

Chapter 14

Metagenomics and Pyrosequencing Approach

Jean Carlos Vencioneck Dutra and Maria do Carmo Pimentel Batitucci

Abstract

Phytopathogens control is necessary for crop biosafety and food security. The control of plant pathogens interferes with agriculture, and the development of molecular analysis has contributed to the detection and identification of phytopathogenic microorganisms. In this context, the use of platforms capable to provide quick and accurate diagnosis helps to reveal the relationships on the pathogen biome and to provide data for disease prediction models and strategies to combat these agents. Among the available methods for metagenomic studies, pyrosequencing has proved to be effective for studies with endophytic microorganisms. On this hand, this chapter is aimed to present the main steps to develop pyrosequencing and metagenomics techniques applied to investigations of endophytic microorganisms.

Key words Endophytic microorganisms, Phytopathogens, Metagenomics, Pyrosequencing, Molecular tools

1 Introduction

Crop biosafety and food security for the population involves controlling phytopathogens and, consequently, the progression of plant diseases. The control of these plant pathogens positively interferes with agriculture, and the advancement of molecular methods applied to studies of phytopathology and microbiology has increased the possibility of more sensitive, rapid detection and identification of them. Thus, conditions that favor the management of different cultures and the control or detection of pathogens reflect on the well-being of the population and in the consequent increase of plant consumption [1].

In view of all losses caused by pathogenic organisms in plant crops, there is a need to develop platforms capable of ensuring a quick and accurate diagnosis of the pathogenic agents, as well as, providing tools to elucidate microbial profiles, reveal the relationships of the pathogen biome and that allow the development of disease prediction models and strategies to combat them.

A. Sankaranarayanan et al. (eds.), *Endophytic Microbes: Isolation, Identification, and Bioactive Potentials*, Springer Protocols Handbooks, https://doi.org/10.1007/978-1-0716-2827-0_14,

In this context, bioinformatics and metagenomics are fields that have provided faster and highly sensitive protocols when compared to conventional methods of microbiological analysis. Research in phytopathology and microbiology tends to involve approaches focused on elucidating the growth dynamics of the microbial biome instead of studies with isolated or closely related organisms, generating advances in the number and quality of information about the microbial ecosystem in natural environments [1].

Among the various methods available for metagenomic studies are techniques that range from conventional PCR to pyrosequencing, which provide a fast and safe detection and delineation of profiles of the biomes composed by the microorganisms of interest.

The use of molecular techniques has some advantages for conducting studies with endophytic microorganisms. Most of these techniques are independent of the cultivation of microorganisms and are high-yielding techniques, thus enabling the study of a biome as a whole and not just an individual or a related group [2]. In this way, such molecular technologies have been used to detect phytopathogens [3] and to establish phytobiomass by fingerprinting or sequencing [4]. However, it is also necessary to consider that even with the great advancement of these tools, we still need to evaluate the efficiency and application of each one of these techniques.

This chapter intends to present the main steps for the development of pyrosequencing and metagenomics techniques applied to investigations of endophytic microorganisms, presenting a critical analysis on the application of these discussed methods.

Pyrosequencing

Among techniques based on the analysis of genetic material, pyrosequencing shows to be a great and promising technology. Pyrosequencing technology is based on the principle of sequencing by synthesis and is a technology capable of providing sequences related to quantitative data in a short period of time.

The use of this technology in the metagenomic evaluation enables the in-depth study of the natural microbiome and the dynamics of microbial interactions of the plant, since this technology can be used for the detection of single nucleotide polymorphisms (SNP), genetic quantification, sequencing of short DNA or even in the quantification of CpG sites.

In the fields of biology and agronomy, it is clear that DNA sequencing favored the most diverse studies related to the genomes of plants and microorganisms related to them.

Historically, sequencing stood out in Science from the Sanger method, which is still widely used, but as genomic studies have expanded, new methods with different principles have been developed, such as pyrosequencing, the subject of our discussion.

Pyrosequencing has emerged as the first DNA sequencing method alternative to the conventional Sanger methodology [5]. This method is based on the sequencing-by-synthesis principle and includes the use of a series of four enzymes to detect nucleic acid sequences. In pyrosequencing, the detection occurs during the synthesis of nucleic acid sequences and has been a widely used technique due to accuracy, flexibility, parallel processing, and is easily automated [6].

454-Pyrosequencing method allows the evaluation of hypervariable regions of bacterial 16S rRNA and variable regions ITS1/ITS4 or D1/D2 that are targeted to fungal populations [7, 8]. Among sequencing-by-synthesis technologies, this pyrosequencing method stands out as a high-throughput metagenomic method. For example, the application of 454-pyrosequencing has allowed to reveal microbial populations in soil samples, water, controllable vegetable carpoplanes and fruits.

2 Materials

Tissue Surface Sterilization

- Plant material.
- Sterile deionized water (washing purpose).
- Sterile filter paper (for partial removal of moisture).
- Sterile Petri dish.
- Sterile blade (for sample segmentation).
- 70% EtOH.
- 95% EtOH.
- 0.525% NaOCl.

CTAB Storage

- Sample pieces.
- Microtubes (2 mL).
- 1 M Tris-HCl pH 8.
- 5 M NaCl.
- 0.5 M EDTA.
- 20 g CTAB.

Silica Gel Storage

- Sample pieces.
- Sterile Petri dish.
- Silica gel.

Tissue Culture

- Sample pieces.
- 2% malt extract agar (MEA).
- Tryptic soy agar (TSA).

2.1 Estimation of DNA

Real-time PCR Method

- Microtubes (0.2 mL).
- Ultra pure water.
- 96 wells real-time PCR microplate
- Real-time PCR (machine).

NanoDropmethod

- Elution buffer.
- NanoDropfluorospectrometer.

2.2 PCR Amplification and Metabarcoding

- Specific primers (forward and reverse).
- Multiplex Identifier (MID) primers.
- Thermal Cycler.
- Agarose.

2.3 OTU-Based Sequence Analysis

- CLOTU app (available in https://www.bio-portal.uio.no/appinfo/show.php?app=clotu).
- BLAST (available inhttps://blast.ncbi.nlm.nih.gov/Blast.cgi).
- BLASTclust (available inhttps://www.ncbi.nlm.nih.gov/Web/Newsltr/Spring04/blastlab.html).
- GenBank (available in https://www.ncbi.nlm.nih.gov/genbank/).

3 Methods

Tissue Surface Sterilization

1. Collect plant material in a sterile plastic bag, transfer to the lab laboratory and processed within 24 h of collection.
2. Thoroughly wash the collected plant material sample with sterile deionized water to remove dirt (repeat this step 2 times).
3. In a sterile Petri dish, cut the sample into approximately 2 mm^2 segments (use a sterile blade) prior to sterilization.
4. To sterilize the surface of the tissues, proceed to the immersion and agitation series in 95% EtOH for 10s, 0.525% NaOCl for 2 min and 70% EtOH for 2 min.
5. Place the sample on sterile filter paper for partial removal of moisture.

6. For fresh tissue analysis, segmented samples must be processed immediately for DNA extraction.

The tissue sample can be stored or cultured before proceeding to DNA extraction (methods described below).

CTAB Storage

1. As established by Arnold et al. [9] and Higgins et al. [8], prepare a CTAB solution using 1 M Tris-HCl pH 8, 5 M NaCl, 0.5 M EDTA, and 20 g CTAB.
2. After the tissue surface sterilization step, partition a set of 24 sample pieces and store the set into a microtube (2 mL) with 750 μL CTAB.
3. Maintain the samples under these conditions until the DNA extraction.

Silica Gel Storage

1. After the tissue surface sterilization step, the tissue sample should be kept in sterile Petri dishes with silica gel.
2. Maintain the samples under these conditions until the DNA extraction.

Tissue Culture

1. In sterile Petri dishes, pre-add tryptic soy agar (TSA) culture medium for bacterial growth, or 2% malt extract agar (MEA) for fungi growth.
2. After the tissue surface sterilization step, add the sample pieces in sterile Petri dishes with culture medium.
3. Seal the Petri dishes with parafilm and incubate at 28 °C for bacterial growth or at room temperature for fungal growth.
4. Maintain samples incubated until microorganism growth is observed and proceed to DNA extraction.

3.1 DNA Extraction

1. Use standard methodology for extracting DNA from samples [7, 10].
2. After extraction, the genetic material is submitted to a DNA purification system (when using kits, follow the manufacturer's specifications).
3. Store the extracted DNA at −20°Cuntil the moment of analysis.

3.2 Estimation of DNA

Real-Time PCR Method

1. In a 0.2 mL microtube, dilute the total DNA 1:10 using ultra pure water.
2. Transfer 10uL of the diluted sample to the well of a real-time PCR microplate (perform the experiment in duplicate or triplicate).

3. In a real-time PCR, read the samples and take the cycle threshold (Ct) values.
4. Calculate the amount of DNA from the taken Ct values, using the standard curves [10].

NanoDropmethod

1. Dilute the total DNA in 100 μL of elution buffer (reserve the sample dilution factor).
2. In a Nanodropfluorospectrometer, read the samples using the 260 nm wavelength and take the absorbance values.
3. Calculate the amount of DNA from the taken absorbance values, using the followed formula [7, 11]:

$$\text{DNA concentration } (\mu g/\mu L) = A_{260nm} \times 0.05 \times \text{dilution factor}$$

where A_{260nm} corresponds to the absorbance of the sample.

3.3 PCR Amplification and Metabarcoding

1. To perform the 454-Pyrosequencing, in a Thermal Cycler, DNA sample should be subjected to the PCR with specific primers (forward and reverse) to generate amplicons*.
2. After clustering, to identify sample molecular specificities, all amplicons obtained and Multiplex Identifier (MID) primers**-must be conducted on a Thermal Cycler***.
3. Following DNA quantification, submit the PCR product to electrophoresis in 1.5% agarose gel****.
4. Purify the pool of amplicons (a gel extraction kit can be used according to the manufacturer's instructions) and the pools generated must be sequenced.

*Amplicons must correspond to the internal transcribed spacers and 5.8S region (ITS rDNA) or fragments of 16S rRNA and nifH genes.

**Select about 30 primer markers with 10 nucleotide MID primer. It can be randomly performed from a list of recommended MID primer markers and the primers should be synthesized to HPLC specification to ensure the quality of results.

***Specifications used in amplification steps, such as temperature and time for DNA denaturation, number of cycles, and final elongation vary according to the investigated sample.

****Labeled PCR amplicons should be pooled in approximately the same amounts and bulk samples should be randomized against the MID sequencing set and primer tags.

3.4 OTU-Based Sequence Analysis

1. Using the CLOTU app, perform filtering, grouping, and BLAST searches of string reads.
2. Discard the readings that are less than 150 bp and readings where the primers and the tag cannot be identified.
3. After discarding, group the remaining readings using BLASTclust (apply 97% similarity and 90% coverage).
4. Subsequently, omit the singletons from the dataset.
5. Construct an OTU table based on the BLASTclust analysis (it must contain the abundance of readings in each OTU of each sample).
6. Normalize the abundance data with the total DNA in each sample as quantified by real-time PCR.
7. After data normalization, unless otherwise indicated, reserve these data for the subsequent analyses.
8. Use a set of readings selected randomly from each OTU to consult manually GenBank for BLAST searches.
9. Using software for scientific data analysis, calculate rarefaction, species accumulation curves, and the non-parametric diversity estimator α Chao1.
10. Finally, deposit the sequence files and metadata in the NCBI sequence reading file.

4 Note

In pyrosequencing, intrinsic biases and insufficient quality control may compromise metagenomic analysis and lead to difficult to detect errors in the estimation of microbial community richness, distribution, and composition.

Tissue storage methods commonly used can be suitable for recovering high-quality plant or mycobiont DNA. However, some techniques can compromise the quality of the pyrosequencing amplicon. For minute fraction of symbiotrophic fungal DNA, silica gel desiccation may not effectively and downstream PCR amplification and pyrosequencing. Furthermore, to verify low-abundance pyrosequencing reads, it is recommended to use genetic material obtained from pure culture, for taxonomic inferences and downstream phylogenetic studies, the full-length ITS rDNA sequences and multilocus sequences.

References

1. Jongman M, Carmichael PC, Bill M (2020) Technological advances in phytopathogen detection and metagenome profiling techniques. Curr Microbiol 77(4):675–681
2. Jongman M, Chidamba L, Korsten L (2017) Bacterial biomes and potential human pathogens in irrigation water and leafy greens from different production systems described using pyrosequencing. J Appl Microbiol 123(4): 1043–1053
3. Carmichael PC, Siyoum N, Jongman M, Korsten L (2018) Prevalence of Botrytis cinerea at different phenological stages of table grapes grown in the northern region of South Africa. Sci Hortic (Amsterdam) 239:57–63
4. Gorni C, Allemand D, Rossi D, Mariani P (2015) Microbiome profiling in fresh-cut products. Trends Food Sci Technol 46(2): 295–301
5. Ronaghi M (2001) Pyrosequencing sheds light on DNA sequencing. Genome Res 11(1):3–11
6. Fakruddin M, Chowdhury A, Hossain MN, Mannan KS, Mazumda RM (2012) Pyrosequencing-principles and applications. Int J Life Sci Pharma Res 2(1):L–65
7. U'Ren JM, Riddle JM, Monacell JT, Carbone I, Miadlikowska J, Arnold AE (2014) Tissue storage and primer selection influence pyrosequencing-based inferences of diversity and community composition of endolichenic and endophytic fungi. Mol Ecol Resour 14(5):1032–1048
8. Higgins KL, Coley PD, Kursar TA, Arnold AE (2011) Culturing and direct PCR suggest prevalent host generalism among diverse fungal endophytes of tropical forest grasses. Mycologia 103(2):247–260
9. Arnold AE, Henk DA, Eells RL, Lutzoni F, Vilgalys R (2007) Diversity and phylogenetic affinities of foliar fungal endophytes in loblolly pine inferred by culturing and environmental PCR. Mycologia 99(2):185–206
10. Martellacci L, Quaranta G, Patini R, Isola G, Gallenzi P, Masucci L (2019) A literature review of metagenomics and culturomics of the peri-implant microbiome: current evidence and future perspectives. Materials (Basel) 12(18):3010
11. Li O, Xiao R, Sun L, Guan C, Kong D, Hu X (2017) Bacterial and diazotrophic diversities of endophytes in *Dendrobium catenatum* determined through barcoded pyrosequencing. PLoS One 12(9):e0184717

Part IV

Methods to Proof Endophytism

Chapter 15

Cultivation-Based Studies of Endophytic Fungi in Plants

Lizzy A. Mwamburi

Abstract

Fungal endophytes live within plants without any symptoms and confer several advantages to the host that may be exploited for use in agriculture and medicine. Studies of the species population dynamics, diversity, and relationship of endophytic fungi and host plants have gained popularity in recent decades. Investigations on the species ecology and diversity of fungal endophytes have been mainly approached by cultivation-based methods. This method is effective, revealing a great number of endophytic fungal species richness and diversity from host plants. The current chapter explores the cultivation-based technique as one of the ways of studying endophytic fungi from plant tissues.

Key words Endophytes, Fungi, Plant tissues, Cultivation, Rapid recovery, Species diversity

1 Introduction

Fungal endophytes are plant mutualists that colonize plant healthy tissues of almost all plants and thrive asymptomatically in different healthy tissues of living plants [1] above or under the ground, including leaves, stems, and roots. Each plant species may host one or more endophytic fungi species. Faeth and Fagan [2] estimated that approximately 1 million endophytic fungal species occur in nature.

The colonization of endophytes in plants offers several benefits to the host including enhanced growth [3, 4] and competitiveness of the host plants against herbivores [5–8] and disease-causing pathogens [9]. They also increase the adaptability and tolerance of the host plants against abiotic stresses such as drought [10, 11], nutrient deficiency [12, 13], heavy metals [14], and salinity [15, 16] in exchange for shelter and nutrition from host plants [17]. Furthermore, some endophytic fungi have been reported to promote the accumulation of secondary metabolites [18] that protect the host plant against epiphytic organisms and may additionally be exploited for agricultural and medicinal purposes.

A. Sankaranarayanan et al. (eds.), *Endophytic Microbes: Isolation, Identification, and Bioactive Potentials*, Springer Protocols Handbooks, https://doi.org/10.1007/978-1-0716-2827-0_15,

It is therefore necessary to isolate endophytic fungi for purposes of morphological characterization, studies of species population diversity, and dynamics of economically important species [19, 20] and for screening secondary metabolites for application in agriculture and medicine.

Different techniques have been used to study and identify endophytes including staining techniques [21], cultivation-based techniques [22, 23], cultivation-independent or molecular identification methods [24]. Despite advances in molecular identification of endophytic fungi, the cultivation-dependent method still remains justifiable for the classification of taxonomic units, since a large number of isolates often remain sterile until the end of the study [25, 26].

The cultivation-dependent technique is a method-dependent process that is effective for the rapid recovery of a large number of endophytic fungal species from plant tissues. This method is most commonly applied, as it is easy and simple, and yields large fungal diversity [27]. The technique involves a series of methods where the living plant tissues are subjected to surface sterilization of the plant, isolation of endophytic fungi from plant samples placed on agar, purification and sporulation of endophytic isolates under various incubation conditions, and identification of the endophytic fungi based on morphological characteristics in cultures [28–30]. Since surface sterilization may affect the fungal endophyte population dynamics, specific isolation procedures need to be adapted for different plant species. Additionally, total surface sterilization should be confirmed by no fungal growth from cultures obtained by imprinting the surface-sterilized plant tissues or aliquots of water from the last rinsing. Finally, the sporulating and non-sporulating structures are morphologically identified. Isolation of endophytic fungi in plant tissues is a critical process, and specific incubation conditions depending on plant tissue are required [31]. Isolated fungal endophytes may then be further used to extract DNA and other molecular techniques for further analysis.

Different types of media should be used in the isolation procedure, including malt extract agar (MEA), potato dextrose agar (PDA), and Sabouraud agar (SA) in order to promote isolate sporulation of different endophytes. Further inclusion of sterile host tissue in cultures has also been reported to increase the number of sporulating isolates [32]. As a major limitation of cultivation-dependent studies, analysis of fungal diversity is often biased toward fast-growing, ubiquitous species, whereas rare species with minor competitive strength and specialized requirements may remain undiscovered. They might lose in competitive interactions while growing toward the agar and during the incubation process, unless different media are used [33] and different detection methods are applied [34, 35].

The isolated strains may then be used for morphotype, morphological identification, or DNA sequencing [36, 37].

This chapter describes the cultivation-dependent technique also referred to as the fragment plating method used to identify endophytes from plant tissues. It requires a series of procedures, and morphological identification can be easily done.

2 Materials

2.1 Collection of Plant Material

- Whole plants at different stages of growth.
- Sterile polythene bags (for sample collection).
- Refrigerator (4 °C) for storage.

2.2 Surface Sterilization of Plant Material

- Running tap water.
- Ethanol (70%).
- Sodium hypochlorite (4% active chlorine).
- Sterile demineralized water.
- Filter paper.
- Spatula.
- Measuring cylinder.
- 1 L conical flasks for media preparation.
- Biomalt agar (Biomalt, 120 g/L BactoTM Agar, pH 5.6 with the antibiotics Penicillin G Na 60 mg/L, Streptomycin sulfate 80 mg/L, and (Oxy)-tetracycline-HCl 50 mg/L) for sterility check.
- pH meter.
- Sterile Petri dishes.
- Sterile blades.
- Sterile plastic bags (for storage).
- 500 mL Erlenmeyer flask.
- Forceps.
- Alcohol lamp.

2.3 Endophytic Fungi Isolation and Cultivation

- Aqueous agar (distilled water and 1.5% agar-agar).
- Surface sterilized stem, root, and leaf sections.
- Measuring cylinder.
- 1 L conical flasks for media preparation.
- Spatula.
- Streptomycin (3 mg/100 mL).
- Sterile Petri dishes,

- Incubator.
- Mycological medium, (malt extract agar (MEA), potato dextrose agar (PDA), or Sabouraud agar (SA)).
- Chloramphe.nicol (0.2 g/L).
- pH meter.
- Biomalt agar.
- Sterile slant tubes.
- For description of morphological characteristics refer to fungal identification keys.

3 Materials and Methods

3.1 Collection of Plant Material

1. Collect plant samples at different growth stages.
2. Collect five to seven samples of plants with and their roots and adhering soil particles.
3. Store plants at 4 °C until later.

3.2 Surface Sterilization

1. Inspect the plants macroscopically and discard areas with discolorations and lesions so that only healthy plant material is used.
2. Wash plants should be surface-sterilized by first thoroughly washing under running tap water to remove soil particles and the surface epiphytes [38] and then immersed in ethanol (70%) for 1 min, followed by soaking in 4.5 min sodium hypochlorite (4% active chlorine) followed by three washings for at least 5 min with sterile demineralized water [39] and then drying on a sterile filter paper.
3. Excise leaves, stem, and roots of plants using a sterile blade.
4. Gently press the leaf, stem, and root samples on filter paper imprinted with Biomalt agar 50 g/L (Biomalt, 120 g/L BactoTM Agar, pH 5.6 with the antibiotics Penicillin G Na 60 mg/L, Streptomycin sulfate 80 mg/L, and (Oxy)-tetracycline-HCl 50 mg/L) as a sterility check [40]. These cultures serve as control plates to check the surface sterilization procedure.
5. After the plant parts impression remove the sections carefully with the help of spatula from the surface of the medium and discard it.
6. Plate the final sterile water rinse on the same medium and observe post-incubation period.
7. Place in separate labeled sterile plastic bags for storage at 4 °C until use.

3.3 Isolation and Cultivation of Endophytic Fungi

1. Cut plants into small pieces using a sterile blade.

   (Note: Tissue fragments should be small and thin as colonization may be sparse [41]).

2. Place plant pieces on aqueous agar (distilled water and 1.5% agar-agar) supplemented with antibiotic streptomycin (3 mg/100 mL) in Petri plates and incubated at 28 ± 2 °C until fungal growth is initiated. Surface-disinfected, and non-disinfected pieces that constitute the controls should also be placed on the same agar to check for fungi.

   (Note: The selected medium is supplemented with antibiotics to prevent any unwanted microbial growth. Antifungal substances may also be added to delay the development of fast-growing fungal contaminants [40]).

3. Remove the tips of the fungal hyphae from the aqueous agar (1–2 mm from the edge) and plate on mycological medium, potato dextrose agar (PDA) or the Sabouraud agar (SA) supplemented with chloramphenicol (0.2 g/L), pH of 7.8 and incubate for 7 days at 27 °C.

   (Note: If the endophytic fungi are slow growing then it is important to include a soaking process [42] to ensure that the endophytic invaders and colonizers of the root cortex or vascular tissue, both intercellular intracellular are trapped).

4. After incubation, the purity of each fungal culture is assessed by examination of colony morphology.
5. Monitor each culture for up to 6 weeks for the outgrowth of mycelium, and then immediately inoculate on fresh biomalt medium [40].
6. After purifying the isolates several times as described in step 4 above, transfer the final pure cultures to PDA slant tubes.
7. Group the pure cultures according to colony morphology and characteristics.
8. Use culture morphological characteristics to identify fungi.

References

1. Bacon CW, White J (2000) Microbial endophytes. CRC, New York
2. Faeth SH, Fagan WF (2002) Fungal endophytes: common host plant symbionts but uncommon mutualists. Integr Comp Biol 42: 360–368
3. White JF, Kingsley KL, Zhang Q, Verma R, Obi N, Dvinskikh S et al (2019) Review: endophytic microbes and their potential applications in crop management. Pest Manag Sci 75:2558–2565
4. Xia Y, Sahib MR, Amna A, Opiyo SO, Zhao Z, Gao YG (2019) Culturable endophytic fungal communities associated with plants in organic and conventional farming systems and their effects on plant growth. Sci Rep 9:1669
5. Vega FE (2008) Insect pathology and fungal endophytes. J Invertebr Pathol 98:277–279
6. Estrada C, Wcislo WT, Van Bael S (2013) Symbiotic fungi alter plant chemistry that discourages leaf-cutting ants. New Phytol 198: 241–251

7. Vidal S, Jaber LR (2015) Entomopathogenic fungi as endophytes: plant-endophyte-herbivore interactions and prospects for use in biological control. Curr Sci 109:46–54
8. Mwamburi LA (2021) Endophytic fungi, *Beauveria bassiana* and*Metarhizium anisopliae*, confer control of the fall armyworm, *Spodoptera frugiperda* (J. E. Smith) (Lepidoptera: Noctuidae), in two tomato varieties. Egypt J Biol Pest Co:31:7
9. Bae H, Roberts DP, Lim H-S, Strem MD, Park S-C, Ryu C-M et al (2011) Endophytic *Trichoderma* isolates from tropical environments delay disease onset and induce resistance against *Phytophthora capsici* in hot pepper using multiple mechanisms. Mol Plant-Microbe Interact 24:336–351
10. Ghaffari MR, Mirzaei M, Ghabooli M, Khatabi B, Wu Y, Zabet-Moghaddam M et al (2019) Root endophytic fungus *Piriformospora indica* improves drought stress adaptation in barley by metabolic and proteomic reprogramming. Environ Exp Bot 157:197–210
11. Qiang X, Ding J, Lin W, Li Q, Xu C, Zheng Q et al (2019) Alleviation of the detrimental effect of water deficit on wheat (*Triticum aestivum* L.) growth by an indole acetic acid-producing endophytic fungus. Plant Soil 439: 73–391
12. Bajaj R, Huang Y, Gebrechristos S, Mikolajczyk B, Brown H, Prasad R et al (2018) Transcriptional responses of soybean roots to colonization with the root endophytic fungus *Piriformospora indica* reveals altered phenylpropanoid and secondary metabolism. Sci Rep 8:10227
13. Aslam MM, Karanja J, Bello SK (2019) *Piriformospora indica* colonization reprograms plants to improved P-uptake, enhanced crop performance, and biotic/abiotic stress tolerance. Physiol Mol Plant Pathol 106:232–237
14. Bilal S, Shahzad R, Imran M, Jan R, Kim KM, Lee IJ (2020) Synergistic association of endophytic fungi enhances *Glycine max* L. resilience to combined abiotic stresses: heavy metals, high temperature and drought stress. Ind Crop Prod 143(111931)
15. Manasa KM, Vasanthakumari MM, Nataraja KN, Uma SR (2020) Endophytic fungi of salt adapted *Ipomeapes-caprae* L. R. Br: their possible role in inducing salinity tolerance in paddy (*Oryza sativa* L.). Curr Sci 118:1448–1453
16. Sampangi-Ramaiah MH, Jagadheesh DP, Jambagi S, Vasantha KMM, Oelmüller R et al (2020) An endophyte from salt-adapted Pokkali rice confers salt-tolerance to a salt-sensitive rice variety and targets a unique pattern of genes in its new host. Sci Rep 10:3237
17. Saikkonen K, Wäli P, Helander M, Faeth SH (2004) Evolution of endophyte–plant symbioses. Trends Plant Sci 9:275–280
18. Jia M, Chen L, Xin H-L, Zheng C-J, Rahman K, Han T, Qin L-P (2016) A friendly relationship between endophytic fungi and medicinal plants: a systematic review. Front Microbiol 7:906
19. Li J, Li L, Si YK, Jiang XJ, Guo LD, Che YS (2011) Virgatolides A–C, benzannulated spiroketals from the plant endophytic fungus *Pestalotiopsis virgatula*. Org Lett 13:2670–2673
20. Tejesvi MV, Kajula M, Mattila S, Pirttilä AM (2011) Bioactivity and genetic diversity of endophytic fungi in *Rhododendron tomentosum* Harmaja. Fungal Divers 47:97–107
21. Johnston PR, Sutherland PW, Joshee S (2006) Visualising endophytic fungi within leaves by detection of (1/3)-ß-d-glucans in fungal cell walls. Mycologist 20:159–162
22. Sieber T, Hugentobler C (1987) Endophytic fungi in leaves and twigs of healthy and diseased beech trees (*Fagus sylvatica* L.). Eur J Forest Pathol 17:411–425
23. Fröhlich J, Hyde KD (1999) Biodiversity of palm fungi in the tropics: are global fungal diversity estimates realistic? Biodivers Conserv 8:977–1004
24. Gao X-X, Zhou H, Xu D-Y, Yu C-H, Chen Y-Q, Qu L-H (2005) High diversity of endophytic fungi from the pharmaceutical plant, *Heterosmilax japonica* Kunth revealed by cultivation-independent approach. FEMS Microbiol Lett 249:255–266
25. Lacap DC, Hyde KD, Liew ECY (2003) An evaluation of the fungal 'morphotype' concept based on ribosomal DNA sequences. Fungal Divers 12:53–66
26. Unterseher M, Reiher A, Finstermeier K, Otto P, Morawetz W (2007) Species richness and distribution patterns of leaf-inhabiting endophytic fungi in a temperate forest canopy. Mycol Prog 6:201–212
27. Arnold AE, Henk DA, Eells RL, Lutzoni F, Vilgalys R (2007) Diversity and phylogenetic affinities of foliar fungal endophytesin loblolly pine inferred by culturing and environmental PCR. Mycologia 99:185–206
28. Guo LD, Huang GR, Wang Y (2008) Seasonal and tissue age influences on endophytic fungi of *Pinus tabulaeformis* (Pinaceae) in the Dongling Mountains. Beijing J Integr Plant Biol 50:997–1003
29. Su YY, Guo LD, Hyde KD (2010) Response of endophytic fungi of *Stipa grandis* to experimental plant function group removal in Inner

Mongolia steppe. China Fungal Divers 43:93–101

30. Sun X, Guo LD, Hyde KD (2011) Community composition of endophytic fungi in *Acer truncatum* and their role in decomposition. Fungal Divers 47:85–95
31. Hallmann J, Berg G, Schulz B (2006) Isolation procedures for endophytic microorganisms. In: Schulz BJE, Boyle CJC, Sieber TN (eds) Microbial root endophytes. Springer, Berlin, pp 299–319
32. Guo LD, Hyde KD, Liew ECY (1998) A method to promote sporulation in palm endophytic fungi. Fungal Divers 1:109–113
33. Bills GF (1995) Analyses of microfungal diversity from a user's perspective. Can J Bot 73: S33–S41
34. Arnold AE, Henk DA, Eells RL, Lutzoni F, Vilgalys R (2007) Diversity and phylogenetic affinities of foliar fungal endophytes inloblolly pine inferred by culturing and environmental PCR. Mycologia 99:185–206
35. Collado J, Platas G, Paulus B, Bills GF (2007) High-throughput culturing of fungi from plant litter by a dilution-to-extinction technique. FEMS Microbiol Ecol 60:521–533
36. Guo LD, Hyde KD, Liew ECY (2000) Identification of endophytic fungi from *Livistona chinensis* based on morphology and rDNA sequences. New Phytol 147:617–630
37. Higgins KL, Arnold AE, Miadlikowska J, Sarvate SD, Lutzoni F (2007) Phylogenetic relationships, host affinity, and geographicstructure of boreal and arctic endophytes from three majorplant lineages. Mol Phylogenet Evol 42:543–555
38. Saad MMG, Ghareeb RY, Saeed AA (2019) The potential of endophytic fungi as bio–control agents against the cotton leafworm, *Spodoptera littoralis* (Boisd.) (Lepidoptera: Noctuidae). Egypt J Biol Pest Control 29:1–7
39. Mat-Jalil M, Ibrahim D (2018) Fungal endophytes isolated from the leaves of a medicinal plant, *Ocimum sanctum* Linn and evaluation of their antimicrobial activities. Afr J Microbiol Res 12:616–622
40. Kjer J, Debbab A, Aly AH, Proksch P (2010) Methods for isolation of marine-derived endophytic fungi and their bioactive secondary products. Nat Prot 5:479–490
41. Anjum N, Chandra R (2019) Endophytic bacteria of *Catharanthus roseus* as an alternative source of vindoline and application of response surface methodology to enhance its production. Arch Biol Sci 71:27–38
42. Sieber T (2002) Fungal root endophytes. In: Eshel A, Beeckman T (eds) The hidden half edition, 4th edn. CRC Press/Taylor & Francis Group, Boca Raton

Chapter 16

Microscopy-Based Studies

C. Brígido and A. Paço

Abstract

Successful colonization of plants by endophytes is the basic requirement for plant-microbe interactions. Thus, the study of plant colonization routes and niches has always been a central aspect in research on plant-microbe interaction. Endophytic microorganisms can be detected by several methods based on microscopy, thus allowing to investigate their behavior, their niches, and how they colonize their hosts. Different techniques can be applied, depending on the study objectives, to single strains or to more complex plant-microbe systems involving full or synthetic microbiomes. Furthermore, the use of microscopy-based tools allows us to track endophytes in plant tissues and to determine how they interact with each other and with their host plants, as well as how they are associated with various plant organs under different plant growth and environmental conditions either from natural or artificial environments. This chapter describes the step-by-step protocol of distinct microscopy-based methods that may be used to visualize and track endophytic microorganisms within plant tissues.

Key words Colonization, Microscopy, Fluorescence in situ hybridization (FISH), Confocal laser scanning microscopy, Fluorescent proteins, Endosphere, Plant-microbe interaction, Dyes

1 Introduction

Endophytic microorganisms maintain intimate associations within their host plants without causing apparent symptoms of disease [1]. A key factor for beneficial plant-microbe interactions relies on the successful colonization of host plant tissues by endophytic microbes. For successful colonization, a series of phenomena such as attachment, entry, motility, transmission, and multiplication of endophytic populations within the host plant is necessary. Therefore, the localization and visualization of the endophytic microorganisms and/or their molecules or enzymes with special function are important methods to understand their behavior, their niches, and their interactions with the host plant.

In this regard, a potent tool that allows the detection of endophytic colonization within plant tissues is the combination of microscopy with distinctive fluorochromes or fluorescent dyes

A. Sankaranarayanan et al. (eds.), *Endophytic Microbes: Isolation, Identification, and Bioactive Potentials*, Springer Protocols Handbooks, https://doi.org/10.1007/978-1-0716-2827-0_16,

used to mark or stain specific microbes or endophytic communities [2]. Fluorescence in situ hybridization (FISH) associated with microscopy is a powerful technique to visualize microbes or specific microbial groups allowing to screen endophytic communities and to elucidate the dynamics of colonization in different plant parts, which helps to better understand complex microbial communities. FISH using fluorescently labeled oligonucleotide probes has been widely utilized to localize different endophytic bacterial species [3–9] or fungi [10] in plant tissues. Moreover, FISH can be widely applied in microbial ecology since it allows the detection, enumeration, and identification of cultivable and non-cultivable microorganisms within their natural microhabitat [4, 5, 8].

Another strategy for tracking microbial colonization inside the plants is the use of broad host range plasmids containing constitutively expressed *gfp* genes [11–15]. GFP-tagged endophytes express the fluorescent proteins in situ, thus allowing to visualize the emission of fluorescent green light in the presence of near-ultraviolet light or blue light. This technique has been widely used in bacterial and fungal endophytes in laboratory/greenhouse conditions to assess their colonization success, to determine their entry into the host plants, and to investigate the colonized microhabitat within plant tissues [13–16]. With this reporter gene system, plant tissues do not need to be fixed and no substrate or cofactor is necessary. GUS (β-glucuronidase) reporter system is another reporter gene system developed for analyzing the time course and tissue infection specificity by endophytic microbes, which is detectable by histochemical staining [17]. Both these systems are suitable for transcriptional fusions of *gusA* and *gfp* to a gene of interest, which allows to study and quantify the expression of a specific bacterial gene *in planta*. For instance, transcriptional fusions of *gusA* and *gfp* to the *nifH* gene of *Azoarcus* sp. strain BH72 revealed high nitrogenase gene expression levels of this endophyte within infected rice roots (*Oryza sativa*) under laboratory conditions [18]. In addition to all the previous approaches to observe the microbial colonization in plants, bacteria can also be observed in planta by fluorogenic dye staining using different dyes, such as 4′,6′ diamino-2-fenil-indol (DAPI), acridine orange (AO) [19], and propidium iodide (PI) [20]. Viable cells do not permeate DAPI and PI due to the integrity of their membrane, however they go through damaged cell membranes and stain the nuclei. DAPI binds to the specific AT sequence of bacterial DNA [21] and is commonly used in combination with Acridine Orange (AO) which stains viable cells. PI is frequently used in combination with fluorogenic dye carboxyfluorescein diacetate (CFDA) [22] or calcofluor white [20]. Altogether, colonization studies can benefit from a combination of different techniques such as FISH, GFP labeling, and GUS staining or other fluorogenic dyes [7, 13, 23].

2 Materials

2.1 Materials for FISH (Fluorescence In Situ Hybridization)

Chemical Reagents

- 5 M NaCl.
- 1 M Tris/HCl pH 8.0.
- Sterile dH_2O or sterile milli-Q water.
- Formamide.
- 10% (w/v) SDS.
- 0.5 M EDTA pH 8.0.
- 1× Phosphate-buffered saline (PBS).
- 25%, 50%, 75%ethanol (diluted in PBS).
- 96% ethanol.
- Oligonucleotide probes (singly labeled or DOPE-labeled) (Table 1).
- 4% paraformaldehyde solution (freshly prepared in PBS).

Other Supplies and Equipment

- Falcon or Eppendorf tubes.
- Tube supports.
- Parafilm.
- Glass slides.
- Glass coverslips.
- Timer.
- Ice.
- Nail polish.
- Micropipettes.
- Incubator.
- Shaker platform.
- Fluorescence microscope (e.g., confocal microscope or epifluorescence microscope).

2.2 Materials for GFP Tagging

- Growth media (liquid and plates), e.g., Luria Bertani (LB); Tryptic soy broth (TSB), etc., according to the strains used.
- Shaker.
- Sterile glass tubes.
- Spectrophotometer.
- Centrifuge.
- Incubator.
- Micropipettes.
- Electroporator and cuvettes.

Table 1
Examples of rRNA-targeted oligonucleotide probes used in FISH assays to detect bacteria or fungi in plant tissues

Probe	Probe sequence (5′-3′)	Target molecule/ position	Reference
EUB338mix (for all bacteria) In equal amounts:			
EUB338	GCTGCCTCCCGTAGGAGT	16S rRNA/ 338–355	[29]
EUB338II	GCAGCCACCCGTAGGTGT	16S rRNA/ 338–355	[30]
EUB338III	GCTGCCACCCGTAGG TGT	16S rRNA/ 338–355	[30]
LGC354MIX (specific for *Firmicutes*) In equal amounts:			
LGC354A	TGGAAGATTCCCTACTGC	16S rRNA/ 354–371	[31]
LGC354B	CGGAAGATTCCCTACTGC	16S rRNA/ 354–371	[31]
LGC354C	CCGAAGATTCCCTACTGC	16S rRNA/ 354–371	[31]
BET42a specific for *Betaproteobacteria*	GCCTTCCCACTTCGTTT	23S rRNA/ 1027–1043	[32]
GAM42a (specific for *Gammaproteobacteria*)	GCCTTCCCACATCGTTT	23S rRNA/ 1027–1043	[32]
ALF969 (for some Alphaproteobacteria)	TGGTAAGGTTCTGCGCGT	16S rRNA/ 969–986	[33]
HGC236 (specific for *Actinobacteria*)	AACAAGCTGATAGGCCGC	16S rRNA/ 235–253	[34]
EUKb310 (for almost all Eukaryota)	TCA GGC BCC YTC TCC G	18S rRNA/ 310–325	[35]
MY1574 (for fungi, Eumycota)	TCCTCGTTGAAGAGC	18S rRNA/ 1474–1489	[36]

- Electrophoresis apparatus.
- Ice.
- Antibiotics.
- 10–30% Glycerol solution.
- Broad-host plasmid containing *gfp* gene (e.g., pHC60 [24]).
- Eppendorf 1.5 mL or 2.0 mL.
- Plates.

2.3 Materials for GUS Staining

- Growth media (liquid and plates), e.g., Luria Bertani (LB); Tryptic soy broth (TSB), etc., according to the strains used.
- Antibiotics.
- 10–30% Glycerol solution.
- Broad-host plasmid containing *gusA* gene (e.g., pGUS3 [25]).
- Donor and helper *E. coli* strains for mating.
- Shaker.
- Sterile glass tubes.
- Microscope.
- Centrifuge.
- Incubator.
- Micropipettes.
- X-gluc ((5-bromo-4-chloro-3-indolyl-beta-D-glucuronic acid, cyclohexylammonium salt) stock solution (50 mg/mL).
- Eppendorf 1.5 mL or 2.0 mL.
- Phosphate buffer (8.1 g NaH_2PO_4, 32.8 g NaH_2HPO_4 $12H_2O$, 0.87 g EDTA, 4.5 g SDS) [26].

2.4 Materials for Fluorogenic Dye Staining

- Inoculated plant material.
- 1× PBS (10 mM Na_2HPO_4, 10 mM NaH_2PO_4, 150 mM NaCl, pH 7.2).
- 4.0% paraformaldehyde in 1× PBS.
- 4% formaldehyde solution in 50 mM PIPES buffer.
- Sterile dH_2O.
- Shaker.
- Paraffin.
- Micrometer.
- Glass slides.
- Coverslips.
- 70% glycerol.
- Ethanol.
- Micropipette.
- DAPI aqueous solution (1 mg/mL) at −20 °C.
- AO aqueous solution (1 mg/mL) at −20 °C.
- 20 μM propidium iodide (PI) solution at 4 °C.
- 0.01% calcofluor white solution.
- Microtome.
- 6% low melting agarose.
- Fluorescence microscope.

3 Methods

3.1 Methods for FISH (Fluorescence In Situ Hybridization)

1. Collect the plants and gently detach the tissues of interest, then rinse them with PBS.
2. Cut the plant tissues into small pieces ($\leq$ 1 cm length) and fix with 4% paraformaldehyde solution for at least 3 h up to overnight, depending on the plant tissue, at 4 °C.

   Optional: Cryosection can be used to obtain transversal sections of the plant tissues embedded in commercial tissue-freezing medium using a low-temperature constant-cooling cryostat. See [7, 8] for details.
3. Wash the samples three times with ice-cold PBS at 4 °C. If long-term storage is required, keep samples at - 20 °C in ice-cold 1:1 (v:v) PBS: 96% ethanol until further processing.

   Optional: Depending on the targeted endophytes and plant tissues, permeabilization of cells and tissues may be needed to be performed before FISH staining [7].
4. Hydrate the samples in an ethanol series of 25%, 50%, 75%, and 99% for 15–30 min each step at 4 °C.
5. For hybridization, transfer the plant tissues to glass slides and apply to each sample the hybridization solution containing each labeled probe (use equimolar mixture of different probes when applicable). Incubate at 42–46 °C for 2 h in the dark.

   Notes: The fluorochromes of the oligonucleotide probes should be chosen taken into consideration the natural background of the autofluorescence of the plant tissues. Prepare each hybridization solution containing as final concentration 20 mM Tris-HCl pH 8.0, 0.01% SDS, 0.9 M NaCl, 20% formamide* and 5–15 ng/μL of each probe). *Check and use the percent formamide recommended for each probe in the hybridization buffer for optimal hybridization conditions in FISH experiments at probeBase (https://probebase.csb.univie.ac.at/) [27] or mathFISH (http://mathfish.cee.wisc.edu/) [28].
6. Wash the unbinding probes with a washing solution containing a final concentration 20 mM Tris-HCl pH 8.0, 0.01% SDS, 5 mM EDTA pH 8.0, and 225 mM NaCl) at 43–48 °C for 10–30 min in the dark.

   Note: The NaCl concentration in the washing solution should be taken into consideration the % formamide used in the hybridization solution [29].
7. Finally, rinse the samples with distilled water and dry out with soft compressed air.
8. Immediately, add antifade reagent and cover the slide with a coverslip, and seal with nail polish.

9. Observe the stained samples with a confocal microscope or other fluorescence microscope equipped with the necessary lasers for the excitation of each of the fluorophores used for probe labeling.

   Note: The occurrence of false positive signals derived from unspecific adhesion of FISH probes must be checked by staining a subsample with a mixture of negative control probe.

3.2 Methods for GFP Tagging

3.2.1 Transformation of Gram-Negative Bacteria by Electroporation

Isolation of Plasmid DNA

1. Grow the *Escherichia coli* cells carrying the plasmid pHC60 (or other) on LB medium supplemented with 10 μg/mL tetracycline.
2. Isolate the plasmid DNA by using standard procedures or commercial plasmid DNA isolation kits (e.g., Zymo Research, Qiagen, Thermo Fisher Scientific).
3. Check the quality and purity of the plasmid DNA by agarose-gel electrophoresis or NanoDropspectrophotometer.
4. Store plasmid DNA at −20 °C until further use.

Preparation of Competent Cells

1. Pick a colony with a toothpick of the bacterial strain (e.g., *Pseudomonas* spp.) to be tagged and put it to grow in appropriate liquid medium (e.g., TSB), at 30 °C overnight (or until it reaches exponential/late exponential phase).
2. Inoculate fresh liquid medium with 1/100th volume of the overnight culture and incubate again at optimal growth temperature, 30 °C.
3. Collect the bacterial cells at exponential, late exponential/early stationary, and stationary phase by centrifugation (3000 g, 10 min, 4 °C).
4. Discard the supernatant carefully.
5. Wash gently the cells pellet three times with 1000 μL sterile ice-cold distilled water and centrifuge again.
6. Discard the supernatant carefully and gently resuspend the cells in 1000 μL of ice-cold 10% glycerol solution ($\pm 10^9$–10^{10} cells/mL).
7. Carefully distribute the culture suspension into chilled Eppendorf 1.5 or 2.0 mL in aliquots of 100 μL.
8. Store the competent cells aliquots at −80 °C until further processing.

3.2.2 Bacteria Electro Transformation

1. Add 10–100 ng of plasmid DNA (pHC60) to 100 μL competent cells aliquots.
2. Mix gently by flicking the tube and incubate the mixture on ice for 20 min.

3. Transfer the entire mixture to a chilled electroporation cuvette (2 mm electrode gap).
4. Dry off the outside of the electroporation cuvette with absorbent paper and place it in the cuvette holder provided with the electroporator.
5. Apply an exponential pulse of 2.5 kV and 25μFm and use 200 Ω.
6. Immediately after pulsing, add to the electropulsed volume 500–1000 μL of culture medium at room temperature and mix by shaking the cuvette.
7. Transfer the cells to a new Eppendorf 1.5 mL or 2.0 mL and incubate to the optimal growth temperature 30 °C for 40 to 3 h with shaking.
8. After regeneration, plate the cells on solid medium containing the appropriate selecting antibiotic 10 μg/mL tetracycline, and incubate at optimal growth temperature, 30 °C until well-defined colonies are formed.

Note: The number of resulting transformants depends on the quantity of DNA put in the cuvette, the dilution used and the volume of aliquot spread on the plate (classically 100 μL, but 300 or 400 μL can be used when expected efficiency is low).

9. Pick several colonies and streak them on duplicate fresh plates containing the selective media and use one of the plates to confirm the selected bacterial transconjugants by detecting fluorescent colonies on agar plates under UV irradiation (e.g., UVP-transilluminator).
10. Or confirm positive transconjugants by resuspending the colony cells in a drop of distilled water placed in the center of a glass slide and cover it with coverslip to observe under near-ultraviolet light or blue light in a fluorescence microscopy.
11. Preserve the positive transconjugants in 20–30% glycerol solution at −80 °C until further processing.

3.2.3 Transformation of Bacteria by Conjugation

1. Grow the *E. coli* S17.1 (pHC60) as a donor strain (for biparental mating) overnight in LB with appropriate antibiotics 10 μg/mL tetracycline for plasmid selection at 37 °C.
2. In other glass tube, grow the recipient strain (e.g., *Mesorhizobium*) in the appropriate liquid medium, Tryptone-Yeast (TY) medium, at 28 °C for 48 h (or until it reaches exponential/late exponential phase).
3. Collect the cells by centrifugation (3000 g, 10 min, room temperature) and wash the donor cells twice with LB medium without antibiotics.

4. Then, resuspend the cells in 0.5 mL of fresh medium without any antibiotics.
5. In a TY plate (or other medium appropriate for the recipient strain) without antibiotics, mix 0.5 mL of recipient cells with 0.2 mL of donor cells into the center of the plate.
6. Incubate the TY plate with the mixture at the optimal growth condition of the recipient strain, 28 °C for 24 h.
7. Collect the mixture of the plate with TY liquid medium and prepare serial dilutions (10^{-1},10^{-2}, 10^{-3},10^{-4}, 10^{-5}).
8. Spread 100–200 μL of the dilutions on TY plates containing selective antibiotics, 10 μg/mL tetracycline for plasmid selection and 5 μg/mL gentamicin for *Mesorhizobium* strain selection.
9. Incubate the plates at optimal growth temperature, 30 °C until well-defined colonies are formed.
10. Pick several colonies and streak them in a fresh TY plate with antibiotics.
11. Confirm positive transconjugants by observing fluorescence under near-ultraviolet light or blue light in a fluorescence microscopy.
12. Preserve the positive transconjugants in 20–30% glycerol solution at −80 °C until further processing.

    Note: Other marker plasmids carrying Enhanced GFP derivatives such as egfp (green), ecfp (cyan), eyfp (yellow), ebfp (blue) [37–40] or the red fluorescent protein, rfp (red), isolated from the coral Discosoma striata [41] can also be used to successfully visualize endophytic microbes in plant tissues [42, 43].

3.2.4 Plant Growth Conditions and GFP-Bacterial Inoculations

GFP-tagged Bacterial Inoculum Preparation

1. Grow the GFP-tagged bacterial strains (e.g., *Mesorhizobium* strain) on TY medium supplemented with appropriate antibiotics (10 μg/mL tetracycline) at 28 °C for 48 h.
2. Wash the GFP-tagged cells twice and resuspend in sterile saline solution (0.85% *w*/*v* NaCl).
3. Measure the optical density (OD) at 600 nm of the cell suspension in a spectrophotometer and adjust the cell suspension to the desired OD.
4. Use this cell suspension for bacterization of the surface-sterilized seeds.

Seed Surface-sterilization and Germination

5. Surface-sterilize the seeds of the plant (e.g. *Cicer arietinum* L.) by soaking the seeds in 14% calcium hypochlorite solution for

45 min followed by rinse the seeds ten times in sterilized distilled water and incubate the seeds for 2 h at 4 °C.

Note: Use the surface-sterilization method appropriate for the plant seeds used.

6. Transfer the seeds to0.75% agar plates and incubate in the dark at 28 °C for 48 h.

Plant Experiments Under Gnotobiotic Conditions

7. Depending on the purposes, transfer the seedings to plasticpots filled with a sterilized substrate (e.g. vermiculite or sand or mixture of both for long-term plant growth assays or to containers with nutrient solution for hydroponic conditions or to the surface of 1.5% agar plates containing appropriatenutrient solution and overlaid with wetted Whatman sterile filter for analysis of early stages of plant-endophytic bacteria interaction.

   Note: Include negative controls, i.e., uninoculated plants in the experiment.
8. Inoculate the seeds with the cell suspension obtained above.
9. Transfer the pots or containers or incubate plates vertically in a growth chamber with the light cycle, temperature, and relative humidity adjusted for the optimal plant growth for the time required for the experiment.
10. When necessary, water the plants with an appropriate nutrient solution.
11. At the end of the experiment, harvest the plants and excise the plant roots to a microscope slide.
12. Examine the roots (or other plant tissues) on glass slides by fluorescence microscopy using *fluorescence filter set* for GFP (Green*Fluorescence*Protein) for visualization of GFP fluorescence of the GFP-tagged bacterial strains.

3.3 Methods for GUS Staining

3.3.1 Transformation of Bacteria by Conjugation (or by Electroporation as Described Earlier)

1. Grow the donor *E. coli* DH5α (containing the pGUS3), helper *E. coli* MT616 (carrying pRK600) [44] overnight in LB agar plates with appropriate antibiotics, 50 μg/mL kanamycin and 25 μg/mL chloramphenicol, respectively, at 37 °C, for triparental mating.
2. Grow your recipient cells (e.g., *Rhizobium leguminosarum*) in the appropriate medium, Tryptone-Yeast (TY) agar plate, at 28 °C for 24–48 h.
3. Transfer a loop of recipient cells to the center of a new TY agar plate and add ½ loop of donor *E. coli* cells followed by ½ loop of helper *E. coli* cells.
4. Mix the cells very well with the help of the loop.
5. Incubate the TY plate with the mixture at the optimal growth temperature of the recipient strain, 28 °C for 24 h.

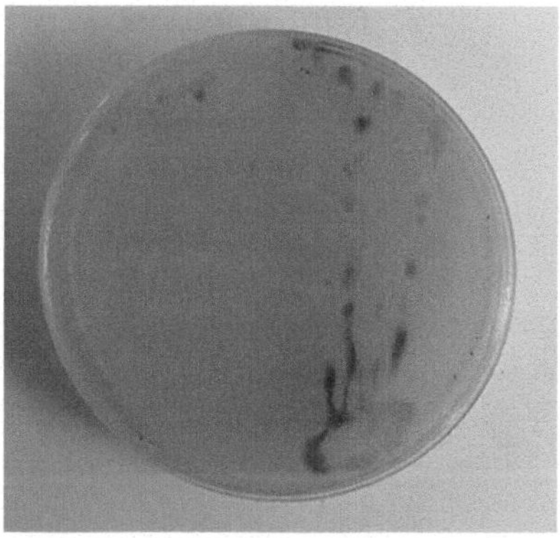

Fig. 1 Colonies of the *R. leguminosarum bv. viciae* 128C53K, transformed with a broad host plasmid containing the *gusA*, in TY plates supplemented with 50 μg/ml X-gluc. A developed green/blue color indicates enzyme β-glucuronidase (GUS) activity

6. Then, resuspend the cells in 0.5 mL of fresh medium without any antibiotics.
7. Collect the mixture of the plate with TY plate and prepare serial dilutions (10^{-1}, 10^{-2}, 10^{-3}, 10^{-4}, etc.).
8. Spread 100–200 μL of the dilutions onto fresh TY plates containing selective antibiotics, 50 μg/mL tetracycline for plasmid selection and another antibiotic for *R. leguminosarum* strain selection.
9. Incubate the plates at optimal growth temperature, 28 °C until well-defined colonies are formed.
10. Pick several colonies and streak them in a fresh TY plate with antibiotics.
11. Confirm putative positive transconjugants by growing the transconjugants on new TY plates containing the selective antibiotic and 50 μg/ml X-gluc (Fig. 1).
12. Preserve the positive transconjugants in 20–30% glycerol solution at −80 °C until further processing.

3.3.2 *Detection of* gusA-*tagged Bacteria Activity* in planta

1. Collect the plants inoculated with the *gusA*-tagged bacterial strain and detach the whole inoculated roots.
2. Wash repeatedly the roots with saline solution (0.85% NaCl) to remove substrate particles and loosely associated bacteria from the roots. Optional: Surface-sterilization of the roots may be performed before histochemical detection.
3. Incubate the roots in phosphate buffer containing X-gluc (50 μg/ml) for 6 h to overnight at 37 °C.
4. Dry the roots under vacuum for 20 min to facilitate penetration of the substrate into the roots.

5. Rinse the roots three times in phosphate buffer and increase the transparency of plant cells by clearing using 50% (V/V) bleach (1.5% final concentration of hypochlorite) for 10 min, followed by washing with deionized water.
6. Visualize the roots (or small pieces of roots) for internal GUS expression under bright field microscopy.

3.4 Methods for Fluorogenic Dye Staining

3.4.1 DAPI/AO Staining

1. Collect the plant material and detach the roots followed by washing three timesin sterile dH_2O to remove unbound bacteria.
2. Fix theroot samples with 4.0% paraformaldehyde during 12–24 h, with gentle agitation at room temperature.
3. Rinse the root samples in 1× PBS and dehydrated through an ethanol series to 95% ethanol.
4. Incorporate the root samples in paraffin for at least 24 h.
5. Cut root segments of ~5 mm in length with a micrometer, or in case of root nodules usewhole nodulessections (40–50 μm).
6. Transfer the sample sections to a clean glass slide and dyed with DAPIand AOdiluted 1:100 or 1:25in 1× PBS, respectively. Apply DAPI and AO droplets directly to the sections and incubate for 5 min at room temperature.
7. Remove the excess of stain by wash three times the sections with PBS (5 min each wash). After, mount in a new slide the sections with 70% glycerol or commercial mounting solution under a coverslip.
8. Analyze the stained sections using a fluorescence microscope (e.g. confocal microscope) with 360 nm excitation and 460 nm emission filters for DAPI fluorescence and with 500 nm excitation and 530 nm emission filtersfor AO fluorescence.

3.4.2 Calcofluor White/Propidium Iodide (PI) Staining

1. Collect the plant material and fix the samples with a 4% formaldehyde solution in 50 mM PIPES buffer by 30 min vacuum infiltration and then kept at room temperature for 45 min.
2. Incorporate the fixed samples in 6% low melting agarose and cut the samples into thin sections (30–50 μm) with microtome.
3. Stain the sample sections with a fresh 0.01% calcofluor white solution for 10 min at room temperature.
4. Then, stain with a 20 μM propidium iodide (PI) solution for 10 min at room temperature.
5. Mount and cover the sample slides with a coverslip.
6. Examine the stained sections using a fluorescence microscope (e.g. confocal microscope). Use a UV laser line for excitation of calcofluor white and detect the emission at 405–450 nm while for PI use an argon laser line at 488 nm and a detection filter at 600–650 nm.

Acknowledgements

This work was funded by National Funds through FCT - Foundation for Science and Technology under the Project UIDB/05183/2020. C. Brígido acknowledges a research contract from the Individual Call to Scientific Employment Stimulus 2018 (CEECIND/00093/2018) from FCT.

References

1. Hallmann J, Quadt-Hallmann A, Mahaffee WF, Kloepper JW (1997) Bacterial endophytes in agricultural crops. Can J Microbiol 43:895–914
2. Kandel SL, Joubert PM, Doty SL (2017) Bacterial endophyte colonization and distribution within plants. Microorganisms 5
3. Lo Piccolo S, Ferraro V, Alfonzo A, Settanni L, Ercolini D, Burruano S, Moschetti G (2010) Presence of endophytic bacteria in *Vitis vinifera* leaves as detected by fluorescence *in situ* hybridization. Ann Microbiol 60:161–167
4. Bulgari D, Casati P, Faoro F (2011) Fluorescence *in situ* hybridization for phytoplasma and endophytic bacteria localization in plant tissues. J Microbiol Methods 87:220–223
5. Lundberg DS, Lebeis SL, Paredes SH, Yourstone S, Gehring J, Malfatti S, Tremblay J, Engelbrektson A, Kunin V, del Rio TG, Edgar RC, Eickhorst T, Ley RE, Hugenholtz P, Tringe SG, Dangl JL (2012) Defining the core *Arabidopsis thaliana* root microbiome. Nature 488:86–90
6. Wagner M, Haider S (2012) New trends in fluorescence *in situ* hybridization for identification and functional analyses of microbes. Curr Opin Biotechnol 23:96–102
7. Castanheira NL, Dourado AC, Pais I, Semedo J, Scotti-Campos P, Borges N, Carvalho G, Barreto Crespo MT, Fareleira P (2017) Colonization and beneficial effects on annual ryegrass by mixed inoculation with plant growth promoting bacteria. Microbiol Res 198:47–55
8. Faddetta T, Abbate L, Alibrandi P, Arancio W, Siino D, Strati F, De Filippo C, Fatta Del Bosco S, Carimi F, Puglia AM, Cardinale M, Gallo G, Mercati F (2021) The endophytic microbiota of *Citrus limon* is transmitted from seed to shoot highlighting differences of bacterial and fungal community structures. Sci Rep 11:7078
9. Thomas P, Franco CMM (2021) Intracellular bacteria in plants: elucidation of abundant and diverse cytoplasmic bacteria in healthy plant cells using in vitro cell and callus cultures. Microorganisms 9
10. Vágí P, Knapp DG, Kósa A, Seress D, Horváth ÁN, Kovács GM (2014) Simultaneous specific *in planta* visualization of root-colonizing fungi using fluorescence *in situ* hybridization (FISH). Mycorrhiza 24:259–266
11. Coombs JT, Franco CMM (2003) Visualization of an endophytic *Streptomyces* species in wheat seed. Appl Environ Microbiol 69: 4260–4262
12. Chi F, Shen S-H, Cheng H-P, Jing Y-X, Yanni YG, Dazzo FB (2005) Ascending migration of endophytic rhizobia, from roots to leaves, inside rice plants and assessment of benefits to rice growth physiology. Appl Environ Microbiol 71:7271–7278
13. Rouws LFM, Meneses CHSG, Guedes HV, Vidal MS, Baldani JI, Schwab S (2010) Monitoring the colonization of sugarcane and rice plants by the endophytic diazotrophic bacterium *Gluconacetobacter diazotrophicus* marked with *gfp* and gusA reporter genes. Lett Appl Microbiol 51:325–330
14. Brígido C, Robledo M, Menéndez E, Mateos PF, Oliveira S (2012) A ClpB chaperone knockout mutant of *Mesorhizobium ciceri* shows a delay in the root nodulation of chickpea plants. Mol Plant-Microbe Interact 25: 1594–1604
15. Wang X, Zhang X, Liu L, Xiang M, Wang W, Sun X, Che Y, Guo L, Liu G, Guo L, Wang C, Yin W-B, Stadler M, Zhang X, Liu X (2015) Genomic and transcriptomic analysis of the endophytic fungus *Pestalotiopsis fici* reveals its lifestyle and high potential for synthesis of natural products. BMC Genomics 16:28
16. Krzyzanowska D, Obuchowski M, Bikowski M, Rychlowski M, Jafra S (2012) Colonization of potato rhizosphere by GFP-tagged *Bacillus subtilis* MB73/2, *Pseudomonas* sp. P482 and *Ochrobactrum* sp. A44 shown on large sections of roots using

enrichment sample preparation and confocal laser scanning microscopy. Sensors 12:17608–17619

17. Yates IE, Hiett KL, Kapczynski DR, Smart W, Glenn AE, Hinton DM, Bacon CW, Meinersmann R, Liu S, Jaworski AJ (1999) GUS transformation of the maize fungal endophyte *Fusarium moniliforme*. Mycol Res 103: 129–136
18. Egener T, Hurek T, Reinhold-Hurek B (1999) Endophytic expression of *nif* genes of *Azoarcus* sp. strain BH72 in Rice roots. Mol Plant-Microbe Interact 12:813–819
19. Dudley ME, Jacobs TW, Long SR (1987) Microscopic studies of cell divisions induced in alfalfa roots by *Rhizobium meliloti*. Planta 171:289–301
20. Liang J, Klingl A, Lin Y-Y, Boul E, Thomas-Oates J, Marín M (2019) A subcompatible rhizobium strain reveals infection duality in lotus. J Exp Bot 70:1903–1913
21. Glavin DP, Cleaves HJ, Schubert M, Aubrey A, Bada JL (2004) New method for estimating bacterial cell abundances in natural samples by use of sublimation. Appl Environ Microbiol 70:5923–5928
22. Kolek J, Branska B, Drahokoupil M, Patakova P, Melzoch K (2016) Evaluation of viability, metabolic activity and spore quantity in clostridial cultures during ABE fermentation. FEMS Microbiol Lett 363
23. Rothballer M, Eckert B, Schmid M, Fekete A, Schloter M, Lehner A, Pollmann S, Hartmann A (2008) Endophytic root colonization of gramineous plants by *Herbaspirillum frisingense*. FEMS Microbiol Ecol 66:85–95
24. Cheng H-P, Walker GC (1998) Succinoglycan is required for initiation and elongation of infection threads during nodulation of alfalfa by *Rhizobium meliloti*. J Bacteriol 180:5183–5191
25. Garcia-Rodriguez FM, Toro N (2000) *Sinorhizobium melilotinfe* (nodulation formation efficiency) genes exhibit temporal and spatial expression patterns similar to those of genes involved in symbiotic nitrogen fixation. Mol Plant-Microbe Interact 13:583–591
26. Wilson KJ, Sessitsch A, Corbo JC, Giller KE, Akkermans AD, Jefferson RA (1995) Beta-Glucuronidase (GUS) transposons for ecological and genetic studies of rhizobia and other gram-negative bacteria. Microbiology 141 (Pt 7):1691–1705
27. Greuter D, Loy A, Horn M, Rattei T (2016) probeBase – an online resource for rRNA-targeted oligonucleotide probes and primers: new features. Nucleic Acids Res 44:D586–D589
28. Yilmaz LS, Parnerkar S, Noguera DR (2011) mathFISH, a web tool that uses thermodynamics-based mathematical models for in silico evaluation of oligonucleotide probes for fluorescence in situ hybridization. Appl Environ Microbiol 77:1118–1122
29. Amann RI, Binder BJ, Olson RJ, Chisholm SW, Devereux R, Stahl DA (1990) Combination of 16S rRNA-targeted oligonucleotide probes with flow cytometry for analyzing mixed microbial populations. Appl Environ Microbiol 56:1919–1925
30. Daims H, Brühl A, Amann R, Schleifer KH, Wagner M (1999) The domain-specific probe EUB338 is insufficient for the detection of all Bacteria: development and evaluation of a more comprehensive probe set. Syst Appl Microbiol 22:434–444
31. Meier H, Amann R, Ludwig W, Schleifer KH (1999) Specific oligonucleotide probes for in situ detection of a major group of gram-positive bacteria with low DNA G + C content. Syst Appl Microbiol 22:186–196
32. Manz W, Amann R, Ludwig W, Wagner M, Schleifer K-H (1992) Phylogenetic Oligodeoxynucleotide probes for the major subclasses of Proteobacteria: problems and solutions. Syst Appl Microbiol 15:593–600
33. Oehmen A, Zeng RJ, Saunders AM, Blackall LL, Keller J, Yuan Z (2006) Anaerobic and aerobic metabolism of glycogen-accumulating organisms selected with propionate as the sole carbon source. Microbiology 152:2767–2778
34. Erhart R, Bradford D, Seviour RJ, Amann R, Blackall LL (1997) Development and use of fluorescent in situ hybridization probes for the detection and identification of "Microthrix parvicella" in activated sludge. Syst Appl Microbiol 20:310–318
35. Baker BJ, Hugenholtz P, Dawson SC, Banfield JF (2003) Extremely acidophilic Protists from acid mine drainage host Rickettsiales-lineage endosymbionts that have intervening sequences in their 16S rRNA genes. Appl Environ Microbiol 69:5512–5518
36. Baschien C, Manz W, Neu TR, Marvanová L, Szewzyk U (2008) In situ detection of freshwater fungi in an alpine stream by new taxon-specific fluorescence *in situ* hybridization probes. Appl Environ Microbiol 74:6427–6436
37. Yang TT, Sinai P, Green G, Kitts PA, Chen YT, Lybarger L, Chervenak R, Patterson GH, Piston DW, Kain SR (1998) Improved

fluorescence and dual color detection with enhanced blue and green variants of the green fluorescent protein. J Biol Chem 273:8212–8216

38. Tsien RY (1998) The green fluorescent protein. Annu Rev Biochem 67:509–544
39. Ellenberg J, Lippincott-Schwartz J, Presley JF (1999) Dual-colour imaging with GFP variants. Trends Cell Biol 9:52–56
40. Matus A (1999) GFP in Motion CD-ROM: introduction: GFP illuminates everything. Trends Cell Biol 9:43
41. Matz MV, Fradkov AF, Labas YA, Savitsky AP, Zaraisky AG, Markelov ML, Lukyanov SA (1999) Fluorescent proteins from nonbioluminescent Anthozoa species. Nat Biotechnol 17: 969–973
42. Stuurman N, Bras CP, Schlaman HRM, Wijfjes AHM, Bloemberg G, Spaink HP (2000) Use of green fluorescent protein color variants expressed on stable broad-host-range vectors to visualize rhizobia interacting with plants. Mol Plant-Microbe Interact 13:1163–1169
43. Bloemberg GV, Wijfjes AH, Lamers GE, Stuurman N, Lugtenberg BJ (2000) Simultaneous imaging of *Pseudomonas fluorescens* WCS365 populations expressing three different autofluorescent proteins in the rhizosphere: new perspectives for studying microbial communities. Mol Plant-Microbe Interact 13:1170–1176
44. Finan TM, Kunkel B, De Vos GF, Signer ER (1986) Second symbiotic megaplasmid in *Rhizobium meliloti* carrying exopolysaccharide and thiamine synthesis genes. J Bacteriol 167:66–72

Chapter 17

Surrogate Host Test

Harshida A. Gamit and Natarajan Amaresan

Abstract

Endophytic microbes colonize plant tissues and play a crucial role in growth, development, and maintaining the health of the host plant. To understand the role of endophytes or the effect of endophytes, their host plants should have the ability to grow under given condition such as laboratory conditions. In some cases, the host plant is not suitable for the research; therefore, a surrogate host for endophytes is needed. This protocol discusses the methods used to investigate surrogate hosts for endophytic microbes.

Key words Host, Bacteria, Water agar, Plastic pots, Sodium hypochlorite

1 Introduction

The study of endophytic microorganism surrogate tests is important for carrying out experiments that may not be conducted with their host plant, due to inappropriate growth of the host plant under laboratory conditions or the desired environment, and the inability to obtain sufficient plant materials. Therefore, it is necessary to determine the surrogate host to investigate the impact of endophytes on the host. In the surrogate test, endophyte-free seeds are germinated and fresh cultures are inoculated into seedlings after determining the surrogate host [1].

2 Materials

- Seeds.
- Petri plates.
- Water agar.
- Scalpel.
- Fresh inoculum.

A. Sankaranarayanan et al. (eds.), *Endophytic Microbes: Isolation, Identification, and Bioactive Potentials*,
Springer Protocols Handbooks, https://doi.org/10.1007/978-1-0716-2827-0_17,

- Sterile sand pot mixture.
- Sterile distilled water.
- 50% sulfuric acid.
- Sodium hypochlorite.
- Clean plastic pots.

3 Protocol [2]

1. Surface sterilizes the seeds by soaking in 50% sulfuric acid for 20 min then rinse with sterile distilled water, followed by 20 min in sodium hypochlorite and subsequently wash with sterile distilled water.
2. Place the sterilized seeds in a Petri plate containing water agar (five seeds per plate) and incubate in dark condition at 20–23 ° C for 5–16 days (or according to your experiment).
3. After the incubation period, inoculate the test culture through the following methods. Use fresh culture of inoculum to be tested.
   - Place culture at the junction of the mesocotyl and coleoptiles of the undamaged shoot or on the coleoptiles just below the first leaf.
   - Using a scalpel, cut 3–5 mm long section of tissue from the junction of the mesocotyl and coleoptiles then place culture in the cut area.
   - Cut 3–5 mm longitudinal slit right through the tissue at the junction of the mesocotyl and coleoptiles and test culture placed in the slit.
4. After inoculation, further incubate the seedlings for 10–15 days in dark condition.
5. Evaluate the seedlings condition and inoculum degree after 7–17 days.
6. After that prepare 10 cm plastic pots with sterile sand pit mixture and transfer the seedlings into the pots.
7. Maintain the control seedling without inoculation of test culture.
8. Evaluate the endophyte infection in 4–6 weeks through fluorescence microscopy and reisolate the endophytes from the treated plants and compare with the original culture.

References

1. Verma SK, Kharwar RN, Gond SK, Kingsley KL, White JF (2019) Exploring endophytic communities of plants: methods for assessing diversity, effects on host development and potential biotechnological applications. In: Verma SK, White JF (eds) Seed endophytes. Springer, Cham, pp 55–82
2. Latch GCM, Christensen MJ (1985) Artificial infection of grasses with endophytes. Ann Appl Biol 107:17–24

Chapter 18

Hood and Shew Staining

Harshida A. Gamit and Natarajan Amaresan

Abstract

Endophytic microorganisms are present in the healthy tissues of almost all plants. Generally, endophytes are detected through plant tissue inoculation using agar plate; KOH-aniline blue staining is used to observe the presence of endophytic fungal mycelia distribution inside healthy plant. This staining technique has a higher resolution and can differentiate between hyphae and host plant tissues.

Key words KOH-aniline, Hood and Shew, Fungi, Gamborg's B-5 media

1 Introduction

Hood and Shew developed a KOH-aniline blue technique for fluorescent staining of endophytic organisms or plant tissue-associated fungi [1]. Histological techniques to investigate tissues and observe all types of endophytes (bacteria and fungi) and characteristics of host colonization or colonization processes, documentation of host-penetration events, fungal reproduction, and detection of inoculums. The aniline blue fluorescence staining method has been widely used to stain callose plugs in the phloem and this method provides a high degree of resolution, and contrast between hyphae and host plant tissue.

Some techniques have been used to evaluate endophytic distribution in plants, and Schulz introduced tissue blotting technique to check sterility for the isolation of endophytes from healthy plant tissues [2]. Furthermore, endophytism has been evaluated by microdissection and histological observation of plant tissue using light microscope, confocal microscope, or electron microscope [3–5]. The KOH-aniline blue technique is a fluorescent staining method for clearing tissue for direct visualization of endophytic fungi [3] and endophytic microbes [1]. Some problems arise

A. Sankaranarayanan et al. (eds.), *Endophytic Microbes: Isolation, Identification, and Bioactive Potentials*, Springer Protocols Handbooks, https://doi.org/10.1007/978-1-0716-2827-0_18,

during staining with fluorescent dye, autofluorescence of plant tissue, and nonspecific binding of fluorochromes. The resulting fluorescence intensity increased, and it may not provide clear differentiation of hyphae from the host plant tissues.

2 Materials

2.1 KOH–Aniline Blue Fluorescence Staining [1]

- Fresh plant specimen.
- 1 M KOH.
- Deionized water.
- 4% NaOCl.
- KOH-aniline blue solution (*see* **Note 1**).

2.2 Preparation of Specimen for Live Staining

- Plant seeds.
- Pathogen culture.
- Gamborg's B-5 media.
- Fluorescent light.
- 1 M and 10 M KOH.
- 2% glutaraldehyde.
- 0.05 M potassium phosphate buffer.

3 Methods

3.1 KOH–Aniline Blue Fluorescence Staining

1. Collect fresh plant specimens and autoclave them for 15 min at 121 °C in 50 mL of 1 M KOH (chlorophylls will be removed).
2. Clear the tissues by rinsing the specimen for three times using deionized water.
3. Wash the tissues 4% NaOCl for 3 min and finally rinse twice with distilled water.
4. Mount a section of the specimen on the slide with a small drop of stain solution (KOH-aniline blue solution) and examine under a compound light microscope for the presence of mycelia.

3.2 Plate Method [1]

1. Inoculate surface-sterilized seeds on Gamborg's B-5 basal salt medium with minimal organic substances.
2. Grow the seeds under continuous fluorescent light at a suitable temperature for 3 weeks.
3. Apply a drop or 50 μL pathogen sporidial (to be tested) suspensions on plant leaves.

4. After 5 days of inoculum observe the leaf infection (medium will contain pathogen hyphae).
5. Place hyphae at 5–10 mm from root and grow hyphae connected to root tissue through the medium.
6. After 24 h incubation observe the root infection by excising the specimen.
7. Treat agar-embedded specimen with 10 M KOH at room temperature for 4 h.
8. Then perform standard staining protocol KOH-aniline blue technique.

3.3 Greenhouse Plant

1. Grow plants under greenhouse conditions for 2 weeks.
2. Inoculate pathogen sporidial suspension (50 mL) by soil drenching and incubate for 72 h (2×10^6 per mL).
3. Rinse the infected plant roots and observe through staining protocol KOH-aniline blue technique (*see* **Note 2**).

4 Notes

1. KOH-aniline blue solution: Prepare before 2–3 h prior to use as 0.05% aniline blue dye in 0.067 M K_2HPO_4 at pH 9.0. (Aniline blue is used predominantly.)
2. Infected plant roots can be preserved in 1 M KOH for 5 months at room temperature. Autoclave the preserved specimen and perform standard staining protocol.

References

1. Hood ME, Shew HD (1996) Applications of KOH-aniline blue fluorescence in the study of plant-fungal interactions. Phytopathology 86: 704–708
2. Schulz B, Guske S, Dammann U (1998) Endophyte-host interactions. II. Defining symbiosis of the endophyte-host interaction. Symbiosis 25:213–227
3. Stone JK, Polishook JD, White JF (2004) Endophytic fungi. In: Foster M, Bills G (eds) Biodiversity of fungi. Elsevier Academic Press, Burlington, pp 241–270
4. Lucero ME, Unc A, Cooke P, Dowd S, Sun S (2011) Endophytemicrobiome diversity in micropropagated *Atriplexcanescens* and *Atriplextorreyi* var *griffithsii*. PLoS One 6:e17693
5. Sun X, Guo LD (2012) Endophytic fungal diversity: review of traditional and molecular techniques. Mycology 3:65–76

Part V

Endophytes for Bioactive Metabolites

Chapter 19

Screening of Endophytes for Plant Growth-Promoting Metabolites

Rinka S. Pramanik, Sapna Chandwani, and Natarajan Amaresan

Abstract

Endophytes live symbiotically with the host plant without harm; instead, they benefit them by making nutrients available for plants and promoting plant growth. After isolating endophytes, it is essential to screen them for plant growth-promoting properties such as production of phytohormones, phosphate solubilization, nitrogen fixation, siderophore production, etc. In this chapter, the protocol for the screening of endophytes for the above-mentioned plant growth-promoting traits is provided. This will help in finding the best plant growth-promoting endophytes so that they can be used as an environmentally friendly replacement for chemical fertilizers and pesticides.

Key words Endophytes, Plant growth promoters, Siderophore, Phosphate solubilization, IAA, Nitrogen fixation, ACC deaminase, HCN

1 Introduction

Endophytes are bacteria or fungi that can live in the plant endosphere during all or part of their life cycle without harming the plant [1]. Plants require macro and micronutrients for their growth and development, which are available in complex forms in soil. Endophytes have the ability to solubilize complex compounds present in the soil and transfer the nutrients to their host plants. Plant growth-promoting endophytes have been the prime focus of research on increasing the ability of plants to grow better under changing environmental biotic and abiotic conditions. These plant growth-promoting endophytes increase plant growth and productivity directly or indirectly through several mechanisms, including biological nitrogen fixation [2], phosphate solubilization [3], siderophores production [4], HCN and ammonium production [5, 6], and synthesis of plant growth hormones [7]. In addition, bacterial

A. Sankaranarayanan et al. (eds.), *Endophytic Microbes: Isolation, Identification, and Bioactive Potentials*, Springer Protocols Handbooks, https://doi.org/10.1007/978-1-0716-2827-0_19,

endophytes can tolerate abiotic stress by lowering host ethylene levels by 1-aminocyclopropane-1-carboxylate (ACC) deaminase activity [8]. The effective use of endophytic microbes for plant growth promotion lessens the dependency of different crops on chemical fertilizers.

2 Materials

2.1 Indole Acetic Acid (IAA) Detection

- Culture media (LB, Nutrient Broth, YEM Broth).
- L-tryptophan.
- Shaker incubator (28–37 °C).
- Indole-3-Acetic acid (IAA).
- Centrifuge.
- Salkowski reagent (in 50 mL of 35% perchloric acid, add 1 mL of 0.5 M ferric chloride, mix it well and store in brown bottle).

2.2 Phosphate Solubilization

- Pikovskaya's Agar: Glucose – 10 g; $Ca_3(PO_4)_2$ – 5 g; $(NH_4)_2SO_4$ – 0.5 g; KCl – 0.2 g; $MgSO_4$ – 0.1 g; $MnSO_4$ – traces, $FeSO_4$ – traces, Yeast Extract – 0.5 g; Agar – 15 g; Distilled water – 1 L, pH – 7.0).
- Petri plates.
- Incubator.

2.3 Nitrogen Fixation Assay

- 0.5% Bromophenol blue in alcohol solution.
- NFB medium (KH_2PO_4 – 0.4 g, K_2HPO_4 – 0.1 g, $MgSO_4$.7H2O – 0.2 g, NaCl – 0.1 g, $CaCl_2$ – 0.02 g, $FeCl_3$ – 0.01 g, Mo.O_4Na.$2H_2O$ – 0.002 g, sodium malate – 5 g, Agar Powder – 15 g, bromothymol blue – 5 mL, pH – 7).
- Petri plates.

2.4 Siderophores Production

- Solution I.
  - Dissolve 0.06 g of chrome azurol S (CAS) in 50 mL of distilled water.
- Solution II.
  - Dissolve 0.0027 g of $FeCl_3$-$6H_2O$ in 10 mL of 10 mM HCl.
- Solution III.
  - Dissolve 0.073 g of hexadecyltrimethylammonium bromide (HDTMA) in 40 mL of distilled water.
- Blue Dye.
  - Mix 50 mL solution I with 9 mL solution II, then mix with 40 mL of solution III.

- Minimal Media 9 (MM9) Salt Solution Stock.
  - Dissolve 50 g NH_4Cl, 25 g NaCl, and 15 g KH_2PO_4 in 500 mL of distilled water.
- 20% Glucose Stock
  - Dissolve 20 g glucose in 100 mL of distilled water.
- NaOH Stock.
  - Dissolve 25 g of NaOH in 150 mL distilled water.
- Casamino acid solution.
  - Dissolve 3 g of casamino acid in 27 mL of distilled water.
  - Extract with 3% 8-hydroxyquinoline in chloroform to remove any trace iron.
  - Filter sterilize the casamino acid solution.
- Petri plates.
- 250 mL Erlenmeyer Flask.
- Incubator 30–37 °C.

2.5 ACC Deaminase Detection

- Culture media containing 5 mM ACC.
- Luria Bertani medium.
- DF salt medium.
- Shaker incubator (150–200 rpm; 30–37 °C).
- Centrifuge.
- Tris–HCl buffer (0.1 M) (pH = 8.5).
- Vortex.
- $(NH_4)_2SO_4$.
- Toluene.
- ACC (0.5 M).
- 0.56 N HCl.
- 0.2% 2, 4-dinitrophenylhydrazine reagent (in 2 N HCl).
- NaOH (2 N).

2.6 HCN Production

- Test culture.
- Picric acid solution.
- Glycine.
- Sterile Whatman filter paper.
- Parafilm.

2.7 Ammonium Production

- Peptone water.
- Nessler's reagent (K_2HgI_4; 1.4%).
- Sterile test tubes.

3 Methods

3.1 Indole Acetic Acid (IAA) Detection [9]

3.1.1 Qualitative Method

1. Inoculate the bacteria in the culture media (LB, Nutrient Broth, YEM Broth) supplemented with 0.1% tryptophan.
2. Incubate the culture in a shaker incubator at 28–37 °C for 48 h.
3. After incubation centrifuge the bacterial culture at 4000 rpm for 20 min.
4. Take 1 mL of supernatant and mix with 2 mL of Salkowski's Reagent.
5. Keep the mixture for 20 min and observe the color change.
6. If it turns pink then it is considered a positive for IAA Production (Fig. 1).

3.1.2 IAA Standard Preparation

1. Prepare the indole-3-acetic acid (IAA) standard in acetone or ethanol ranging from 10 to 100 μg per mL of concentration.
2. Take 1 mL of each standard of IAA in a separate tube and add 1 mL Salkowski's reagent.
3. Prepare blank by using 1 mL solvent without IAA and 1 mL Salkowski reagent.
4. Mix the reaction mixture by gentle vortex and incubate it in the dark for 30 min.
5. After incubation read the absorbance at 536 nm wavelength and prepare an IAA standard calibration graph by plotting a graph of absorbance versus IAA concentration.

3.1.3 Estimation of IAA Production from Endophytic Bacteria

1. Inoculate the fresh bacterial culture in the culture media (LB, Nutrient Broth, YEM Broth) supplemented with 0.1% tryptophan.
2. Incubate the inoculated culture in a dark condition (by wrapping the container with aluminum foil) in shaking condition at 120 rpm and 30 °C temperature for 48 h.
3. After incubation, centrifuge it at 10,000 rpm for 5 min.

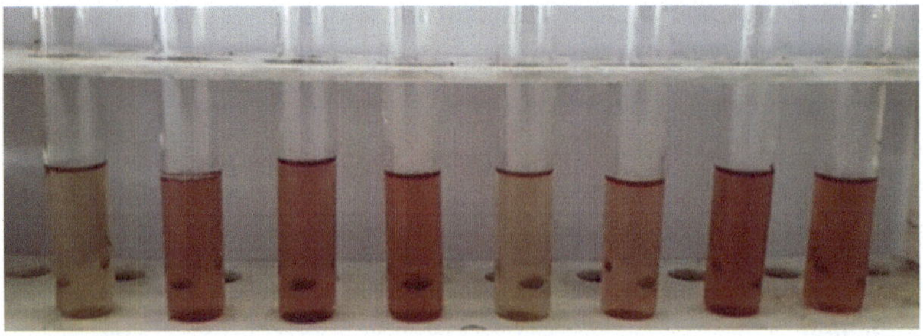

Fig. 1 Qualitative determination of IAA production. Pink color indicates the production of IAA

4. Transfer the supernatant into new test tube.
5. Add equal volume of Salkowski reagent, vortex gently and incubate the reaction at 30 °C in a dark condition for 30 min.
6. Prepare blank by replacing the supernatant with 1 mL of uninoculated medium and mix with 1 mL of Salkowski reagent.
7. Measure the color intensity spectrophotometrically at 536 nm and use uninoculated media as bank.
8. Compare the optical density of the test sample with a standard IAA curve (10–100 $\mu g{\cdot}mL^{-1}$) to calculate the IAA concentration.

3.2 Phosphate Solubilization

1. Prepare Pikovskaya's agar medium [10] and autoclave it at 121 °C and 15 psi for 15 min.
2. Prepare the plates by pouring the media in sterile Petri plates.
3. Spot/streak inoculates the 10 μL of culture in the Pikovskaya's agar medium plate containing tricalcium phosphate as a sole source of phosphate.
4. Incubate the inoculated plates at 28–37 °C for 48–96 h.
5. Observe the plate for the presence of clear a zone around the colony, which indicates the solubilization of phosphate (Fig. 2).
6. Calculate the solubilization efficiency (SE) by [11].

$$\text{Solubilization efficiency} = \frac{\text{Solubilization diameter (S)}}{\text{Growth diameter (G)}} \times 100$$

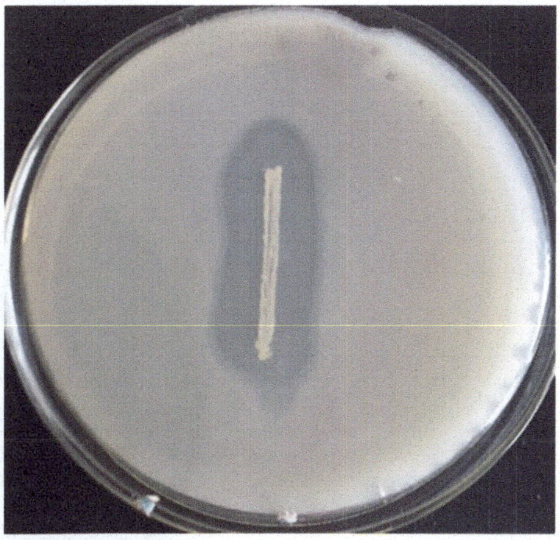

Fig. 2 Phosphate solubilization by endophytic bacteria on Pikovskaya's agar medium plate amended with tricalcium phosphate

3.3 Nitrogen Fixation Assay

1. Prepare the *nitrogen-free* bromothymol blue *malate (NFB) medium* [12] *and autoclave at 121 °C and 15 psi for 15 min.*
2. *Aseptically transfer the NFB medium into sterile Petri plates.*
3. *Inoculate the culture at the center of the plates and inoculate the plates at 35 °C for 10 days.*
4. *The presence of a blue-colored zone around the colony will be considered as nitrogen-fixing bacteria.*

3.4 Siderophore Production [13]

1. Clean all glassware with 6 M HCl for removing any trace elements followed by rinsing with distilled water.
2. Dissolve 100 mL of MM9 salt solution to 750 mL of distilled water.
3. Dissolve 32.24 g piperazine-N,N′-bis(2-ethanesulfonic acid) PIPES.

   (Note: PIPES will not dissolve below pH of 5. Bring pH up to 6 and slowly add PIPES while stirring. The pH will drop as PIPES dissolves. Bring the pH up to 6.8 while stirring but the pH should not exceed 6.8 otherwise it will turn green.)
4. Add 15 g agar powder to the MM9/PIPES mixture.
5. Autoclave the mixture and cool to 50 °C.
6. Add 10 mL of sterile 20% glucose solution and 30 mL of sterile casamino acid solution to MM9/PIPES mixture.
7. Slowly add 100 mL of blue dye solution along the glass wall with enough agitation to mix thoroughly aseptically.
8. Inoculate the culture at the center of the plate and incubate at 28 °C for 48–72 °C.
9. The appearance of an orange halo around the colony indicates the siderophore synthesis (Fig. 3).

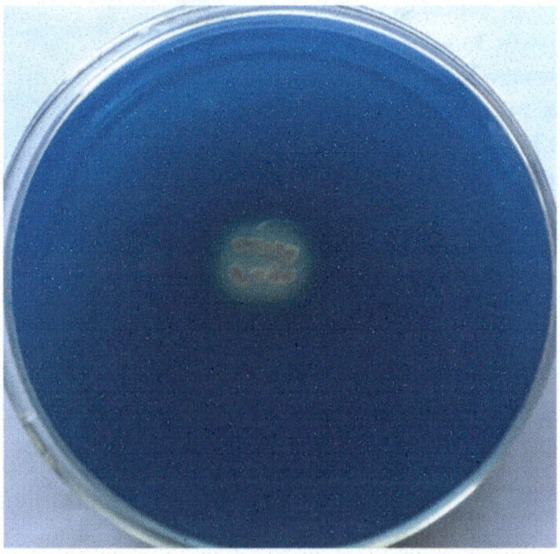

Fig. 3 Siderophore production by endophytic bacteria on CAS medium

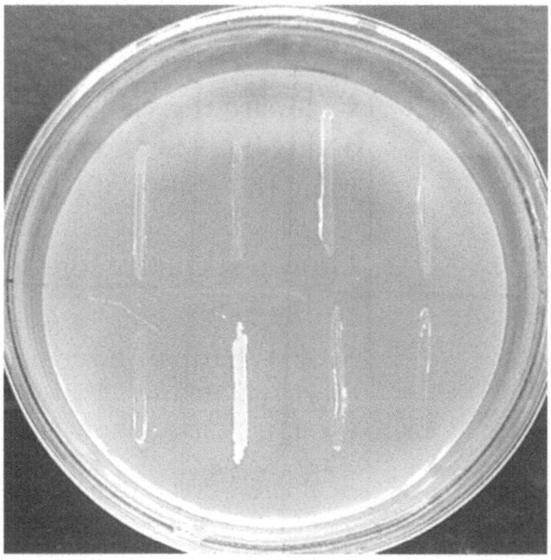

Fig. 4 Endophytic bacteria showing the growth on minimal medium amended with 3 mM ACC

3.5 ACC Deaminase Production [14]

3.5.1 Qualitative Method [14]

1. Grow the bacterial isolates in LB medium for 24 h at 28 °C in a rotary condition at 150 rpm.
2. Centrifuge at 8000 rpm for 5 min, collect the cell pellet, and perform the washing step with sterile distilled water.
3. Resuspend the pellet in 1 mL of sterile distilled water and spot inoculate it on Petri plates containing DF salt medium/minimal salt medium supplemented with 3 mM ACC as nitrogen source.
4. Compare the growth of isolates on ACC supplemented plates with positive control [$(NH_4)_2SO_4$) as nitrogen source] and negative control (DF salt medium/minimal salt medium without ACC) after 3–4 days of incubation at 28 °C.
5. Isolates grown well on ACC supplemented plates were considered as positive for ACC deaminase production (Fig. 4).

3.5.2 Quantitative Method [15]

1. Inoculate the isolates/bacterial culture in Luria broth.
2. Incubate the cultures at room temperature for 48 h with constant shaking.
3. In a 96-well micro-titer plate, add 122 μL of DF medium containing $(NH_4)_2SO_4$ in lane 5 and 6 and fill 6 μL of thawed ACC with DF medium without any nitrogen source in lane 1 and 2.
4. Fill the two other lanes 3 and 4 with DF medium without $(NH_4)_2SO_4$ or substrate ACC.
5. Inoculate each well with 22 μL of bacterial culture except in uninoculated control wells where add 22 μL of distilled water in place of bacteria.

6. Measure the optical density after 0, 24, 48, 72, and 96 h at 496 nm.
7. Compare the value of ACC without ACC/ammonium sulfate and $(NH_4)_2SO_4$ well along with distilled water wells to determine the ability of bacteria to metabolize ACC.

3.6 HCN Production [16]

1. Grow test culture for 48 h on a suitable medium supplemented with glycine.
2. Soak Whatman filter paper into the picric acid solution (0.5% picric acid and 2% of sodium carbonate) and place inside the lids.
3. Seal the Petri pates using parafilm and incubate at 26–28 °C for 7 days.
4. After incubation period, observe the color of Whatman filter paper.
5. The color changes orange or red, brown indicates the microbial HCN production (Fig. 5).

3.7 Ammonia Production [17]

1. Grow the test culture in a peptone water filled test tubes and incubate at 30 °C for 72 h.
2. After incubation period, add 1 mL of Nessler's reagent in cultured tubes.
3. Observe the color change – faint yellow color will indicate small amount of ammonia and deep yellow to brownish color will indicate maximum ammonium production (Fig. 6).

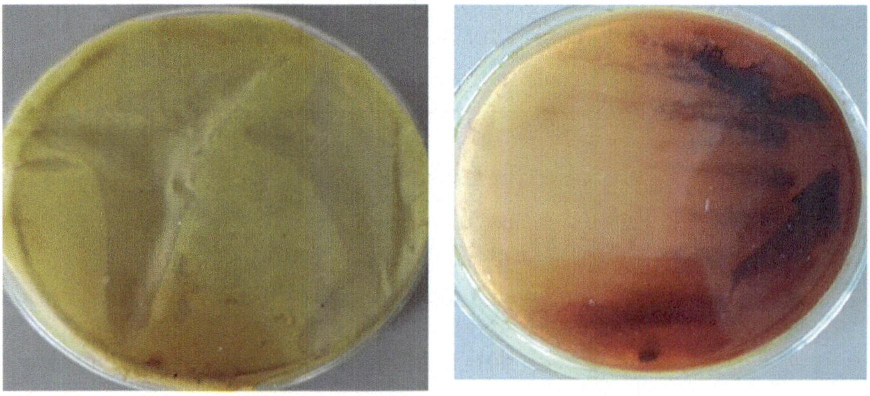

Fig. 5 Endophytic bacteria showing the production of HCN

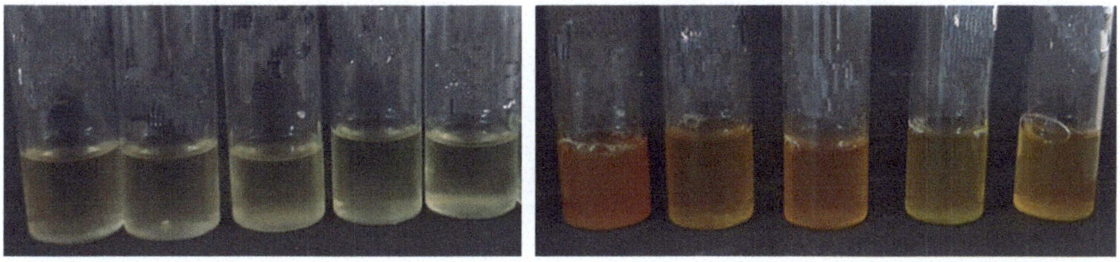

Fig. 6 Endophytic bacteria showing the development of faint yellow to deep yellow to brownish color after adding Nessler's reagent

References

1. Wilson D (1995) Endophyte: the evolution of a term, and clarification of its use and definition. Oikos 73:274
2. Lowman S, Kim-Dura S, Mei C, Nowak J (2016) Strategies for enhancement of switchgrass (Panicum virgatum L.) performance under limited nitrogen supply based on utilization of N-fixing bacterial endophytes. Plant Soil 405:47–63
3. De Abreu CS, Figueiredo JEF, Oliveira CA, Dos Santos VL, Gomes EA, Ribeiro VP, Barros BA, Lana UGP, Marriel IE (2017) Maize endophytic bacteria as mineral phosphate solubilizers. Genet Mol Res 16
4. Rosconi F, Davyt D, Martínez V, Martínez M, Abin-Carriquiry JA, Zane H, Butler A, de Souza EM, Fabiano E (2013) Identification and structural characterization of serobactins, a suite of lipopeptidesiderophores produced by the grass endophyte *Herbaspirillumseropedicae*: Siderophores of *Herbaspirillumseropedicae*. Environ Microbiol 15:916–927
5. Lorck H (1948) Production of hydrocyanic acid by bacteria. Physiol Plant 1:142–146
6. Castric PA (1977) Glycine metabolism by *Pseudomonas aeruginosa*: hydrogen cyanide biosynthesis. J Bacteriol 130:826–831
7. Defez R, Andreozzi A, Bianco C (2017) The overproduction of Indole-3-Acetic Acid (IAA) in endophyte supregulates nitrogen fixation in both bacterial cultures and inoculated rice plants. Microb Ecol 74:441–452
8. Glick BR (2014) Bacteria with ACC deaminase can promote plant growth and help to feed the world. Microbiol Res 169:30–39
9. Gordon SA, Weber RP (1951) Colorimetric estimation of indoleacetic acid. Plant Physiol 26:192–195
10. Pikovskaya RI (1948) Mobilization of phosphorus in soil in connection with the vital activity of some microbial species. Microbiol 17:362–370
11. Nguyen C, Yan W, Le Tacon F, Lapayrie F (1992) Genetic variability of phosphate solubilizing activity by monocaryotic and dicaryotic mycelia of the ectomycorrhizal fungus *Laccariabicolor* (Maire) PD Orton. Plant Soil 143: 193–199
12. Döbereiner J, Day JM (1976) Associative symbioses in tropical grasses: characterization of microorganisms and dinitrogen-fixing sites. In: Newton WE, Nyman CJ (eds) Proceedings of the first international symposium on N_2-fixation, vol 2. Washington University Press, Washington, DC, pp 518–538
13. Schwyn B, Neilands JB (1987) Universal chemical assay for the detection and determination of siderophores. Anal Biochem 160:47–56
14. Govindasamy V, Senthilkumar M, Mageshwaran V, Annapurna K (2009) Detection and characterization of ACC deaminase in plant growth promoting rhizobacteria. J Plant Biochem Biotechnol 18:71–76
15. Jacobson CB, Pasternak JJ, Glick BR (1994) Partial purification and characterization of 1-aminocyclopropane-1-carboxylate deaminase from the plant growth promoting rhizobacterium *Pseudomonas putida* GR12-2. Can J Microbiol 40:1019–1025
16. Bakker AW, Schipper B (1987) Microbial cyanide production in the rhizosphere in relation to potato yield reduction and *Pseudomonas*spp mediated plant growth stimulation. Soil Biol Biochem 19:451–457
17. Cappuccino JG, Sherman N (1992) Biochemical activitiesof microorganisms. In: Cappuccino JG, Welsh CT (eds) Microbiology, a laboratory manual, 1st edn. The Benjamin/Cummings Publishing Co, San Francisco, pp 105–300

Chapter 20

Screening of Endophytes for Biocontrol Properties

Shah Rushabh, Sapna Chandwani, and Natarajan Amaresan

Abstract

Endophytes are widely present in plants, and findings are constantly being made on their ability to protect plants from disease-causing pathogens, particularly in agriculture. Furthermore, they have an advantage compared to other biocontrol agents in that they can colonize the internal plant tissues. This chapter aims to describe the methods for evaluating the biocontrol properties of individual strains.

Key words Endophyte, Biocontrol agents, Plant pathogens, Antagonistic activity, Plant diseases

1 Introduction

Endophytes are microorganisms that colonize inside the plant tissues. The use of antagonistic endophytes as biocontrol agents is gaining popularity as a promising alternative for controlling various plant diseases. Biological control of plant diseases is an eco-friendly and practical approach to prevent or reduce the disease-producing activities of pathogens by using microorganisms [1]. The mechanisms of biocontrol may be competition for nutrients and space, production of antibiotics, enzymes, and volatile organic compounds, induced systemic resistance in plants, and prevent the colonization of pathogens to the host tissues [2]. Biocontrol may also enhance the production of biochemical compounds related to host defense, including massive accumulation of phenolic compounds and phytoalexins; defense enzymes and transcripts; improved lignifications, and increases in the activities of pathogenesis-related (PR) proteins, which may all contribute to increased resistance [3]. Biocontrol agents may use more than one of these mechanisms; therefore, the biocontrol of plant pathogens is the overall result of multiple mechanisms that may be activated concurrently.

A. Sankaranarayanan et al. (eds.), *Endophytic Microbes: Isolation, Identification, and Bioactive Potentials*,
Springer Protocols Handbooks, https://doi.org/10.1007/978-1-0716-2827-0_20,

2 Materials

- Potato dextrose agar.
- Nutrient broth.
- Fungal plant pathogens.
- Bacterial plant pathogens.
- Endophytic bacteria.
- Petri plates.
- Nichrome wire.
- Autoclave.
- Cork borer.
- Laminar flow cabinet.

3 Methods

3.1 Dual Culture Method (Fungal Plant Pathogens)

1. Isolate the fungal plant pathogens (*Fusarium oxysporum, Colletotrichum falcatum, Sclerotium rolfsii, Rhizoctonia solani,* etc.) or obtain the strain from any culture collection center (*see* **Note 1**).
2. Prepare potato dextrose agar (PDA) for fungal growth, inoculate the culture, and incubate for 7–10 days at 28 ± 2 °C.
3. Prepare the nutrient broth for bacterial growth, inoculate each endophytic bacterial strain, and incubate the tubes at 37 ± 2 °C for 24 h.
4. Using a cork borer, take 5 mm agar plugs with the actively growing culture of fungal strain and place them on PDA for antagonistic activity.
5. Streak the endophytic bacteria 2.5 cm apart from the fungal disk and incubate the plates for 7–10 days at 28 ± 2 °C. Examine and record any changes in the morphology of each colony regularly.
6. Maintain the control plate without endophytic bacteria.
7. On the bottom of the control and test plates, note where the pathogen colony's margin is located every day until the growth ceases.
8. Inhibition of mycelia growth toward the direction of the endophytic bacteria indicates antagonistic activity (Fig. 1).
9. At the end of the experiment, calculate the percentage of radial mycelia growth inhibition by the following formula:

$$\text{Inhibition }(\%) = \frac{(C - T) * 100}{C}$$

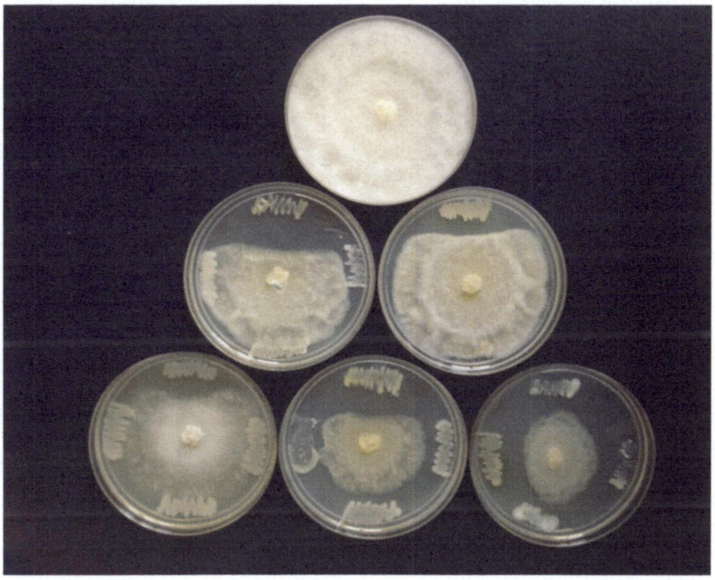

Fig. 1 Endophytic bacteria showing mycelial inhibition of *Pythium aphanidermatum*

where C – diameter of radial growth of pathogenic fungi in control; T – diameter of radial growth of pathogenic fungi in with antagonistic bacteria.

3.2 Agar Well Diffusion Method (Bacterial Plant Pathogens)

1. Streak the bacterial plant pathogens (Ralstonia solanacearum, Pseudomonas syringae pathovars, Xanthomonas oryzae pv. Oryzae, Agrobacterium tumefaciens, etc.) onto a nutrient agar plate to produce single colonies (*see* **Note 1**).
2. Incubate plates upside down overnight in an incubator at 37 ± 2 °C. These plates can be stored at 4 °C.
3. Pour 5 mL nutrient broth into a 15 mL tube. Inoculate the broth with picked bacterial plant pathogen colonies using an inoculation loop and incubate the culture at 37 ± 2 °C with shaking at ~150 rpm overnight.
4. Prepare the melting agar tube and place the tube in a 50 °C water bath to prevent solidification.
5. Once the top agar reaches 40–45 °C, add the bacterial plant pathogen (the bacteria in the sample will be killed if the agar is too hot).
6. Gently invert the tube two times after covering it with a sterile lid.
7. Pour the agar quickly into the Petri dish containing the solidified base agar and gently swirl the Petri dish to spread it evenly. Allow the agar to solidify in the Petri dish under the laminar flow cabinet with the lid partially open.

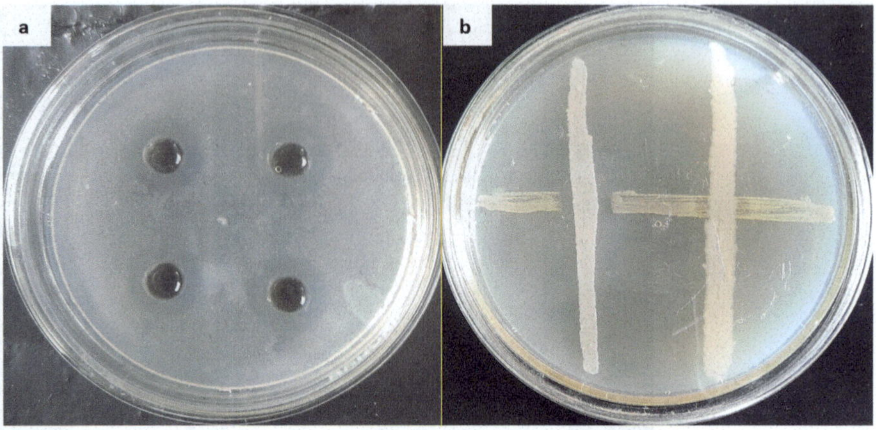

Fig. 2 Endophytic bacteria showing inhibition of *Xanthomonas campestris:* (**a**) Well-diffusion method, (**b**) Streaking method

8. With the help of a cork borer, make a well in nutrient agar plate previously seeded with culture.
9. Add the endophytic bacterial culture into the well with the help of a sterile pipette or using the 0.22 μm filter, remove the cell, and add the filtrate into the well. Cross streaking of both endophytes and pathogens can also be performed.
10. Incubate the plate for 2–3 days at 37 ± 2 °C until a clear zone on the agar is visible at the center of the plate where the endophytic bacterial suspension is applied. Measure the inhibition zone (mm) at least three values for each sample (Fig. 2).

4 Note

1. Take a specific medium that allows the growth of bacterial/ fungal plant pathogens

References

1. Heimpel GE, Mills NJ (2017) Biological control. Cambridge University Press, Cambridge/ New York
2. Narayanasamy P (2013) Mechanisms of action of fungal biological control agents. In: Narayanasamy P (ed) Biological management of diseases of crops. Springer, Dordrecht, pp 99–200
3. Reddy PP (2014) Mechanisms of Biocontrol. In: Reddy PP (ed) Plant growth promoting rhizobacteria for horticultural crop protection. Springer, New Delhi, pp 55–68

Chapter 21

Screening of Endophytes for Biocatalytic Tools

David Gonzalez, Emiliana Botto, Sonia Rodríguez Giordano, and Paula Rodríguez Bonnecarrere

Abstract

Endophytic communities thrive in environments that are rich in biologically active compounds. To survive, endophytes have adapted to produce secondary metabolites mimicking their host plant and synthesize the required enzymes which are used for cell wall penetration and colonization of the host plant. Therefore, endophytic microorganisms could be a source of novel enzymes able to catalyze the organic transformation of a wide range of valuable fine chemicals derived from natural products.

This chapter presents various strategies to discover new biocatalyst within the endophytic community. The chemical entities naturally present in the plant suggest that the associated endophytes could be able to synthesize, metabolize, or modify a portion of the chemical structure. Enrichment and isolation in the presence of the metabolite of interest is a useful strategy to selectively isolate endophytes with the metabolic capacity to modify or degrade the compound. In addition, the use of different growing media, including one that mimics the internal environment of the plant, has also proven advantageous.

Key words Endophytes, Screening, Carbonyl reductase, Xylanase, Cellulase

1 Introduction

Biocatalytic transformations play an important role in the manufacturing of optically pure chiral fine chemicals for use in the pharmaceutical, or other high technology industries.

Biocatalysis provides an interesting alternative for the synthesis of chiral molecules since the reaction conditions are efficient, performed under mild conditions, purification procedures are often simple, and the regio- and stereoselectivity achieved is extremely high. These virtues make biocatalysis a valid green alternative to conventional organic transformations [1].

The development of biocatalysis requires the discovery or design of novel biocatalysts suitable for a wide range of applications. Isolation of novel microorganisms from different habitats is an interesting approach for expanding the biocatalytic repertoire. The diversity of microbial life is enormous and the niches in

A. Sankaranarayanan et al. (eds.), *Endophytic Microbes: Isolation, Identification, and Bioactive Potentials*, Springer Protocols Handbooks, https://doi.org/10.1007/978-1-0716-2827-0_21,

which microbes live at are of amazing diversity; among them, an under-explored source of biocatalytic diversity is probably represented by the endophytic community [2].

Endophytic microorganisms comprise special features that make them an appealing source of novel biocatalysts [3]. They have adapted to survive in an environment rich in biologically active compounds and they have likely developed an enzymatic arsenal to deal with a diverse array of toxic compounds [4–6]. Furthermore, several endophytes are able to produce secondary metabolites identical to those from their host plant [6–8]. Additionally, these microorganisms can synthesize the required enzymes which are used for cell wall penetration and colonization of the host plant [9]. Endophytic fungi in association with the hosts secrete enzymes which are presumed to aid in growth, nourishment, and defense [10]. Likewise, there is ample evidence that endophytes can carry out chemical transformations of natural products [10–12]. In addition, these microorganisms have recently been used in biocatalytic transformation of exobiotic substrates, particularly in redox reactions [1, 2, 13–16]. These characteristics indicate that endophytic microorganisms can be a rich source of novel enzymes [14].

The search for novel biocatalysts among the endophytic community could be guided by previous knowledge of the host plant secondary metabolism. The hypothesis behind this strategy is that endophytes have developed adaptation strategies to cope with their host chemical arsenal, and thus they will likely biotransform the natural products present in their host plant [3, 4, 6, 17]. An alternative but related approach is to search for a particular enzymatic activity. In this case, information on the plant secondary metabolism can be used as a guide to find compounds that can be modified by that enzyme. Our group has followed this strategy using carvone (*4R*)-(−)-carvone) as model substrate since this monoterpene allowed exploring different biocatalytic activities within the same molecule [14]. *Mentha pulegium* plant was selected as a potential source of endophytic biocatalysts since its essential oil is rich in monoterpenes structurally similar to carvone [18]. The hypothesis proposed by us was that pennyroyal endophytes must be adapted to the presence of these terpenes, and therefore could exhibit interesting enzymatic activity for the biotransformation of (*4R*)-(−)-carvone and (*4S*)-(+)-carvone. Following this strategy, two biocatalysts were isolated with very good enoate reductase activity and complementary substrate specificity [14].

Another approach in the search for biocatalysts among endophytes is aimed to the community present in plants with technologically valuable enzymatic activity like amylases, chitinases, lipases, proteases, cellulases, and xylanases [10]. Particularly cellulolytic, hemicellulolytic, and ligninolytic extracellular enzymes produced by endophytic microorganisms are useful for obtaining C5 and C6 sugars from the lignocellulosic biomass. Xylose and glucose are used as substrates in fermentation processes to produce biofuels, building blocks, and high value-added compounds. In this context, we investigated the ability of endophytic microorganisms that populate the leaves, bark, and branches of a *Eucalyptus globulus* tree to produce xylanase and cellulase enzymes. In this investigation, we isolated two endophytic microorganisms with high enzymatic activity. A *Bacillus* strain was found to produce xylanase and a *Preussia* strain synthesized cellulase enzymes [19].

A third and slightly different strategy on the identification of novel endophytic biocatalyst with a particular enzymatic activity is to focus on a plant that has shown a particular enzymatic action. This strategy was applied by our group in the search for biocatalyst with carbonyl reductase activity. The isolation of endophytes from plants with reported Prelog carbonyl reductase activity like carrot (*D. carota*) [20, 21], fennel (*F. vulgare*) [22, 23], and squash (*C. maxima*) [22, 23] yielded interesting biocatalysts with the same activity [15, 24]. On the other hand, the isolation of endophytes from radish (*R. sativus*), a plant with recognized anti-Prelog reductase activity [25], yielded some isolates that presented the same enantioselectivity [2].

2 Materials

2.1 General Procedure for the Isolation of Endophytes to Obtain New Biocatalysts

2.1.1 Sampling

- Sterile scissors.
- Sterile sealed bags (for sample collection).

2.1.2 Disinfection

- 10% Sodium hypochlorite.
- Deionized sterile water.
- 70% Ethanol.
- Sterilized filter paper (dry plant material purpose).
- Sterile stainless-Steel tweezer.
- Sterile 200 μL tips.
- Micropipette 200 μL.
- Drigalski spatula.
- Inoculating loop.
- Petri dishes with Tryptic Soy Agar (TSA) and Potato Dextrose Agar (PDA).
- Incubator (28 and 35 °C).
- Autoclave.

- Laminar flow chamber (if not available, disinfected lab bench with two Bunsen burners).
- Sterilization oven.

2.1.3 Isolation of Endophytic Microorganisms

- Erlenmeyer with 50 mL appropriate sterilized broth.
- Petri dishes with appropriate media: Tryptic Soy Agar (TSA), Potato Dextrose Agar (PDA), M9 mineral medium, Potato Dextrose Broth (PDB), Yeast Nitrogen Base (YNB). For customized plant-medium: Suspend groundplant material from the same plant that is used to make the isolates of the endophytes in distilled water.
- Agar.
- Sterile mortar.
- Sterile knife.
- Laboratory balance.
- Orbital shaker (at 28 and 35 °C) and incubator (28 and 35 °C).
- Optical microscopy.
- General microbiology lab equipment as listed in Subheading 2.1.2.

2.1.4 Conservation of Endophytic Microbial Strains

- 2 mL Cryotubes.
- 30% Glycerol (bacteria and yeast conservation purpose).
- Petroleum jelly (fungi conservation purpose).
- Vials with sterile distilled water (fungi conservation purpose).
- Freezer (−70 °C and −20 °C).
- Refrigerator (4 °C).
- General microbiology lab equipment.

2.1.5 Identification of Bacteria Endophytic Strains

- Gram stain dyes: crystal violet, iodine, ethanol, and safranin.
- N, N-dimethyl-p-phenylenediamine for oxidase activity assay.
- 1% H_2O_2 for catalase activity assay.
- OF Basal Medium supplemented with glucose.
- Thioglycolate broth.
- Bacterial genomic DNAs extraction and purification commercial kit.
- Eubacterial primers: primers 27F (5′-AGAGTTTGATCCTGGCTCAG-3′) and 1492R (5′-GGTTACCTTGTTACGACTT-3′).
- PCR reaction fine chemicals as per standard protocols.
- Thermocycler.
- UV transilluminator.

2.1.6 Identification of Fungal Endophytic Strains

- YES media culture.
- Eppendorf tubes.
- Liquid nitrogen.
- Sterile glass rod.
- Chemicals as described by Griffiths et al. standard to fungi DNA extraction [26].
- Primers: ITS1 (TCCGTAGGTGAACCTGCGG) and ITS4 (TCCTCCCGCTTATTGATATGC).
- Primers: ITS1-F (5′-TCCGTAGGTGAACCTGCGG-3′) and NL-4 (5′-TCCTCCGCTTATTGATATGC-3′).
- General molecular biology reagents and equipment as mentioned in Subheading 2.1.5.

***2.2 Isolation of Endophytes with Biocatalytic Activity from* Mentha pulegium**

a. **Isolation of Endophytic Microorganisms**
   - 10 mM *R*-carvone or *S*-carvona as substrate for selective pressure.
   - *Mentha pulegium* leaves.
   - 50 mL Erlenmeyer flasks with 10 mL of appropriate media.
   - Petri dishes with appropriate media.
   - Sterile knife.
   - Sterile mortar.
   - Orbital shaker (at 28 and 35 °C) and incubator (28 and 35 °C).

b. **Screening for Biocatalytic Activity Among Endophytic Strains**
   - mM *R*-carvone or *S*-carvona as substrate model.

   (i) Screening of biocatalytic activity by bacteria.
      - Inoculating loop.
      - Petri dishes with TSA.
      - Test tube with 5 mL TSB.
      - Sterile 1000 μL tips.
      - Micropipette 1000 μL.
      - 25 mL Erlenmeyer flask with 5 mL of TSB.
      - 2 mL Eppendorf.
      - Orbital shaker and incubator at 35 °C.

   (ii) Screening of biocatalytic activity by yeast.
      - Inoculating loop.
      - Petri dishes with PDA.
      - 500 mL Erlenmeyer flask with 100 mL of YM Broth.
      - Laboratory centrifuge.
      - 150 mL sterilized physiological serum.

- Laboratory balance.
- 25 mL Erlenmeyer flask with 5 mL of 10% dextrose.
- Orbital shaker and incubator at 28 °C.

(iii) Screening of biocatalytic activity by fungi.

- Inoculating loop.
- Petri dishes with PDA.
- 50 mL sterile physiological serum.
- Neubauer chamber to count fungi spores.
- 100 mL Erlenmeyer flask with 50 mL of PDB.
- 2 mL Eppendorf.
- Orbital shaker and incubator at 28 °C.

(iv) Analysis of the biotransformation products.

- Laboratory centrifuge.
- Tubes to collect biotransformation reaction.
- Ethyl acetate.
- Na_2SO_4 (anh.).
- Filter paper.
- Rotary evaporator.
- CH_2Cl_2.
- Gas chromatograph (GC).

2.3 Isolation of Endophytes with Xylanase and Cellulase Activity

(a) **Isolation of Endophytic Microorganisms**

- Beechwood xylan.
- Carboxymethyl cellulose (CMC).
- Leaves, bark, and branches of *Eucalyptus* tree.
- Sterile knife.
- Sterile mortar.
- 50 mL Erlenmeyer flasks with 10 mL of appropriate media.
- Petri dishes with appropriate media.
- Agar.
- Orbital shaker (at 28 and 35 °C) and incubator (28 and 35 °C).

(b) **Screening for Xylanase and Cellulase Activity on Solid Media**

- Inoculating loop.
- Petri dishes with PDA and TSA.
- Petri dishes with YNB/M9 agar with 0.5% xylan/CMC.

- Congo red (0.1% Congo red in distilled water).
- 1 M NaCl for 15 min.
- Incubator (28 and 35 °C).

(c) **Xylanase and Cellulase Production on Liquid Media**

- Inoculating loop.
- Petri dishes with PDA and TSA.
- Erlenmeyer with 20 mL of PDB, YNB, TSB, or M9 with addition of 0.5% beechwood xylan/CMC.
- Orbital shaker (at 28 and 35 °C) and Incubator (28 and 35 °C).

(d) **Xylanase and Cellulase Quantification**

- Suspension of beechwood xylan in 50 mM sodium citrate buffer at pH 4.8.
- 3,5-dinitrosalicylic acid (DNS).
- mL of sodium citrate buffer (50 mM, pH 4.8) with 50 mg filter paper (Whatman no. 1, 0.25 mm pore size).
- UV-spectrophotometer.
- Incubator.

2.4 Isolation Endophytes with Carbonyl Reductase Activity

(a) **Isolation Endophytic Microorganisms**

- 10 mM of ethyl 2-methylacetoacetate or similar substrate for selective pressure.
- Fresh carrot (*Daucus carota*) root.
- Fennel (*F. vulgare*) leaves.
- Endocarp and mesocarp of squash (*C. maxima*).
- Materials describe above to disinfect plant tissue.
- 50 mL Erlenmeyer flasks with 10 mL of appropriate media.
- Petri dishes with appropriate media.
- Sterile knife.
- Sterile mortar.
- Orbital shaker (at 28 and 35 °C) and incubator (28 and 35 °C).

(b) **b. Screening Biocatalytic Activity from Endophytic Strains**

- 10 mM *R*-carvone or *S*-carvona as substrate model.
- Same materials as mentioned in Subheading 2.2.b.
- To identify the different products obtained in the biotransformation **2** or **3** compounds proceed as described by Rodríguez et al. [15].

3 Methods

3.1 General Procedure for the Isolation of Endophytes to Obtain New Biocatalysts According to the Strategies Mentioned above

3.1.1 Collect healthy plant material in a sterile plastic bag and transport it to the lab.

3.1.2 Surface disinfection of each plant material must be done under aseptic conditions as follows:

(i) Immerse plant material in a 10% sodium hypochlorite solution for 3 min.

(ii) Rinse three times with sterile distilled water.

(iii) Immerse in 70% ethanol for 3 min.

(iv) Rinse three times with sterile distilled water.

(v) Dry the plant material by draining it on filter paper.

3.1.3 To confirm that the surface disinfection process was successful, perform:

(i) Press a sample of disinfected plant material onto the surface of the culture media Tryptic Soy Agar (TSA) and Potato Dextrose Agar (PDA). After the impression remove the plant material carefully with the help of tweezers from the surface of the medium and discard it.

(ii) Sow in the surface on TSA and PDA media 100 uL of the last volume of water used in the wash (step iv of disinfection).

(iii) Incubate the plates at 28–30 °C for 10 days to verify the absence of epiphytic microorganisms.

3.1.4 Inoculate the plant fragments in different media to isolate endophytic microorganisms. To increase the diversity of isolated endophytes, use liquid and solid media with different compositions:

(i) TSA and Tryptic Soy Broth (TSB). These media support growth of a broad range of bacteria.

(ii) Minimal medium M9 containing the substrate of interest as sole carbon source. This medium is used to enrich the culture with bacteria that could metabolize the substrate of interest. Add agar to 20 g/L for solid media.

(iii) PDA and Potato Dextrose Broth. These media support growth of a broad range of fungi.

(iv) YNB supplemented with the substrate of interest as sole carbon source. This medium is used to enrich the culture with yeast and filamentous fungi that could metabolize the substrate of interest. Add agar to 20 g/L for solid media.

(v) Customized plant-medium: this medium is used to mimic plant host internal environment. To prepare this culture medium, plant material from the same plant that is being assayed for endophyte isolation is used. This medium is prepared by grinding the plant material and suspending it in the appropriate amount of distilled water. Agar is added to 20 g/L for solid media. Sterilization is performed in autoclave, at 121 °C and 1.05 atm. For 15 min [15].

To cultivate endophytic microorganisms on liquid media, place the disinfected plant material into a sterile mortar under aseptic conditions and macerate. Use this macerate to inoculate each of liquid media proposed above. Incubate at 28 °C and 35 °C and 150 rpm for 24–96 h. Follow with three rounds of serial culturing (1:10 dilutions) incubating at the same conditions. In each step, spread an aliquot of cell culture onto PDA, TSA and plant-medium agar and incubate at 28 °C and 35 °C. Inspect the plates daily. Isolate all colonies found and maintain for later experiments.

To cultivate endophytic microorganisms on solid media, cut aseptically surface sterilized plant material into small and thin segments and plate by pressing directly onto each of solid media proposed above. Incubate the plates at 28 °C and 35 °C for 24 h to 10 days. Inspect the plates daily. Transfer endophytic microorganisms that emerge from the plant tissue to fresh media to isolate them.

3.1.5 Characterize isolated colonies as bacteria, yeast, or filamentous fungi by macro and microscopic observation.

3.1.6 Conservation of endophytic microbial strains.

(i) For bacteria and yeasts conservation use a cryopreservation procedure. Make microbial suspensions in glycerol at 30% and store them in a freezer at −70 °C.

(ii) For filamentous fungi, grow isolate strains in PDA tubes and cover with sterile petroleum jelly to preserve them in an oxygen-free atmosphere. The obtained filamentous fungi are stored at a temperature of 4 °C by subculturing them every 8 months.

3.1.7 Identification of endophytic strains.

3.1.7.1 The identification of bacterial strains is done based on classical phenotypic analysis, and phylogenetic analysis of the 16S RNA gene.

Conduct classical phenotypic analysis for microbial identification at genus level based on the morphological characteristics and selected standard physiological and biochemical tests, according to the Bergey's Manual of Systematic Bacteriology [27] and Cowan and Steel's Manual for the identification of medical bacteria [28]. Perform colony morphology, Gram-staining, oxidase reaction, and catalase activity tests on 24 h grown cultures on TSA. Perform oxidase reaction with a 1% aqueous solution of N, N-dimethyl-p-phenylenediamine, and determine catalase activity by adding a 1% H_2O_2 solution to a culture grown on TSA for 24 h [29]. Determine oxidation and fermentation of glucose in OF medium [30]. Analyze the oxygen requirements of the isolated bacterial strains by Thioglycolate broth.

For phylogenetic analysis of the 16S RNA gene amplify of almost full-length 16S rRNA gene fragments using eubacterial primers DNA. First, extract genomic DNA from cells growing overnight on TSB. Performed the PCR using the following conditions: initial denaturation at 94 °C for 5 min, followed by 30 cycles of denaturation (1 min at 94 °C), annealing (1 min at 55 °C) and extension (3 min at 72 °C) and elongation step (7 min at 72 °C). Compare the 16S rRNA gene sequence obtained with other released sequences in the GenBank database using BLAST search (National Center for Biotechnology Information) and the Ribosomal Database Project II (release 9.47) [31]. Perform phylogenetic analysis using the software package MEGA version 2.1 [32] as described by Menes et al. [29].

3.1.7.2 Identification of fungal endophytic strains.

Perform phenotypic identification to the genus level according to Pitt and Hocking [33] and at the species level by sequence analysis of the ITS1–5.8S-ITS2 genomic region of the nuclear rRNA gene and the D1/D2 variable domains of the 26S rRNA gene at the 5'end of the large subunit rRNA gene [34]. For genomic DNA extraction, inoculate fungi isolates in 10 mL of liquid media composed of 2% yeast extract and 15% sucrose (YES) and incubate for 5 days at 28 °C. Recover mycelium or yeast cells and dry using sterile paper towels. Place approximately 200 mg of biomass in a 2 mL Eppendorf tube, frozen in liquid nitrogen, and grind to a fine powder with a sterile glass rod. Extract genomic DNA according to Griffiths et al. [26]. Use genomic DNA to amplify the ITS1–5.8SrDNA-ITS2 region using primer pair ITS1–ITS4 and D1/D2 domain of the LSU rRNA gene with primers ITS1-F and NL-4. Perform the PCR using the following conditions: initial denaturation at 96 °C for 2 min, followed by 35 cycles of denaturation (30 s at 96 °C), annealing (45 s at 51 °C), extension (120 s at 72 °C), and a final elongation step (7 min at 72 °C). Sequence the resulting PCR products and compare with other released sequences in the GenBank database using BLAST program (National Center for Biotechnology Information) (http://www.ncbi.nlm.nih.gov).

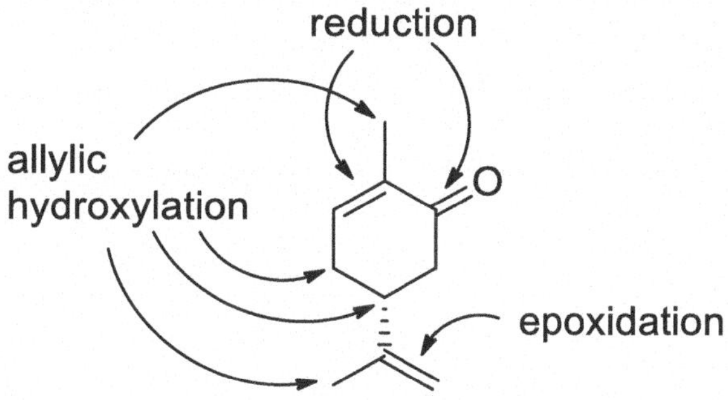

4R-(-)-carvone (**1**)

Fig. 1 Use of *4R*-(-)-carvoneas substrate to explore biocatalytic activity

3.2 Particular Cases

3.2.1 Isolation of Endophytes with Biocatalytic Activity from Mentha pulegium

(i) Use *4R*-(-)-carvone (**1**) as model substrate since it allows to explore allylic hydroxylation, epoxidation, carbonyl reduction, and enoate reduction activities within the same molecule (Fig. 1) to obtain novel biocatalyst.

(ii) Use *Mentha pulegium* leaves as plant material to isolate endophytic biocatalyst since it produces an essential oil rich in monoterpenes structurally like **1**.

(iii) Follow steps 3.1.1 to 3.1.3.

(iv) To isolate endophytic microorganism from liquid culture place 10 mL of the liquid media proposed in 3.1.4 in a 50 mL Erlenmeyer flask. Macerate 2 G of disinfected leaves and inoculate the medium. Add carvone to a 10 mM concentration to provide selective pressure. To isolate endophytic microorganism from solid culture, use a cut and disinfected leave as inoculum on solid media as proposed in 3.1.4. Incubate as described previously.

(v) Follow steps 3.1.5 to 3.1.7.

(vi) Screening biocatalytic activity from endophytic strains:

a) Screening of biocatalytic activity (bacteria).

Inoculate fresh plates of each isolated strain from the frozen stock in TSA and grow overnight at 35 °C. Use a single colony to inoculate 5 mL TSB. Incubate each culture at 35 °C in a rotary shaker at 150 rpm overnight and use 500 μL of this culture to inoculate 5 mL of fresh TSB containing the appropriate substrate in a concentration of 10 mM, in a 25 mL Erlenmeyer flask. Grow this biotransformation culture under the same conditions and sample periodically for GC analysis.

b) Screening of biocatalytic activity (yeast).

Inoculate fresh plates of each isolated yeast strain from the frozen stock in PDA. Inoculate a single colony to 100 mL of YM Broth. Incubate culture in a rotary shaker at 28 °C and 150 rpm for 48 h. Collect the cells by centrifugation at 4000 rpm and 4 °C, for 15 min. Wash the pellet three times with 50 mL physiological serum. Suspend 0.5 g of yeast cells (wet weight) in 5 mL of 10% dextrose solution and add the appropriate substrate to a final concentration of 10 mM. Incubate the reaction mixture at 28 °C and 150 rpm. Aseptically withdrawn samples from the reaction mixture at different times for GC analysis.

c) Screening of biocatalytic activity (fungi).

Incubate fresh plates of each isolated fungal strain for 5–7 days in PDA at 28 °C. Dilute the spores in physiological serum to a final concentration of 10^7 spores.mL^{-1}. Inoculate spore suspension (5 mL) into 50 mL of PDB and incubate the culture at 28 °C in an orbital shaker at 150 rpm. After 48 h of incubation, add the appropriate substrate at a final concentration of 10 mM. Incubate the culture at 28 °C and 150 rpm for 7–10 days. Aseptically withdrawn samples from the reaction mixture at different times for GC analysis.

Subsequently in all assays, separate incubation cells by centrifugation at 4000 rpm and 4 °C, for 15 min and extract biotransformation media 3 times with ethyl acetate. Dry organic layers over Na_2SO_4 (anh.), filter the desiccant agent, and distil ethyl acetate under vacuum on a rotary evaporator at 30 °C. Take the residue in CH_2Cl_2, and analyze aliquots containing 1 mg/mL of products by GC. Identification of the different products obtained in the biotransformation of **1** is done as described by Marconi et al. [14].

Perform all the assays under aseptic condition and perform blank assays without substrate and without bacterial strain in parallel experiments.

3.2.2 Isolation Endophytes with Xylanase and Cellulase Activity from *Eucalyptus* sp. Leaves, Bark, and Branches

(i) To obtain endophytic biocatalyst with xylanase and cellulase use a plant material with lignocellulosic composition like leaves, bark, and branches of a *Eucalyptus* tree.

(ii) Follow steps 3.1.1 to 3.1.3.

(iii) To isolate endophytic microorganisms from leaves macerate 2 g. of disinfected plant material. To isolate endophytic microorganisms from branches and bark, cut the disinfected branches into 1 cm fragments. Use the three disinfected plant material to inoculate on Erlenmeyer flask of 50 mL volume with 10 mL of different liquid media proposed in 3.1.4 and add beechwood xylan or carboxymethylcellulose (CMC) at 0.5%. For isolate endophytic microorganism on solid culture inoculate cut fragments of 1 cm of disinfected of leaf, bark, and branch on solid media as proposed in 3.1.4. Incubate as described previously.

(iv) Follow steps 3.1.5 to 3.1.7.

(v) Screening for xylanase and cellulase activity on solid media.

(a) Inoculate pure cultures of isolates in YNB agar with 0.5% xylan/CMC for xylanase/cellulase activity screening for fungi and bacterial strain, respectively.

(b) Incubated at 28 °C for fungi and 35 °C for bacterial strain.

(c) When visual growth is observed, flooded the plates with an aqueous solution of Congo red (0.1% Congo red in distilled water).

(d) After 15 min, pour off the Congo red solution and further flood plates with 1 M NaCl for 15 min.

(e) Xylanase/cellulase activity is detected by observation of clear zones around colonies.

(f) Selected strains with a hydrolysis halo radius > 0.5 cm for further analysis.

(vi) Xylanase and cellulase production on liquid media

(a) Use pure cultures of isolates selected by the screening described above (Subheading 3.2.2 vi) to inoculate in two different media with addition of 0.5% beechwood xylan/CMC for xylanase/cellulase activity screening. PDB and YNB are used as nutrient and minimal medium for fungi strains TSB and M9 for bacteria strains. Inoculate a single colony from fresh culture of yeast or bacteria in 100 mL Erlenmeyer flasks containing 20 mL of the corresponding medium. For fungal strains, use as inoculum a piece of 1 cm diameter of mycelial growth in solid media.

(b) Incubate cultures for 24, 48, and 72 h for bacteria at 35 ° C; 24, 48, and 72 h for yeast at 28 °C yeast and 48, 120, and 168 h for filamentous fungi at 28 °C in an orbital shaker at 150 rpm.

(c) At the end of the incubation period collect in clean and sterile tubes the culture filtrates for xylanase quantification. Perform all experiments in triplicates.

(vii) Xylanase quantification.

(a) Incubate at 30 °C for 15 min the reaction mixture of 1.5 mL of a 2.0% (w/v) suspension of beechwood xylan in 50 mM sodium citrate buffer at pH 4.8 and 0.5 mL of the enzyme diluted in the same buffer [35].

(b) Reducing sugars are determined by adding 3.0 mL of 3,5-dinitrosalicylic acid solution [36] and incubate the mixture at 95 °C for 5 min.

(c) Measured absorbance at 540 nm on a UV-spectrophotometer.

(d) One unit of enzyme is defined as the amount of enzyme catalyzing the release of 1 μmol of reducing sugars as xylose per min. Under the specified conditions.

(viii) Cellulase quantification.

(a) Filter paper cellulase activity (FPase) is assayed by incubating 0.5 mL of enzyme solution and 1.0 mL of sodium citrate buffer (50 mM, pH 4.8) with 50 mg filter paper (Whatman no. 1, 0.25 mm pore size); at 50 °C for 60 min.

(b) Reducing sugars are determined as above (Subheading 3.2.2 vii b).

(c) One international unit of FPase activity is the amount of enzyme that forms 1 μmol of reducing sugars as glucose per minute during the hydrolysis reaction.

3.2.3 Isolation Endophytes with Prelog Carbonyl Reductase Activity from *Daucus carota, Curcubita maxima,* and *Foeniculum vulgare*

(i) Use ethyl 2-methylacetoacetate (**2**) or ethyl 2-ethylacetoacetate (**3**) as model substrate (Fig. 2). α-substituted β-keto esters are common substrates to evaluate potential bioreducing agents. The reduction of these compounds not only renders optically active hydroxyesters, but also the presence of two contiguous stereocenters in the products allows to test for enantio and diastereoselectivity in a single experiment (Fig. 2) [15].

(ii) Use plant tissue with Prelog carbonyl reductase activity such as carrot (*Daucus carota*) [20, 21], fennel (*F. vulgare*), and squash (*C. maxima*) [22, 23] as plant material to isolate endophytic with the same biocatalytic activity.

Wash thoroughly with running tap water the carrot and pumpkin. Then remove the external layer with a sterile knife. Proceed as is described in steps 3.1.1 to 3.1.3 with this plant material and leaves of fennel.

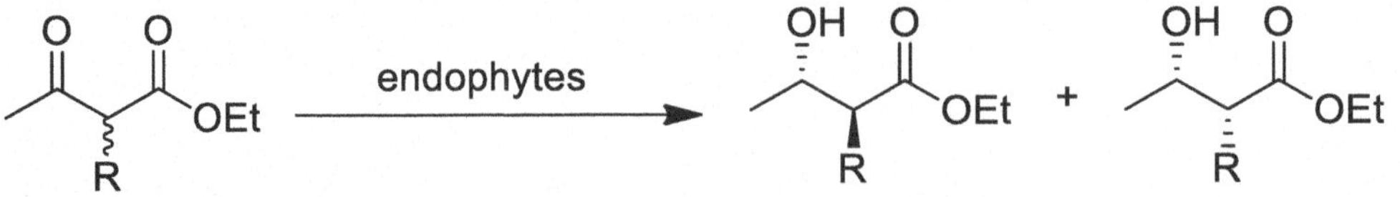

2 R = Me
3 R = Et

2a R = Me
3a R = Et
anti

2b R = Me
3b R = Et
syn

Fig. 2 **2** and **3** as substrates to explore carbonyl reductase activity

(iii) Slice disinfected carrot and pumpkin with a sterile knife and macerate fennel disinfected leaves.

(iv) Use the plant fragments of carrot, pumpkin, and fennel leaves macerated to inoculate the different liquid medium on Erlenmeyer flask of 50 mL volume with 10 mL of different liquid media proposed in 3.1.4 and add **2** or **3** in a concentration of 10 mM for selective pressure. To isolate endophytic microorganism from solid culture, inoculate a cut and disinfected plant material on solid media as proposed in 3.1.4. Incubate as previously described.

(v) Follow steps 3.1.5 to 7.

(vi) To screen carbonyl reductase activity from endophytic strains proceed as described in Subheading 3.2.1 vi. a-c using 2-ethylacetoacetate as substrate. To identify the different products obtained in the biotransformation **2** or **3** compounds proceed as described by Rodríguez et al. [15].

References

1. Rodríguez P, Barton M, Aldabalde V, Onetto S, Panizza P, Menéndez P, Rodríguez S (2007) Are endophytic microorganisms involved in the stereoselective reduction of ketones by *Daucus carota* root? J Mol Catal B Enzym 49:8–11
2. Rodriguez P, Magallanes-Noguera C, Menéndez P, Orden AA, Gonzalez D, Kurina-Sanz M, Rodríguez S (2015) A study of *Raphanus sativus* and its endophytes as carbonyl group bioreducing agents. Biocatal Biotransformation 33:121–129
3. Rodriguez P, Gonzalez D, Rodríguez Giordano S (2016) Endophytic microorganisms: a source of potentially useful biocatalysts. J Mol Catal B Enzym 133:S569–S581
4. Yue Q, Bacon CW, Richardson MD (1998) Biotransformation of 2-benzoxazolinone and 6-methoxy-benzoxazolinone by *Fusarium moniliforme*. Phytochemistry 48:451–454
5. Wang Y, Dai CC (2011) Endophytes: a potential resource for biosynthesis, biotransformation, and biodegradation. Ann Microbiol 61: 207–215
6. Zikmundová M, Drandarov K, Bigler L, Hesse M, Werner C (2002) Biotransformation of 2-benzoxazolinone and 2-hydroxy-1,4-benzoxazin-3-one by endophytic fungi isolated from *Aphelandra tetragona*. Appl Environ Microbiol 68:4863–4870
7. Strobel G, Daisy B (2003) Bioprospecting for microbial endophytes and their natural products. Society 67:491–502
8. Kusari P, Spiteller M, Kayser O, Kusari S (2014) Recent advances in research on Cannabis sativa L. endophytes and their prospect for the pharmaceutical industry. In: Microbial diversity and biotechnology in food security, Springer, New Delhi, pp 3–15
9. Preethi K, Manon Mani V, Lavanya N (2021) Endophytic fungi: a potential source of bioactive compounds for commercial and therapeutic applications in endophytes. Singapore, Springer, pp 247–272
10. Mishra R, Kushveer JS, Revanthbabu P, Sarma VV (2019) Endophytic fungi and their enzymatic potential. In: Advances in endophytic fungal research, Springer, Cham, pp 283–337
11. Huang Q, An H, Song H, Mao H, Shen W, Dong J (2015) Diversity and biotransformative potential of endophytic fungi associated with the medicinal plant *Kadsura angustifolia*. Res Microbiol 166:45–55
12. Qin D, Wang L, Han M, Wang J, Song H, Yan X, Dong J (2018) Effects of an endophytic fungus Umbelopsis dimorpha on the secondary metabolites of host–plant *Kadsura angustifolia*. Front Microbiol 9:1–14
13. Pedrini P, Giovannini PP, Mantovani M, Andreotti E, Colalongo C (2009) Reduction screening with endophytic fungi: synthesis of homochiral secondary alcohols. J Mol Catal B Enzym 60:45–150
14. Marconi F, Umpiérrez ML, Gonzalez D, Giordano SR, Rodriguez P (2018) Endophytic biocatalysts with enoate reductase activity isolated

from *Mentha pulegium*. World J Microbiol Biotechnol 34:50–60

15. Rodríguez P, Reyes B, Barton M, Coronel C, Menéndez P, Gonzalez D, Rodríguez S (2011) Stereoselective biotransformation of α-alkyl-β-keto esters by endophytic bacteria and yeast. J Mol Catal B Enzym 71:90–94
16. Li H, Li Z, Ruan G, Yu Y, Liu X (2016) Asymmetric reduction of acetophenone into R-(+)-1-phenylethanol by endophytic fungus Neofusicoccum parvum BYEF07 isolated from Illicium verum. Biochem Biophys Res Commun 473:874–878
17. Werner C, Petrini O, Hesse M (2006) Degradation of the polyamine alkaloid aphelandrine by endophytic fungi isolated from *Aphelandra tetragona*. FEMS Microbiol Lett 155:147–153
18. Mahboubi M, Haghi G (2008) Antimicrobial activity and chemical composition of *Mentha pulegium* L. essential oil. J Ethnopharmacol 119:325–327
19. Marques NP, de Cassia PJ, Gomes E, da Silva R, Araújo AR, Ferreira H, Bocchini DA (2018) Cellulases and xylanases production by endophytic fungi by solid state fermentation using lignocellulosic substrates and enzymatic saccharification of pretreated sugarcane bagasse. Ind Crop Prod 122:66–75
20. Baldassarre F, Bertoni G, Chiappe C, Marioni F (2000) Preparative synthesis of chiral alcohols by enantioselective reduction with *Daucus carota* root as biocatalyst. J Mol Catal B Enzym 11:55–58
21. Yadav JS, Nanda S, Thirupathi Reddy P, Bhaskar Rao A (2002) Efficient enantioselective reduction of ketones with *Daucus carota* root. J Org Chem 67:3900–3903
22. Aldabalde V, Arcia P, Gonzalez A, Gonzalez D (2007) Enzymatic synthesis of chiral heteroaryl alcohols using plant fragments as the only biocatalyst and reducing agent. Green Chem Let Rev 1:25–30
23. Bruni R, Fantin G, Maietti S, Medici A, Pedrini P, Sacchetti G (2006) Plants-mediated reduction in the synthesis of homochiral secondary alcohols. Tetrahedron Asymmetry 17: 2287–2291
24. Bonnecarrere, PR (2013) Rol de endófitos en reacciones biocatalíticas mediadas por vegetales. Identificación y caracterización de nuevos biocatalizadores In: RIQUIM – Repositorio Institucional de la Facultad de Química – UdelaR, Montevideo
25. Orden AA, Bisogno FR, Giordano OS, Sanz MK (2008) Comparative study in the asymmetric bioreduction of ketones by plant organs and undifferentiated cells. J Mol Catal B Enzym 51:49–55
26. Griffiths RI, Whiteley AS, O'Donnell AG, Bailey MJ (2000) Rapid method for coextraction of DNA and RNA from natural environments for analysis of ribosomal DNA-and rRNA-based microbial community composition. Appl Environ Microbiol 66:5488–5491
27. Brenner DJ, Krieg NR, Staley JT, Garrity G (2005) The Proteobacteria. In: Bergey's manual of systematic bacteriology, Springer, New York
28. Barrow GI, Feltham RKA (1993) Cowan and steel's manual for the identification of medical bacteria
29. Menes RJ, Viera CE, Farí ME, Seufferheld MJ (2011) Halomonas vilamensis sp. nov., isolated from high-altitude Andean lakes. Int J Syst Evol Microbiol 61:1211–1217
30. Hugh R, Leifson E (1953) The taxonomic significance of fermentative versus oxidative metabolism of carbohydrates by various gram-negative bacteria. J Bacteriol 66:24–26
31. Cole JR, Wang Q, Cardenas E, Fish J, Chai B, Farris RJ, Tiedje JM (2009) The ribosomal database project: improved alignments and new tools for rRNA analysis. Nucleic Acids Res 37:D141–D145
32. Kumar S, Tamura K, Jakobsen IB, Nei M (2001) MEGA2: molecular evolutionary genetics analysis software. Bioinformatics 17: 1244–1245
33. Pitt JI, Hocking AD (2009) Fungi and food spoilage, Boston
34. Decarlini MF, Aimar ML, Vázquez AM, Vero S, Rossi LI, Yang P (2017) Fungi isolated from food samples for an efficient stereoselective production of phenylethanols. Biocatal Agric Biotechnol 12:275–285
35. Botto E, Gioia L, Menéndez MP, Rodríguez P (2019) *Pseudozyma* sp. isolation from Eucalyptus leaves and its hydrolytic activity over xylan. Biocatal Agric Biotechnol 21:101282
36. Adney B, Baker J (1996) Measurement of cellulase activities: Laboratory Analytical Procedure (LAP). National Renewable Energy Laboratory

Chapter 22

Screening of Endophytes for Biosurfactant Production

V. Ambikapathy, S. Babu, A. Anbukumaran, and A. S. Shijila Rani

Abstract

The endophytes isolated from different plants are known to produce biosurfactant. The biosurfactant ability of the endophytes could be analyzed in the culture supernatant by performing drop-collapse assay, oil spread assay, emulsification index, and surface tension.

Key words Minimal salt medium, Olive oil, Endophytes, Polystyrene 96-well microplate, K6 tensiometer

1 Introduction

Biosurfactants defined as green synthesized surfactants are surface-active compounds produced by microbes. These products are shown to be efficient in processes of microbial enhanced oil recovery and bioremediation in hydrocarbon contaminated environments. They also possess potential applications in agriculture, raw materials for cosmetics, pharmaceutical products, detergents, personal hygiene products, and food processing [1]. They are more advantageous than synthetically derived ones when considering their biodegradability and low toxicity. Biosurfactants are biologically produced from various substrates such as waste from tropical agronomic cultures, food processing industries, fruit processing industries, petroleum processing, and coffee processing industries [2].

Biosurfactants are produced by endophytic bacteria, yeasts, and filamentous fungi. Endophytes are microorganisms that inhabit the plant tissue without causing any harmful effects on the host plant. According to, endophytic microorganisms live inside the plant tissues, at least for one period of their life cycle instead of the epiphytic microorganisms that live on a plant's surface [3]. Microorganisms are able to grow under environmental stressed condi-

A. Sankaranarayanan et al. (eds.), *Endophytic Microbes: Isolation, Identification, and Bioactive Potentials*, Springer Protocols Handbooks, https://doi.org/10.1007/978-1-0716-2827-0_22,

tions: environments with low pH, deficiency in nutrients, and low water availability media favor their growth [4]. They may still, though more rarely, produce biosurfactants [5].

The host plant may take up hydrocarbon from the soil and translocate it to their different tissues in endophytic microbes [6, 7]. Sometimes phytoremediation of hydrocarbons is mediated by biosurfactants produced by endophytic bacteria [8]. Biosurfactants produced by endophytic bacteria facilitate their biofilm development and cell signaling, which will enhance the plant-endophyte interaction [9].

2 Materials

- Mineral salt medium (gL^{-1}: 0.5 g NH_4Cl, 4 g NaCl, 0.5 g KH_2PO_4,1 g Na_2HPO_4 and 0.5 g $MgSO_4.7H_2O$).
- Olive oil or kerosene.
- Endophytes (*Fusarium* sp., *Pseudomonas* sp., and *Bacillus subtilis*).
- Petri plate.
- Polystyrene 96-well microplate.
- Conical flask.
- Spirit lamp.
- Hot air oven.

3 Methods [10]

3.1 Collection of Biosurfactant from Endophytes

- Take pure culture of endophytes (*Fusarium* sp., *Pseudomonas* sp., *Bacillus subtilis*) isolated from *Centella asiactica*.
- Prepare mineral salt medium by autoclaving at 121 °C for 30 min.
- Inoculate the endophytic microbes in mineral salt medium amended with 10 mL/L of olive oil or kerosene in a conical flask.
- Incubate the conical flask at room temperature (30 ± 4 °C) for 10 days.
- After 10 days, centrifuge the medium was 10,000 rpm for 20 min in 4 °C to harvest biosurfactant.
- Use the collected supernatant for further biosurfactant screening tests.

3.2 Drop Collapse Assay [11]

- Take a polystyrene 96-well micro-plate, to each well add 2 μL of olive oil or kerosene and keep aside for 24 h at 22 °C.
- Add 5 μL of filtered cell-free supernatant to the center of the oil-coated well.
- After 1–2 min observe the results. Positive results indicate the flat oil drop and the rounded oil drops indicate negative.

3.3 Oil Spread Assay [12]

- Take 20 mL of distilled water in a Petri plate with 1 mL of olive oil or kerosene to the water surface.
- Add 20 μL of cell-free culture broth (endophytes supernatant) to the oil surface.
- Observe the plate in the cell-free culture broth. The oil will be displaced and clear zone diameter indicate oil displacement activity.

3.4 Emulsification Index (E24) [13]

- Take 3 mL of the cell-free culture extracts, add 2 mL of olive oil or kerosene to the test tubes.
- Mix the test tube by vortexing for 2 min at 70 rpm and keep aside for 24 h.
- After incubation, compare the emulsified oil height (cm) with the total volume.
- Calculate the emulsification by the following formula:

$$\mathrm{E} - 24\ \text{index}\% = (\text{Height of emulsion}/\text{Total height of solution}) \times 100$$

References

1. Sourav D, Susanta M, Ghosh A, Saha R, Saha B (2015) A review on natural surfactants. RSC Adv 5:65757–65767
2. Bento MF, Camargo FA, Gaylarde CC (2008) Biossurfactantes. In: Melo IS, Azevedo JL (eds) Microbiologia ambiental, 2nd edn. EmbrapaMeioAmbiente, Jaguariuna, p 647
3. Shijila Rani AS, Ambikapathy V, Panneerselvam A (2018) Diversity of endophytic fungi from leaves of *Centellaasiactica* L. and *Melia azedarch* L. Int J Rec Sci Res 9(1):23083–23086
4. Mollea C, Bosco F, Ruggeri B (2005) Fungal biodegradation of naphthalene: microcosms studies. Chemosphere 60:636–643
5. Luz CC, Santos EA, Santos MOS, Mussy MY, Yamashita M, Bastos WR (2011) Biodegradation study of diesel by microbial consortium collected in Porto Velho-RO, Amazonia. Quim Nova 34:775–779
6. Weyens N, Lelie Van Der D, Taghavi S, Vangronsveld J (2009) Phytoremediation: plant – endophyte partnerships take the challenge. Curr Opin Biotechnol 20:248–254
7. Khan K, Naeem M, Javed Arshed M (2006) Extraction and characterization of oil degrading bacteria. J Appl Sci 6:2302–2306
8. Pawlik M, Cania B, Thijs S, Vangronsveld J, Piotrowska Seget Z (2017) Hydrocarbon degradation potential and plant growth-promoting activity of culturableendophytic bacteria of *Lotus corniculatus* and *Oenother abiennis* from a longterm polluted site. Environ Sci Pollut Res 24:19640–19652

9. Mikolajczyk OM, Drozdzynski P, Pietrzyk D, Antczak T (2018) Biosurfactant production and hydrocarbon degradation activity of endophytic bacteria isolated from *Chelidoniummajus* L. Microb Cell Factories 17:1–9
10. Suresh Chander CR, Lohitnath T, Kumar DJM, Kalaichelvan PT (2012) Production and characterization of biosurfactant from *Bacillus subtilis* MTCC441 and its evaluation to use as bioemulsifier for food biopreservative. Adv Appl Sci Res Res Libr 3:1827–1831
11. Jain DK, Collins Thompson DL, Lee H, Trevors JT (1991) A drop-collapsing test for screening surfactant producing microorganisms. J Microbiol Methods 13:271–279
12. Morikawa M, Hirata Y, Imanaka T (2000) A study on the structure – function relationship of lipopeptidebiosurfactants. Biochim Biophys Acta 1488:211–218
13. Ali SR, Chowdhury BR, Mondal P, Rajak S (2013) Screening and characterization of biosurfactants producing microorganism form natural environment (whey spilled soil). J Nat Sci Res 3:53–59

Chapter 23

Screening of Endophytes for Antibacterial Properties

M. Veerapagu, K. R. Jeya, and A. Sankaranarayanan

Abstract

Endophytes are microbes that reside inside the tissues of healthy plants with the varied relationship but not producing any disease symptoms including but not limited to bacteria, actinomycetes, mycoplasma, and fungi. Endophytes are ubiquitous with rich biodiversity, which have been found in almost all species of plant. Endophytes of plant are considered as a vital source of natural bioactive compounds that help the host plant to improve defense against pathogens causing disease to plant can be exploited to produce antibiotics in pharmaceutical industries for the treatment of human diseases caused by microbes. Hence techniques to screen endophytes for antibacterial properties by cross-streak method and agar plug diffusion methods are important in search of new natural antibacterial drugs with improved potential.

Key words Plant, Endophyte, Antibiotics, Cross-streak method, Agar plug diffusion method

1 Introduction

Plants are considered to be a vital source of bioactive compounds having medicinal value from ancient times. Many diseases are cured from the extracts of different plant parts which exhibited distinct remedial properties. Amusingly, in recent years, products of therapeutic potential have been proved to be offered from microbes of plant endophytes and have been increased. Most microorganisms that inhabit the plants exhibit positive interactions. A plant can harbor thousands of microorganisms which can be epiphytes or endophytes depending on their colonization in the plant [1].

Endophytes are a diverse group of endosymbionts mostly bacteria including actinomycetes and mycoplasma and fungi which colonizes the internal plant tissues without causing any type of harm to the host plant [2]. Their whole life cycle or part of their life cycle occurs within the plant host. Endophytes could have entered inside the plants principally through the aerial, subaerial, and underground parts of plants. Many endophytes synthesize broad spectrum of bioactive metabolites [3]. Biopotential metabolites of plant endophytes have inherent applications in agronomy,

A. Sankaranarayanan et al. (eds.), *Endophytic Microbes: Isolation, Identification, and Bioactive Potentials*, Springer Protocols Handbooks, https://doi.org/10.1007/978-1-0716-2827-0_23,

beauty aids, pharmaceutics, food production, and processing technology [4]. Diverse bioactive metabolites from endophytic plant microbes displayed germicidal, antioxidative, antitumoral, insecticidal, antiparasitics, and immunosuppressive properties [5]. Hence the products from microbial endophytes of plants exerted considerable use in a large number of industries.

Demand for new drugs or secondary metabolites is increasing by virtue of the development of resistance of pathogens to antibiotics. Endophytes could be a promising resource for the discovery of new bioactive compounds with possible medicinal properties. Endophytic bacteria, actinomycetes, and fungi appear to be one of the promising sources of valuable antibacterial metabolites [6]. Many endophytes have been identified to exhibit antibacterial activity and some of which cited in recent literatures are listed in Table 1.

Altogether 15 bacterial endophytes from leaves of *Tamarindus indica* revealed antagonistic toward *B. cereus, S. aureus,* and *E. coli* [22]. Endophytic actinomycetes *Aeromicrobium, Actinomadura, Microbacterium, Microbispora, Micrococcus, Sphaerisporangium, Streptomyces,* and *Williamsia* isolated from medicinal plant *Vochysia divergens* disclosed in opposition to *S. aureus* resistance to methicillin [23]. Endophytic fungi *Syncephalastrum rasemosum, Syncephalastrum racemosums328* were isolated from mature healthy leaves and bark of medicinal plant *Saraca asoca* showed antibacterial activity against three pathogenic bacteria, i.e., *Streptococcus mutans, L. acidophilus,* and *E. faecalis* [24].

Endophytes isolated from leaves, stems, and roots of the plant were screened for their antibacterial properties. There are different methods available while cross-streak method and agar plug diffusion methods are simple, efficient, and most widely used for preliminary qualitative screening of antibacterial activities of endophytes. The antimicrobial activity of seven endophytic bacterial isolates from *Moringa peregrina* was screened against 11 human pathogens and were screened qualitatively by the cross-streak method [25]. Syed et al. [26] evaluated the antibacterial activity of three isolates from the root of *Capsicum annum* against *Staphylococcus aureus, E. coli, Pseudomonas* sp., *P.vulgaris,* and *Klebsiella* sp., through the cross-streak method.

Agar plug diffusion method was used for the preliminary screening of antibacterial properties of 12 endophytic fungal isolates from the bark of *Cinnamomum mercadoi* antagonistic to *S. aureus, B. cereus, E. coli,* and *P. aeruginosa* [27]. Endophytic fungi, a total of sixty five isolates from leaves of *Indigofera suffruticosa* were preliminarily screened for antibacterial activity by agar plug method against *S. aureus, B. subtilis, E. coli, K. pneumoniae,* and *P. aeruginosa* [28]. Several literatures are available for the screening of antibacterial property of endophytes by cross-streak method [29, 30] and agar plug diffusion method [31, 32] (Table 1).

Table 1
Antibacterial properties of endophytes

Endophyte	Host plant	Activity against bacterium	Reference
Alternaria alternata	*Picorrhiza kurroa*	*B. subtilis, S. aureus.*	Chandra et al. [7]
Raffaele sp.	*Cocos nucifera (L.)*	*E. coli, B. subtilis, S. aureus*	Fuego et al. [8]
Bacillus amyloliquefaciens	*Salicornia brachiata*	*S. aureus* and *Xanthomonas campestri*	Singh et al. [9]
*Diaporthe*sp	*Ligustrum obtusifolium.*	*S.aureus,E.coli,Salmonella enterica, B.cereus, Listeria monocytogenes*	Saravanakumar et al. [10]
Pseudomonas putida and *Metarhizium anisopliae*	*Oryza sativa* L	*Xanthomonas oryzae* up. *oryzae*	Saikia & Bora et al. [11]
Fusarium sp.	*Falcataria moluccana*	*B. subtilis, P. aeruginosa, E. coli*	Rumidatul et al. [12]
Aspergillus tubenginses	*Hyoscyamus muticus*	*P. aeruginosa, S. aureus, E.coli, Candida albicans*, and *B. subtilis*	Elkhouly et al. [13]
Bacillus amyloliquefaciens	*Drosera burmannii*	*B. cereus, S. haemolyticus,S.epidermidis,P. aeruginosa, K. pneumonia*	Chaudhuri et al. [14]
P. chrysogenum	*Artemisia sieberi*	*A. baumannii, E. aerogenes, E. coli, K. oxytoca, K. pneumoniae, P. aeruginosa, Streptococcus agalactiae, S. epidermis, S. aureus*	Ababutain et al. [15]
Streptomyces armeniacus	*Thymus roseus*	*S.aureus,Bacillus cereus, Salmonella enteritidis*	Musa et al. [16]
Bacillus subtilis	*Pseudobrickellia brasiliensis*	*E. coli, P. aeruginoa, S. aureus*	Cardoso et al. [17]
Burkholderia sp. WYAT7	*Artemisia nilagirica* (Clarke) Pamp.	*P. aeruginosa* (MTCC 2453), *P. aeruginosa*(MTCC 1610), *Salmonella paratyphi, B. subtilis*	Ashitha et al. [18]
Bacillus velezensis	*Tamarindus indica* L.	*P. aeruginosa, N. gonorrhoeae*	Chigurupati et al. [19]
Streptomyces sp.	*Neesia altissima*	*Bacillus cereus* (ATCC 10876), *taphylococcus aureus* (ATCC 25923)	Pratiwi et al. [20]
Aspergillus cejpii	*Nelumbo nucifera*	Methicillin-resistant *Staphylococcus aureus* DMST20651 (MRSA)	Techaoei et al. [21]

2 Materials

2.1 Screening of Endophytes for Antibacterial Property by Cross-Streak Method (Fig. 1)

- 70% v/v ethanol.
- Double distilled water.
- Erlenmeyer flask.
- Measuring cylinder.
- pH meter.
- Nutrient agar medium.
- Petri plates.
- Microbial culture: Pure culture of endophytic bacteria, actinomycetes, and target bacteria.
- Bunsen burner.
- Inoculation loop.
- Digital balance.

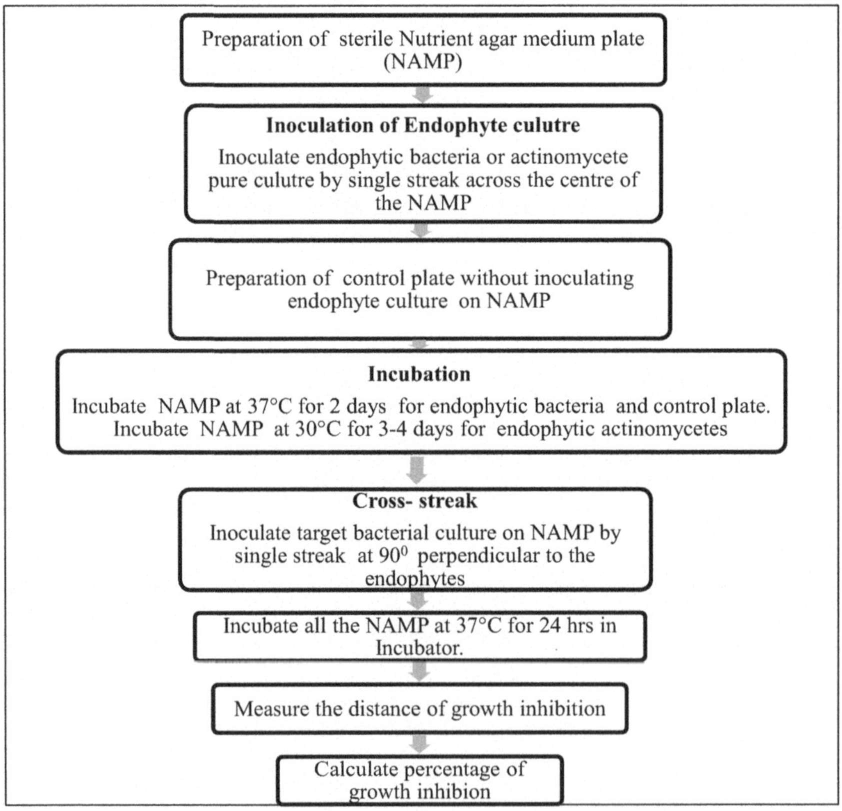

Fig. 1 Screening of endophytes for antibacterial properties by cross-streak method

- Hot air oven.
- Water bath.
- Autoclave.
- Laminar airflow hood.
- Incubator.
- Non-absorbent cotton.

2.2 Screening of Endophytes for Antibacterial Property Agar Plug Diffusion Method

- 70% v/v ethanol.
- Double distilled water.
- Erlenmeyer flask.
- Measuring cylinder.
- pH meter.
- Potato dextrose agar (PDA).
- Yeast malt agar (YMA).
- Muller Hinton agar (MHA).
- Microbial culture: Pure culture of endophytic fungus, Actinomycetes, and target bacteria.
- 0.5McFarland standard (Himedia).
- Antibiotic zone scale (Himedia).
- Petri plate 100 x 15 mm.
- Bunsen burner.
- Inoculation loop.
- Cork borer.
- Forceps.
- Knife.
- Scalpel blade.
- Digital balance.
- Hot air oven.
- Water bath.
- Autoclave.
- Laminar airflow hood.
- Incubator.
- Sterile Cotton swab.

3 Methodology

3.1 Screening of Endophytes for Antibacterial Property by Cross-Streak Method

3.1.1 Inoculation of Endophyte Culture by Single Streak

1. Prepare the nutrient agar medium and sterilize it [33].
2. Take the pure culture of endophytic bacteria or actinomycetes from the nutrient agar slant aseptically.
3. Inoculate the endophytes across the center of the nutrient agar medium plate by single streak [34].
4. Prepare the control plate without streaking endophytic culture on the nutrient agar medium.
5. Incubate the endophytic bacteria at 37 °C for 2 days and at 30 °C for 3–4 days for endophytic actinomycetes.

3.1.2 Inoculation of Target Bacterial Culture by Cross-Streak

1. Inoculate the target bacteria by single streak at an angle of 90° perpendicular to the endophyte without touching the line of endophytes including control plates.
2. Incubate the plates at 37 °C for 24 h in an incubator.
3. Measure the distance of inhibition growth after incubation (Fig. 3).
4. Calculate the percentage of inhibition by the following formula [35] (Table 2):

$$\text{Growth Inhibition\%} = \frac{\text{length of target bacterial streak–length of target bacterial growth} \times 100}{\text{length of target bacterial streak}}$$

5. Note the percentage of inhibition as

   Strong inhibition [+4, ≥ 90%]

   Good inhibition [+3, ≥ 75%]

   Moderate inhibition [+ 2, ≥ 50%]

   Weak inhibition [+, ≥ 15%]

   No inhibition [−]

Table 2
Percentage of growth inhibition of target bacteria by endophyte using the cross-streak method

Target bacteria	Length of streak (mm) A	Length of growth (mm) B	Growth inhibition % (A-B)/ AX100
Target bacteria 1			
Target bacteria 2			
Target bacteria 3			

3.2 Screening of Endophytes for Antibacterial Property Agar Plug Diffusion Method (Fig. 2)

3.2.1 Preparation of Agar Plug

1. Select the well-isolated fungal endophytes or actinomycetes colony from the growing edge with the aid of inoculation loop.
2. Spot inoculate fungal colony at the center of the PDA agar medium or actinomycetes on yeast malt agar medium (YMA).
3. Incubate the PDA plates for 21 days at 25 ± 2 °C [36] or YMA at 28 ± 2 °C for 7–10 days [37].
4. With the help of sterile cork borer cut an agar plug of 6 mm diameter from the actively growing margin of endophyte [38].

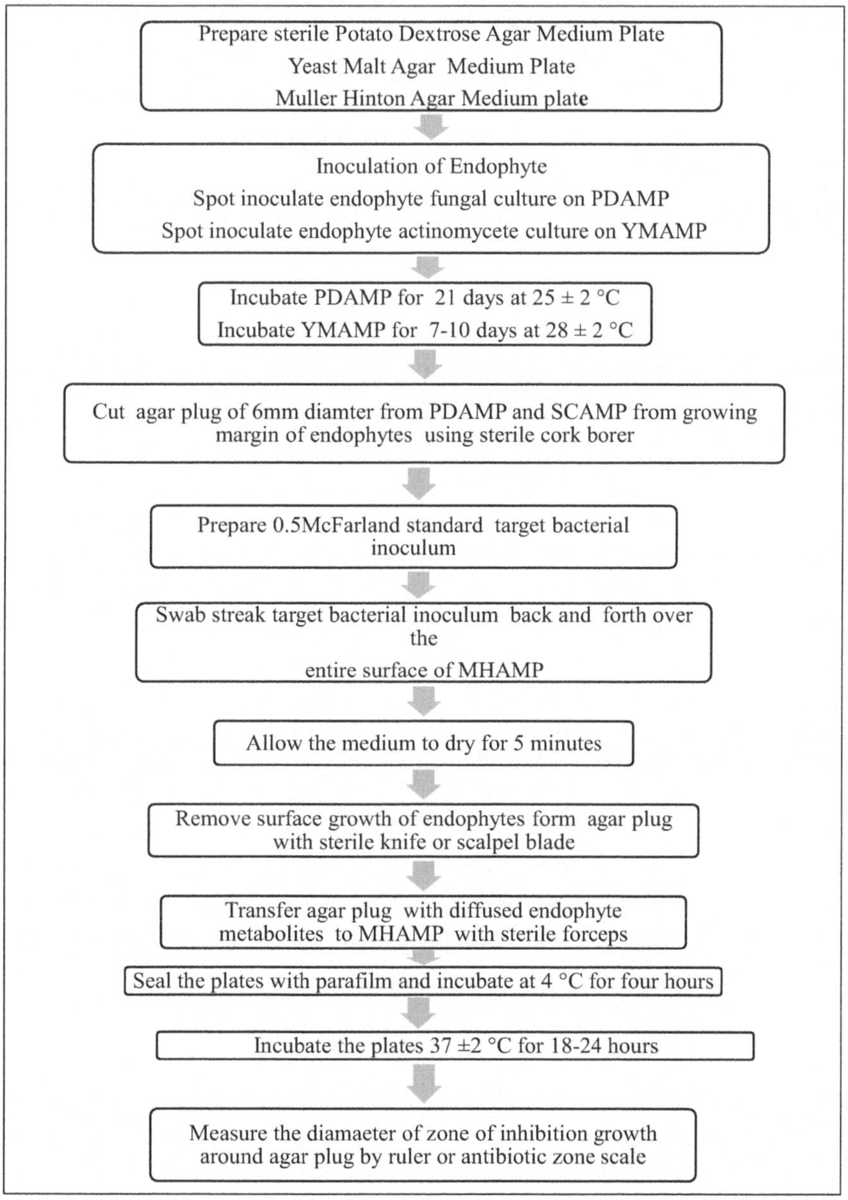

Fig. 2 Screening of endophytes for antibacterial properties by agar plug diffusion method

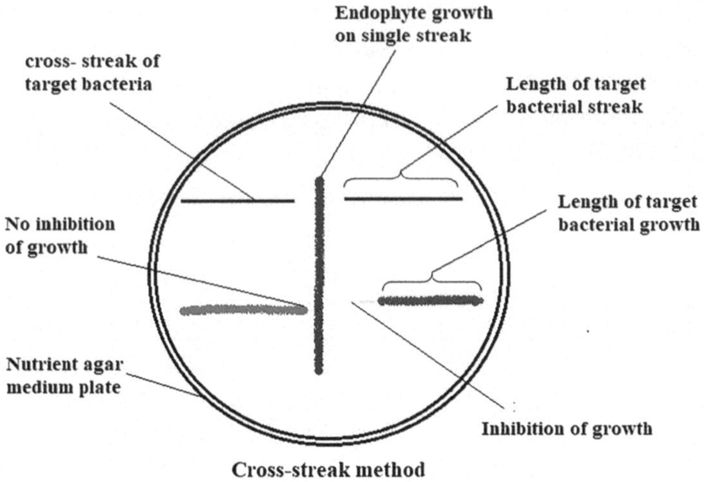

Fig. 3 Diagrammatic representation of cross-streak method for screening of antibacterial property of endophytes

3.2.2 Preparation of 0.5McFarland Standard Target Bacterial Inoculum

1. Take four or five colonies of target bacteria from the pure culture of the organism (not more than 48 h old except slow-growing organisms).
2. Transfer the bacterial colonies into 2 mL of sterile saline and gently vortex it.
3. Add more colonies or dilute with sterile suspension to make up the turbidity of bacterial suspension as per 0.5McFarland standard [39].
4. Use this inoculum suspension (1–2 × 10^8 CFU/mL) within 15 min of preparation.

3.2.3 Inoculation of MHA Plate

1. Inoculate the bacteria on MHA medium and leave the plates for 5–15 mins.

3.2.4 Transfer of Agar Plug

1. Remove the growth of endophyte from the surface of agar plug with a sterile knife or scalpel blade carefully without damaging agar plug [40].
2. Hold the agar plug containing only diffused metabolites of endophytes with the help of sterile forceps.
3. Retain the agar plug in the middle of the MHA plates pre-seeded with target bacteria.
4. Seal the plates with parafilm.
5. Keep the plate in an ice box at 4 °C for 4 h for metabolite diffusion [41].

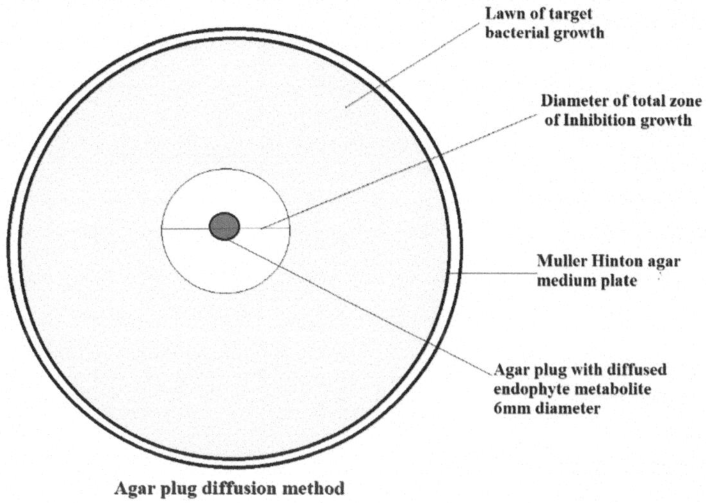

Fig. 4 Diagrammatic representation of agar plug diffusion method for screening of endophytes for antibacterial property

Table 3
Zone of growth inhibition of target bacteria by endophyte using agar plug diffusion method

Target bacteria	Total zone of inhibition growth diameter (mm) A	Agar plug diameter (mm) B	Zone of inhibition growth (mm) A-B
Target bacteria 1			
Target bacteria 2			
Target bacteria 3			

6. Incubate the MHA plates at 37 ± 2 °C for 18–24 h in an incubator.
7. Measure the zone of inhibition around the agar plug following incubation (Fig. 4).
8. Calculate zone of inhibition growth by decreasing the diameter of the agar plug from the total zone of inhibition growth [42] (Table 3).

Zone of inhibition growth
= Diameter of total zone of Inhibition growth (mm)–Diameter of agar plug (6 mm).

4 Observations

4.1 Screening of Endophytes for Antibacterial Properties by Cross-Streak Method

4.2 Screening of Endophytes for Antibacterial Properties by Agar Plug Diffusion Method

References

1. Gouda S, Das G, Sen SK, Shin HS, Patra JK (2016) Endophytes: a treasure house of bioactive compounds of medicinal importance. Front Microbiol 7:1538
2. Tadych M, White JF (2019) Endophytic microbes, encyclopedia of microbiology, 4th edn. Academic Press, pp 123–136
3. Mohana NC, Rakshith D, Ramesha KP, Nuthan BR, Harini BP, Satish S (2021) TLC directed isolation and in silico analysis of antimicrobial metabolite from *Nigrospora sphaerica* inhabiting *Croton bonplandianus* Baill. South Afr J Bot 139:106–113
4. Nazir A, Rahman HA (2018) Secrets of plants: endophytes. Int J Plant Biol 9(1). https://doi.org/10.4081/pb.2018.7810
5. Singh A, Singh DK, Kharwar RN, White JF, Gond SK (2021) Fungal endophytes as efficient sources of plant-derived bioactive compounds and their prospective applications in natural product drug discovery: insights, avenues, and challenges. Microorganisms 9(1): 197
6. Sharma D, Pramanik A, Agrawal PK (2016) Evaluation of bioactive secondary metabolites from endophytic fungus *Pestalotiopsis* neglecta BAB-5510 isolated from leaves of *Cupressus torulosa* D. Don. 3 Biotech 6:210
7. Chandra H, Kumari P, Prasad R, Gupta SC, Yadav S (2021) Antioxidant and antimicrobial activity displayed by a fungal endophyte *Alternaria alternata isolated* from *Picrorhiza kurroa* from Garhwal Himalayas, India. Biocatal Agric Biotechnol:101955
8. Fuego B, Romano K, Pinlac CLirio G (2021) Evaluation of the antimicrobial activity of endophytic fungus isolated from *Cocos nucifera* (L.) cotyledon against medically-important pathogens. J Biosci Med 9:86–97
9. Singh S, Ghadge VA, Kumar P, Mathew DE, Dhimmar A, Sahastrabudhe H, Nalli Y, Rathod MR, Shinde PB (2021) Biodiversity and antimicrobial potential of bacterial endophytes from halophyte *Salicornia brachiate*. Antonie Van Leeuwenhoek:1–18. https://doi.org/10.1007/s10482-021-01544-4
10. Saravanakumar K, Sriram B, Sathiyaseelan A, Hu X, Mariadoss AVA, MubarakAli D, Wang MH (2021) Molecular identification, volatile metabolites profiling, and bioactivities of an indigenous endophytic fungus (*Diaporthe* sp.). Process Biochem 102:72–81
11. Saikia K, Bora LC (2021) Exploring actinomycetes and endophytes of rice ecosystem for induction of disease resistance against bacterial blight of rice. Eur J Plant Pathol 159(1):67–79
12. Rumidatul A, Rahmawati N, Sunarya S (2021) Production of secondary metabolites and its antibacterial and antioxidant activity during the growth period of endophytic fungi isolated from gall rust *Sengon*. Plants Pharma J 13(2): 325–331

13. Elkhouly HI, Hamed AA, El Hosainy AM, Ghareeb MA, Sidkey NM (2021) Bioactive secondary metabolite from endophytic *Aspergillus tubenginses* ASH4 isolated from *Hyoscyamus muticus*: antimicrobial, antibiofilm, antioxidant and anticancer activity. Pharmacog J 13(2):434–442
14. Chaudhuri M, Paul AK, Arundhati P (2021) Isolation of potential antimicrobial metabolite from endophytic *Bacillus amyloliquefaciens*DL06 of carnivorous plant *Drosera burmannii* VAHL. Asian J Pharm Clin Res 14(1):163–170
15. Ababutain IM, Aldosary SK, Aljuraifani AA, Alghamdi AI, Alabdalall AH, Al-Khaldi EM, Borgio JF (2021) Identification and antibacterial characterization of endophytic fungi from *Artemisia sieberi*. Int J Microbiol:Article ID 6651020. https://doi.org/10.1155/2021/6651020
16. Musa Z, Ma J, Egamberdieva D, Abdelshafy Mohamad OA, Abaydulla G, Liu Y, Li WJ, Li L (2020) Diversity and antimicrobial potential of cultivable endophytic actinobacteria associated with the medicinal plant *Thymus roseus*. Front Microbiol 11:191
17. Cardoso VM, Campos FF, Santos ARO, Ottoni MHF, Rosa CA, Almeida VG, Grael CFF (2020) Biotechnological applications of the medicinal plant *Pseudobrickellia brasiliensis* and its isolated endophytic bacteria. J Appl Microbiol 129(4):926–934
18. Ashitha A, Radhakrishnan EK, Mathew J (2020) Characterization of biosurfactant produced by the endophyte *Burkholderia* sp. WYAT7 and evaluation of its antibacterial and antibiofilm potentials. J Biotechnol 313:1–10
19. Chigurupati S, Vijayabalan S, Selvarajan KK, Aldubayan M, Alhowail A, Mani V, Das S (2020) Antimicrobial exploration between counterpart endosymbiont and host plant (*Tamarindus indica* Linn.). Curr Pharm Biotechnol 21(5):384–389
20. Pratiwi RH, Hidayat I, Hanafi M, Mangunwardoyo W (2020) Isolation and structure elucidation of phenazine derivative from *Streptomyces* sp. strain UICC B-92 isolated from *Neesia altissima* (Malvaceae). Iran J Microbiol 12(2):127–137
21. Techaoei S, Jirayuthcharoenkul C, Jarmkom K, Dumrongphuttidecha T, Khobjai W (2020) Chemical evaluation and antibacterial activity of novel bioactive compounds from endophytic fungi in *Nelumbo nucifera*. Saudi J Biol Sci 27(11):2883–2889
22. Sawarkar A, Sharma RK, Gautam V, Shraman K, Jain S (2021) Antibacterial activity of endophytic bacteria from leaves of *Tamarindus indica*. Pharma Innov J 10(2):472–475
23. Gos FM, Savi DC, Shaaban KA, Thorson JS, Aluizio R, Possiede YM, Glienke C (2017) Antibacterial activity of endophytic actinomycetes isolated from the medicinal plant *Vochysia divergens* (Pantanal, Brazil). Front Microbiol 8:1642
24. Tilwari A, Dixit P (2018) Isolation and screening of endophytic fungus from medicinal plant *Saraca asoca* for antibacterial activity. J Pharmacogn Phytother. SP2: 351–354
25. Aljuraifani A, Aldosary S, Ababutain I (2019) In vitro antimicrobial activity of endophytes, isolated from *Moringa peregrina* growing in eastern region of Saudi Arabia. Natl Acad Sci Lett 42(1):75–80
26. Syed CS, Mounika PPN, Mounika Y, Kumar SS, Bai VT, Audipudi AV (2017) Evaluation of antimicrobial and antibiotic sensitivity of chilli root endophytic bacteria for eco-friendly biofertilizer. Int J Curr Microbiol App Sci 5:45–53
27. Marcellano JP, Collanto AS, Fuentes RG (2017) Antibacterial activity of endophytic fungi isolated from the bark of *Cinnamomum mercadoi*. Pharm J 9(3):405–409
28. Santos IPD, Silva LCND, Silva MVD, Araújo JMD, Cavalcanti MDS, Lima VLDM (2015) Antibacterial activity of endophytic fungi from leaves of *Indigofera suffruticosa* Miller (Fabaceae). Front Microbiol 6:350
29. Chandrakar S, Gupta AK (2017) Antibiotic potential of endophytic actinomycetes of medicinal herbs against human pathogenic bacteria. Proc Natl Acad Sci India B Biol Sci 87(3):905–915
30. Shukla S, Naik G, Mishra SK (2021) Potential antimicrobial activity of bacterial endophytes isolated from *Flacourtia jangomas* (Lour.) Raeusch, a less explored medicinal plant. J Microbiol Biotechnol Food Sci:473–477
31. Muazzam KAAR, Taufig MMJ, Darah I (2015) Isolation and screening of antibacterial activity of endophytic fungi isolates from *Ocimum citriodorum* Vis. leaves. Aust J Basic Appl Sci 9(28):155–161
32. Sadrati N, Zerroug A, Demirel R, Bakli S, Harzallah D (2020) Antimicrobial activity of secondary metabolites produced by *Aspergillus neobridgeri* isolated from *Pistacia lentiscus* against multi-drug resistant bacteria. Bangladesh J Pharmacol 15(3):82–95
33. AhmedZ AS, Alauddin S, Mohiuddin SG, Javed A, Ahmed R, Bitar AN, Sheikh Ghadzi SM (2020) In vitro cytotoxic and antimicrobial activities of *Erythrina suberosa* (Roxb) bark. J Pharm Bioallied Sci 12(2):210–216

34. Mukherjee A, Das R, Pal A, Paul AK (2017) Antibacterial activity of *endophytic Bacillus safensis*isolated from *Ophioglossum reticulatum* L. Microbiol Res J Int 18(5):1–12
35. Kampapongsa D, Kaewkla O (2016) Biodiversity of endophytic actinobacteria from jasmine rice (*Oryza sativa* L. KDML 105) grown in Roi-Et Province, Thailand and their antimicrobial activity against rice pathogens. Ann Microbiol 66(2):587–595
36. Ramesha KP, Mohana NC, Nuthan BR, Rakshith D, Satish S (2020) Antimicrobial metabolite profiling of *Nigrospora sphaerica* from *Adiantum philippense* L. J Gene Eng Biotechnol 18(1):1–9
37. Radhakrishnan M, Vijayalakshmi G, Gopikrishnan V, Jerrine J (2016) Bioactive potential of actinobacteria isolated from certain understudied regions in. India J Appl Pharm Sci 6(08):151–155
38. Mahadevamurthy M, Puttaswamy H, Channappa TM, Sidappa M, Madegowda P, Chikkamanchegowda JS, Nagaraj AK (2016) Antibacterial potential of fungal endophytes isolated from *Boerhaavia diffusa* L. J Appl Pharm Sci 6:216–221
39. Debalke D, Birhan M, Kinubeh A, Yayeh M (2018) Assessments of antibacterial effects of aqueous-ethanolic extracts of *Sida rhombifolia's* aerial part. Sci World J. Article ID 8429809. https://doi.org/10.1155/2018/8429809
40. Uma V, Arunadevi S, Elizabeth JP, Vidhya A (2017) Bioprospecting of actinobacteria from Senganatham Hill Anunexplored ecosystem. Ind J Appl Microbiol 20(1):31–41
41. Manikkam R, Ganesan V, Kaari M, Venugopal G, Arumugam S, Joseph J (2021) Antibacterial efficacy of *Streptomyces maritimus* SACC-E6 against carbapenem-resistant *Klebsiella pneumoniae* ATCC BAA-1705. J Appl Pharm Sci 11(01):089–094
42. Habtom S, Gebrehiwot S (2019) In-vitro antimicrobial activities of crude extracts of *Vernonia amygdalina* and *Croton macrostachyus* against some bacterial and fungal test pathogens. J Phytopharmacol 8(2):57–62

Chapter 24

Screening of Endophytes for Nematicidal Activities

Soma Barman, Ratan Chowdhury, and Satya Sundar Bhattacharya

Abstract

Endophytes are symbiotic microorganisms found in almost all living plant species and hold a very special place due to the fact that it colonizes in the plant tissues, thereby plant gets multiple benefits from it. One of such utilization as an alternative to nematicides is the enhancement of plant health for providing sustainable agriculture. Endophytes have been reported to produce some specific compounds such as antibiotics, siderophores which kill different pathogens either fungi, bacteria, or different insects and nematodes. Endophytes are identified based on the isolation, culturing, microscopic as well as molecular identification methods from different plant parts. The present objective of the work is to screen the potent endophytes for nematicidal activities.

Key words Endophytes, Isolation, Screening, Identification, Nematicidal

1 Introduction

Endophytes are groups of microbial species that usually harbor within the living plant tissues for shelter and nutrition without harming their host. These endophytic microbes have been employed in vitro experiments as well as in field conditions as a biocontrol agent to control different types of phytopathogenic fungi [1], insects [2, 3], and nematodes [4]. Gall formation in roots, lesions in roots, disproportionate root branching, wounded root tips, and underdeveloped root systems were caused by root-knot nematodes the most common example is *Meloidogyne incognita*, plant wilting even in the presence of abundant soil moisture, foliage symptoms viz, yellowing were some of the phyllospheric plant parts which were indicated by slow growth of the whole plant, also this leads to reduction, smaller, curled, and discoloration of leaves, which finally ends on the margin of leaves showing brown spots while the leaflets size and number also reduce drastically, disturbing the photosynthetic machinery of the plant. Other nematodes include *Heterodera* spp., *Globodera* spp. which causes cyst; root lesions caused by *Pratylenchus* spp., burrowing disease in

A. Sankaranarayanan et al. (eds.), *Endophytic Microbes: Isolation, Identification, and Bioactive Potentials*, Springer Protocols Handbooks, https://doi.org/10.1007/978-1-0716-2827-0_24,

plants by *Radopholus similis*; bulb and stem disease caused by *Ditylenchus dipsaci*; spiral disease caused by *Helicotylenchus* spp.; reniform nematode, e.g., *Rotylenchulus reniformis* causes impaired root system in below ground part and malformed fruits and seeds in above ground part; dagger disease by *Xiphinema* spp.; bud and leaf disease caused by *Aphelenchoides* spp.; pine wilt disease caused by *Bursaphelenchus xylophilus*, etc. These pathogenic nematodes infect the plants and cause severe damages leading to severe agricultural losses [5]. Symptoms that are caused by the invasion of this organism are basically a gall at the root. The gall can create severe impairment in the host plant. Plants usually derive their food from their root cells. But whenever there is severe impairment, in the root cells this leads to the blockage of the plant roots for up taking the water and nutrients from the soil [6].

Endophytes act as a biological control microorganism in case of the pathogenic nematodes as they have the ability to significantly control these pests. Although there are several agronomic measures (viz, crop rotations, using resistant varieties, and applying chemical nematicidal agents) for controlling nematode infections in crops but these practices are inadequate to combat severe infestations [2, 7]. Application of chemical nematicidal agents for controlling nematodes is harmful due to their high toxicity. Due to which the end results of the residuary can be visible throughout the chain of food cycle which ultimately affects plants and human health [8, 9]. This gives a preference for using endophytes as perfect biocontrol nematicides. The endophytes use antagonistic property or mechanism of removal of pathogens by niche exclusion principle, and thus can promote a healthy disease-free plant growth [10]. Apart from these, endophytes share a close relationship with the host plants at the cellular and molecular levels. They produce a large number of antimicrobial compounds like antibiotics which help the host to resist the plant against these pathogenic nematodes [11]. They secrete different kinds of naturally occurring molecules like the siderophores and chitinolytic enzymes [2]. Siderophores help in the binding and transport of iron into the plants. On the other hand, the secretion of chitinolytic enzymes helps in the induction of plant resistance against pathogen infestation, thus they co-exist and control the pests from harming the plants [12, 13]. They include mostly the *Bacillus* spp., *Streptomyces* spp., *Herbaspirillum* spp., etc. belonging to Gram-positive and Gram-negative categories. Other examples of endophytic bacterial communities are *Pseudomonas fluorescens*, *Serratia marcescens*, etc. [14]. Different endophytic fungi (*Alternaria*, *Colletotrichum*, *Cladosporium*, *Fusarium*, *Pestalotiopsis*, and *Trichoderma* species) provide a fair choice for some of the disease management in plants that can be useful as a biocontrol agent [15, 16].

Table 1
Endophytic microbes, their host plant, and range of nematicidal activity

Endophytes	Host plant	Nematicidal activity	References
Pseudomonas sp. (strain 273), *Bacillus* spp. (strain MPB04), *Methylobacterium* sp.	*Abelmoschus esculentus* L.	*Meloidogyne incognita*	[27]
B. subtilis (strain 168)	*Musa paradisiaca* L.	*Radopholus similis*, *Pratylenchus coffeae*	[27]
Kocuria sp.	*Pinus densiflora* Siebold & Zucc.	*Bursaphelenchus xylophi lus*	[28]
Pseudoxanthomonas sp.	*Pinus koraiensis* Siebold & Zu cc.	*B. Xylophilus*	[28]
Stenotrophomonas sp.	*P. rigida* mill.	*B. Xylophilus*	[28]
Bacillus sp.	*P. thunbergia* Parl.	*B. Xylophilus*	[28]
Paenarthrobacter sp.	*Pinus* species	*B. Xylophilus*	[28]
Okibacterium sp., *Flexivirga* sp., *Naumannella* sp.	*Pinus* species	*B. Xylophilus*	[28]
Exiguobacterium aurantiacum, *E. mexicanum*	*P. rigida* mill.	*M. incognita*	[28]
Chaetomium globosum	*Ginkgo biloba* L.	*M. javanica*	[29]
Burkholderia ambifaria	*Zea mays* L.	*M. incognita*	[30]

Some of the endophytic fungal isolates which showed nematicidal activity were *Fusarium*, *Trichoderma*, *Chaetomium*, *Acremonium*, *Paecilomyces*, and *Phyllosticta*. Some *Fusarium* isolates can induce systemic acquired resistance (SAR) in banana plant against burrowing nematodes *Radopholus similis* [17, 18]. Other studies found that fungi from the genera *Acremonium*, *Paecilomyces*, and *Chaetomium* produced nematicidal compounds [19–22]. Some other endophytes, their host range, and nematicidal activity are listed in Table 1. One potent isolate from the genus *Phyllosticta* was found to be antagonistic to *M. incognita*. They suppressed nematodes mainly before the penetration into the host plant [23]. Dababat and Sikora (2007) [24, 25] reported that non-pathogenic *Fusarium* sp. Fo162 altered the chemical composition of root exudates of tomato. Thus, reduced the attraction and invasion of *M. incognita* and induced SAR against *M. incognita*. Endophytic fungi produced active nematicidal compounds, and thus trigger plant defense. These techniques are considered to be most likely to be used as seed treatment agents for the biocontrol of root-knot nematodes. At the young seedlings stage, the direct contact between small root, fungi, and nematodes is enough. Nematicidal compounds from the fungi can play the major role suppressing nematode penetration [26].

Endophytes were identified based on the staining methods, biochemical assays, and molecular techniques. In the present endeavor, a user-friendly protocol for screening and isolation of endophytes with nematicidal activities from the natural environment is elaborated.

2 Materials

2.1 Isolation of Endophytic Microbes from Roots, Shoots, and Leaves of Plant Samples

- Root/shoot samples collected.
- Sterilized polythene bag (for sample collection).
- Sterilized scalpel (for cutting the root samples).
- Autoclaved distilled water (washing purpose).
- Sterilized mortar and pestle.
- Sodium hypochlorite (NaClO) solution (0.01%).
- Triton X-100 (0.05%).
- Sterilized potassium (K) phosphate buffer pH 7 (0.02 M).
- Sodium (Na) phosphate buffer [pH 7 (0.02 M)].
- King's B medium (KB).
- Tryptic Soya Broth (TSA).
- Agar powder.
- Nutrient agar (NA).
- Potato dextrose broth (PDB).
- Petri dishes.
- Griseofulvin (100 μg/mL).
- Streptomycin sulfate (100 μg/mL).
- Micropipettes.
- BOD incubator with shaker (30 °C).
- Autoclave.
- Laminar air flow.
- Ethanol 70% (v/v).
- Glass slides Light Microscope.

2.2 Nematicidal Efficacy of Endophytic Microbes In Vitro

- Bacterial culture.
- Shaker (at 160 rpm) and incubator (28 °C).
- Centrifuge.
- Millipore filter (0.22 μm), Microcentrifuge tube, Microtiter plates.
- Sterile distilled water (SDW).

- Population of *M. incognita* and any other nematodes [second stage juvenile nematodes (J2s) suspension].
- Highly susceptible plant variety.
- Greenhouse.
- Nematode eggs (NE).
- NaOCl solution (15 g/L).
- Extraction trays.
- Agar discs.
- 300 mL flasks.
- Potato Dextrose Broth medium (PDB).
- Filter paper (Whatman no. 1).
- Petri dishes.

2.2.1 Preparation of Fungal Culture Filtrates (FCF)

- 250 mL flasks.
- BOD incubator with shaker (150 rpm speed).
- Filter paper (Whatman No.1).
- Syringe filter (0.45 μm).

2.3 Screening of Endophytic Fungi for Nematicidal Activities (In Vivo)

- Incubator.
- Sterilized water.
- Pipette.
- Fine mesh screen (0.15 mm).
- Neubauer hemacytometer.
- Soil mixture (sand:loam = 1:1 v/v).
- Pots (size 400 cm^3).
- Surface-sterilized plant seeds.
- Greenhouse (temperature 25 °C, continuous 16 h artificial light).
- Poly fertisol (N:P: K = 14:10:14).
- Computer installed with SPSS 16.0 software.

3 Methods

3.1 Isolation of Endophytic Microbes from Roots, Shoots, and Leaves of Plant Samples

The procedure is represented in Fig. 1. Molecular characterization and identification of the bacterial isolates is done according to the method of Ghosh et al. (2016) [31].

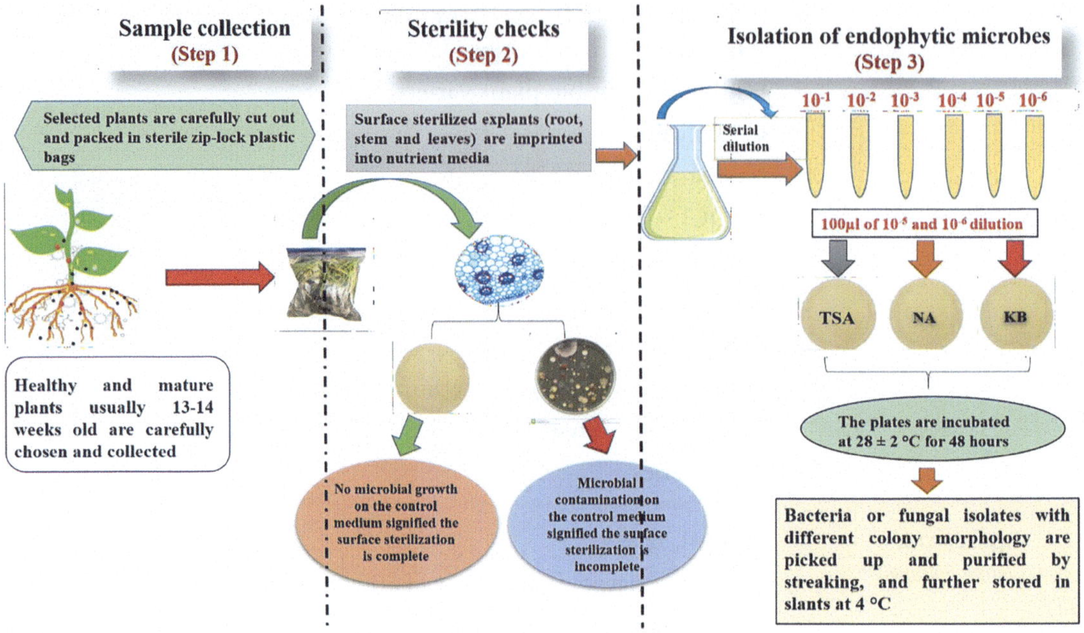

Fig. 1 Procedure for isolation of endophytic microbes

3.2 Nematicidal Efficacy of Endophytic Microbes In Vitro

For finding out the nematicidal activities among the microbes, there are various tests which can be carried out in vitro. These tests are carried out using sterile water and bacterial cells immersed in it [32]. Endophytic bacteria were tested for their efficacy against burrowing nematodes.

3.3 For Endophytic Bacteria

1. The purified bacterial strains are cultured in 100 mL of nutrient broth (for bacteria) or PDB (for fungi) culture medium at 28 °C for 24 h at 160 rpm.
2. Then the bacterial broth is centrifuged (8000 rpm for 5 min).
3. Now, in the third step, the cells are removed from the supernatant after centrifugation and filtered with a 0.22 µm membrane filter.
4. 100 hatched pre-parasitic second stage juveniles of nematode are added to the microcentrifuge tube containing 1 mL of liquid medium and centrifuged at 2500 rpm for 1 min.
5. The obtained supernatant is incubated at 28 °C.
6. A highly susceptible host plant variety is used in the greenhouse with a population of *M. incognita* or any other nematode.
7. Extraction of NE is done from diseased plant roots using 15 g/L NaOCl solution [33].
8. NE are incubated (25 °C for 14 days) in extraction trays to get J2s. The J2s suspension is accustomed at 200 nematodes/mL.

The in vitro test studied by Chen and Dickson (2000) [34] is as follows:

- 200 μL of bacterial suspension (10^9CFU/mL) in sterile distilled water (SDW) is added to each well in 24-well microtiter plates.
- Surface sterilized nematode suspension in SDW (50 mL containing 30 nematodes is then added to each well).
- Each treatment is replicated thrice. Wells containing only SDW are considered as controls.
- The plates are incubated at 28 °C.
- The number of live and dead nematodes was counted after 72 h under a stereomicroscope by adding 20 mL of 1 N NaOH.
- The results were recorded and the percentage mortality of nematodes over the control was calculated.
- The data are subjected to angular transformation and analyzed by ANOVA.

3.4 For Endophytic Fungi

1. 300 mL Erlenmeyer flasks are used for inoculation of agar cakes from the 14 days old PDA plates into 200 mL PDB and incubated at 25 °C for 7 days at 100 rpm.
2. Whatman no.1 filter paper and millipore filter (0.22 μm) are used for obtaining of the FCF. It is obtained by the separation of the liquid from the mycelia.
3. To test any toxicity of the FCF (in vitro) against J2s, 5 mL FCF of each isolate and 1 mL J2s suspension (200 J2s/mL) are pipetted into each Petri dishes. 5 mL PDB and 1 mL J2s suspension are pipetted into the control Petri dishes. The presence of nonmotile nematodes (percentage) is calculated after 48 h. They are allowed to be recovered in tap water for 5 h.
4. All of those that remained insensitive are considered as dead.
5. A total of 6 replicates (both treatment and control) are used. Significant differences are calculated using one-way ANOVA and Least Significant Difference (LSD) post-hoc test in SPSS software.

3.5 Preparation of FCF

1. In 250 mL flasks the fungi are cultured containing 100 mL of PDB medium (pH 5.0).
2. The flasks are shaken at 150 rpm for 7 days in darkness at 28 ± 2 °C.
3. Then the culture broth was passed through double layers of Whatman No.1 filter paper (Fig. 2).

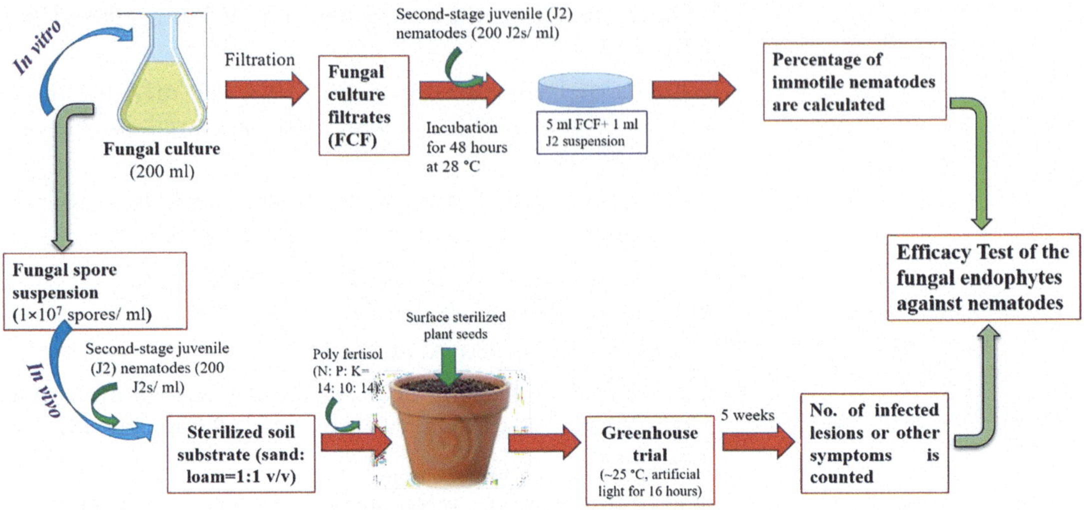

Fig. 2 Screening and efficacy of fungal endophytes as nematicidal agent

4. The culture broths are passed through a filter of 0.45 μm membrane filter to remove fungal spores.
5. The filtrates are used at concentrations of the original preparation.

3.6 Screening of Endophytic Fungi for Nematicidal Activities (In Vivo)

1. The fungal isolates are incubated at 28 ± 2 °C on PDA plates for 15 days.
2. Then the surface of the plates is flooded with 5 mL of sterilized water, and the spores as well as conidia are scrapped from the Petri dishes.
3. Fine mesh screen (0.15 mm) is used for the filtration of the fungal spore suspension to separate spores from hyphae.
4. Spore suspension concentration is adjusted to 1×10^7 spores/ mL with the help of Neubauer hemacytometer and is kept at 4 ° C for future use.
5. The mixture of soil substrate is autoclaved at 121 °C for 2 h and taken into 400 cm^3 pots.
6. 3 mL of J2s suspension is mixed properly with the soil (Fig. 2).
7. Surface sterilized plant seeds are placed into 1 cm deep in the soil mixture.
8. On top of the seed, 1 mL spore suspension is pipetted of the tested isolate and covered with top soil.
9. Then 1 mL water is pipetted around the control seeds. All together six replicates are prepared for each fungal endophyte including the control sets. The plants are watered daily and kept in the greenhouse (~25 °C, continuous 16 h artificial light). They are also manured every 7 days interval with 3 g/ L of polyfertisol in the mentioned ratio.

10. The pots are organized as random block design technique.
11. The plants are harvested after 5 weeks of seed germination, and the number of infected lesions or other symptoms (e.g., galls) is calculated.
12. All the data are analyzed using ANOVA in SPSS software.
13. LSD values are calculated and their control efficacies are determined by selecting 10 isolates with the highest significant differences.

3.7 Efficacy Test and Colonization Ratio of Selected Fungal Isolates

1. Among 10 potent fungal endophytes the control efficiencies are tested repeatedly following the same procedure as the first screening (Fig. 2).
2. The measurement of symptoms (e.g., gall counting or any others) is evaluated to assess the colonization ability of the endophyte. 3 plant replicates of each endophyte are taken for further isolation. From each plant replicate nearly about 30 underground and phyllospheric fragments are impregnated into 5–6 PDA containing plates.
3. The number of plant fragments (N) with visible fungal growth (colonization ratio) is calculated after 14 days with the formula $N/30 \times 100\%$ [24].

4 Observation

Acknowledgments

The authors thankfully acknowledged the Scientific and Engineering Research Board (SERB), National Post-Doctoral Fellowship (NPDF), India (file no. PDF/2017/002639) for financial assistance.

References

1. Guetsky R, Shtienberg D, Elad Y, Dinoor A (2001) Combining biocontrol agents to reduce the variability of biological control. Phytopathology 91:621–627
2. Verma SK, White JF (2018) Indigenous endophytic seed bacteria promote seedling development and defend against fungal disease in brown top millet *Urochloa ramosa* L. J Appl Microbiol 124:764–778
3. Azevedo JL, Maccheroni JW, Pereira JO, Luiz de Araujo W (2000) Endophytic microorganisms: a review on insect control and recent advances in tropical plants. Electron J Biotechnol 3(1):40–65
4. Spadaro D, Gullino ML (2005) Improving the efficacy of biocontrol agents against soilborne pathogens. Crop Prot 24:601–613
5. Bernard GC, Egnin M, Bonsi C (2017) The impact of plant-parasitic nematodes on agriculture and methods of control. In: Shah MM (ed) Nematology: concepts, diagnosis and control. Intech Open, London, pp 121–151
6. Griffiths BS (1986) Mineralization of nitrogen and phosphorus by mixed cultures of the ciliate

protozoan *Colpoda steinii*, the nematode *Rhabditis* sp. and the bacterium *Pseudomonas fluorescens*. Soil Biol Biochem 18:637–641

7. Roberts PA (1992) Current status of the availability, development and use of host plant resistance to nematodes. J Nematol 24:213–227
8. Nico AI, Jimenez-Diaz RM, Castilla P (2004) Control of root knot nematodes by composed agro-industrial wastes in potting mixtures. Crop Prot 23:581–587
9. Carriger J, Rand G, Grdinali P, Perry W, Tompkins M, Fernanndes A (2006) Pesticides of potential ecological concern in sediment from South Florida canals: an ecological prioritization for aquatic arthropods. Soil Sed Contam 15:21–45
10. Kusari S, Hertweck C, Spiteller M (2012) Chemical ecology of endophytic fungi: origins of secondary metabolites. Chem Biol 19:792–798
11. Raaijmakers JM, Mazzola M (2012) Diversity and natural functions of antibiotics produced by beneficial and plant pathogenic bacteria. Annu Rev Phytopathol 50:403–424
12. Hallmann J (2001) Plant interactions with endophytic bacteria. In: Jeger MJ, Spence NJ (eds) Biotic interactions in plant-pathogen associations. CABI Publishing, New York, pp 87–119
13. Bacon CW, Hinton DM (2006) Bacterial endophytes: the endophytic niche, its occupants, and its utility. In: Gnanamanickam SS (ed) Plant-associated bacteria. Springer, Dordrecht, pp 155–194
14. Adesemoye A, Egamberdieva D (2013) Beneficial effects of plant growth-promoting rhizobacteria on improved crop production: prospects for developing economies. In: Maheshwari D, Saraf M, Aeron A (eds) Bacteria in agrobiology: crop productivity. Springer, Berlin, pp 45–63
15. Prihanto AA, Firdaus M, Nurdiani R (2011) Endophytic fungi isolated from mangrove (*Rhizophora mucronata*) and its antibacterial activity on *Staphylococcus aureus* and *Escherichia coli*. J Food Sci Eng 1:386–389
16. Barman S, Ghosh R, Mandal N, Gond SK, Mandal NC (2016) Assessment of antimicrobial activity of endophytic fungi isolated from *Justicia adhatoda* L. UGB J Plant Biol Biotech 1(1):19–23
17. Yan XN, Sikora RA, Zheng JW (2011) Potential use of cucumber (*Cucumis sativus* L.) endophytic fungi as seed treatment agents against root-knot nematode Meloidogyne incognita. J Zhejiang Univ Sci B 12(3): 219–225. https://doi.org/10.1631/jzus.B1000165
18. Vu T, Sikora R, Hauschild R (2006) *Fusarium oxysporum* endophytes induced systemic resistance against *Radopholus similis* on banana. J Nematol 8(6):847–852. https://doi.org/10.1163/156854106779799259
19. Sharon E, Bar-Eyal M, Chet I, Herrera-Estrella A, Kleifeld O, Spiegel Y (2001) Biological control of the root-knot nematode *Meloidogyne javanica* by *Trichoderma harzianum*. Phytopathology 91(7):687–693. https://doi.org/10.1094/PHYTO.2001.91.7.687
20. Nitao JK, Meyer SLF, Oliver JE, Schmidt WF, Chitwood DJ (2002) Isolation of flavipin, a fungus compound antagonistic to plant-parasitic nematodes. Nematology 4(1):55–63. https://doi.org/10.1163/156854102760082203
21. Goswamia J, Pandey RK, Tewaria JP, Goswami BK (2008) Management of root knot nematode on tomato through application of fungal antagonists, *Acremonium strictum* and *Trichoderma harzianum*. J Environ Sci Health B 43(3):237–240. https://doi.org/10.1080/03601230701771164
22. Kalele DN, Affokpon A, Coosemans J (2007) Efficacy of *Paecilomyces lilacinus* strain 251 against root knot nematodes in tomato under greenhouse conditions. Commun Agric Appl Biol Sci 72(1):209–213
23. Kiewnick S, Sikora RA (2006) Biological control of the root-knot nematode *Meloidogyne incognita* by *Paecilomyces lilacinus* strain 251. Biol Control 38(2):179–187. https://doi.org/10.1016/j.biocontrol.2005.12.006
24. Dababat AA, Sikora RA (2007) Induced resistance by the mutualistic endophyte, *fusarium oxysporum* 162, toward *Meloidogyne incognita* on tomato. Biocontrol Sci Tech 17(9): 969–975. https://doi.org/10.1080/09583150701582057
25. Dababat AA, Sikora RA (2007) Influence of the mutualistic endophyte *fusarium oxysporum* 162 on *Meloidogyne incognita* attraction and invasion. Nematology 9(6):771–776. https://doi.org/10.1163/156854107782331225
26. Tian B, Yang J, Zhang KQ (2007) Bacteria used in the biological control of plant-parasitic nematodes: populations, mechanisms of action, and future prospects. FEMS Microbiol Ecol 61(2):197–213. https://doi.org/10.1111/j.1574-6941.2007.00349.x
27. Vetrivelkalai P, Sivakumar M, Jonathan EI (2010) Biocontrol potential of endophytic

bacteria on *Meloidogyne incognita* and its effect on plant growth in bhendi. J Biopest 3(2): 452–457

28. Ponpandian LN, Rim SO, Shanmugam G, Jeon J, Park YH, Lee SK, Bae H (2019) Phylogenetic characterization of bacterial endophytes from four *Pinus* species and their nematicidal activity against the pine wood nematode. Sci Rep 9(1):1–11. https://doi.org/10.1038/s41598-019-48745-6
29. Khan B, Yan W, Wei S, Wang Z, Zhao S, Cao L, Rajput NA, Ye Y (2019) Nematicidal metabolites from endophytic fungus *Chaetomium globosum* YSC5. FEMS Microbiol Lett 366(14). https://doi.org/10.1093/femsle/fnz169
30. Pavithra G, Bindal S, Rana M, Srivastava S (2020) Role of endophytic microbes against plant pathogens: a review. Asian J Plant Sci 19(1):54–62. https://doi.org/10.3923/ajps.2020.54.62
31. Ghosh R, Barman S, Mukherjee R, Mandal NC (2016) Role of phosphate solubilizing *Burkholderia* spp. for successful colonization and growth promotion of *Lycopodium cernuum* L. (Lycopodiaceae) in lateritic belt of Birbhum district of West Bengal, India. Microbiol Res 83:80–91
32. Tian H, Riggs RD, Crippen DL (2000) Control of soybean cyst nematode by chitinolytic bacteria with chitin substrate. J Nematol 32: 370–376
33. Hussey RS, Barker KR (1973) A comparison of methods of collecting inocula *Meloidogyne* spp. including a new technique. Plant Dis Rep 57(6):1025–1028
34. Chen SY, Dickson DW (2000) A technique for determining live second-stage juveniles of *Heterodera glycines*. J Nematol 32:117–121

Chapter 25

Screening of Endophytes for Virucidal Activity

L. Manjunatha, G. Narasa Reddy, LeelaSaisree Uppala, B. M. Ravikumara, and R. Ranjini

Abstract

Currently, crop production is at risk due to the global warming effect with increased biotic and abiotic stresses. Viral diseases of plants cause enormous economic losses particularly in the tropics and semi-tropics. Assessments of losses as a result of pest or pathogen attack in crop plants are mostly lacking. In particular, viral diseases cause serious economic losses destroying crops and reducing agronomic productivity, they become the limiting factor production of both open field and under greenhouse cultivation systems. This is because plant viruses are obligate parasites and require living tissue for their multiplication and spread. Therefore, they are able to interfere with plant metabolism and compete for host plant resources, so determining a decrease of plant growth and productivity. Many diverse approaches have been tried to minimize the losses caused by these diseases. In such a scenario, it is necessary to introduce new strategies for controlling plant viruses and parasites in order to help maintain ecosystems and to boost sustainable agriculture. The work on virucidal activity of endophytes against plant viruses is very negligible and absent. Therefore, based on available literature and experience, we have formulated the protocols for the testing of plant endophytes against plant viruses.

Key words Endophyte, Virus, Screening, Isolates, Microbes

1 Mechanism of Endophytes in Antiviral Activity

The information on mechanism behind antiviral properties of endophytes is very less. The discovery of the possible antiviral compounds from endophytic fungi or bacteria is still novel and in an infancy [1]. There is only limited number of compounds reported as antiviral agents from fungal and bacteria endophytes. The main drawback in discovering antiviral compounds is perhaps due to the absence of antiviral screening methods against plant viruses.

Many endophytic bacteria secrete enzymes including RNases, which are involved immobilization of organic phosphates for adaptation to changing environmental conditions. *Bacillus licheniformis*

A. Sankaranarayanan et al. (eds.), *Endophytic Microbes: Isolation, Identification, and Bioactive Potentials*,
Springer Protocols Handbooks, https://doi.org/10.1007/978-1-0716-2827-0_25,

Table 1
Some of the endophytes reported to be virucidal activity against plant viruses

Endophytes	Antiviral components	Antiviral against plant virus	Reference
Pseudomonas putida A3	RNase	TMV-infected tobacco plants	[5]
Bacillus pumilus7P/3-19	RNase	TMV-infected tobacco plants	[6]
–	RNase	PVX, PVY, PVM, and PVS in potato	[7]
Schizosaccharomyces pombe	RNase gene PAC1	China soybean mosaic virus (SMV) in soybean	[8]
B. amyloliquefaciens	baRNase gene	Tomato leaf curl virus in tobacco	[9]

produces extracellular RNases [2, 3] which stimulate plant growth and resistance to a wide spectrum of stress elements, and high levels of them display antiviral activity against viral RNA. The *Pantoea, Cronobacter, Microbacterium,* and *Staphylococcus* spp. originated from Cucurbitaceae plants and as 73% of Bacillus, 30% of Enterobacteriaceae, and 27% of *Paenibacillus* isolates produce nucleases [4] Therefore, utilization of endophytic *Bacillus* can protect crop plants against plant virus by producing RNases which directly breaks the viral RNA particles in infected plant tissues (Table 1).

2 Isolation of Endophytes

Endophytes can be effortlessly isolated on any microbial media, potato dextrose agar, and any nitrogen- or carbon-containing media. The most common method used to detect and count endophytes includes isolation from surface-sterilized tissue of the host plant. Only healthy and disease-free plants should be collected to minimize the presence of pathogenic and saprophytic microbes. The plants once collected should be stored in distinct plastic bags at 4 °C in an ice box till isolation processes started. Surface-sterilize the leaf samples and isolate the endophytic bacteria as the method described by Yan et al. [10]. Various protocols are available for isolation of entophytic microbes. Here, we have outlined the three methods of testing endophytic microbes against plant viruses for their virucidal properties. They are (a) ribonuclease activity estimation method using endophytic bacteria, (b) screening of endophytes for virucidal activity by sap transmission method, and (c) screening of endophytic microbes for virucidal activity using insect vector.

2.1 Ribonuclease Activity Estimation Method Using Endophytic Bacteria

The quantitative estimation of extracellular RNase activity in liquid culture medium can be carried out according to the modified method of Sokurenko et al. [11, 12].

Protocol

1. Grow the bacteria strains and isolates on LP medium (low phosphate peptone, 2%;glucose, 1%; Na2HPO4, 0.04%; CaCl2, 0.01%; MgSO47H2O,0.03%; MnSO4, 0.01%; NaCl, 0.3%, 120 mg/mL of phosphorus).
2. Cultivate the bacterial endophyte for its multiplication at 37 °C in a laboratory shaker for 24–48 h at 120 rpm.
3. Measure the bacterial growth spectrophotometrically at 590 nm and express its number as optical density units (OD590).
4. Centrifuge the bacterial density of 10^8 cells/mL cultures for 15 min at 8000 rpm in a refrigerated centrifuge (Eppendorf, Germany).
5. Homogenize the potato leaves in sterile bags in 1 mL of 0.05 M tris–HCl buffer (pH 8.5) using blender Bag Mixer 400 W (Interscience, France) and incubate for 60 min at 4 °C.
6. This homogenate is subjected to centrifuge for 15 min at 8000 rpm in a centrifuge (Eppendorf, Germany) and measure the pH using pH meter.
7. Add 20 μL of homogenates into 1.98 mL of 50 mg/mL of torula yeast RNA solution in 0.05 M of tris–HCl buffer (pH 8.5) (Sigma, United States) and incubate for 1 h at 25 ° C and measure its absorbance at 260 nm comparative to the control (reaction mixture without leaf extract or bacterial medium) at 260 nm on spectrophotometer.
8. Consider the unit of nuclease activity as an amount of enzyme causing an increase in adsorption by 1.0 optical unit at 260 nm for 1 h at 25 °C [13].
9. Express the RNase activity as units/min/mL of liquid culture medium (extracellular *Bacillus* RNase activity) or units/min/mg of protein (RNase activity in plants). Protein concentration in plants can be measured using the Bradford assay.

2.2 Screening of Endophytes for Virucidal Activity by Sap Transmission Method

2.2.1 Materials Required

Plants, virus isolates, endophytic fungi or bacteria (in the form of endospores) and virus-maintaining hosts such as *Nicotiana benthamiana*, *Nicotiana rusticator*, and *N. tabacum*. Plant growth conditions such as insect-proof greenhouse with controlled environmental conditions at 25 °C/20 °C temperature and 12 h photoperiod with light intensity of 20,000 lux and relative humidity 85% (day) and 90% (night). Plants grown in pots of an appropriate size.

2.2.2 Protocol for Screening of Endophytic Microbes for Virucidal Activity

1. Dilute the viable spores of endophyte in sterile water @ 5.5×10^7 cfu/mL (Fungi) and 5.0×10^5cfu/mL bacteria [14].
2. Apply the endophytic spores either as drenching (60 mL per pot) or foliar spray (spraying until run-off at a maximum rate estimated up to 15 mL of suspension per plant, depending on age of the plant).
3. Drench or spray at a weekly interval for 2–3 times. The first application of endophytes at the developmental stage of the second leaf on the main shoot unfolded or 1–2 leaf stage or cotyledonary stage that depends on the type of virus we are investigating.
4. Inoculate the plant with plant virus one day after the second application of endophyte spores. (Foliar application experiments must be conducted only in the growth chamber which could compromise the viability of spores on leaf surface due to low relative humidity conditions lower than 50% in the greenhouse) (use of air-conditioners for temperature control).
5. Furthermore, test a single drench application of spores in air-dried potting medium at the time of sowing in pots (prior germination).
6. In general, mix the 1 kg of dry potting medium with 1–1.5 liter of 5.5×10^7 cfu/mL suspension and use the resulting mixture for seeds sowing.
7. Afterward, inoculate the virus on plants at the developmental stage of the third leaf on main shoot similarly to triple drench or foliar application experiments.
8. The chemical analogs of salicylic acid (SA) have been reported to induce resistance against TSWV [15] [16]. Therefore, Benzol [1–3] thiadiazol-7-carbothioic acid-S-methyl ester (BTH, Sigma), a chemical analog of SA, can be used as a positive control for activation of systemic acquired resistance (SAR).
9. Irrespective of spore application method in all the experiments, use BTH at 100 mM in foliar application experiments three times at weekly interval identical to that of endophyte applications. Besides, include controls as water treatment of virus-challenged plants and mock-inoculated plants.
10. For all sap transmitted plant viruses, prepare the viral inoculum from virus-infected *Nicotiana* sp. leaf tissues and ground in 0.01 M sodium phosphate or 0.05Mor 0.01 M potassium phosphate buffer of pH 7 containing 0.02 M β-mercaptoethanol in 1:3 w/v dilution with 3% w/v active carborundum 600 mesh [17] to create sub-lethal wounds on the leaf surface for virus entry. Inoculate the leaf surface with plant virus inoculum, onto the youngest fully expanded composite leaf and keep for 24 h in the dark at 22 °C to 24 °C.

11. The virus inoculum should also be inoculated onto local lesion hosts such as *Chenopodium quinoa* and *C. amaranticolor*, *Petunia hybrid* to quantify the infectivity of each inoculum.
12. To examine the action of endophytic spores against plant virus, test the different treatment combinations in the greenhouse using water, BTH, spores, triple or single drench or foliar application and in independent experiments. Follow a randomized complete block design with three replications consisting of 10 plants per pot (30 plants per treatment).
13. Assess the disease progress by visual observation of symptoms induced by the virus characteristic to the disease and also by double-antibody sandwich enzyme-linked immunosorbent assay [18] (DAS-ELISA) by utilizing commercial antibodies of the respective virus.
14. Perform ELISA in both challenged inoculated and the apical non-inoculated leaves (to check systemic virus movement) at three different time points, 14, 19, and 30 dpi for TSWV, and 7, 14, and 30 dpi for PVY [14]. Consider the samples as positive when their absorbance at 405 nm is higher than three times the mean absorbance value at 405 nm of three healthy control samples.
15. If the endophytic fungi and bacteria spores treated plant samples show absence of disease/significantly less disease and three-fold low absorbance value compared to control are considered as endophyte has virucidal activity against that plant virus.
16. For further validation, assess the height, fresh and dry weight, gene expression studies through quantitative PCR analysis [19] can be performed as described by Beris et al. [14].

2.3 Screening of Endophytic Microbes for Virucidal Activity Using Insect Vector

Lehtonen et al. [20] established the screening protocol for endophytic microbes for virucidal activity against barley yellow dwarf virus (BYDV) transmitted by bird cherry-oat aphid (*Rhopalosiphum padi*), which may be deterred by endophyte-origin alkaloids within the meadow ryegrass (*Lolium pratense* = *Festuca pratensis*). The percentage of BYDV infections and number of aphid vectors were lower in endophyte-infected plants compared to endophyte-free plants, signifying that endophyte infection could safeguard meadow ryegrass from infections of BYDV. Generally, meadow ryegrass is commonly infected by systemic and severely transmitted fungal endophyte *Neotyphodium uncinatum* and its frequencies vary from 0% to 100%. Therefore, the protocol is modified and outlined below. This method can be possibly used for screening of endophytes against various viruses such as Tospoviruses (Groundnut bud necrosis virus), Geminiviruses (Tomato leaf curl virus), Emaravirus (Pigeonpea sterility mosaic virus transmitted by eriophyid mite - *Aceriacajani*), All potyviruses such as papaya ring spot virus, Zucchini yellow mosaic virus (ZYMV), Chilli veinal mottle virus (ChVMV), Watermelon mosaic virus (WMV), Cucumber mosaic virus of bromoviridae, etc.

Protocol

1. The isolated endophytic bacteria or fungi from the host plants are subjected for multiplication.
2. Multiply the endophytic fungi or bacteria and dilute the viable spores of endophyte in sterile water at 5.5×10^7 cfu/mL (Fungi) and 5.0×10^5cfu/mL bacteria [14].
3. Raise the host seedlings of the respective plant virus in insect proof greenhouse containing four insect rearing cages of 5 pots each with 25 plants in each cage.
4. Apply the endophytic spores either as drenching (60 mL per pot) or foliar spray (spraying until run-off at a maximum rate estimated up to 15 mL of suspension per plant, depending on the age of the plant).
5. Drench or spray at a weekly interval for 2–3 times. The first application of endophytes at the developmental stage of the second leaf on main shoot unfolded or 1–2 leaf stage or cotyledonary stage that depends on type of virus we are investigating.
6. The insect vector (aphids, Whitefly, thrips, and leaf hoppers) were fed on virus-infected host plant leaves for 1 week before the experiment started.
7. Release the 30–50 insect vectors carrying virus into each cage and leave to reproduce for 2 months. The cages should be strictly insect proof (1.5 m × 1.5 m × 1 m (height)) to prevent viral infections from spreading to the neighboring cages.
8. The pots containing soil should be nutrient-rich with high in organic matter.
9. Harvest the plants and count the insect vector and samples for virus analysis. Calculate fresh and dry weight of the plant sample.
10. Perform the statistical analyses using a standardized statistical package.
11. Take 0.5 g of leaf samples for virus analysis before and after the insect vector experiment and frozen the same leaves at –20 °C for 1–3 months till ELISA test to carry out.
12. Grind the 0.1 g of each leaf sample in 1 mL of the sample extraction buffer [18] and conduct DAS-ELISA assay for virus presence.
13. Use the respective virus-specific antisera, conjugate, and the alkaline phosphate substrate according to the manufacturer's instructions (available in all the countries).

14. Add substrate and incubate for 1 h of incubation. Measure the absorbance at the optical density of 620 nm. Absorbance values must be at least twice as high as in healthy leaves for positive virus infection.
15. The leaf samples can also be used for RT-PCR-based molecular method using respective virus-specific primers for the detection of virus.
16. The endophytic bacteria or fungi could be considered as virucidal or antiviral activity based on negative reaction of endophyte-treated samples in ELISA or PCR test.

References

1. Rajamanikyam M, Vadlapudi V, Upadhyayula SM (2017) Endophytic fungi as novel resources of natural therapeutics. Brazil Arch Biol Technol 60
2. Ulyanova V, Mahmud RS, Dudkina E, Vershinina V, Domann E, Ilinskaya O (2016) Phylogenetic distribution of extracellu-Lar guanylpreferring ribonucleases renews taxonomic status of two bacillus strains. J Gen Appl Microbiol 62:181–188. https://doi.org/10.2323/jgam.2016.02.005
3. Ilinskaya O, Ulyanova V, Lisevich I, Dudkina E, Zakharchenko N, Kusova A (2018) The native monomer of *Bacillus pumilus ribonuclease* doesnot exist extra-cellularly. BioMed Res Int:ID4837623. https://doi.org/10.1155/2018/4837623
4. Khalaf EM, Raizada MN (2018) Bacterial seed endophytes of domesticated cucurbits antagonize fungal and oomycete pathogens including powdery mildew. Front Microbiol 9:42. https://doi.org/10.3389/fmicb.2018.00042
5. Yang J, Guo C, Zhai X, Shen L, Qian Y, Wang F (2012) Inactivation of tobacco mosaic virus in soil by pseudomonas putida A3-m strain to prevent virus mosaic disease. Afr J Microbiol Res 6:6300–6307. https://doi.org/10.5897/AJMR12.1123
6. Sharipova M, Rockstroh A, Balaban N, Mardanova A, Toymentseva A, Tikhonova A (2015) Antiviral effect of ribonuclease from Bacilluspumilusagainst phytopathogenic RNA-viruses. Agric Sci 6:1357–1366. https://doi.org/10.4236/as.2015.611130
7. Trifonova EA, Ibragimova SM, Volkova OA, Shumny VK, Kochetov AV (2018) Ribonuclease activity as a new prospective disease resistance marker in potato. Vavilov J Genet. Breed 22:987–991. https://doi.org/10.18699/VJ18.441
8. Yang X, Niu L, Zhang W, He H, Yang J, Xing G et al (2019) Increased multiple virus resistance in transgenic soybean overexpressing the double strand RNA-specific ribonuclease gene PAC1. Trans Res 28:129–140. https://doi.org/10.1007/s11248-018-0108-8
9. Pakniat-Jahromy A, Behjatnia SA, Dry IB, Izadpanah K, Rezaian MA (2010) A new strategy for generating geminivirus resistant plants using a DNA betasatellite/split baRNase construct. J Virol Methods 170:57–66
10. Yan X, Wang Z, Mei Y, Wang L, Wang X, Xu Q, Peng S, Zhou Y, Wei C (2018) Isolation, diversity, and growth-promoting activities of endophytic bacteria from tea cultivars of Zijuan and Yunkang-10. Front Microbiol 9:1848. https://doi.org/10.3389/fmicb.2018.01848
11. Sokurenko Y, Nadyrova A, Ulyanova V, Ilinskaya O (2016) Extracellular ribonuclease from *bacillus licheniformis* (balifase), a new member of the N1/T1 RNase superfamily. Biomed Res Int 4239375. https://doi.org/10.1155/2016/4239375
12. Sorokan A, Cherepanova E, Burkhanova G, Veselova S, Rumyantsev S, Alekseev V, Mardanshin I, Sarvarova E, Khairullin R, Benkovskaya G, Maksimov I (2020) Endophytic bacillus spp. as a prospective biological tool for control of viral diseases and non-vector Leptinotarsadecemlineata say.in Solanumtuberosum L. Front Microbiol 11:569457. https://doi.org/10.3389/fmicb.2020.569457
13. Ilinskaya ON, Karamova NS, Ivanchenco OB, Kipenskaya LV (1996) SOS-inducing ability of native and mutant microbial ribonucleases. Mutation Res 354:203–209. https://doi.org/10.1016/0027-5107(96)00012-7
14. Beris D, Theologidis I, Skandalis (2018) *Bacillus amyloliquefaciens* strain MBI600 induces

salicylic acid dependent resistance in tomato plants against *tomato spotted wilt virus* and *potato virus Y*. Sci Rep 8:10320. https://doi.org/10.1038/s41598-018-28677-3

15. Mandal B, Mandal S, Csinos AS, Martinez N, Culbreath AK, Pappu HR (2008) Biological and molecular analyses of the acibenzolar-S-methyl-induced systemic acquired resistance in flue-cured tobacco against tomato spotted wilt virus. Phytopathology 98(2):196–204. https://doi.org/10.1094/PHYTO-98-2-0196
16. Nie X (2006) Salicylic acid suppresses potato virus Y isolate N: O-induced symptoms in tobacco plants. Phytopathology 96:255–263
17. Manjunatha L, Patil MS, Kavitha TR, Vanitha LS, VijayaMahantesha SR (2010) Mechanical transmission of groundnut bud necrosis virus. Environ Ecol 28(4A):2456–2458
18. Clark MF, Adams AN (1977) Characteristics of the microplate method of enzyme-linked immunosorbent assay for the detection of plant viruses. J Gen Virol 34:475–483
19. Steibel JP, Poletto R, Coussens PM, Rosa GJ (2009) A powerful and flexible linear mixed model framework for the analysis of relative quantification RT-PCR data. Genomics 94: 146–152
20. Lehtonen PT, Helander M, Siddiqui SA, Lehto K, Saikkonen K (2006) Endophytic fungus decreases plant virus infections in meadow ryegrass (Loliumpratense). Biol Lett 2(4): 620–623

Chapter 26

Screening of Endophytes for Fungicidal Activities

K. R. Jeya, M. Veerapagu, and A. Sankaranarayanan

Abstract

Fungal diseases pose a significant threat to many plants and animals besides public health. Many fungal pathogens invade plants and affect sustainable food production. However, only a few antifungal antibiotics are currently available for the treatment of fungal diseases. Hence the use of replacement medicines has increased popularly over the past few decades. Endophytes are ample and reliable source of unusual bioactive compounds with antifungal potential for development in medical and agricultural industries. Dual culture method is a significant approach to screen antifungal properties of endophytes in pursuit of contemporary natural drugs with enhanced prospective.

Key words Fungal pathogen, Fungal disease, Antifungal antibiotic, Plant, Endophyte, Dual culture method

1 Introduction

In general, plants are conjoined with variegated microbes. Microorganisms which cause infection of the plant and reside within internal plant tissues usually don't affect plants and produce no or any disease symptoms are often called as endophytes. An uncommon environment for existence and initiation of endophytes is usually afforded almost by the plants. Endophytes are omnipresent with affluent biological variety. They are present in almost all plant species and each plant contains one or more endophytes [1]. Endophyte comprises a diverse range of microbes including bacteria, fungi, and actinomycetes, composed of a multifarious microecosystem [2].

Endophytes help in the adaptation of plants to the environmental stress and increase their tolerance to drought, salinity, and high temperature. They produce several bioactive secondary metabolites and protect plant hosts against predators and resistance to disease-causing pathogens [3–5]. Several endophytes of plant

A. Sankaranarayanan et al. (eds.), *Endophytic Microbes: Isolation, Identification, and Bioactive Potentials*,
Springer Protocols Handbooks, https://doi.org/10.1007/978-1-0716-2827-0_26,

exhibited activity against pathogenic fungus [6]. Antifungal properties of endophytes in the recent literatures are listed in Tables 1 and 2.

The metabolites of endophytic bacterium *Bacillus amyloliquefaciens* DL06 from *Drosera burmannii* plant leaf exhibited activity against *Alternaria solani*, *Cochliobolus lunatus*, *Rhizoctonia solani*, *Fusarium oxysporum*, *Sclerotium rolfsii*, and *Saccharomyces boulardii* [22]. Endophytic actinomycetes *Streptomyces lydicus* from *Oryza sativa* L. showed strongly antagonistic against rice pathogenic fungi, *Fusarium oxysporum*, *Rhizoctonia solani*, and *Helminthosporium oryzae* [23].

Screening of antifungal properties of endophytes is performed after its isolation from varied parts of plants such as leaves, stems, and roots. Dual culture method is principal and foremost among the other methods widely used for the evaluation of antifungal properties of endophytic isolates. The antifungal activity of two endophytic fungal isolates *Alternaria* sp. from *Lithraeacaustica* and *Aureobasidium* sp. from *Echinopsis chiloensis* against phytopathogenic fungus *Botrytis cinerea* causing graymold disease was screened qualitatively by dual culture method [24]. Khan et al. [25] employed a dual culture method for the evaluation of antifungal activity of endophyte *Bacillus velezensis* from *Lilium leucanthum* against *Botryosphaeria dothidea*, *Fusarium oxysporum*, *Botrytis cinerea*, and *Fusarium fujikuroi*. Many authors investigated the antifungal activity of endophytes by dual culture methods [26–28].

2 Materials

- 70% v/v ethanol.
- Distilled water.
- Conical flask.
- Measuring cylinder.
- Bunsen burner.
- Inoculation loop.
- Digital balance.
- Hot air oven.
- Water bath.
- Hot plate.
- Autoclave.
- Laminar airflow hood.
- Cork borer.
- Forceps.

Table 1
Fungicidal activities of endophytes

Endophytes	Host plant	Activity against fungi	Reference
Bacillus sp.	*Theobroma cacao*	*Moniliophthora roreri, Moniliophthora perniciosa*	Serrano et al. [7]
Bacillus velezensis	*Populusdavidiana* × *P. alba var. pyramidalis (PdPap)*	*Alternaria alternata*	Zhang et al. [8]
*Bacillus*sp, *Enterobacter* sp	Tomato plant	*Botrytis cinerea*	Chaouachi et al. [9]
Aspergillus, fusarium	*Sceletium tortuosum*	*Fusarium oxysporum, fusarium graminearum, Colletotrichum gleosporioides. Borytiscinerea*	Sishuba et al. [10]
Trichoderma sp.	*Leucas aspera, Pterocarpus santalinus, Juniperuslutchuensis*	*Sclerotinia sclerotiorum, Sclerotium rolfsii, fusarium oxysporum*	Rajani et al. [11]
Trichoderma longibrachiatum, T. asperellum, T. atroviride	Soybean (*Glycine max* L.)	*Rhizoctonia solani*	Sallam et al. [12]
Serratia sp.	*Mimosa pudica*	*Fusarium* sp., *Alternaria solani,*	Méndez-Santiago et al. [13]
Streptomyces sp.	*Pinus dabeshanensis*	*Fusarium oxysporum, Rhizoctonia solani, Phytophthora infestan, Candida albicans.*	Peng et al. [14]
Streptomyces polyantibioticus	Thymus roseus	*Alternaria solani, Valsamalicola, Valsamali*	Musa et al. [15]
Streptomyces sp.	*Artemisia herba-alba*	*Botrytis cinerea*	El-Shatoury et al. [16]
Aplosporellajaveedii	*Orychophragmus violaceus*	*Candida albicans*	Gao et al. [17]
Trichoderma atroviride, Trichoderma koningii	*Cupressaceae*	*A. niger, A. fumigatus*	Erfandoust et al. [18]
Fusarium verticilloides	*Mentha piperita* L.	*A. austroafricanus, A. versicolor, Phomafungicola, A. turbingensis, C. albicans*	Ebrahim et al. [19]

(continued)

Table 1
(continued)

Endophytes	Host plant	Activity against fungi	Reference
Rhizopus oryzae, fusarium fujikuroi	*Debregeasia salicifolia*	*Aspergillus flavus, aspergillus Niger*	Nisa et al. [20]
Trichoderma longibrachiatum, Macrophomina phaseolina	*Helianthus tuberosus*	*Sclerotium rolfsii*	Suebrasri et al. [21]

Table 2
Fungicidal activities of endophytes against target fungus by dual culture method

Endophytes	Radius of target fungal mycelium on control plate (mm) R1	Radius of target fungal mycelium on dual culture plate (mm) R2	Mycelial growth Inhibition % [(R1–R2)/R1] × 100
Fungus			
Actinomycetes			
Bacteria			

- Incubator.
- Digital caliper/Verner Scale.
- Microbial culture: Pure cultures of endophytic bacteria, fungi, actinomycetes, and target fungi.
- Potato Dextrose Agar: (Potato infusion from – 200 gm, Dextrose – 20 g, Agar – 15 g, distilled water – 1000 mL, pH – 5.6 ± 0.2 at 25 °C).
- Yeast Malt Agar: (peptone – 5gm, yeast extract – 3 g, malt extract – 3 g, dextrose – 10 g, Agar – 20 g, distilled water – 1000 mL, pH – 6.2 ± 0.2 at 25 °C).
- Petri plate 90 mm × 15 mm.

3 Methodology

3.1 Preparation of Potato Dextrose Agar Medium Plate (PDAMP)

1. Measure the required quantity of PDA medium according to the manufacturer's recommendation.
2. Suspend it in distilled water in a conical flask.

3. Boil the medium for complete dissolution in a water bath or hot plate, then adjust the pH of the medium to 5.6 ± 0.2.

3.2 Preparation of Yeast Malt Agar Medium Plate (YMAMP)

Prepare yeast malt agar medium plate by following procedures as mentioned in Subheading 3.1.

3.3 Preparation of Endophytic Mycelial Disk (Fungus & Actinomycetes)

1. Inoculate fungal endophyte as spot at the center of PDA agar medium plate PDAMP or actinomycetes on yeast malt agar medium plate YMAMP.
2. Incubate the PDAMP for 21 days at 25 ± 2 °C [29] or YMAMP at 28 ± 2 °C for 7–10 days [30].
3. Prepare mycelial disk of 5mm^2 diameter from actively growing margin of endophytes with the aid of sterile cork borer [31]

3.4 Preparation of Target Fungal Mycelial Disk

Prepare target fungal mycelial disk as mentioned in Subheading 3.3.

3.5 Inoculation of Dual Culture (Figs. 1 and 2)

3.5.1 Inoculation of Target Fungus and Endophytes (Fungus & Actinomycetes)

1. Place the target fungal mycelial disk onto PDAMP approximately 25 mm away on the 90 mm PDAMP edge on one side [32].
2. Hold the endophytic fungal mycelial disk or actinomycetal disk of 5 mm^2 diameter with sterile forceps.
3. Place them 25 mm away from the margin of PDAMP opposite to target fungus.

3.5.2 Inoculation of Endophytic Bacteria

1. Take a single colony of bacterial endophyte from pure culture and make a streak opposite to target fungus at 25 mm away from the margin of PDAMP [33]
2. The distance between the target fungus and endophytes is around 3 cm away from each other

3.6 Incubation

1. Prepare a control plate with target fungal mycelial disk 5 mm^2 diameter and replace endophytes by 5 mm^2 diameter PDA agar plug on PDAMP [34].
2. Incubate all the PDAMP plates at 28 ± 2 °C for 5–7 days [35].
3. At a regular interval observe the growth of target fungus until the colony spreads over the entire surface of PDAMP in control.
4. Measure the radius of target fungal mycelium (mm) on the control (R1) and on the dual culture plate (R2) with aid of vernier caliper [36] or digital caliper [37].

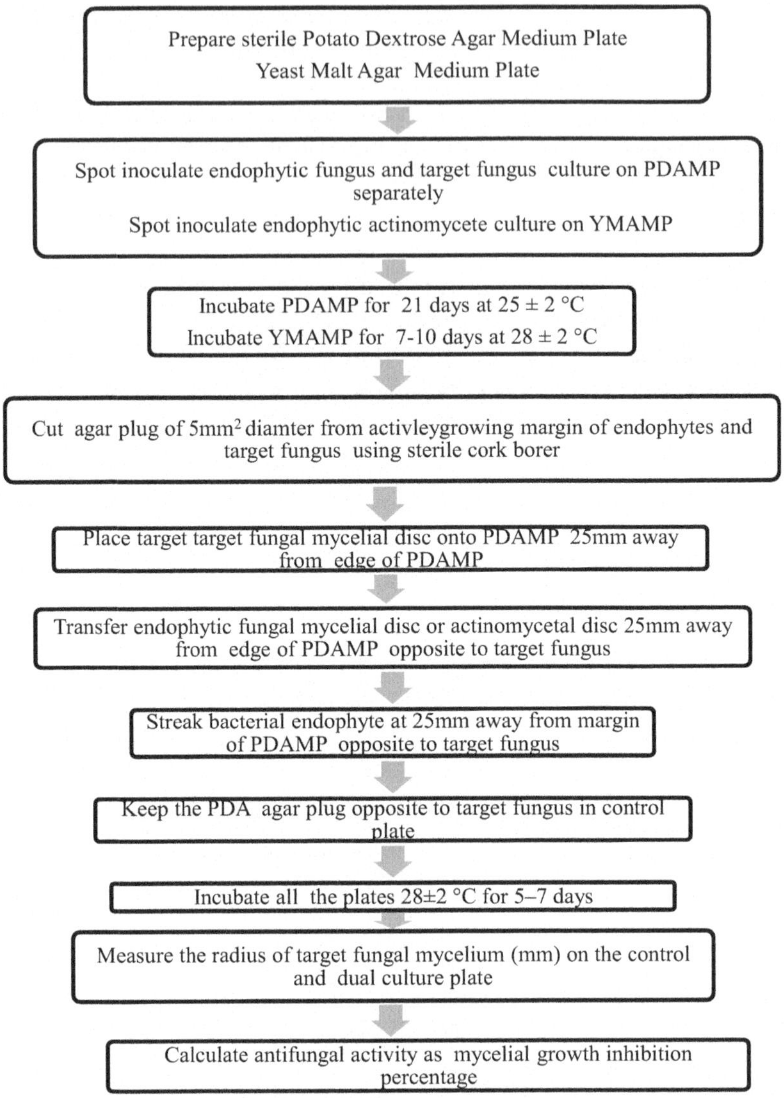

Fig. 1 Various steps for the screening of endophytes for fungicidal activities by dual culture method

5. Calculate the antifungal activity of endophytes as mycelial growth inhibition percentage by the following formula [38]:

$$MGI\% = [(R1 - R2)/R1] \times 100$$

4 Observations

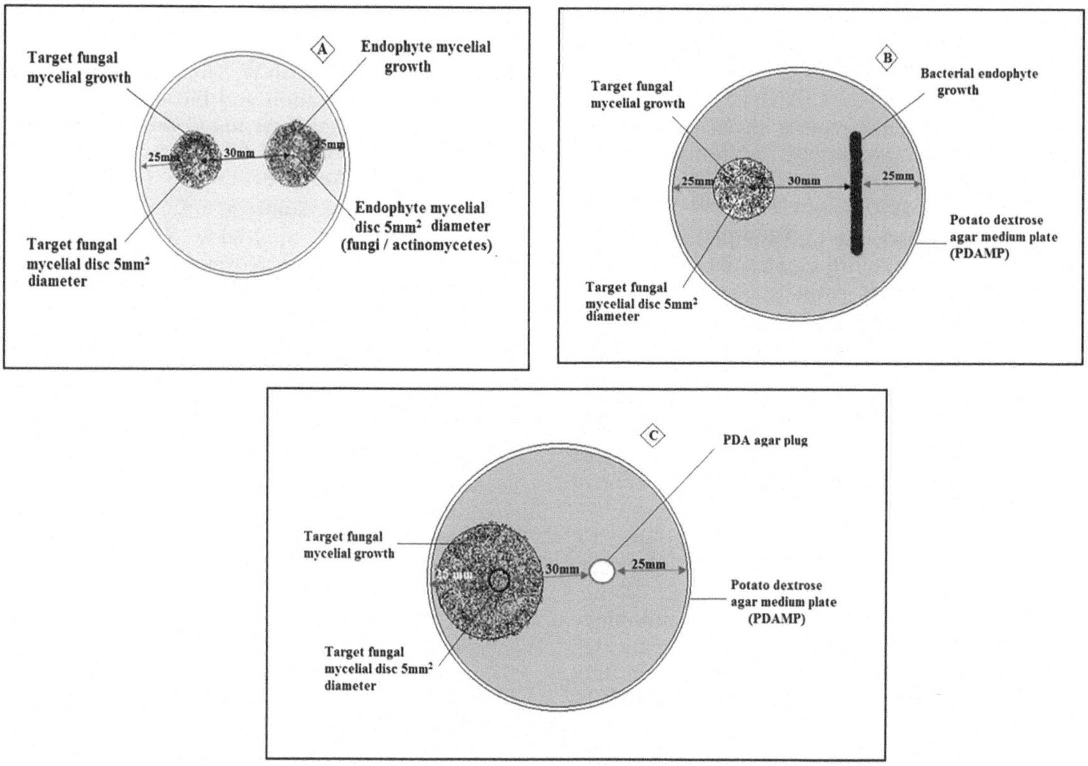

Fig. 2 Diagrammatic representation of screening of endophytes for fungicidal activities by dual culture method. (**a**) Dual culture plate with target fungus and endophytes (fungus/actinomycetes). (**b**) Dual culture plate with target fungus and endophytic bacteria. (**c**) Control plate with target fungus and PDA agar plug

References

1. Kouipou Toghueo RM, Boyom FF (2019) Endophytic fungi from *Terminalia* species: a comprehensive review. J Fungi 5(2):43
2. Golinska P, Wypij M, Agarkar G, Rathod D, Dahm H, Rai M (2015) Endophytic actinobacteria of medicinal plants: diversity and bioactivity. Antonie van Leeuwenhoek 108(2): 267–289
3. Deng Z, Cao L (2017) Fungal endophytes and their interactions with plants in phytoremediation: a review. Chemosphere 168:1100–1106
4. Hassan SE (2017) Plant growth-promoting activities for bacterial and fungal endophytes isolated from medicinal plant of *Teucrium polium* L. J Adv Res 8:687–695
5. Pandey SS, Singh S, Babu CS, Shanker K, Srivastava NK, Kalra A (2016) Endophytes of opium poppy differentially modulate host plant productivity and genes for the biosynthetic pathway of benzylisoquinoline alkaloids. Planta 243:1097–1114
6. Alice FSH, David EW, Charles LC, Camilea RC, Zhiqiang P, Rita MM, Victor LM, Luiz HR (2015) Diversity and antifungal activity of the endophytic fungi associated with the native medicinal cactus *Opuntia humifusa* (*Cactaceae*) from the United States. Microbiol Res 175:67–77
7. Serrano L, Moreno AS, Castillo DSD, Bonilla J, Romero CA, Galarza LL, Coronel–León JR (2021) Biosurfactants synthesized by endophytic *Bacillus* strains as control of *Moniliophthoraperniciosa* and *Moniliophthoraroreri*. Scientia Agricola 78:e20200172
8. Zhang P, Diao J, Xie G, Ma L, Wang L (2021) A complete genome sequence of the wood stem endophyte *Bacillus velezensis* BY6 strain possessing plant growth-promoting and antifungal activities. BioMed Res Int 3904120
9. Chaouachi M, Marzouk T, Jallouli S, Elkahoui S, Gentzbittel L, Ben C, Djébali N (2021) Activity assessment of tomato

endophytic bacteria bioactive compounds for the postharvest biocontrol of *Botrytis cinerea*. Postharvest Biol Technol 172:111389

10. Sishuba A, Leboko J, Ateba CN, Manganyi MC (2021) First report: diversity of endophytic fungi possessing antifungal activity isolated from native Kougoed (*Sceletiumtortuosum* L.). Mycobiology 49(1):89–94
11. Rajani P, Rajasekaran C, Vasanthakumari MM, Olsson SB, Ravikanth G, Shaanker RU (2021) Inhibition of plant pathogenic fungi by endophytic *Trichoderma*sp. through mycoparasitism and volatile organic compounds. Microbiol Res 242:126595
12. SallamN AEF, Seleim MA, Bagy HMK (2021) Endophytic fungi associated with soybean plants and their antagonistic activity against *Rhizoctonia solani*. Egypt J Biol Pest Control 31(1):1–9
13. Méndez-Santiago EW, Gómez-Rodríguez O, Sánchez-Cruz R, Folch-Mallol JL, Hernández-Velázquez VM, Villar-Luna E, Wong-Villarreal A (2021) *Serratia* sp., an endophyte of *Mimosa pudica* nodules with nematicidal, antifungal activity and growth-promoting characteristics. Arch Microbiol 203(2): 549–559
14. PengC AD, Ding WX, Zhu YX, Ye L, Li J (2020) Fungichromin production by *Streptomyces*sp.WP-1,an endophyte from *Pinus dabeshanensis*, and its antifungal activity against *Fusarium oxysporum*. Appl Microbiol Biotechnol 104(24):10437–10449
15. Musa Z, Ma J, Egamberdieva D, Mohamad OAA, Abaydulla G, Liu Y, Li L (2020) Diversity and antimicrobial potential of cultivable endophytic actinobacteria associated with the medicinal plant *Thymus roseus*. Front Microbiol 11:191
16. El-Shatoury SA, Ameen F, Moussa H, Wahid OA, Dewedar A, AlNadhari S (2020) Biocontrol of chocolate spot disease (*Botrytis cinerea*) in faba bean using endophytic actinomycetes *Streptomyces*: a field study to compare application techniques. Peer J 8:e8582
17. Gao Y, Wang L, Kalscheuer R, Liu Z, Proksch P (2020) Antifungal polyketide derivatives from the endophytic fungus *Aplosporellajaveedii*. Bioorg Med Chem 28(10):115456
18. Erfandoust R, Habibipour R, Soltani J (2020) Antifungal activity of endophytic fungi from *Cupressaceae* against human pathogenic *aspergillus fumigatus* and *aspergillus Niger*. Journal de Mycologie Médicale 30(3):100987
19. Ebrahim W, Özkaya FC, Ebada SS (2020) Antifungal metabolites from endophytic fungus *fusarium verticillioides* strain WF18. S Afr J Bot 133:40–44
20. Nisa S, Khan N, Shah W, Sabir M, Khan W et al (2020) Identification and bioactivities of two endophytic fungi *fusarium fujikuroi* and *aspergillus tubingensis* from foliar parts of *Debregeasiasalicifolia*. Arab J Sci Eng 45:4477–4487
21. Suebrasri T, Somteds A, Harada H, Kanokmedhakul S, Jogloy S, Ekprasert J, Boonlue S (2020) Novel endophytic fungi with fungicidal metabolites suppress sclerotium disease. Rhizosphere 16:100250
22. Chaudhuri M, Paul AK, Pal A (2021) Isolation of potential antimicrobial metabolite from endophytic *bacillus amyloliquefaciens*DL06 of carnivorous plant *Droseraburmannii*vahl. Asian J Pharm Clin Res 14(1):163–170
23. Mingma R, Duangmal K (2018) Characterization, antifungal activity and plant growth promoting potential of endophytic actinomycetes isolated from Rice (*Oryza sativa* L.). Chiang Mai J Sci 45(7):2652–2665
24. Vidal A, Parada R, Mendoza L, Cotoras M (2020) Endophytic fungi isolated from plants growing in central Andean Precordillera of Chile with antifungal activity against *Botrytis cinerea*. J Fungi 6(3):149
25. Khan MS, Gao J, Chen X, Zhang M, Yang F, Du Y, Zhang X (2020) The endophytic bacteria *bacillus velezensis* Lle-9, isolated from *Lilium leucanthum*, harborsantifungal activity and plant growth-promoting effects. J Microbiol Biotechnol 30(5):668–680
26. Ali A, Junda M, Karim H, Nuryani S (2017) Characterization and in vitro antifungal assay against *fusarium oxysporum*f. Sp. *passiflorae* of endophytic actinomycetes from purple passion fruit plants of South Sulawesi, Indonesia. Asian J Microbiol Biotech Environ Sci 19(2): 344–353
27. Verma SK, Kingsley KL, Bergen MS, Kowalski KP, White JF (2018) Fungal disease prevention in seedlings of rice (*Oryza sativa*) and other grasses by growth-promoting seed-associated endophytic bacteria from invasive *Phragmites australis*. Microorganisms 6(1):21
28. Cao P, Li C, Wang H, Yu Z, Xu X, Wang X, Xiang W (2020) Community structures and antifungal activity of root-associated endophytic actinobacteria in healthy and diseased cucumber plants and *Streptomyces* sp. HAAG3-15 as a promising biocontrol agent. Microorganisms 8(2):236
29. Ramesha KP, Mohana NC, Nuthan BR, Rakshith D, Satish S (2020) Antimicrobial metabolite profiling of *Nigrosporasphaerica*

from *Adiantum philippense* L. J Genet Eng Biotechnol 18:66

30. Radhakrishnan M, Vijayalakshmi G, Gopikrishnan V, Jerrine J (2016) Bioactive potential of actinobacteria isolated from certain understudied regions in India. J Appl Pharm Sci 6(8):151–155
31. Li ZF, Wang LF, Feng ZL, Zhao LH, Shi YQ, Zhu HQ (2014) Diversity of endophytic fungi from different *Verticillium*-wilt-resistant *Gossypium hirsutum* and evaluation of antifungal activity against *Verticillium dahliae* in vitro. J Microbiol Biotechnol 24(9):1149–1161
32. Mohamad OA, Li L, Ma JB, Hatab S, Xu L, Guo JW, Li WJ (2018) Evaluation of the antimicrobial activity of endophytic bacterial populations from Chinese traditional medicinal plant *licorice* and characterization of the bioactive secondary metabolites produced by *bacillus atrophaeus*against *Verticillium dahliae.* Front Microbiol 9:924
33. Luu T, Phi Q, Nguyen, Van Dinh M, Pham BN, Do QT (2021) Antagonistic activity of endophytic bacteria isolated from weed plant against stem end rot pathogen of *pitaya* in Vietnam. Egypt J Biol Pest Control 31:14
34. Huang LQ, Niu YC, Su L, Deng H, Lyu H (2020) The potential of endophytic fungi isolated from cucurbit plants for biocontrol of soilborne fungal diseases of cucumber. Microbiol Res 231:126369
35. Ali S, Hameed S, Shahid M, Iqbal M, Lazarovits G, Imran A (2020) Functional characterization of potential PGPR exhibiting broad-spectrum antifungal activity. Microbiol Res 232:126389
36. Landum MC, Do Rosário FM, Alho J, Garcia R, Cabrita MJ, Rei F, Varanda CM (2016) Antagonistic activity of fungi of *Olea europaea* L. against *Colletotrichum acutatum.* Microbiol Res 183:100–108
37. Barra-Bucarei L, France Iglesias A, Gerding González M, Silva Aguayo G, Carrasco-Fernández J, Castro JF, Ortiz Campos J (2020) Antifungal activity of *Beauveria bassiana* endophyte against *Botrytis cinerea* in two *Solanaceae* crops. Microorganisms 8(1):65
38. Nguyen MH, Yong JH, Sung HJ, Lee JK (2020) Screening of endophytic fungal isolates against *Raffaeleaquercus-mongolicae* causing oak wilt disease in Korea. Mycobiology 48(6): 484–494

Chapter 27

Screening of Endophytes for Anticancer Compounds

**Drielly Dayanne Monteiro dos Santos Baliza,
Juliana Fonseca Moreira da Silva, Eskálath Morganna Silva Ferreira,
Irlon M. Ferreira, Edilene Oliveira da Silva,
José Luiz Martins do Nascimento, and Raphael Sanzio Pimenta**

Abstract

Endophytes are microorganisms that colonize plant tissues and participate in the production of several compounds with biotechnological interest. Available treatments against cancer have inefficiencies, prompting studies on potential alternative sources to the synthesis of metabolites, such as those from fungi. Isolation and study of endophytes is an alternative method to extremely low amounts of active principles extracted from plants, deforestation and difficulty in re-isolation of active ingredients due to changes in environment, season, and geography. The evolution of different morphological, biochemical, and molecular techniques provides a vision capable of directing the biological activity of new secondary metabolites of endophytes, monitoring the progression of cell death under any conditions, demonstrating how these metabolites work.

Key words Endophytic, Bioactive compounds, Cancer cell lines, Cytotoxicity, In vitro techniques

1 Introduction

Endophytes are microorganisms that colonize internal tissues of plants (roots, stem, fruits, leaves, and seeds) without causing immediate negative effects, which can bring several benefits to the host plant, such as protection and growth promotion [1]. Within medicinal plants, they influence the formation of secondary metabolites, being reported the ability to produce distinctly aliphatic compounds, alkaloids, benzopyranones, chinones, phenolic acids, phytohormones, quinones, steroids, saponins, tannins, terpenoids, tetralones, xanthones, and others compounds with biotechnological interest [2–6].

Among the current methods of cancer treatment, chemotherapy is the most effective for the treatment of metastatic tumors, but it has characteristics of multi-resistance of cancer cells and side

A. Sankaranarayanan et al. (eds.), *Endophytic Microbes: Isolation, Identification, and Bioactive Potentials*,
Springer Protocols Handbooks, https://doi.org/10.1007/978-1-0716-2827-0_27,

effects, triggering treatment inefficiency [7]. Hartwell [8] listed more than 3000 species of plants used in the treatment of cancer and there are several anticancer drugs with active ingredients of plant origin, but for the preparation of these drugs there are some difficulties such as extremely low amounts of active ingredients, severe deforestation and the difficulty in re-isolation of active ingredients due to changes in environment, season, and geography [9].

In some cases, the difficulty with isolation of bioactive compounds from plants has made the extraction of new chemical compounds from endophytic derivatives an important object of study in recent years [10]. Thus, several studies involving higher plants have shown the association between endophytes and active metabolites, as well as their potential for medicinal use [11]. These potential endophytic microorganisms are favorable alternative sources for the synthesis of mimetic metabolites of interest to the host and prevent the overexploitation of host plants [12, 13].

The evolution of different morphological, biochemical, and molecular techniques provides insight capable of targeting biological activity of novel secondary metabolites from endophytes monitoring the progression of cell death in any conditions. These techniques resulted in an enormous advantage in knowledge on the regulatory network it takes for the cell to die [14–19]. In addition, in vitro methods are important to show how these secondary metabolites work [20].

2 Materials

2.1 Obtaining Active Fungal Colonies

- Endophytic fungi in Castellani, glycerol, or other stock culture.
- Petri dishes with Potato Dextrose Agar (PDA).
- Inoculation loop.
- B.O.D. incubator.

2.2 Obtaining Fungal Colonies in Liquid Medium

- Petri plate (90 mm) with pure colonies of endophytic fungi.
- YM broth (0.3% yeast extract, 0.3% malt extract, 0.5% peptone, 1% glucose).
- Flask Erlenmeyer (250 mL).
- Plastic straw (4 mm in diameter).
- Toothpicks.
- Shaker incubator.

2.3 Extraction of Metabolites from Endophytic Fungi

- Extraction funnel.
- Fungus grown in YM broth.
- Dichloromethane P. A. ACS.

- Rotary evaporator.
- Small glass bottles.
- Exhaust fume hood.
- Flat bottom balloon.

2.4 Processing of Extracts and Verification of Metabolites

- Gas Chromatography – Mass Spectrometry (GC-MS).
- Methanol/water solution (1:1).
- Hexane.
- Chloroform.
- Ethyl acetate.
- Magnesium sulfate anhydrous.
- Helium gas.

2.5 Cancer Cell Lines Culture, Reagent, and Equipment

- Cancer cell lines can be obtained from purchased from a cell bank.
- Appropriated culture media, sera, and reagents.
- Cell Culture Plates (multiwell plates).
- PEN STREP - Penicillin Streptomycin solution.
- CMF – Calcium medium free.
- TRYPSIN/EDTA solution.
- Wash bath.
- pH meter.
- Automatic pipettes.
- Centrifuge.
- Refrigerate and freeze (−20 °C).
- Liquid Nitrogen (N_2) containers.
- Sterilizer.
- CO_2 incubator.
- Laminar flow.
- Inverted microscope.

2.6 Cytotoxicity Assays in Cancer Cell Lines

2.6.1 Trypan Blue Protocol

- Cancer cell lines.
- Pasteur pipette.
- PBS (Phosphate buffered saline).
- Trypan blue solution.
- Hemocytometer chamber.
- Optical microscope.

2.6.2 MTT Colorimetric Assay

- Cancer cell lines.
- MTT (3-(4,5-dimethylthiazol-2-yl)-2,5-diphenyl tetrazolium bromide).
- Dimethyl sulfoxide (DMSO).
- Isopropanol.
- HCl.
- Microplate Reader.

2.6.3 LDH

- Cancer cells lines.
- 96 well flat-bottom plate.
- Sodium Pyruvate.
- NADH (Nicotinamide adenine dinucleotide hydrogen).
- KPO_4 (Potassium phosphate buffer).
- Microplate Reader.

2.6.4 Alamar Blue

- Cancer cell lines.
- Alamar blue solution (10%).
- Medium free-serum.
- Excitation wavelength 530 nm.
- Emission wavelength 590 nm.
- Fluorimeter of microplates.

2.6.5 Colony Formation Assay

- Cancer cell lines.
- Trypsin solution.
- Dimethyl sulfoxide (DMSO).
- Ethanol.
- Crystal violet.
- Optical Microscope.

2.6.6 Electron Microscopic Analysis

- Cancer cell lines.
- PHEM buffer ($MgCl_2$-5 mM; KCl-70 mM; EDTA-10 mM; HEPES-20 mM; PIPES-60 mM).
- Glutaraldehyde.
- Formaldehyde.
- OsO_4 (Osmium tetroxide).
- PBS (Phosphate saline buffer).
- $K_4Fe(CN)_6 \cdot 3H_2O$ (Potassium ferrocyanide).
- $CaCl_2$ (Calcium chloride).

- EponResin.
- Uranyl acetate.
- Lead citrate.
- Ultramicrotome.
- Nickel Grids 300 mesh.
- Fume hood.
- Transmission Electron Microscopy.

2.6.7 Colony Formation Assay

- Cancer cell lines.
- Trypsin solution.
- Dimethyl sulfoxide (DMSO).
- Ethanol.
- Crystal violet.
- Optical microscope.

2.6.8 Annexin V-FITC-propidium Iodide (PI) Assay

- Cancer cell lines.
- Annexin VFITC Apoptosis KIT.
- Propidium Iodide (PI).
- Binding Buffer (50 mM HEPES, 700 mM NaCl, 12,5 mM $CaCl_2$, pH = 7.4).
- FACS Tubes.
- Flow Cytometer.

2.6.9 Analysis of Mitochondrial Transmembrane Potential ($\Delta\Psi m$)

- Cancer cell lines.
- JC1 – fluorescent carbocyanine dye.
- 12, 24, or 96-well plates.
- Dimethyl sulfoxide (DMSO).
- PBS.
- Flow Cytometer.

2.6.10 Cell Cycle Analysis

- Cancer cell lines.
- Dimethyl sulfoxide (DMSO).
- PBS (Phosphate saline buffer).
- Ethanol.
- CO_2 Incubator.
- RNaseA.
- Propide Iodide.
- Flow Cytometer.

2.6.11 Western Blotting

- Cellular lysates.
- Electrophoresis SDS PAGE gel.
- Nitrocellulose membranes.
- Primary and secondary antibody.
- Nonfat milk.
- Semidry transfer system.
- Chemiluminescent substrate Kit.
- Images captured by chemiDoc system.

2.6.12 Caspase-3 Activity

- Cell-culture 96-well, flat-bottom microplate.
- CHAPS buffer.
- Chromogenic substrates acetyl-Asp-Glu-Val-Asp p-nitroanilide (Ac-DEVD-pNA).
- p-nitroaniline (pNA).
- Phosphate buffer.
- Tris–HCl.
- Microplate reader.

3 Methods

3.1 Obtaining Active Fungal Colonies

- With the aid of a sterile microbiological loop, remove one or two disks (1 cm diameter) of the fungus (depending on the size of the colony grown), stored in Castellani, glycerol, or other storage methods.
- Insert the disk in the center of the plate containing previously sterilized solid culture medium (Potato Dextrose Agar) and incubate at 25 °C in B.O.D. for 5 days, or until pattern growth of the inserted colony is observed, as shown in Fig. 1.

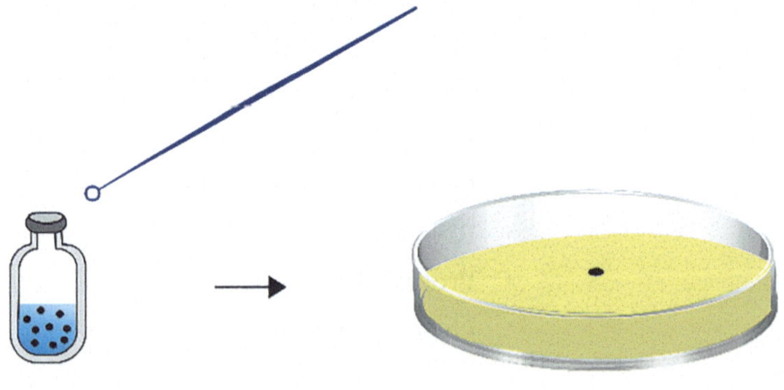

Fig. 1 Obtaining a fungal colony from stored endophytes

3.2 Obtaining Fungal Colonies in Liquid Medium

- With the plates with the purified fungi, remove five disks with the aid of a straw and toothpick, previously sterilized, and transfer them to an Erlenmeyer containing 200 mL of previously autoclaved YM broth, as shown in Fig. 2.
- Incubate at 25 °C in Shaker, at a rotation of 120 rpm, for 14 days.

3.3 Extraction of Metabolites from Endophytic Fungi

- Pour the YM broth with the fungal growth into the extraction funnel and top up with 1:1 dichloromethane. Shake to homogenize.
- Allow to decant until the water/dichloromethane separation, as shown in Fig. 3.

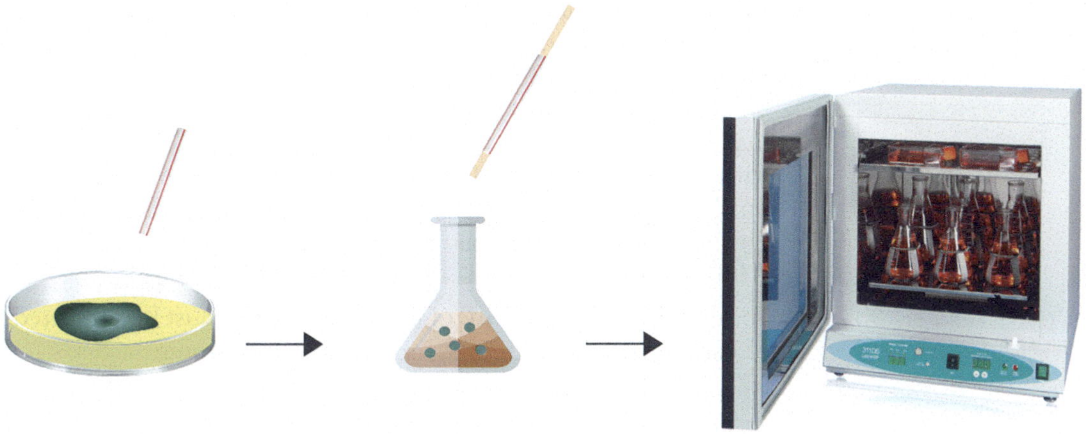

Fig. 2 Obtaining fungal colonies in liquid medium

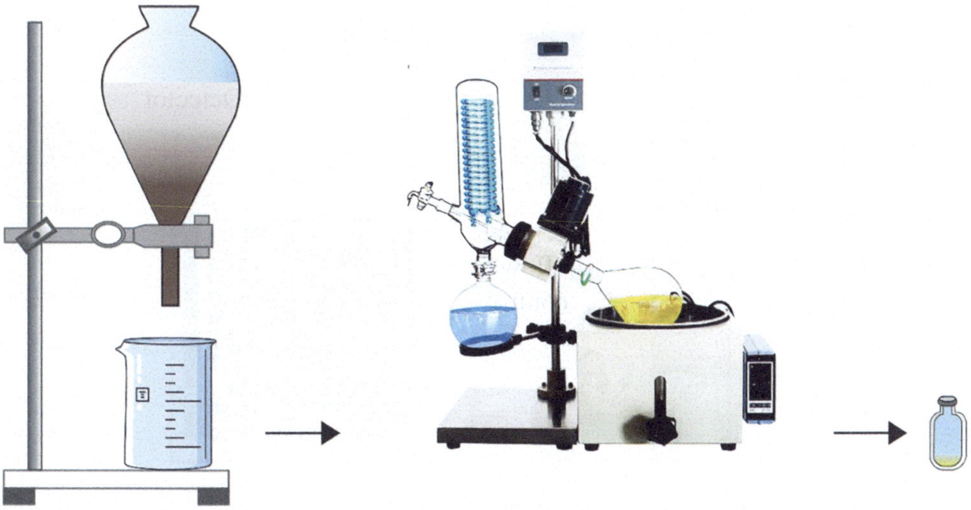

Fig. 3 Scheme for extracting metabolites from endophytic fungi

- Remove the fraction of dichloromethane, which is at the bottom of the bottle, and put it to rotaevaporate.
- The rotating process evaporates the solvent, concentrating only the extract in the flask. Wash the flask with dichloromethane to loosen the evaporated and distribute in smaller flasks.
- Place the flasks to evaporate again in the exhaust hood.

3.4 Processing of Extracts and Verification of Metabolites

- Resuspend approximately 2.7 g of the lyophilized material in 50 mL of a methanol/water (1:1) solution and then submit to a liquid/liquid partition (2 × 50 mL) using the solvents hexane, chloroform, and ethyl acetate, respectively.
- In each fraction add anhydrous magnesium sulfate and filter again.
- Rotaevaporate excess solvent at reduced pressure and analyze by Gas Chromatography – Mass Spectrometry (GC-MS): Shimadzu/CGMS-PQ2010SE apparatus, coupled to Shimadzu/AOC-5000 autoinjector and a mass detector (Shimadzu MS2010 Plus) with electron impact (70 eV) equipped with a 30 m long RTX-5MS fused silica column, 0.25 mm outside diameter and 0.25 μm inside diameter, as shown in Fig. 4, using helium as a carrier gas, supporting a flow 1.03 mL/min.
- Soluble the sample in 2 μL of hexane and 1 μL must be subjected to experimental conditions, for example, as follows:
  - Injector temperature at 210 °C.
  - Detector temperature at 250 °C.
  - Helium gas as a drag in a flow of 3 mL/min.

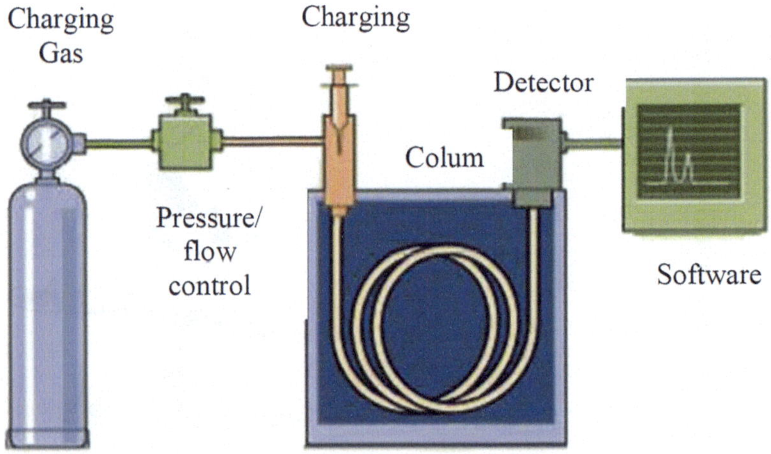

Fig. 4 Scheme of Gas Chromatograph coupled to Mass Spectrometer (GC-MS)

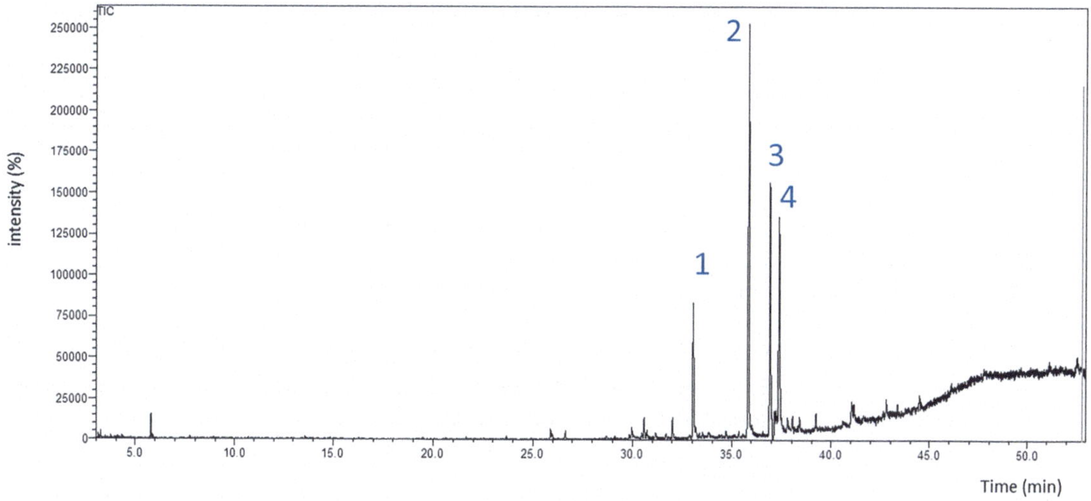

Fig. 5 Example of chromatogram of the hexane fraction from the ethanol extract of the roots of an Amazonian medicinal plant

  - Injection divided in the proportion of 1/15.
  - Set the column temperature from 90 °C, with an increase of 6 °C/min to 250 °C, ending with 5 min isothermal at this temperature.
- The substances identified in the fraction will show predominance peaks, as shown in Fig. 5.

3.5 Anticancer Analysis of Bioactive Compounds from Endophytes

3.5.1 Cancer Cell Lines

- Maintain cancer cell line at 37 °C, 5% CO_2 in 95% relative humidity in multiwell plates.
- Use media and sera appropriated for cancer cell line growth.
- View the cultures under a phase contrast inverted microscope.
- Remove small aliquots of the cells for determination of cell numbers.
- Add the appropriate cell density for each experiment.
- Change the medium every third day for 14 days to obtain well-differentiated monolayers of cancer cell line.

3.5.2 Perform Bioactive Compounds Treatment

- Remove the existing culture medium from the cells and replace with new medium.
- Incubate cancer cell line culture with different concentrations of bioactive compounds.
- Untreated cells as control.
- Discard media from cell cultures after treatment with bioactive compounds.

- Incubate cancer cell line cultures in the presence of bioactive compounds for 6, 12, 24, 48, and 72 h.
- At the end of the incubation period, determine the cell number by cytotoxicity methods and methods to detect apoptotic cell death.

3.5.3 Trypan Blue Assay Protocol

- Determine the cell density using a hemocytometer.
- Prepare a 0.4% solution of trypan blue in buffered isotonic salt solution, pH 7.4 phosphate-buffered saline.
- Add 0.1 mL of trypan blue stock solution to 0.1 mL of cells.
- Load a hemocytometer and examine immediately under a microscope at low magnification.
- Count the number of blue staining cells and the number of total cells. Cell viability should be at least 95% for healthy log-phase culture.
- Take images with an optical microscope.

3.5.4 MTT (3-(4,5-Dimethylthiazol-2-yl)-2,5-Diphenyltetrazolium Bromide) Assay

- 10^6 cancer cells line in culture.
- Remove the existing culture medium from the cells and replace with new medium.
- Incubate the cells according to the standard treatment protocol.
- Discard media from cell cultures after treatment with bioactive compounds.
- Add 100 μL of serum-free media and 20 μL of MTT solution (5 mg/mL) into each well.
- Incubate for 2 h at room temperature in the dark until purple precipitate.
- Add 20 μL DMSO or acid-isopropanol (100 μL 0,04 N HCl in Isopropanolol) to dissolve the purple crystals.
- Measure the absorbance at 570 nm using a microplate reader.

3.5.5 LDH Assay

- 10^6 cancer cell lines in culture.
- Remove the existing culture medium from the cells and replace with new medium.
- Incubate the cells according to the standard treatment protocol.
- Discard media from cell cultures after treatment with bioactive compounds.
- Add 200 μL of serum-free media.
- 2,3 μMol of sodium pyruvate.
- 0,2 mg NADH in 0,1 M KPO_4 into each well pH 7,0.

- Recorded automatically at 2 s interval.
- Measure absorbance at 340 nm using a microplate reader.
- Enzyme activity expressed in (LDH U/mL).

3.5.6 Alamar Blue Assay Method

- 10^6 cancer cell in culture.
- Remove the existing culture medium from the cells and replace with new medium.
- Incubate the cells according to the standard treatment protocol.
- Discard media from cell cultures after treatment with bioactive compounds.
- Dilute Alamar blue stock solution to 1:1 with serum-free medium.
- Volume of 50 μL/well.
- Final assay volume of 250 μL.
- Final concentration of 10% alamar blue.
- Read fluorescence at excitation 530 nm, emission at 590 nm.

3.5.7 Colony Formation Assay

- 2.5×10^5 cancer cell in culture.
- Remove the existing culture medium from the cells and replace with new medium.
- Incubate the cells according to the standard treatment protocol.
- Trypsinize cells and plate 1×10^3 of viable cells.
- After 14 days, stain at least 50% of the cells during 30 min with 0.25% crystal violet.
- Take images with phase contrast microscope.

3.5.8 Annexin V-FITC-Propidium Iodide (PI) Assay

- 100 uL of cell suspension containing 10^4 cells culture.
- Remove the existing culture medium from the cells and replace with new medium.
- Incubate the cells according to the standard treatment protocol.
- Add 5 μL of Annexin V-FITC.
- Add 5 μL of PI.
- 15 min at room temperature in the darkness.
- 400 uL of binding buffer.
- Annexin V-FITC.
- Measure 100,000 events using the green channel FL1 for Annexin V-FITC and red channel FL3 for PI.
- Analyze by Flow Cytometer.

3.5.9 Analysis of Mitochondrial Transmembrane Potential (ΔΨm)

- Seed 10^4 or 10^5 cells/mL in 12, 24 or 96-well plates.
- Washing with cold PBS.
- Incubate the cells according to the standard treatment protocol.
- After incubation, centrifuge cells for 3–4 min at 400× g at 4 ° Cand aspirate the supernatant.
- Add 500 μL PBS to suspend cells.
- Add 20 μM JC-1 (250 μL) for 15 min at room temperature.
- Measure a total of 100,000 events with the capture emissions set to 590 nm (red) and 530 nm (green).
- Analysis by Flow Cytometer.

3.5.10 Transmission Electron Microscopy (TEM) Analysis

- Seed line cells (2 × 10^6) in 75 cm^2 flasks.
- Incubate cells with different concentration of the bioactive compounds.
- After 24 h, wash cells with PBS and fix them with 1% glutaraldehyde, 4% freshly prepared formaldehyde, and 2.5% sucrose in 0,1 M PHEM buffer, at pH 7.2.
- Postfix in a 1% osmium tetroxide, 0.8% ferrocyanide potassium, and 5 mM calcium chloride solution for 1 h.
- Wash cells with 0.1 PHEM buffer.
- Dehydrate cells in graded acetone solutions (50, 70, 90, and two changes in 100% for 10 min).
- After dehydration, embed cells in Epon® resin.
- Cut thin sections on a ultramicrotome.
- Stain ultrathin sections with a saturated solution of uranyl acetate and lead citrate.
- Examine samples at 80 kV accelerating Voltage on a TEM.

3.5.11 Cell Cycle Analysis

- 1. 5 × 10^5 cancer cells in culture.
- Remove the existing culture medium from the cells and replace with new medium.
- Incubate the cells according to the standard treatment protocol.
- Fix cells in 5 mL of cold 70% ethanol and stored at 4 °C for subsequent cell cycle analysis.
- Wash cells with PBS twice, incubate with 50 μL of RNase A/PBS (100 μg/mL) at 37 °C for 30 min and stain with 200 μL of PI (50 μg/mL) at room temperature in the dark for 15 min.
- Analysis by Flow cytometer.

3.5.12 Western Blotting

- Remove the existing culture medium from the cells and replace with new medium.
- Incubate cancer cell line with control or different concentration of bioactive compounds.
- After the induction of apoptosis, scrape cells and collect in 15 mL tube.
- Spin at 250 g for 10 min at 4 °C.
- Resuspend pellet in 6 mL 1× PBS and spin again at 250 g for 10 min at 4 °C to wash.
- Remove as much PBS as possible without disrupting pellet and then add 100 μLlysis buffer with appropriate protease inhibitor.
- Vortex and keep on ice for 20 min vortexing periodically.
- Spin at 1100 g for 15 min at 4 °C and collect supernatant.
- Keep lysates at −40 °C until further use.
- Cell lysates (30 μg/sample) were subjected to electrophoresis on a 10% SDS-PAGE gel.
- Electrotransfer gels onto nitrocellulose membranes.
- Semidry transfer system.
- Block the membranes with 5% nonfat milk at room temperature for 1 h.
- Incubate with primary antibody used at different dilution at 4 °C overnight and after use a horseradish peroxidase (HRP)-conjugated secondary antibody at 1:5000 for 1 h.
- Detect immunoblotting using a chemiluminescent substrate kit.
- Capture the images by a ChemiDoc system.

3.5.13 Caspase Activity Assay

- After the induction of apoptosis with bioactive compounds or control, scrape cells and collect in 15 mL tube.
- Spin at 250 g for 10 min at 4 °C.
- Resuspend pellet in 6 mL 1× PBS and spin again at 250 g for 10 min at 4 °C.
- Homogenized in Buffer containing 25 mM Hepes, 5 mM $MgCl_2$, 5 mM EDTA, 2 mM DTT and 0.1% CHAPS (pH 7.5).
- Centrifuges at 12,000 × g for 10 min (4 °C).
- Total protein (100 μg) from the supernatants.
- Determine activities of caspase-3 using the chromogenic substrates, namely 200 μM Ac-DEVD-pNA, PH 7.5.
- Absorbance at 405 nm determined in a microplate reader after incubation at 37 °C for 1 h.
- Specific activities of caspase-3 (nmol of *p*NA formed/min/mg) calculated using standard curve of *p*-nitroaniline (*p*NA), using 100 μL volume in a 96-well plate using an ELISA reader.

References

1. Jia M, Chen L, Xin HL, Zheng CJ, Rahman K, Han T, Qin LPA (2016) Friendly relationship between endophytic fungi and medicinal plants. Front Microbiol 7:907
2. Saikkonen K, Wäli P, Helander M, Faeth SH (2004) Evolution of endophyte-plant symbioses. Trends Plant Sci 9(6):275–280
3. Tan RX, Zou WX (2001) Endophytes: a rich source of functional metabolites. Nat Prod Rep 18(4):448–459
4. Strobel G, Daisy B (2003) Bioprospecting for microbial endophytes and their natural products. Microbiol Mol Biol Rev 67(4):491–502
5. Gouda S, Das G, Sen SK, Shin HS, Patra JK (2016) Endophytes: a treasure house of bioactive compounds of medicinal importance. Front Microbiol 29(7):1538
6. El-Hawary SS, Sayed AM, Rateb ME, Bakeer W, AbouZid SF, Mohammed R (2017) Secondary metabolites from fungal endophytes of *Solanum nigrum*. Nat Prod Res 31(21):2568–2571
7. Gottesman MM, Fojo T, Bates SE (2002) Multidrug resistance in cancer: role of ATP-dependent transporters. Nat Rev Cancer 2(1):48–58
8. Hartwell JL (1982) Plants used against cancer: a survey. Quarterman, Lawrence
9. Kharwar RN, Mishra A, Gond SK, Stierle A, Stierle D (2011) Anticancer compounds derived from fungal endophytes: their importance and future challenges. Nat Prod Rep 28(7):1208–1228
10. Li SJ, Zhang X, Wang XH, Zhao CQ (2018) Novel natural compounds from endophytic fungi with anticancer activity. Eur J Med Chem 5(156):316–343
11. Chandra H, Kumari P, Prasad R, Gupta SC, Yadav S (2021) Antioxidant and antimicrobial activity displayed by a fungal endophyte *Alternariaalternata* isolated from *Picrorhizakurroa* from Garhwal Himalayas, India. Biocatal Agric Biotechnol 33:101955
12. Verma VC, Kharwar RN, Strobel GA (2009) Chemical and functional diversity of natural products from plant associated endophytic fungi. Nat Prod Commun 4(11):1511–1532
13. Su LJ, Mahabir S, Ellison GL, McGuinn LA, Reid BC (2012) Epigenetic contributions to the relationship between cancer and dietary intake of nutrients, bioactive food components, and environmental toxicants. Front Genet 9(2):91
14. Gao H, Li G, Lou HX (2018) Structural diversity and biological activities of novel secondary metabolites from endophytes. Molecules 23: 646–677
15. Fox R, Aubert M (2008) Flow cytometric detection of activated Caspase-3. In: Mor G, Alvero AB (eds) Methods in molecular biology, apoptosis and cancer: methods and protocols. Humana Press Inc, Totowa, pp 47–56
16. Koopman G, Reutelingsperger COM, Kuijten GAM, Keehnen RMJ, Pals ST, van Oers MHJ (1994) Annexin V for flow cytometric detection of phosphatidylserine expression on B cells undergoing apoptosis. Blood 84:1415–1142
17. Maues LAL, Alves GM, Couto NMG, da Silva BJM, Arruda MSP, Macchi BM, Senac CBC, Pradoc AF, Crespo-Lopez ME, Silva EO, do Nascimento JLM (2019) Flavonoids from the Amazon plant *Brosimumacutifolium* induce C6 glioma cell line apoptosis by disrupting mitochondrial membrane potential andreducing AKT phosphorylation. Biomed Pharmacother 13:1.08728
18. McCarthy N, Evan G (1998) Methods for detecting and quantifying apoptosis. Curr Top Dev Biol 36:258–278
19. Mosmann T (1983) Rapid colorimetric assay for cellular growth and survival: application to proliferation and cytotoxicity assays. J Immunol Methods 65:55–63
20. Obeng E (2021) Apoptosis (programmed cell death) and its signals – a review Braz. J Biol 81: 1133–1143

INDEX

A. Sankaranarayanan et al. (eds.), *Endophytic Microbes: Isolation, Identification, and Bioactive Potentials*, Springer Protocols Handbooks, https://doi.org/10.1007/978-1-0716-2827-0,

V

W

Y

The manufacturer's authorised representative in the EU is Springer Nature Customer Service Centre GmbH, Europaplatz 3, 69115 Heidelberg, Germany. If you have any concerns regarding our products, please contact ProductSafety@springernature.com

Printed and bound by CPI Group (UK) Ltd, Croydon, CR0 4YY

30/10/2026

02242634-0007